NONLINEAR PHENOMENA
IN POWER ELECTRONICS

NONLINEAR PHENOMENA IN POWER ELECTRONICS

Attractors, Bifurcations, Chaos, and Nonlinear Control

Soumitro Banerjee
Indian Institute of Technology
Kharagpur, India

George C. Verghese
Massachusetts Institute of Technology
Cambridge, MA, USA

WILEY

Published by John Wiley & Sons, Hoboken, NJ

For general information on our other products and services, please contact our Customer Care Department within the United States at 800-762-2974, outside the United States at 317-572-3993 or fax 317-572-4002.

For more information about Wiley products, visit our website at www.wiley.com.

ISBN 0-7803-5383-8

Library of Congress Cataloging-in-Publication Data

Banerjee, Soumitro
  Nonlinear phenomena in power electronics : attractors, bifurcations, chaos, and nonlinear control/Soumitro Banerjee, George C. Verghese.
    p.cm.
  Includes bibliographical references and index.
  ISBN 0-7803-5383-8
  1. Power electronics. 2. Nonlinear theories. I. Verghese, George C. II. Title.

TK7881.15 B36 2001
621.31'7—dc21   2001016864

CONTENTS

CHAPTER 8   NONLINEAR CONTROL AND CONTROL OF CHAOS

Contents   **xiii**

PREFACE

Power electronics is a relatively new and fast-growing area of electronics with wide practical application. It is concerned with the efficient conversion of electrical power from one form to another, at power levels ranging from fractions of a watt or a few watts (e.g., for on board dc/dc converters), to kilowatts (in industrial drives), to megawatts (in ac transmission systems). To achieve high-efficiency power conversion, the active semiconductor devices (thyristors, power transistors) are always used in a switching mode in combination with passive components (power diodes, inductors, capacitors, and transformers).

Power converters exhibit a wealth of nonlinear phenomena. The prime source of nonlinearity is the switching element present in all power electronic circuits. Nonlinear components (e.g., the power diodes) and control methods (e.g., pulse-width modulation) are further sources of nonlinearity. Therefore, it is hardly surprising that feedback-controlled power converters routinely exhibit various types of nonlinear phenomena. Although first studied in depth for dc/dc converters, nonlinear phenomena have become significant throughout the field of power electronics.

The nonlinear phenomena of interest include bifurcations (sudden changes in operating mode), coexisting attractors (alternative stable operating modes), and chaos (apparently random behavior). If reliable power converters are to be designed, an appreciation that these possibilities exist is vital, together with a knowledge of how to investigate them, use them, or avoid them.

Unfortunately, the academic training of engineers has traditionally emphasized linear methods (e.g., Fourier and Laplace transforms, matrices, and transfer functions). By their very nature, linear methods alone give little insight into nonlinear phenomena and are of limited value in predicting and analyzing them. It is therefore our belief that every power electronics engineer can benefit from some background in nonlinear dynamics.

In the past few years, nonlinear phenomena have been thoroughly investigated in many power converters, but the results are available only in research papers, often written in language accessible only to specialists in the field. At present no book is available that presents the required foundations in nonlinear dynamics in a form matched to the typical background of power electronics engineers. Although some excellent introductory texts on nonlinear dynamics have become available in recent years, most of them emphasize mathematics and examples from physics, appeal little to the intuition of electrical engineers, and have few engineering applications. None of them includes examples from power electronics.

This book attempts to bridge the gap by providing an introduction to nonlinear dynamics and its applications, specifically aimed at power electronics professionals.

With contributions from many leading workers in the field, this book builds upon the requisite foundation and presents a range of recent advances in our understanding of the nonlinear dynamics of power electronic circuits.

The book is organized as follows.

Chapter 1 sets the background for the book, and introduces the main problems to be treated. It presents several examples of nonlinear phenomena in power electronics circuits to whet the reader's appetite, and explains the need for understanding nonlinear dynamics. It also identifies the chief sources of nonlinearity in typical power electronic circuits.

Chapter 2 treats various methods of modeling for power electronic circuits. The starting point for dynamic modeling in power electronics is a continuous-time state-space representation with switched dynamics. From this, aggregated models are obtained by sampling or averaging. Section 2.1 gives an overview of these modeling techniques. Sampled-data models are well adapted to studies of bifurcation and chaos, and to cycle-by-cycle digital control schemes. Since this approach is adopted in most studies in nonlinear dynamics, a detailed account of the method is presented in Section 2.2.

In Chapter 3, the theory of bifurcation and chaos is presented in a form easily accessible to the power electronics engineer. No prior background in nonlinear dynamics is assumed. While the theory of nonlinear dynamics constitutes a vast body of knowledge, only the issues relevant to the phenomena observed in power electronics circuits are presented in this chapter. Starting with a relatively informal introduction to the theory of nonlinear dynamics, this chapter explains various types of bifurcations in smooth and piecewise-smooth maps, coexisting attractors, basins of attraction, crises, and the analytical method of Schwarzian derivatives.

Experimental investigation of nonlinear phenomena involves special techniques developed in the past few years. Section 4.1 introduces the reader to techniques for sampling switching waveforms, for displaying chaotic waveforms or phase portraits or Poincaré sections on the oscilloscope, and for obtaining bifurcation diagrams. The remainder of Chapter 4 deals with computational issues related to determining fixed points, eigenvalues and associated manifolds, Lyapunov exponents, operating mode boundaries, average values of state variables, and spectral characteristics.

A major line of investigation over the past few years has been the nonlinear dynamics of feedback-controlled dc/dc converters. Chapter 5 presents the analysis of the dynamics of various dc/dc converter topologies, including the current-mode-controlled boost converter, Cúk converter, and converters in discontinuous conduction mode. The voltage-controlled buck converter has been the subject of intensive investigation in recent times, because of its rich dynamical behaviors. This converter is treated in two parts: experimental investigations and possible pathways to chaos for the converter with the latch (preventing multiple pulsing) is dealt with in Section 5.2, and analytical investigation of the converter without the latch (exhibiting richer dynamics) is discussed in Section 5.3.

The switching characteristics of thyristors introduce some special types of nonlinearities. Thyristors are widely used in high-power applications (up to megawatts). In Chapter 6 these phenomena and their theoretical analyses are presented.

One source of nonlinearity in a power electronic circuit is the inductor because of the magnetic nonidealities and hysteretic effects. Section 7.1 shows how to model such nonlinearities and demonstrates that these effects can give rise to bifurcation phenomena even in very simple systems. The presence of noise, ubiquitous in all power electronic circuits, can also lead to nonlinear effects, and Section 7.2 discusses nonlinear noise interactions in converters and inverters. Nonlinear phenomena in three high-power systems, namely inverters under tolerance band control (Section 7.3), induction motors under current control (Section 7.4), and induction motor drive systems (Section 7.5), are also treated in this chapter.

Although most feedback control schemes in power electronics have nonlinear aspects, the conscious application of methods developed in nonlinear control theory to power converters is somewhat more recent. Nevertheless, a single chapter cannot hope to cover this area comprehensively. Chapter 8 simply aims to convey the flavor of such nonlinear control approaches. PWM rectifiers, inverters, and cycloconverters are presented as nonlinear modulators, and the basis of sliding modes, hysteretic control, energy-based control, and boundary control are presented in the context of dc/dc converters, rectifiers, and inverters. In systems exhibiting nonlinear phenomena and chaos, one might endeavor to control chaos by using feedback to enforce periodic orbits or delay the onset of bifurcations that may lead to chaotic behavior. However, it has been shown in recent years that operation of a converter in chaos may offer specific advantages due to the spreading of the output spectrum, so there is also motivation to develop control strategies that permit a converter to operate reliably in the chaotic mode. Both these approaches to control of chaos are discussed in Chapter 8.

Soumitro Banerjee
Indian Institute of Technology
Kharagpur, India

George C. Verghese
Massachusetts Institute of Technology
Cambridge, MA, USA

ACKNOWLEDGMENTS

The Editors are very grateful to each of the authors for their enthusiastic participation in this project and for their prompt and cheerful cooperation with our various suggestions and requests along the way. If the value of this book to its readers is commensurate with the collective effort of the authors, then this project will have been worthwhile. Special thanks go to Lynne Hamill for helping create the book's webpage and to author (Michael) Chi K. Tse for taking on the added burden of hosting and maintaining the "Authors' Exchange" homepage, which were crucial to coordinating the efforts of our world-wide web of authors. We also appreciate the technical assistance and good humor of author José Luis R. Marrero during critical stages of the editorial process, and the logistical and administrative assistance of Vivian Mizuno at MIT. Additional thanks go to the staff at IEEE Press, for their patience and help in seeing this project to completion. Finally, the Editors are enormously grateful to their respective families for their forbearance and cheerful support – in hundreds of ways, big and small – during the preparation of this book.

Soumitro Banerjee
Indian Institute of Technology
Kharagpur, India

George C. Verghese
Massachusetts Institute of Technology
Cambridge, MA, USA

LIST OF CONTRIBUTING AUTHORS

Eyad Abed
Department of Electrical Engineering
University of Maryland
College Park, MD 20742
U.S.A.
E-mail: abed@isr.umd.edu

Soumitro Banerjee
Department of Electrical Engineering
Indian Institute of Technology
Kharagpur – 721302
INDIA
E-mail: soumitro@ee.iitkgp.ernet.in

Carles Batlle
Department de Matematica Aplicada
i Telematica
E.U.P.V.G. Universitat Politecnica
de Catalunya
Av. Victor Balaguer, s/n
E-08800 Vilanova i la Geltru
SPAIN
E-mail: carles@mat.upc.es

Mario di Bernardo
Applied Nonlinear Mathematics Group
Department of Engineering Mathematics
University of Bristol
Bristol BS8 1TR
U.K.
E-mail: M.diBernardo@bristol.ac.uk

Roberto Santos Bueno
Departamento de Electrónica y
Automática
ETS de Ingeniería ICAI
Universidad Pontificia Comillas
Alberto Aguilera 23
28015 Madrid
SPAIN
E-mail: rsantos@ dea.icai.upco.es

Santanu Das
Department of Microsystems Engineering
FHF, University of Applied Science
D-78120 Furtwangen
GERMANY
E-mail: das@alpha.fh-furtwangen.de

Jonathan H. B. Deane
Department of Electrical & Electronic
Engg.
University of Surrey
Guildford GU2 5XH
U.K.
E-mail: J.Deane@ee.surrey.ac.uk

Ian Dobson
Department of Electrical
and Computer Engineering
University of Wisconsin
Madison, WI 53706
U.S.A.
E-mail: dobson@ece.wisc.edu

Gerardo Escobar
Northeastern University
Department of Electrical and
Computer Engineering, 442DA
Boston, MA 02115
U.S.A.
E-mail: gescobar@ece.neu.edu

Chung-Chieh Fang
Taiwan Semiconductor
Manufacturing Company
Science-Based Industrial Park
Hsinchu 300
TAIWAN
E-mail: ccfangb@tmsc.com.tw

Enric Fossas
Department of Applied Mathematics
and Telematics
Polytechnical University of Catalonia
Barcelona 08034
SPAIN
E-mail: fossas@mat.upc.es

Celso Grebogi
Instituto de Fisica
Universidade de Sao Paulo
Caixa Postal 66318
05315-970 Sao Paulo
BRAZIL

David C. Hamill
Surrey Space Centre
University of Surrey
Guildford GU2 5XH
U.K.
E-mail: D.Hamill@surrey.ac.uk

Steven Isabelle
Intel Corporation
77 Reed Road
Hudson, MA 01749
U.S.A.
E-mail: steve.lisabelle@intel.com

Debaprasad Kastha
Department of Electrical Engineering
Indian Institute of Technology
Kharagpur – 721302
INDIA
E-mail: kastha@ee.iitkgp.ernet.in

Toshiji Kato
Department of Electrical Engineering
Doshisha University
Kyotanabe, Kyoto 610-0321
JAPAN
E-mail: kato@kairo.doshisha.ac.jp

Philip T. Krein
Department of Electrical Engineering
University of Illinois at Urbana-Champaign
U.S.A.
E-mail: krein@ece.uiuc.edu

Yasuaki Kuroe
Department of Electronics
and Information Science
Kyoto Institute of Technology
Matsugasaki, Sakyo-ku
Kyoto 606-8585
JAPAN
E-mail: kuroe@dj.kit.ac.jp

Yuk-Ming Lai
Department of Electronic
and Information Engineering
Hong Kong Polytechnic University
Hung Hom, Kowloon, Hong Kong
CHINA
E-mail: enymlai@polyu.edu.hk

Andreas Magauer
Abteilung für Elektronik und Informatik
Höhere Technische Bundeslehranstalt
Salzburg
Salzburg – A-5020
AUSTRIA
E-mail: magauer@alpin.or.at

José Luis Rodriguez Marrero
Department of Electronics – ICAI
Univ. Pontifica Comillas
Alberto Aguilera 23
28015 Madrid
SPAIN
E-mail: marrero@dea.icai.upco.es

Pallab Midya
Motorola Corporate Research and
Development Laboratories
1301 E. Algonquin Rd.
Schaumburg, Illinois 60196
U.S.A.
E-mail: apm016@email.mot.com

I. Nagy
Technical University of Budapest
Department of Automation
1111 Budapest
Budafoki ut.8. Fep.II.lp.mf.
HUNGARY
E-mail: nagy@elektro.get.bme.hu

Gerard Olivar
Department de Matematica Aplicada i
Telematica
E.U.P.V.G. Universitat Politecnica de
Catalunya
Av. Victor Balaguer, s/n
E-08800 Vilanova i la Geltru
SPAIN
E-mail: gerard@mat.upc.es

Romeo Ortega
Laboratory for Signals and Systems
CNRS – SUPELEC
Gif sur Yvette
FRANCE
E-mail: Romeo.Ortega@lss.supelec.fr

Priya Ranjan
Department of Electrical Engineering
University of Maryland
College Park, MD 20742
U.S.A.
E-mail: priya@glue.umd.edu

Seth Sanders
Department of Electrical Engineering
and Computer Science
University of California – Berkeley
Berkeley, CA 94720
U.S.A.
E-mail: sanders@eecs.berkeley.edu

Alex Stanković
Department of Electrical
and Computer Engineering
Northeastern University
Boston, MA 02110
U.S.A.
E-mail: astankov@hilbert.cdsp.neu.edu

Zoltán Sütő
Department of Automation
Technical University of Budapest
XI., Budafoki ut 8
Budapest H-1111
HUNGARY
E-mail: suto@elektro.get.bme.hu

Chi K. Tse
Department of Electronic
and Information Engineering
Hong Kong Polytechnic University
Hung Hom, Hong Kong
CHINA
E-mail: encktse@polyu.edu.hk

Francesco Vasca
Universita' del Sannio
Facolta' di Ingegneria
Corso Garibaldi 107,
Palazzo Bosco,
Benevento 82100
ITALY
E-mail: vasca@disna.dis.unina.it

George C. Verghese
Department of Electrical Engineering
and Computer Science
Room 10-093
Massachusetts Institute of Technology
Cambridge, Massachusetts 02139
U.S.A.
E-mail: verghese@mit.edu

Chapter 1

INTRODUCTION

David C. Hamill
Soumitro Banerjee
George C. Verghese

In all chaos there is a cosmos, in all disorder a secret order, in all caprice a fixed law.
—Carl Gustav Jung (1875–1961)[1]

1.1 INTRODUCTION TO POWER ELECTRONICS

Most branches of electronics are concerned with processing information or signals; in contrast, power electronics deals with the processing of electrical energy. Power converters do not have an end of their own, but are always an intermediary between an energy producer and an energy consumer. The field is one of growing importance: it is estimated that during the twenty-first century, 90% of the electrical energy generated in developed countries will be processed by power electronics before its final consumption.

Power electronics is a "green" technology, with three main aims:

- To convert electrical energy from one form to another, facilitating its regulation and control
- To achieve high conversion efficiency and therefore low waste heat
- To minimize the mass of power converters and the equipment (such as motors) that they drive.

Unlike other areas of analog electronics, power electronics uses semiconductor devices as *switches*. Since electrical power supplies can be either dc or ac, there are four basic types of power converter: ac/dc converters, dc/ac converters, dc/dc converters, and ac/ac converters. Here *ac* typically refers to nominally *sinusoidal* voltage waveforms, while *dc* refers to nominally *constant* voltage waveforms. Small deviations from nominal are tolerable. An ac/dc converter (which has an ac power source and a dc load) is also called a *rectifier*, and a dc/ac converter is called an *inverter*.

1. *The archetypes and the collective unconscious,* in *Collected Works,* 2nd ed., vol. 9, part 1, Routledge & Kegan Paul, London, 1968, p. 32.

Power electronics technology is increasingly found in the home and workplace [1,2,3,4]. Familiar examples are the domestic light dimmer, switched-mode power supplies in personal computers, heating and lighting controls, electronic ballasts for fluorescent lamps, drives for industrial motion control, induction heating, battery chargers, traction applications such as locomotives, solid-state relays and circuit breakers, off-line dc power supplies, spacecraft power systems, uninterruptible power supplies (UPSs), conditioning for alternative energy sources, automobile electronics, and electric vehicles.

High efficiency is needed in order to reduce energy costs, but also because it reduces the amount of dissipated heat that must be removed from the power converter. Efficiencies of higher than 99% can be obtained in large, high-power systems, while small, low-power systems may have efficiencies closer to 80%. The goal of high efficiency dictates that the power processing components in the circuit be close to lossless. Switches, capacitors, inductors, and transformers are therefore the typical components in a power electronic converter.

The switches are operated cyclically, and serve to vary the circuit interconnections—or the *topological state* of the circuit—over the course of a cycle. The capacitors and inductors perform filtering actions, regulating power flows by temporarily storing or supplying energy. The transformers scale voltages and currents, and also provide electrical isolation between the source and load. *Ideal* switches, capacitors, inductors and transformers do not dissipate power, and circuits comprising only such elements do not dissipate power either (provided that the switching operations do not result in impulsive currents or voltages, a constraint that is respected by power converters).

In particular, an *ideal switch* has zero voltage across itself in its *on* (or closed, or conducting) state, zero current through itself in its *off* (or open, or blocking) state, and requires zero time to make a transition between these two states. Its power dissipation is therefore always zero. Of course, practical components depart from ideal behavior, resulting in some power dissipation. However, for the types of dynamic behavior examined in this book, it suffices to assume ideal switches.

1.1.1 Power Switching Devices

The key to power electronics is the availability of suitable switching devices. The main types are listed below.

- *Diode*: Diodes may be thought of as passive switches, or *non-return valves*. Ideally the resistance is zero for current in the *forward* direction (the forward-biased case), so the diode functions as a closed switch under this condition; and the resistance is infinite for voltages applied in the *reverse* direction (the reverse-biased case), so the diode then functions as an open switch. The types currently available include fast recovery *pn* junction, *p-i-n*, and Schottky diodes. The latter have low conduction loss and negligible charge storage, and are widely used at low voltages.

- *Thyristor or SCR*: The SCR is a *pnpn* device that may be thought of as a diode with an additional *gate* terminal. When reverse biased, it blocks the flow of current; when forward biased, conduction is inhibited until a trigger pulse is applied to the gate. The SCR then conducts until the current through it falls to zero, whereupon it resumes blocking. Modern variants include asymmetric SCRs, reverse conducting thyristors, and gate turn-off thyristors. Being rugged devices available in high ratings, thyristors have been widely applied up to extremely high power levels (e.g., in the 2GW England–France HVDC link). Most types are rather slow, limiting their applications to low frequencies.

- *Bipolar junction transistor*: Silicon bipolar junction transistors (BJTs) were developed during the 1960s, and by the 1970s were employed in switched-mode power supplies. These are controlled via an appropriate drive at the base or gate, which can cause the transistor to act as an open switch (in *cutoff*) or closed switch (in *saturation*). BJTs are minority carrier devices, so speed is a limitation: practical switching frequencies are limited to around 40kHz.

- *Power MOSFET*: The power MOSFET (metal oxide semiconductor field effect transistor) became a commercial proposition in the early 1980s. A majority carrier device, it is capable of switching at megahertz frequencies, but contains a slow parasitic body-drain diode. MOSFETs have largely replaced BJTs in low-power applications such as switched-mode power supplies. The MOSFET's construction is not suitable for very high powers, and voltage ratings are lower than for competing devices.

- *IGBT*: The insulated gate bipolar transistor (IGBT) became a commercial reality in the late 1980s. It acts like a MOSFET driving a power BJT, and has some of the advantages of both: ease of drive, high ratings, and low conduction loss. But minority carrier charge storage makes the IGBT turn off with a long *current tail*, causing high switching loss. One of the main application areas of IGBTs is in multikilowatt motor drives, where they are the dominant switching device. Switching frequencies in the tens of kilohertz are used.

Each type of semiconductor switch is amenable to a characteristic mode of control. Diodes are at one extreme, as they *cannot* be controlled; they conduct or block as a function solely of the current through them or the voltage across them, so neither their turn-on nor turn-off can be directly commanded by a control action. For thyristors, the turn-off happens as for a diode, but the turn-on is by command, under appropriate circuit conditions. For transistors, both the turn-on and turn-off occur in response to control actions, provided circuit conditions are appropriate. The power loss associated with real switches comes from a nonzero voltage drop when they are closed, a nonzero leakage current when they are open, and a finite transition time from closed to open or vice versa, during which time both the voltage and current may be significant simultaneously.

A higher switching frequency generally implies a more compact converter, since smaller capacitors, inductors, and transformers can be used to meet the specified circuit characteristics. However, the higher frequency also means higher switching losses associated with the increased frequency of switch transitions, as well as other losses and limitations associated with high-frequency operation of the various components. Switching frequencies above the audible range are desirable for many applications. The choice of switch implementation depends on the requirements of each particular converter.

1.1.2 Sources of Nonlinearity in Power Electronics

Power Converters

Since the object is to convert electrical energy at high efficiency, the ideal power converter would contain only lossless components. Two basic groups that can be approximated by real components are available:

- *Switching components, such as transistors and diodes.* Active switches such as transistors or MOSFETS turn on and off in response to an applied signal, and in feedback-controlled systems the switching signal depends on the state variables. Passive switches (diodes) have a highly nonlinear *v-i* characteristic.
- *Reactive (energy storing) components, such as inductors and capacitors.* They are characterized by differential equations, $v = L\, di/dt$ for an inductor, $i = C\, dv/dt$ for a capacitor. They absorb energy from a circuit, store it, and return it.

Power converters employ components from both groups. Energy is steered around the circuit by the switching components, while the reactive components act as intermediate energy stores and input/output reservoirs. The presence of both types of component implies that the circuits are nonlinear, time-varying dynamical systems. This has two implications: power converters are difficult to analyze, and they are likely to show a wealth of unusual behavior.

There are also several unavoidable sources of unwanted nonlinearity in practical power converters:

1. The semiconductor switching devices have intrinsically nonlinear dc characteristics. They also have nonlinear capacitances, and most suffer from minority carrier charge storage.
2. Nonlinear inductances abound: transformers, chokes, magnetic amplifiers, and saturable inductors used in snubbers.
3. The control circuits usually involve nonlinear components: comparators, PWMs, multipliers, phase-locked loops, monostables, and digital controllers.

Electrical Machines and Drives

Adjustable-speed drives constitute a rapidly growing market for power electronics. Here, power converters are combined with electric motors and sophisticated control electronics. The main thrust of current work is to replace conventional dc drives with ac drives. The dc motors are easy to control for a good dynamic response, but have a complex physical construction and a poor power-to-weight ratio. They utilize commutators and brushes, which cause sparking and radio interference, and are subject to mechanical wear. Much research has been done into supplying and controlling ac machines such as squirrel-cage induction motors, permanent-magnet synchronous motors, *brushless dc* motors, and switched reluctance motors. These machines are mechanically simple and are therefore inexpensive and reliable, but they are difficult to control if variable speed and rapid dynamic response are required. Power electronics and digital control techniques are now being applied to obtain speed variation and good dynamic response in these machines. Unfortunately, ac motors are themselves inherently nonlinear. The induction motor, for example, may be modeled by a non-linear and highly interactive multivariable structure. It is the task of vector control techniques to unravel this model, decouple the flux and torque variables, and allow a relatively simple outer control loop. Another example is the switched reluctance motor, in which the self and mutual inductances vary not only with the shaft rotation, but also with saturation of the magnetic path—which itself depends on the shaft position as well as the drive waveform. As a final example, the permanent magnet stepper motor, operated open loop with an inertia load, exhibits bifurcation from steady rotation to irregular back-and-forth juddering, a phenomenon that has been well known for many years but little studied. Combining switched circuits and nonlinear electromechanical devices, adjustable-speed drives would seem to be a rich source of nonlinear behavior and, because of their importance to industry, an appropriate subject for detailed study.

Power Systems

The field of electric power systems deals with the generation, transmission, and distribution of 50/60Hz power. Bifurcation theory has been applied successfully to simple models of power systems [5,6,7,8,9], and it has been shown that the theory of nonlinear dynamics can be used to explain undesirable low-frequency oscillations (subsynchronous resonances) and voltage collapses.

Power systems are seeing increased use of power electronics. In developed countries, about 60% of electricity generated is used to power motors, and a further 20% is consumed by lighting; as power electronics penetrates these areas, more and more power converters will be connected to the ac supply.

Furthermore, power electronics is increasingly being used by the utilities themselves to process power on a large scale. Widespread use of multimegawatt power converters in flexible ac transmission systems (FACTS) is anticipated. In order to maximize the capacity and cost effectiveness of existing power systems as demand rises, progressive interlinking is taking place on a continental level. Undesirable nonlinear effects are likely, unless measures are taken to understand them. It is to be hoped that catastrophic bifurcations can be avoided with proper understanding of the phenomena involved.

1.2 AN EXAMPLE: THE BUCK DC/DC CONVERTER

A concrete example of a power converter, the buck dc/dc converter, is now presented. In this example, the conventional modeling approach is contrasted with one derived from nonlinear dynamics. The buck converter is one of the simplest but most useful power converters: a chopper circuit that converts a dc input to a dc output at a lower voltage. Many switched-mode power supplies employ circuits closely related to it. An application of current importance is conversion of the standard 5V dc supply used in computers to the 3.3V or less needed by processor chips, such as those of the Pentium family. A buck converter for this purpose can achieve a practical efficiency of 92%, whereas a linear regulator would be only 66% efficient—producing four times as much waste heat. Although this example is at a low power level, buck converters are also used at several kilowatts.

The basic open-loop buck converter is shown in Figure 1.1. The switch S opens and closes periodically at the switching frequency f_s, with a duty ratio d (the fraction of

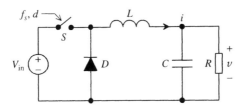

Figure 1.1 The open-loop buck dc/dc converter.

time the switch is on). When S is closed, the input voltage V_{in} is transferred to the LC low-pass filter. When S is open, the inductor maintains its current flow, forcing the diode D to conduct and ground the input of the LC filter. Thus the filter sees a square wave between 0 and V_{in}. The cutoff frequency of the filter is much lower than f_s, removing most of the switching ripple and delivering a relatively smooth output voltage v to the load resistance R. The output voltage can be varied by changing the duty ratio d (i.e., by pulse-width modulation (PWM)).

The operation described is known as *continuous-conduction mode* (CCM), since the inductor passes current without a break. However, if the output is only lightly loaded, the inductor current can become zero for part of the cycle as D comes out of conduction. This situation is called the *discontinuous-conduction mode* (DCM). (The terms *continuous* and *discontinuous* are used in a nonmathematical sense here.)

In practice, it is necessary to regulate the output voltage v against changes in the input voltage and the load current, by adding a feedback control loop as in Figure 1.2. In this simple proportional controller, a constant reference voltage V_{ref} is subtracted from the output voltage, and the error v_e is amplified with gain A to form a control signal, $v_{con} = A(v - V_{ref})$. This signal feeds a simple PWM circuit comprising a ramp (sawtooth) oscillator of frequency f_s and voltage v_{ramp} between levels V_l and V_u, and a comparator that drives the switch. This switch conducts whenever $v_{con} < v_{ramp}$; thus v_{con} determines d. The intended mode of operation is a steady state in which the output voltage stays close to V_{ref}.

1.2.1 Conventional Model of the Buck Converter

The conventional way of modeling this type of circuit is to take an average over a switching cycle, an approach first proposed by Wester [10]. Since conventional control theory requires a linear model, the averaged circuit is generally linearized about a

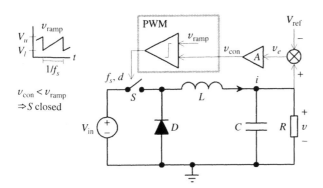

Figure 1.2 The buck converter with proportional closed-loop controller.

suitable operating point. State space averaging, developed by Ćuk [11,12], operates on the state equations of the circuit. An alternative method uses "injected and absorbed" currents [13]. Vorperian [14,15] suggested a method of treating the switch-diode combination in isolation from the converter circuit. Regardless of the details, these methods have the same aim: to replace the nonlinear, time-varying dynamical system with an averaged, linearized one. The justification is that when designing the control circuit, one need no longer be concerned with the microscopic details of the power switching. Clearly, something must be lost in the process.

Continuous Conduction Mode

In order to demonstrate the shortcomings of the conventional modeling technique, we illustrate here the state space averaging approach for the buck converter. A more detailed account of averaging techniques will be presented in Chapter 2.

The state equations are:

$$\frac{di}{dt} = \begin{cases} (v_{\text{in}} - v)/L & S \text{ conducting, } D \text{ blocking} & (a) \\ -v/L & S \text{ blocking, } D \text{ conducting} & (b) \\ 0 & S \text{ and } D \text{ both blocking} & (c) \end{cases} \tag{1.1}$$

and

$$\frac{dv}{dt} = \frac{i - v/R}{C} \tag{1.2}$$

where, for notational simplicity the time argument t has been dropped from $i(t)$, $v(t)$ and $v_{\text{in}}(t)$.

Averaging. In CCM, S conducts for a fraction d of each cycle and D conducts for the remainder, $1 - d$; (1.1c) is not involved. The averaged equations are found by multiplying (1.1a) by d and (1.2b) by $1 - d$, and summing:

$$\frac{di}{dt} = \frac{dv_{\text{in}} - v}{L} \tag{1.3}$$

$$\frac{dv}{dt} = \frac{i - v/R}{C} \tag{1.4}$$

In this simple example, only one of the state equations is affected: (1.2) comes through the averaging process unaltered. This may not be the case with other converters. Note, however, that the variables in (1.3) and (1.4) now actually represent local averages of the instantaneous variables in (1.1) and (1.2).

Perturbation. Let each quantity comprise a constant (dc) nominal component, represented by an uppercase symbol, and a small perturbation component, represented by a lowercase symbol with a circumflex. Thus, for instance, let $i(t) = I + \hat{i}(t)$. Doing this for i, v and d, substituting into (1.3) and (1.4), and using the fact that $dI/dt = 0$, $dV/dt = 0$ (I and V being constants), the following expressions are obtained:

$$\frac{d\hat{i}}{dt} = \frac{(\overline{D} + \hat{d})(V_{\text{in}} + \hat{v}_{\text{in}}) - (V + \hat{v})}{L} \tag{1.5}$$

$$\frac{d\hat{v}}{dt} = \frac{I + \hat{i} - (V + \hat{v})/R}{C} \tag{1.6}$$

where $\overline{D}$ denotes the nominal value of d (we have already used D to denote the diode).

Steady State. To find the steady state (the equilibrium point of the averaged dynamical system), all perturbation terms are set to zero, and the LHS of each state equation is also set to zero. This results in

$$V = \overline{D}V_{\text{in}} \quad \text{and} \quad I = V/R \tag{1.7}$$

which accord with an intuitive understanding of the circuit's operation.

Linearization. Finally, the system is linearized about this steady-state operating point. Expanding (1.5) and (1.6), neglecting second-order perturbation terms, and subtracting away the respective steady-state equations of (1.7), the following are obtained:

$$\frac{d\hat{i}}{dt} = \frac{\hat{d}V_{\text{in}} + \overline{D}\hat{v}_{\text{in}} - \hat{v}}{L} \tag{1.8}$$

$$\frac{d\hat{v}}{dt} = \frac{\hat{i} - \hat{v}/R}{C} \tag{1.9}$$

These linear differential equations represent the small signal (ac) behavior of the buck converter.

Transfer Functions. Laplace transforms of (1.8) and (1.9) are taken by writing s for d/dt. Eliminating $\hat{i}$ between the two transformed equations with $\hat{v}_{\text{in}} = 0$ yields the control-to-output transfer function:

$$\frac{\hat{v}}{\hat{d}} = \frac{V_{\text{in}}}{1 + sL/R + s^2LC} \tag{1.10}$$

where $\hat{v} = \hat{v}(s)$ now means the Laplace transform of $\hat{v}(t)$, etc. This transfer function forms part of the feedback loop and determines the closed-loop stability. Using a similar averaging approach, the transfer function of the error amplifier and PWM is easily found as

$$\frac{\hat{d}}{\hat{v}_e} = \frac{A}{V_u - V_l} \tag{1.11}$$

Hence the overall loop gain is

$$G(s) = \frac{\hat{v}}{\hat{v}_e} = \frac{AV_{\text{in}}}{V_u - V_l} \cdot \frac{1}{1 + sL/R + s^2LC} \tag{1.12}$$

Stability. Equation (1.12) describes a standard second-order system, with dc gain $AV_{\text{in}}/(V_u - V_l)$, undamped natural frequency $\omega_n = 1/\sqrt{LC}$, and damping factor $\zeta = \sqrt{L/4CR^2}$. Given R, the values of L and C are chosen by the designer on power considerations: L is made large enough to ensure CCM operation, and C is chosen to give an acceptably small output voltage ripple. This generally results in an under-damped response ($\zeta < 1$) with $\omega_n/2\pi << f_s$.

Consider the example of a buck converter designed to accept an input voltage of 15V to 40V and produce a regulated output voltage close to 12V [16,17,18]. The following parameter values apply: $f_s = 2.5\text{kHz}$, $A = 8.4$, $V_u = 8.2\text{V}$, $V_l = 3.8\text{V}$, $V_{\text{ref}} = 12\text{V}$, $L = 20\text{mH}$, $C = 47\mu\text{F}$ and $R = 22\Omega$; therefore $\omega_n/2\pi = 164\text{Hz}$ and $\zeta = 0.47$. The system's phase margin can be calculated from (1.12), and varies from $10.2°$ at the minimum input voltage of 15V to $6.2°$ at the maximum, 40V. These figures are rather small: a phase margin of $45°$ or more would be desirable. Nevertheless, according to the average model, the closed-loop converter is stable over the entire input voltage range.

1.2.2 Actual System Behavior

If we actually study the system behavior as V_{in} is varied, we find that for low values of the parameter the system waveforms are periodic, and the periodicity is same as that of the ramp waveform. But at $V_{\text{in}} \approx 24.4\text{V}$, this periodic behavior or "orbit" becomes unstable and is replaced by the behavior that repeats every two cycles of the ramp (Figure 1.3). This is the *period-2* subharmonic behavior.

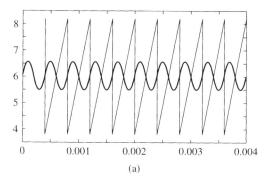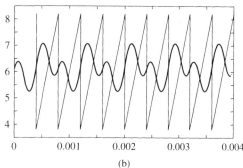

| (a) | (b) |

Figure 1.3 (a) Period-1 orbit of the buck converter at $V_{\text{in}} = 24\text{V}$ and (b) Period-2 orbit of the buck converter at $V_{\text{in}} = 25\text{V}$. The v_{ramp} and v_{con} waveforms are shown.

If the parameter V_{in} is increased even further, the behavior changes to an apparently random, erratic, and aperiodic waveform. This situation is illustrated in Figure 1.4. Such a bounded aperiodic behavior is known as *chaos*.

The standard averaging method of analysis predicts that the buck converter will be stable over the whole operating range of input voltage and load resistance. But it is evident from numerical simulation as well as experiment (see Section 5.2) that this converter can exhibit subharmonics and chaos over a significant range of parameter values. No method that relies upon linearization is able to predict such effects, which are peculiar to nonlinear systems. In addition, the process of averaging may suppress behavior that a more detailed model might display.

1.2.3 Nonlinear Map-Based Model of the Buck Converter

An alternative modeling approach is to move into the discrete domain, by means of sampled-data modeling [17]. Here the full details of the switching operations are retained, so the model is likely to be more accurate; but inevitably, the description will

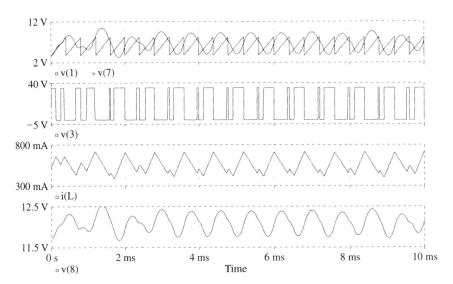

Figure 1.4 Simulated chaotic waveforms for the buck converter in CCM with $V_{in} = 35$V. Top to bottom: v_{ramp} and v_{con}; voltage across D; inductor current i; output voltage v.

be relatively complex. The full details of this method are given in Chapter 2; here we offer only the flavor of how the problem can be probed with the tools of nonlinear dynamics.

We first introduce the idea of a *mapping* or transformation. Put simply, a mapping is a mathematical function that takes each point of a given space to another point (see Figure 1.5). A mapping **F** that converts a point in the n-dimensional real space $I\!\!R^n$ to another point in the same space can be written $\mathbf{F} : I\!\!R^n \mapsto I\!\!R^n$. This notation is used throughout nonlinear dynamics, where $I\!\!R^n$ is treated as an n-dimensional space of real numbers and **F** is a nonlinear transformation. If a certain point in the space maps to itself (i.e., is invariant under the mapping), it is said to be a *fixed point*. In functional notation, $\mathbf{x}^* = \mathbf{F}(\mathbf{x}^*)$, where $\mathbf{x}^* \in I\!\!R^n$ is the fixed point.

Because the buck converter has two state variables, the inductor current i and the capacitor voltage v, it has a two-dimensional state space. Our aim is to find a two-dimensional mapping $\mathbf{F} : I\!\!R^2 \mapsto I\!\!R^2$ that describes how the state vector **x** evolves from

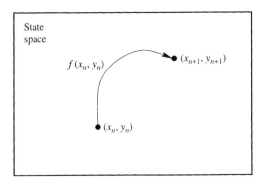

Figure 1.5 Illustration of a discrete mapping.

one ramp cycle to the next: $\mathbf{x}_{m+1} = \mathbf{F}(\mathbf{x}_m)$, where $\mathbf{x}_m$ is a vector with components that are i and v at the end of the mth ramp cycle. Steady-state period-1 operation corresponds to a fixed point of the map, $\mathbf{x}^* = \mathbf{F}(\mathbf{x}^*)$. It will be assumed that the converter always operates in CCM, and that the filter network is underdamped.

With S closed, (1.1a) and (1.2) govern the time evolution of the state vector. Their solutions may be written

$$i = \exp\frac{-t}{2CR}\left(a_1 \sin \omega_d t + b_1 \cos \omega_d t\right) + \frac{V_{\text{in}}}{R} \tag{1.13}$$

$$v = \exp\frac{-t}{2CR}\left(a_2 \sin \omega_d t + b_2 \cos \omega_d t\right) + V_{\text{in}} \tag{1.14}$$

where $\omega_d = \left(\frac{1}{LC} - \frac{L}{C^2 R}\right)^{1/2}$ and a_1, b_1, a_2, b_2 are constants derived from the initial conditions. The state vector follows (1.13) and (1.14) until the switching condition $v_{\text{con}} = v_{\text{ramp}}$ is satisfied. Then S opens at the switching instant $t = t_s$. For this circuit the state vector is continuous (in the mathematical sense), so the final values of i and v for one interval become the initial values for the next. With S open, (1.1b) and (1.2) govern the motion. Resetting t to zero, the solutions are now

$$i = \exp\frac{-t}{2CR}\left(a_1' \sin \omega_d t + b_1' \cos \omega_d t\right) \tag{1.15}$$

$$v = \exp\frac{-t}{2CR}\left(a_2' \sin \omega_d t + b_2' \cos \omega_d t\right) \tag{1.16}$$

where the new constants a_1', b_1', a_2', b_2' can be calculated from a_1, b_1, a_2, b_2.

This process of alternating switch transitions, applied over the ramp cycle $t \in [0, 1/f_s)$, where f_s is the switching frequency, defines the mapping $\mathbf{F}$ that takes $\mathbf{x}_m$ to $\mathbf{x}_{m+1}$. Unfortunately, there is a snag: it is not straightforward to find the set of switching instants t_s. Switching occurs whenever $A(v - V_{\text{ref}}) = v_{\text{ramp}}$, and this introduces two problems. First, because $v(t)$ is a damped sinusoid, finding the switching instants involves solving a transcendental equation, which must be done numerically.[2] Second, if a latch is not included in the control loop, there is no guarantee that the switch will close and reopen exactly once in every ramp cycle. In fact, it turns out that the switch can operate any number of times, from zero to infinity. (In practice there is an upper bound, set by parasitic effects.) However, it is possible to write a subroutine that gives the state vector at a clock instant in terms of the state vector at the previous clock instant.

The discrete model of this converter can thus be expressed as a deterministic algorithm that allows numerical investigations. This is presented in detail in Section 5.3.

There are a few methods and tools in the theory of nonlinear dynamics that have proved very useful in studying the behavior of power electronic converters. Details of these will be presented in subsequent chapters; here we offer a brief introduction.

Suppose that the discrete model of such a system is available. For the buck converter in question, it is a two-dimensional map obtained by sampling the state variables once every ramp cycle. Now one can iterate the map starting from any initial condition and can plot the discrete-time evolution in the 2-D state space. The picture thus

---

2. For some other control schemes—current mode control, for example—this problem does not arise, and the map can be obtained in closed form.

obtained is called a phase-portrait (or Poincaré section) of the system. The phase portrait of the buck converter at $V_{in} = 35V$ is shown in Figure 1.6. The fact that it has an infinite number of points tells us that the waveform is aperiodic. We also find that the points are bounded within a definite region of the state space. If an initial condition is placed outside this region, in subsequent iterations of the map it moves to the set of points shown in the figure. It is as if points in the state space are *attracted* to this region in the state space and in this sense the region shown in the figure is called an *attractor*.

Attractors occur in stable periodic systems also—where initial conditions are attracted to a single stable fixed point of the map. There can also be stable behaviors where initial conditions are attracted to *two* points in the state space, and in the steady state it toggles between the two. This is then called a *period-2 attractor*. If an attractor contains an infinite number of points bounded within a definite region of the state space, the resulting behavior is called *chaos*. Thus chaos is a bounded aperiodic behavior of a system. Figure 1.6 suggests that at $V_{in} = 35V$, the buck converter behaves chaotically. Definitive tests of chaos involve the Lyapunov exponent. This is discussed in subsequent chapters.

The question then is, How did the system change from periodic behavior (which would mean iterates falling on the same point in the discrete state space) to that shown in Figure 1.6? Such changes in the behavior of the system occur when a parameter is changed. In this case at $V_{in} = 15V$ the system exhibited a nice periodic ripple, while at $V_{in} = 35V$ the ripple waveform was chaotic. To study such changes, one plots what is known as a *bifurcation diagram*. One parameter (in this case V_{in}) is varied while the others are kept fixed. The value of this parameter is plotted along the x-axis and the asymptotic steady-state behavior of one of the discrete state variables is plotted along the y-axis. The experimentally obtained bifurcation diagram of the buck converter is given in Figure 1.7. The method of obtaining such bifurcation diagrams is presented in Section 4.1.

This bifurcation diagram tells us that the periodic behavior was first transformed to the period-2 subharmonic, which subsequently changed to chaos. Such a qualitative change in the system behavior is called a *bifurcation*. The mathematics of bifurcation theory has been a subject of intense study over the past few years. Physicists and mathematicians have developed a theory of bifurcations that has proved very useful

Figure 1.6 The phase portrait of the buck converter at $V_{in} = 35V$ obtained from simulation.

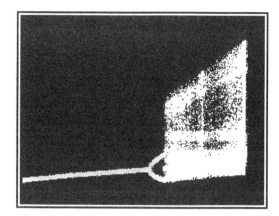

Figure 1.7 Experimental bifurcation diagram of the buck converter while V_{in} was varied from 20V to 35V. The sampled inductor current is plotted in the y-axis.

in studying nonlinear phenomena in power electronics. On the other hand, the peculiar features of power electronic systems have demanded a further development of bifurcation theory and thus spurred renewed research in that area [19,20,21]. With this theoretical understanding one can tell *why* one system behavior changes to the other as a parameter is varied.

1.2.4 Discontinuous Conduction Mode

Analysis of operation in DCM is more complicated, because there are now three circuit configurations during a cycle: S conducts for a fraction d of each switching cycle; D conducts for a time that depends on the circuit action and ceases when $i = 0$; and both S and D block for the remaining time. Thus equations (1.1a)–(1.1c) are all involved, together with a condition determining D's conduction interval. Despite the increased complexity, similar principles can be applied as for CCM. The averaging technique gives the linearized control-to-output transfer function as

$$\frac{\hat{v}}{\hat{d}} = \frac{A_0}{1 + s\tau} \tag{1.17}$$

where the dc gain A_0 is a function of f_s, L, R, and D, and the time constant τ is

$$\tau = \frac{CR}{2}\left(1 - \frac{1}{\sqrt{1 + 8f_s L/RD^2}}\right) \tag{1.18}$$

Note that the DCM model is of first order, not second order as might be expected. An explanation is that the inductor does not really enter into the long-run dynamics of the system. By definition i is zero at both the start and the finish of every cycle; the role of L is simply to set the amount of charge transferred from V_{in} to C. The change of order can be seen in the simulation of Figure 1.8, in which the load resistance R is stepped so that the CCM/DCM boundary is crossed. The pole at $s = -1/\tau$ is not fixed, but varies with the operating point.

Since a first-order system with proportional control has a phase margin greater than $90°$, its stability is expected to be extremely good. However, in experiment as well as in numerical simulation with the exact state equations, it is found that the DCM-operated buck converter also exhibits dynamical instabilities, bifurcations, and chaos.

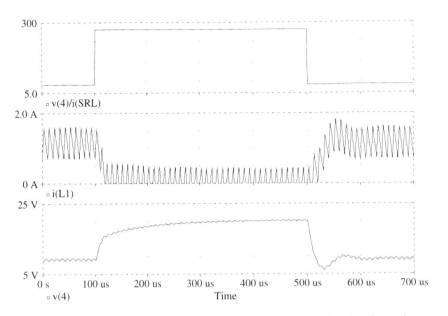

Figure 1.8 Transient response of open-loop buck converter, showing first-order
characteristics in DCM and second-order characteristics in CCM. Top
to bottom: variation of the load resistance R; inductor current i; output
voltage v.

In the nonlinear map-based modeling technique, if the state vector is sampled at the
start of each ramp cycle, the discrete system becomes truly one-dimensional: since $i = 0$
at every sample (assuming the converter stays in DCM), v is the only state variable.
From an approximate analysis, Tse [22] found a map $F : \mathbb{R} \mapsto \mathbb{R}$ of the form

$$v_{m+1} = \alpha v_m + \frac{\beta V_{\text{in}}(V_{\text{in}} - v_m)[\text{sat}(d_m)]^2}{v_m} \tag{1.19}$$

where α and β are constants involving f_s, L, C, and R, and sat($\cdot$) is a saturation function
that limits the duty ratio so that $d_m \in [0, 1]$. The value of d_m was set by a proportional
feedback scheme to $d_m = \overline{D} - A(v_m - V)$, where $\overline{D}$ and V are the steady-state (dc)
components of d and v respectively. Using the gain A as the bifurcation parameter, a
period-doubling route to chaos (see Sections 3.2 and 3.5 for explanation) was predicted,
and confirmed by simulation using the exact equations. Experimental results supported
the prediction.

1.2.5 Limitations and Extensions of Average Models

The above shows that the averaging process has some evident limitations. First, all
information about operation within a cycle is lost. Furthermore, the switching frequency f_s does not appear in the CCM model, though it must certainly have some
effect. A subtler point is that d is purportedly a continuous-time variable, yet the
duty ratio is defined in terms of discrete time. The averaging process is exact when
the perturbation frequency is zero, but is further in error the higher the perturbation
frequency. In fact the natural sampling inherent in PWM imposes a Nyquist limit of

$f_s/2$, beyond which the model is meaningless. Another problem is that the true duty ratio is constrained to the interval [0, 1], but the averaged variable d is not bounded (at least, not explicitly).

The conventional averaging technique gives a useful representation of the system and allows simple design procedures for operation in certain regimes. Nonlinear averaged models can also predict some cases of instability. For example, the onset of the first instability (Hopf bifurcation) in the autonomous Ćuk converter can be predicted with the nonlinear averaged model [23].

However, the averaged model is of little or no use in predicting and analyzing subharmonics and chaos of the sort exposed in our buck converter example. In the case of the Ćuk converter, where it can successfully predict the first instability, it failed to throw any light on the subsequent bifurcation sequences. More detailed analyses based on other models may therefore be warranted for safe and reliable operation of a power electronic system.

Considerable effort has been expended to validate and improve upon the basic averaging process. Sanders et al. [24] developed a generalized averaging method with greater applicability; Krein et al. [25] considered Bogoliubov averaging; Tymerski applied the theory of time-varying transfer functions [26] and Volterra series [27]; variable structure systems theory (sliding-mode control) was explored by Sira-Ramirez [28] and Bass [29]. These investigations build on sound theoretical bases, and usually recover state space averaging as the zero-order approximation, with higher terms giving more accurate results. Nonetheless, the simplistic averaging technique remains the most popular with practising power electronics engineers: it is easy to understand at some level, and straightforward to apply, particularly for dc/dc converters.

1.3 STUDY OF NONLINEAR DYNAMICS AND CHAOS IN POWER ELECTRONICS

Nonlinear dynamics is an older and broader field than power electronics. It is only relatively recently that power electronics researchers have begun using the techniques and ideas of nonlinear dynamics to analyze power electronic circuits. The history of chaotic dynamics can be traced back to the work of Henri Poincaré on celestial mechanics around 1900. However, the first inkling that chaos might be important in a real physical system was given in 1963 by Lorenz [30], who discovered the extreme sensitivity to initial conditions in a simplified computer model of atmospheric convection. Lorenz's paper, which appeared in an obscure journal, was largely overlooked for some years. Li and Yorke first used the term *chaos* in their 1975 paper "Period three implies chaos" [31]. In 1976, May published an influential article [32] describing how simple nonlinear systems can have complex, chaotic behavior. In the late 1970s, Feigenbaum analyzed the period-doubling cascades that form one of the commonest routes to chaos [33]. Over the past two decades there has been a great deal of advancement in the theory of nonlinear dynamics and it has been found that rich and interesting nonlinear phenomena are very common in a large number of physical systems.

Chaotic effects in electronic circuits were first noted by Van der Pol in 1927 [34,35]. A relaxation oscillator, comprising a battery, a neon bulb, a capacitor, and a resistor, was driven by a 1kHz sinusoidal signal and tuned to obtain subharmonics, but "an irregular noise" was often heard. There was little interest in explaining such spurious oscillations for about 50 years. In 1980 Baillieul, Brockett, and Washburn [36] suggested that chaos might occur in dc/dc converters and other control systems incorporating a pulse-width modulator (PWM). In 1981 Linsay published the first modern experimental report of electronic chaos [37]: a driven resonant circuit, employing a varactor diode as a nonlinear capacitor. The driven resistance-inductance-diode circuit has a close relative in power converters: when a transformer feeds a rectifier diode, the leakage inductance resonates with the diode's nonlinear capacitance to give a chaotic transient when excited by the switches.

In 1983 Chua and Matsumoto synthesized the first autonomous chaotic electronic circuit [38], the *double scroll oscillator*, now usually known simply as *Chua's circuit*, which has been widely studied as the archetypal chaotic electronic circuit [39].

In 1984, Brockett and Wood [40] presented a conference paper describing chaos in a controlled buck dc/dc converter. A 1988 letter by Hamill and Jefferies [41] was the first detailed analysis of chaos in power electronics. Wood further described chaos in a switching converter at a 1989 conference [42], and soon after a paper by Deane and Hamill [43] identified several other ways by which chaos might arise in power electronics. These ideas were further developed in [16,44,45], which are mainly concerned with prediction and experimental confirmation of chaos in dc/dc converters under various control schemes.

The initial investigations in this line were done with the exact differential equation models of these systems, which were integrated to obtain the trajectories [46]. With this kind of system description, it was difficult to go beyond empirical observation of the phenomena. After it was conclusively demonstrated that all feedback-controlled switching circuits are inherently nonlinear and many nonlinear phenomena occur in them, efforts were directed toward developing system models with which one can investigate such phenomena theoretically.

Taking the clue from system descriptions used in the nonlinear dynamics literature, Hamill and Deane proposed nonlinear map-based modeling [17]. Sampled-data modeling techniques of power electronic circuits presented in the textbook by Kassakian, Schlecht, and Verghese [4] helped in this development. In this method, one discretely observes the state variables at specific instants of time. It is clear that the choice of sampling instants is not unique. For example, in the buck converter system described in Section 1.2, one can sample the state variables in three possible ways:

1. At the beginning of each ramp cycle (clock instant)
2. At those clock instants that result in a switching event (i.e., skipped cycles are ignored)
3. At each switching event (i.e., when $v_{con} = v_{ramp}$)

Deane and Hamill proposed the second one in [17] and later applied that method to analyze the current-mode-controlled boost converter [44,45]. Di Bernardo et al. used the third method in analyzing the voltage-controlled buck converter [47,48,49,50]. On the other hand, Banerjee and Chakrabarty [51], Chan and Tse [52], and Marrero et al. [53] applied the first method (known as the stroboscopic sampling) in studying the dynamics of current-mode-controlled converters.

With these tools at their disposal, researchers focused on studying nonlinear phenomena occurring in specific converters and pulse-width modulation (PWM) schemes. The voltage-controlled buck converter and the current-mode-controlled boost converter have received high research attention—the first one because it exhibits a wealth of nonlinear phenomena and the second one because it is easy to obtain a closed form expression of the map, facilitating analytical treatment.

After Deane and Hamill presented the numerical as well as experimental bifurcation diagrams of the voltage-controlled buck converter [17], Fossas and Olivar [18] explored the stability of the periodic solutions and obtained the conditions of instability. Banerjee [54] observed that multiple attractors coexisting with the main attractor are responsible for the sudden expansion of the chaotic attractor through interior crisis. Di Bernardo et al. [49,50] explored the bifurcation sequence in detail and concluded that a period-5 orbit organizes the enlarged attractor in five zones.

Deane first obtained the nonlinear map-based model of the current-mode-controlled boost converter in closed form [45] by observing the state variables at every switch-on instant. Chan and Tse [52] obtained the equivalent stroboscopic map. This spurred research in two directions. Banerjee and Chakrabarty [55] tried to make the model more exact by including parasitic elements like the resistances of the inductor and the capacitor and showed that the model can still be obtained in closed form. On the other hand, it was shown that under certain reasonable assumptions the discrete model becomes a simple one-dimensional piecewise-linear map suitable for analytical treatment. Deane and Hamill [44] obtained the map under switch-on sampling and Banerjee et al. [56] obtained the map under stroboscopic sampling. With these tools, the bifurcation phenomena in this converter were studied in detail [52].

Nonlinear phenomena in other converter configurations were also investigated. Tse [57] showed that the boost converter under discontinuous conduction mode yields a one-dimensional smooth (continuous and everywhere differentiable) map, and the bifurcation phenomena (like repeated period doublings) expected in such maps occur in this system. Tse and Chan [58] also investigated the bifurcation phenomena in current-mode-controlled Ćuk converters.

In the initial years of the study of nonlinear phenomena in power electronics, dc/dc converters received most attention, mainly because this was the class of systems in which such phenomena were first detected. In 1992 and 1993, Dobson and his co-workers [59,60] showed that thyristor circuits used to model Static Var Control exhibit a new kind of bifurcation phenomenon in which switching times change discontinuously as a parameter is varied. The occurrence of this *switching time bifurcation* cannot be predicted from the Jacobian of the fixed point. They also showed that discrete modeling of such systems can yield discontinuous maps with multiple attractors.

The phenomenon of ferroresonance (a tuned circuit involving a saturating inductor) was studied in [61,62,63,64,65]. Its practical relevance is that it is exploited to regulate voltages, but unintended ferroresonance in power systems can cause excessive voltages and currents [66].

Some effort also has gone into investigating nonlinear phenomena in other (high-power) systems. Nagy et al. [67] investigated the current control of an induction motor drive; Magauer and Banerjee [68] reported quasiperiodicity, period doubling, chaos, and various crises in a system controlled by the tolerance-band PWM technique; Kuroe and Hayashi [69] observed and analyzed many interesting bifurcation phenomena in power electronic induction machine drive systems.

In all these studies, the essential method has been to obtain a discrete-time model of the systems under study and to analyze the observed phenomena in terms of the theory of bifurcations in maps developed by mathematicians and physicists. In many cases this approach worked well. In some cases, however, very atypical bifurcation phenomena were observed. For example, direct transition from a periodic orbit to a chaotic orbit was observed in [52,58,70] and nonsmooth period doubling was reported in [45,52]. These phenomena could not be explained in terms of the standard bifurcation theory developed for smooth (everywhere differentiable) maps.

Banerjee et al. showed [56,71] that in most of these systems the discrete model yields piecewise-smooth maps and the atypical bifurcations observed in such systems belong to a new class called *border-collision bifurcation*. Earlier, mathematicians like Nusse and Yorke [72,73] had shown that peculiar bifurcations can occur in piecewise-smooth maps, but at that time no physical examples of such systems were known. In fact, power electronic circuits offered the first examples of physical systems that yield piecewise-smooth maps. This created renewed interest in the theoretical analysis of piecewise-smooth systems; the conceptual framework for understanding and categorizing such bifurcations has recently been developed by Banerjee et al. [74,75], and some work done earlier by Feigin has been brought to the English-speaking world [21]. With this body of knowledge, many empirically observed bifurcation phenomena now have theoretical explanations.

It is now understood that all three basic types of maps occur in power electronics:

1. Smooth maps
2. Piecewise-smooth maps
3. Discontinuous maps

Bifurcation theory developed for these classes of maps is relevant in power electronics and helps us to understand *why* one type of system behavior changes to another as a parameter is varied.

Recently there has been renewed interest in interconnected systems of converters due to the increased demand for better flexibility in high-current, high-power applications. It has been reported that parallel connected systems of dc/dc converters exhibit a range of bifurcation behaviors, such as period doubling, border-collision [76], Neimark-Sacker bifurcation [77], and so on.

Many of the theoretical investigations outlined above have been backed by experiments. Deane and Hamill reported experimental observations of the bifurcations in the voltage-controlled buck converter [43,17]. Tse et al. have backed up their numerical investigation on the current-mode-controlled DCM boost converter [57] and CCM boost converter [78] through experimental results. Chakrabarty, Poddar, and Banerjee have reported experimental studies on the buck converter [46] and the boost converter [51].

Attempts to control chaos into periodic orbits have been a hot pursuit of researchers in nonlinear dynamics ever since Ott, Grebogi, and Yorke published their pioneering work in 1990 [79]. Subsequently, various methods for controlling chaos were developed by physicists and mathematicians, and have been applied in practical systems like lasers. Similar methods also have been developed for power electronic circuits. Poddar, Chakrabarty, and Banerjee reported experimental control of chaos in the buck converter and the boost converter [80,81]. Di Bernardo developed an adaptive control technique [82]. Batlle, Fossas, and Olivar reported the time-delay stabilization of periodic orbits in a current-mode-controlled boost converter [83]. In a 1995 review paper, Hamill [84] conjectured that power electronic converters operating under controlled chaos (instead of a stable periodic orbit) may have a better dynamic response—

just as fighter aircraft are designed to be open-loop unstable but are then stabilized by feedback, making them more agile than conventional designs. Similarly, stabilized chaotic power converters may react more quickly, for instance in moving rapidly from one commanded output voltage to another. However, quantitative understanding of this possibility has yet to emerge.

After a reasonable understanding of the nonlinear phenomena in power electronics is obtained, one has to address the question: Can we make engineering use of them? One possible area of application is in reducing electromagnetic interference (EMI) in switch-mode power supplies, which are notorious generators of both conducted and radiated EMI, owing to the high rates of change of voltage and current which are necessary for efficient operation. The problem is particularly acute in the aviation sector, and a number of electromagnetic compatibility regulations are coming into force.

Efforts have been made to counter the problem by spreading the spectrum of converters through pseudorandom modulation of the clock frequency [85,86]. The first suggestion that this problem can also be attacked with deliberate use of chaos came in Hamill's review paper [84]. Then in [53] Marrero, Font, and Verghese observed that "a potential advantage of chaotic operation is that the switching spectrum is flattened." These were followed up by Deane and Hamill in 1996 [87], who experimentally demonstrated a reduction of the spectral peaks when a converter was operated in chaos.

However, in order to bring this possibility into engineering practice, some theoretical issues needed to be addressed. First, in order to formulate design procedures for chaotically operated converters, one needs a theory for calculating the average values of state variables under chaos. Second, one needs a theory to predict the structure of the power spectrum of a chaotically operated power converter. Third, it is known that in most chaotic systems there are periodic windows in the parameter space, and a slight inadvertent change of a parameter can bring the system out of chaos. Often, coexisting stable orbits occur in such nonlinear systems and ambient noise may knock the system from one orbit to the other. How, then, can we ensure reliable operation of a converter under chaos?

First attacks on all these theoretical problems have been made. Current mode controlled dc/dc converters have been used for these studies since it is known that under certain reasonable assumptions they give rise to piecewise-linear one-dimensional maps. Isabelle [88] argued that these piecewise-linear maps can be approximated to a smaller class known as *Markov maps*—for which calculation of average values is tractable. This idea has been further developed by Marrero et al. and is presented in Section 4.3 of this book.

The second problem has been attacked for dc/dc converters that can be modeled by piecewise-linear one-dimensional maps: Deane et al. [89] developed a method for calculating the line spectrum at the switching frequency and its harmonics, and Baranovski et al. [90,91] extended the analysis to the continuous part of the spectrum.

In attacking the third problem, there have been two approaches. Banerjee, Yorke, and Grebogi developed the theory of robust chaos [19]—the analytical condition under which there would be no periodic window or coexisting attractor in a chaotic system—and demonstrated that such a condition does occur in current-mode-controlled converters. On the other hand, Bueno and Marrero applied the chaos-control technique to stabilize the chaotic regime [92].

Over the past few years there has been a rapid growth of our understanding of nonlinear phenomena in power electronics. The rest of this book reflects our current state of understanding.

1.4 CONCLUSIONS

High-efficiency solid-state power conversion has become possible through the continuing development of high-power semiconductor devices. The operation of these devices as switches, which is necessary for high efficiency, means that power electronic circuits are essentially nonlinear time-varying dynamical systems. Though this makes them difficult to study, the effort is well worthwhile because they have many practical applications and are becoming increasingly important in the delivery and utilization of electrical energy. The conventional modeling approach generally ignores nonlinear effects, and can sometimes mislead the designer into thinking a circuit will perform acceptably when in practice it will not. Thus the traditional approach does not always produce reliable models. Discrete nonlinear modeling offers another way of looking at the circuits, one that is more accurate and able to reproduce nonlinear effects such as subharmonics and chaos. Unfortunately it demands a mental shift on the part of power electronics engineers, away from linear systems thinking and toward the unfamiliar realm of nonlinear dynamics.

REFERENCES

[1] B. K. Bose, *Modern Power Electronics: Evolution, Technology and Applications.* New York: IEEE Press, 1992.

[2] R. E. Tarter, *Solid-State Power Conversion Handbook.* New York: Wiley-Interscience, 1993.

[3] N. Mohan, T. M. Undeland, and W. P. Robbins, *Power Electronics: Converters, Applications and Design.* New York: Wiley, second ed., 1995.

[4] J. G. Kassakian, M. F. Schlecht, and G. C. Verghese, *Principles of Power Electronics.* Reading, MA: Addison-Wesley, 1991.

[5] I. Dobson and H.-D. Chiang, Towards a theory of voltage collapse in electric power systems, *Systems and Control Letters,* vol. 13, pp. 253–262, 1989.

[6] H. G. Kwatny, A. K. Pasrija, and L. H. Bahar, Static bifurcations in electric power networks: Loss of steady-state stability and voltage collapse, *IEEE Trans. on Circuits and Systems,* vol. 33, pp. 981–991, October 1986.

[7] H. D. Chiang, C. W. Liu, P. P. Varaiya, F. F. Wu, and M. G. Lauby, Chaos in a simple power system, *IEEE Trans. on Power Systems,* vol. 8, pp. 1407–1417, November 1993.

[8] B. Lee and V. Ajjarapu, Period-doubling route to chaos in an electrical power system, *IEE Proc., Part C,* vol. 140, no. 6, pp. 490–496, 1993.

[9] H. O. Wang, E. Abed, and A. M. A. Hamdan, Bifurcations, chaos, and crises in voltage collapse of a model power system, *IEEE Trans. on Circuits and Systems, Part I,* vol. 41, pp. 294–302, March 1994.

[10] G. W. Wester and R. D. Middlebrook, Low-frequency characterization of switched dc-to-dc converters, *IEEE Power Processing and Electronics Specialists' Conf.* (Atlantic City), pp. 9–20, May 1972.

[11] R. D. Middlebrook and S. Ćuk, A general unified approach to modeling switching converter power stages, *IEEE Power Electronics Specialists' Conference* (Cleveland), pp. 18–34, 1976.

[12] S. Ćuk and R. D. Middlebrook, A general unified approach to modeling switching dc-to-dc converters in discontinuous conduction mode, *Power Electronics Specialists' Conf.*, pp. 36–57, 1977.

[13] A. S. Kislovski, R. Redl, and N.O. Sokal, *Dynamic Analysis of Switching-Mode DC/DC Converters*. New York: Van Nostrand Reinhold, 1991.

[14] V. Vorperian, Simplified analysis of PWM converters using model of PWM switch—Part I: Continuous conduction mode, *IEEE Trans. on Aero. and Elec. Systems*, vol. 26, pp. 490–496, May 1990.

[15] V. Vorperian, Simplified analysis of PWM converters using model of PWM switch—Part II: Discontinuous conduction mode, *IEEE Trans. on Aero. and Elec. Systems*, vol. 26, pp. 497–505, May 1990.

[16] J. H. B. Deane and D. C. Hamill, Analysis, simulation and experimental study of chaos in the buck converter, *IEEE Power Electronics Specialists' Conference* (San Antonio), pp. 491–498, 1990.

[17] D. C. Hamill and J. H. B. Deane, Modeling of chaotic dc-dc converters by iterated nonlinear mappings, *IEEE Trans. on Power Electronics*, vol. 7, pp. 25–36, January 1992.

[18] E. Fossas and G. Olivar, Study of chaos in the buck converter, *IEEE Trans. on Circuits and Systems—I*, vol. 43, no. 1, pp. 13–25, 1996.

[19] S. Banerjee, J. A. Yorke, and C. Grebogi, Robust chaos, *Physical Review Letters*, vol. 80, pp. 3049–3052, 1998.

[20] S. Banerjee and C. Grebogi, Border collision bifurcations in two-dimensional piecewise smooth maps, *Physical Review E*, vol. 59, no. 4, pp. 4052–4061, 1999.

[21] M. di Bernardo, M. I. Feigin, S. J. Hogan, and M. E. Homer, Local analysis of C-bifurcations in n-dimensional piecewise smooth dynamical systems, *Chaos, Solitons & Fractals*, vol. 10, no. 11, pp. 1881–1908, 1999.

[22] C. K. Tse, Chaos from a buck switching regulator operating in discontinuous mode, *Int. J. Circuit Theory and Applications*, vol. 22, pp. 262–278, July–August 1994.

[23] C. K. Tse, Y. M. Lai, and H. H. C. Iu, Hopf bifurcation and chaos in a free-running autonomous Ćuk converter, *IEEE Trans. on Circuits and Systems, Part I*, vol. 47, pp. 448–457, April 2000.

[24] S. R. Sanders, J. M. Noworolski, X. Z. Liu, and G. C. Verghese, Generalized averaging method for power conversion circuits, *IEEE Trans. on Power Electronics*, vol. 6, no. 2, pp. 251–259, 1991.

[25] P.T. Krein, J. Bentsman, R. M. Bass, and B. C. Lesieutre, On the use of averaging for the analysis of power electronic systems, *IEEE Trans. on Power Electronics*, vol. 5, no. 2, pp. 182–190, 1990.

[26] R. Tymerski, Application of the time-varying transfer function for exact small-signal analysis, *IEEE Trans. on Power Electronics*, vol. 9, no. 2, pp. 196–205, 1994.

[27] R. Tymerski, Volterra series modelling of power conversion systems, *IEEE Trans. on Power Electronics*, vol. 6, no. 4, pp. 712–718, 1991.

[28] H. Sira-Ramirez and M. Rios-Bolivar, Sliding mode control of dc-to-dc power converters via extended linearization, *IEEE Trans. on Circuits and Systems, Part I*, vol. 41, no. 10, pp. 652–661, 1994.

[29] R. M. Bass, B. S. Heck, and R. A. Khan, Average modelling of current-controlled converters: Instability predictions, *Int. J. Electronics*, vol. 77, no. 5, pp. 613–628, 1994.

[30] E. N. Lorenz, Deterministic nonperiodic flow, *J. Atmospheric Sciences*, vol. 20, no. 2, pp. 130–141, 1963.

[31] T. Y. Li and J. A. Yorke, Period three implies chaos, *American Mathematical Monthly*, vol. 82, pp. 985–992, 1975.

[32] R. M. May, Simple mathematical models with very complicated dynamics, *Nature*, vol. 261, pp. 459–467, June 1976.

[33] M. J. Feigenbaum, Universal behavior in nonlinear systems, in *Universality in Chaos* (Predrag Cvitanovic, ed.), Bristol, U.K.: Adam Hilger Ltd., 1984.

[34] B. van der Pol and J. van der Mark, Frequency demultiplication, *Nature*, vol. 120, pp. 363–364, September 1927.

[35] M. P. Kennedy and L. O. Chua, Van der Pol and chaos, *IEEE Trans. on Circuits and Systems*, vol. 33, pp. 974 980, October 1986.

[36] J. Baillieul, R. W. Brockett, and R. B. Washburn, Chaotic motion in nonlinear feedback systems, *IEEE Trans. on Circuits and Systems*, vol. 27, pp. 990–997, November 1980.

[37] P. S. Linsay, Period doubling and chaotic behavior in a driven anharmonic oscillator, *Physical Review Letters*, vol. 47, pp. 1349–1352, November 1981.

[38] L. O. Chua, The genesis of Chua's circuit, *Archiv für Elektronik und Ubertragungstechnik*, vol. 46, no. 4, pp. 250–257, 1992.

[39] L. O. Chua, Chua's circuit 10 years later, *Int. J. Circuit Theory and Appl.*, vol. 22, no. 4, pp. 279–305, 1994.

[40] R. W. Brockett and J. R. Wood, Understanding power converter chaotic behavior in protective and abnormal modes, in *Powercon II*, 1984.

[41] D. C. Hamill and D. J. Jefferies, Subharmonics and chaos in a controlled switched-mode power converter, *IEEE Trans. on Circuits and Systems*, vol. 35, no. 8, pp. 1059–1061, 1988.

[42] J. R. Wood, Chaos: A real phenomenon in power electronics, *Applied Power Electronics Conference* (Baltimore), pp. 115–124, March 1989.

[43] J. H. B. Deane and D. C. Hamill, Instability, subharmonics, and chaos in power electronics circuits, *IEEE Trans. on Power Electronics*, vol. 5, pp. 260–268, July 1990.

[44] J. H. B. Deane and D. C. Hamill, Chaotic behaviour in current-mode controlled dc-dc converter, *Electronics Letters*, vol. 27, pp. 1172–1173, June 1991.

[45] J. H. B. Deane, Chaos in a current-mode controlled boost dc-dc converter, *IEEE Trans. on Circuits and Systems—I*, vol. 39, pp. 680–683, August 1992.

[46] K. Chakrabarty, G. Podder, and S. Banerjee, Bifurcation behavior of buck converter, *IEEE Trans. on Power Electronics*, vol. 11, no. 3, pp. 439–447, 1996.

[47] M. di Bernardo, F. Garofalo, L. Glielmo, and F. Vasca, Quasi-periodic behaviors in dc/dc converters, *IEEE Power Electronics Specialists' Conference*, pp. 1376–1381, 1996.

[48] M. di Bernardo, C. Budd, and A. Champneys, Grazing, skipping and sliding: analysis of the non-smooth dynamics of the dc-dc buck converter, *Nonlinearity*, vol. 11, no. 4, pp. 858–890, 1998.

[49] M. di Bernardo, E. Fossas, G. Olivar, and F. Vasca, Secondary bifurcations and high-periodic orbits in voltage controlled buck converter, *Int. J. Bifurcation and Chaos*, vol. 7, no. 12, pp. 2755–2771, 1997.

[50] M. di Bernardo, F. Garofalo, L. Glielmo, and F. Vasca, Switchings, bifurcations and chaos in dc-dc converters, *IEEE Trans. on Circuits and Systems—I*, vol. 45, no. 2, pp. 133–141, 1998.

[51] S. Banerjee and K. Chakrabarty, Nonlinear modeling and bifurcations in the boost converter, *IEEE Trans. on Power Electronics*, vol. 13, no. 2, pp. 252–260, 1998.

[52] W. C. Y. Chan and C. K. Tse, Study of bifurcations in current programmed dc/dc boost converters: from quasiperiodicity to period doubling, *IEEE Trans. on Circuits and Systems—I*, vol. 44, no. 12, pp. 1129–1142, 1997.

[53] J. L. R. Marrero, J. M. Font, and G. C. Verghese, Analysis of the chaotic regime for dc-dc converters under current mode control, *IEEE Power Electronics Specialists' Conference*, pp. 1477–1483, 1996.

[54] S. Banerjee, Coexisting attractors, chaotic saddles and fractal basins in a power electronic circuit, *IEEE Trans. on Circuits and Systems—I*, vol. 44, no. 9, pp. 847–849, 1997.

[55] S. Banerjee and K. Chakrabarty, Nonlinear modeling and bifurcations in the boost converter, *IEEE Trans. on Power Electronics*, vol. 13, no. 2, pp. 252–260, 1998.

[56] S. Banerjee, E. Ott, J. A. Yorke, and G. H. Yuan, Anomalous bifurcations in dc-dc converters: Borderline collisions in piecewise smooth maps, *IEEE Power Electronics Specialists' Conference*, pp. 1337–1344, 1997.

[57] C. K. Tse, Flip bifurcation and chaos in three-state boost switching regulators, *IEEE Trans. on Circuits and Systems—I*, vol. 41, pp. 16–23, January 1994.

[58] C. K. Tse and W. C. Y. Chan, Instability and chaos in a current-mode controlled Ćuk converter, *IEEE Power Electronics Specialists' Conference*, 1995.

[59] S. G. Jalali, I. Dobson, and R. H. Lasseter, Instabilities due to bifurcation of switching times in a thyristor controlled reactor, *IEEE Power Electronics Specialists' Conference* (Toledo, Spain), pp. 546–552, May 28–31, 1992.

[60] R. Rajaraman, I. Dobson, and S.G. Jalali, Nonlinear dynamics and switching time bifurcations of a thyristor controlled reactor, *IEEE Int. Symposium on Circuits & Systems* (Chicago), pp. 2180–2183, May 1993.

[61] L. O. Chua, M. Hasler, J. Neirynck, and P. Verburgh, Dynamics of a piecewise-linear resonant circuit, *IEEE Trans. on Circuits and Systems*, vol. 29, pp. 535–547, August 1982.

[62] C. Kieny, Application of the bifurcation theory in studying and understanding the global behavior of a ferroresonant electric power circuit, *IEEE Trans. on Power Delivery*, vol. 6, pp. 866–872, April 1991.

[63] A. E. A. Araujo, A. C. Soudack, and J. R. Marti, Ferroresonance in power systems, *IEE Proc., Part C*, vol. 140, pp. 237–240, May 1993.

[64] J. H. B. Deane, Modeling of a chaotic circuit containing a saturating/hysteretic inductor, *Electronics Letters*, vol. 29, pp. 957–958, May 1993.

[65] J. H. B. Deane, Modeling the dynamics of nonlinear inductor circuits, *IEEE Trans. on Magnetics*, vol. 30, pp. 2795–2801, September 1994.

[66] J. A. Mohamed, Existence and stability analysis of ferroresonance using the generalized state-space averaging technique. PhD diss., EECS Dept., MIT, February 2000.

[67] Z. Sütő, I. Nagy, and E. Masada, Avoiding chaotic processes in current control of ac drive, *IEEE Power Electronics Specialists' Conference*, pp. 255–261, 1998.

[68] A. Magauer and S. Banerjee, Bifurcations and chaos in the tolerance band PWM technique, *IEEE Trans. on Circuits and Systems—I*, vol. 47, pp. 254–259, February 2000.

[69] Y. Kuroe and S. Hayashi, Analysis of bifurcation in power electronic induction motor drive systems, *IEEE Power Electronics Specialists' Conference*, pp. 923–930, 1989.

[70] M. Ohnishi and N. Inaba, A singular bifurcation into instant chaos in a piecewise linear circuit, *IEEE Trans. on Circuits and Systems—I*, vol. 41, no. 6, pp. 433–442, 1994.

[71] G. H. Yuan, S. Banerjee, E. Ott, and J. A. Yorke, Border collision bifurcations in the buck converter, *IEEE Trans. on Circuits and Systems—I*, vol. 45, no. 7, pp. 707–716, 1998.

[72] H. E. Nusse and J. A. Yorke, Border-collision bifurcations including "period two to period three" for piecewise smooth maps, *Physica D*, vol. 57, pp. 39–57, 1992.

[73] H. E. Nusse and J. A. Yorke, Border-collision bifurcations for piecewise smooth one dimensional maps, *Int. J. Bifurcation and Chaos*, vol. 5, no. 1, pp. 189–207, 1995.

[74] S. Banerjee, M. S. Karthik, G. H. Yuan, and J. A. Yorke, Bifurcations in one-dimensional piecewise smooth maps—theory and applications in switching circuits, *IEEE Trans. on Circuits and Systems—I*, vol. 47, no. 3, 2000.

[75] S. Banerjee, P. Ranjan, and C. Grebogi, Bifurcations in two-dimensional piecewise smooth maps—theory and applications in switching circuits, *IEEE Trans. on Circuits and Systems—I*, vol. 47, no. 5, 2000.

[76] H. H. C. Iu and C. K. Tse, Instability and bifurcation in parallel-connected buck converters under a master-slave current sharing scheme, *IEEE Power Electronics Specialists' Conference* (Galway, Ireland), pp. 708–713, June 2000.

[77] H. H. C. Iu and C. K. Tse, Bifurcation in parallel-connected boost dc/dc converters, *IEEE Int. Symp. on Circuits and Systems* (Geneva, Switzerland), pp. II-473–476, June 2000.

[78] C. K. Tse and W. C. Y. Chan, Experimental verification of bifurcations in current-programmed dc/dc boost converters: from quasi-periodicity to period-doubling, *European Conf. on Circuit Theory and Design*, (Budapest), pp. 1274–1279, September 1997.

[79] E. Ott, C. Grebogi, and J. A. Yorke, Controlling chaos, *Physical Review Letters*, vol. 64, no. 11, pp. 1196–1199, 1990.

[80] G. Podder, K. Chakrabarty, and S. Banerjee, Experimental control of chaotic behavior of buck converter, *IEEE Trans. on Circuits and Systems—I*, vol. 42, no. 8, pp. 100–101, 1995.

[81] G. Podder, K. Chakrabarty, and S. Banerjee, Control of chaos in the boost converter, *Electronics Letters (IEE)*, vol. 31, no. 11, p. 25, 1995.

[82] M. Di Bernardo, An adaptive approach to the control and synchronization of continuous-time chaotic systems, *Int. J. Bifurcation and Chaos*, vol. 6, pp. 557–568, 1996.

[83] C. Batlle, E. Fossas, and G. Olivar, Time-delay stabilization of periodic orbits of the current mode controlled boost converter, *Linear Time Delay Systems* (L. D. J. Dion and M. Fliess, eds.), pp. 111–116, 1998.

[84] D. C. Hamill, Power electronics: A field rich in nonlinear dynamics, *3rd Int. Specialists' Workshop on Nonlinear Dynamics of Electronic Systems* (University College, Dublin), pp. 165–178, 1995.

[85] A. M. Stanković, G. C. Verghese, and D. J. Perreault, Analysis and synthesis of randomized modulation schemes for power converters, *IEEE Trans. on Power Electronics*, vol. 10, no. 6, pp. 680–693, 1995.

[86] A. M. Stanković, G. C. Verghese, and D. J. Perreault, Randomized modulation of power converters via Markov chains, *IEEE Trans. Control Systems Technology*, vol. 5, pp. 61–73, January 1997.

[87] J. H. B. Deane and D. C. Hamill, Improvement of power supply EMC by chaos, *Electronic Letters*, vol. 32, p. 1045, June 1996.

[88] S. H. Isabelle, A signal processing framework for the analysis and applications of chaotic systems. PhD diss., EECS Department, MIT, 1995.

[89] J. H. B. Deane, P. Ashwin, D. C. Hamill, and D. J. Jefferies, Calculation of the periodic spectral components in a chaotic dc-dc converter, *IEEE Trans. on Circuits and Systems—I*, vol. 46, no. 11, pp. 1313–1319, 1999.

[90] A. L. Baranovski, A. Mögel, W. Schwarz, and O. Woywode, Statistical analysis of a dc-dc converter, *Nonlinear Dynamics of Electronic Systems* (Ronne, Bornholm, Denmark), July 15–18 1999.

[91] A. L. Baranovski, A. Mögel, W. Schwarz, and O. Woywode, Chaotic control of a dc-dc converter, *IEEE Int. Symp. on Circuits & Systems* (Geneva, Switzerland), May 28–31 2000.

[92] R. S. Bueno and J. L. R. Marrero, Control of a boost dc-dc converter in the chaotic regime, *IEEE Int. Conf. on Control Applications*, pp. 832–837, 1998.

Chapter 2

DYNAMIC MODELS OF POWER CONVERTERS

2.1 INTRODUCTION TO POWER ELECTRONIC CONVERTERS AND MODELS

George C. Verghese
Alex M. Stanković

2.1.1 Introduction

This section is aimed at the reader who is not already familiar with the basic types of circuitry that characterize power electronics. The section will also benefit the reader who—though familiar with power electronics—has not been exposed to the principles by which tractable dynamic models are obtained for analysis, simulation, and control design of power converters. A more detailed development may be found in [1] and in the relevant portions of [2].

2.1.2 Types of Power Electronic Converters

This subsection briefly describes the structure and operating principles of some basic power electronic converters; these (or related) converters appear in later chapters of this book. Many power electronic systems involve *combinations* of such basic converters. For instance, a high-quality power supply for electronic equipment might comprise a unity-power-factor, pulse-width-modulated (PWM) rectifier cascaded with a PWM dc/dc converter; a variable-frequency drive for an ac motor might involve a rectifier followed by a variable-frequency inverter.

High-Frequency PWM DC/DC Converters

Given a dc voltage of value V (which can represent an input dc voltage, or an output dc voltage, or a dc difference between input and output voltages), we can easily arrange for a controlled switch to chop the dc waveform into a *pulse waveform* that alternates between the values V and 0 at the switching frequency. This pulse waveform can then be low-pass-filtered with capacitors and/or inductors that are configured to respond to its average value (i.e., its dc component). By controlling the *duty ratio* of the switch (i.e., the fraction of time that the switch is closed in each cycle), we can control the fraction of time that the pulse waveform takes the value V, and thereby control the dc component of this waveform. This control approach is referred to as *pulse-width modulation* (PWM).

An important class of dc/dc converters is based on the above principle. They are referred to as *switching regulators* or *switched-mode* converters or high-frequency PWM dc/dc converters. Switching frequencies in the range of 15–300kHz are common.

The *boost* (or voltage step-up) converter in Figure 2.1 is a high-frequency PWM dc/dc converter of significant practical interest. Here V_{in} denotes the voltage of a dc source, while the voltage across the load (modeled for simplicity as being just a resistor) is essentially a dc voltage V_o, with some small switching-frequency ripple superimposed on it. The values of L and C are chosen such that the ripple in the output voltage is a suitably small percentage (typically $< 5\%$) of the nominal load voltage. The left terminal of the inductor is held at a potential of V_{in} relative to ground, while its right terminal sees a pulse waveform that is switched between 0 (when the transistor is on, with the diode blocking) and the output voltage (when the transistor is off, with the diode conducting). In nominal operation, the transistor is switched periodically with duty ratio D, so the average potential of the inductor's right terminal is approximately $(1 - D)V_o$. A periodic steady state is attained only when the inductor has zero average voltage across itself, i.e., when

$$V_o \approx \frac{V_{in}}{(1 - D)} \tag{2.1}$$

Otherwise the inductor current at the end of a switching cycle would not equal that at the beginning of the cycle, which contradicts the assumption of a periodic steady state. Since $0 < D < 1$, we see from (2.1) that the output dc voltage is *higher* than the input dc voltage, which is why this converter is termed a *boost* converter.

Other High-Frequency PWM Converters

Appropriate control of a high-frequency PWM dc/dc converter also enables conversion between waveforms that are not dc, but that are nevertheless slowly varying relative to the switching frequency. If, for example, the input is a slowly varying unidirectional voltage—such as the waveform obtained by rectifying a 60Hz sinewave—while the converter is switched at a much higher rate, say 50kHz, then we can still arrange for the output of the converter to be essentially dc. The result would be a so-called *PWM rectifier*.

In a high-frequency *PWM inverter*, the situation is reversed. The heart of it is still a dc/dc converter, and the input to it is dc. However, the switching is controlled in such a way that the filtered output is a slowly varying rectified sinusoid at the desired frequency. This rectified sinusoid can then be unfolded into the desired sinusoidal ac

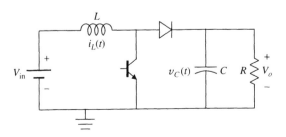

Figure 2.1 The average voltage across the inductor must be zero in the periodic steady state that results when the transistor is switched periodically with duty ratio D. Also, if the switching frequency is high enough, the output voltage is essentially a dc voltage V_o. It follows from these facts that $V_o \approx V_{in}/(1 - D)$ in the steady state, so the boost converter steps up the dc input voltage to a *higher* dc output voltage.

waveform, through the action of additional controllable switches arranged in a bridge configuration. In fact, both the chopping and unfolding functions can be carried out by the bridge switches, and the resulting high-frequency PWM bridge inverter is the most common implementation, available in single-phase and three-phase versions. These inverters are often found in drives for ac servo-motors, such as the permanent-magnet synchronous motors (also called *brushless dc* motors) that are popular in robotic applications. The inductive windings of the motor perform all or part of the electrical low-pass filtering in this case, while the motor inertia provides the additional mechanical filtering that practically removes the switching-frequency component from the mechanical motion.

Other Inverters

Another common approach to constructing inverters again relies on a pulse waveform created by chopping a dc voltage of value V, but with the frequency of the pulse waveform now *equal* to that of the desired ac waveform, rather than much higher. Also, the pulse waveform is now generally caused (again through controllable switches configured in a bridge arrangement) to have a mean value of zero, taking values of V, 0, and $-V$, for instance. Low-pass filtering of this pulse waveform to keep only the fundamental and reject harmonics yields an essentially sinusoidal ac waveform at the switching frequency. The amplitude of the sinusoid can be controlled by varying the duty ratio of the switches that generate the pulse waveform; this may be thought of as low-frequency PWM. It is easy to arrange for the pulse waveform to have no even harmonics, and more elaborate design of the waveform can eliminate designated low-order (e.g., third, fifth, and seventh) harmonics, in order to improve the effectiveness of the low-pass filter. This sort of inverter might be found in variable-frequency drives for large ac motors, operating at power levels where the high-frequency PWM inverters described in the previous paragraph would not be practical (because of limitations on switching frequency that become dominant at higher power levels). The low-pass filtering again involves using the inductive windings and inertia of the motor.

Resonant Converters

There is an alternative approach to controlling the output amplitude of a dc/ac converter such as that presented in the previous paragraph. Rather than varying the duty ratio of the pulse waveform, a resonant inverter uses frequency variations. In such an inverter, a resonant bandpass filter (rather than a low-pass filter) is used to extract the sinewave from the pulse waveform; the pulse waveform no longer needs to have zero mean. The amplitude of the sinewave is strongly dependent on how far the switching frequency is from resonance, so control of the switching frequency can be used to control the amplitude of the output sinewave.

If the sinusoidal waveform produced by a resonant inverter is rectified and low-pass filtered, what is obtained is a resonant dc/dc converter, as opposed to a PWM dc/dc converter. This form of dc/dc converter can have lower switching losses and generate less electromagnetic interference (EMI) than a typical high-frequency PWM dc/dc converter operating at the same switching frequency, but these advantages come at the cost of higher peak currents and voltages, and therefore higher component stresses.

Phase-Controlled Converters

A diode bridge is commonly used to convert an ac waveform into a unidirectional or rectified waveform. Using *controllable* switches instead of the diodes allows us to *partially* rectify a sinusoidal ac waveform, with subsequent low-pass filtering to obtain an essentially dc waveform at a specified level. This is the basis for phase-controlled rectifiers, which are used as drives for dc motors or as battery-charging circuits.

AC/AC Converters

For ac/ac conversion between waveforms of the *same frequency*, we can use switches to window out sections of the source waveform, thereby reducing the fundamental component of the waveform in a controlled way. Subsequent filtering can be used to extract the fundamental of the windowed waveform. More intricate use of switches—in a *cycloconverter*—permits the construction of an approximately sinusoidal waveform at some specified frequency by splicing together appropriate segments of a set of three-phase (or multiphase) sinusoidal waveforms at a *higher* frequency; again, subsequent filtering improves the quality of the output sinusoid. While cycloconverters and matrix converters effect a direct ac/ac conversion, it is also common to construct an ac/ac converter as a cascade of a rectifier and an inverter (generally operating at *different* frequencies), forming a *dc-link converter*.

2.1.3 Averaged and Sampled-Data Models for Analysis, Simulation, and Control of Converter Dynamics

Elementary circuit analysis of a power converter typically produces detailed, continuous-time, nonlinear, time-varying models in state-space form (for details about state-space formalism, see Section 3.1). These models have rather low order, provided one makes approximations that are reasonable from the viewpoint of control-oriented modeling: neglecting dynamics that occur at much higher frequencies than the switching frequency (e.g., dynamics due to parasitics, or to *snubber* elements that are introduced around the switches to temper the switch transitions), and focusing instead on components that are central to the power processing and control functions of the converter.

Such models capture essentially all the effects that are likely to be significant for analysis of energy conversion, but they are generally still too detailed and awkward to work with. The first challenge, therefore, is to extract from such a detailed model a simplified approximate model, preferably time-invariant, that is well matched to the particular analysis or control task for the converter being considered. There are systematic ways to obtain such simplifications, notably through:

- *Averaging*, which blurs out the detailed switching artifacts
- *Sampled-data modeling*, again to suppress the details internal to a switching cycle, focusing instead on cycle-to-cycle behavior

Both methods can produce time-invariant but still nonlinear models. In the remainder of this section, we illustrate the preceding comments through a more detailed examination of the boost converter that was introduced in the previous section. Extensions to other converters can be made along similar lines.

The boost converter of Figure 2.1 is redrawn in Figure 2.2 with some additions. The figure includes a schematic illustration of a typical analog PWM control method that uses output feedback. This control configuration is routinely and widely used, in a single-chip implementation; its operation will be explained shortly. We have allowed the input voltage $v_{in}(t)$ in the figure to be time varying, to allow for a source that is nominally dc at the value V_{in}, but that has some time-varying deviation or ripple around this value. Although a more realistic model of the converter for control design would also, for instance, include the equivalent series resistance—or ESR—of the output capacitor, such refinements can be ignored for our purposes here; they can easily be incorporated once the simpler case is understood. The rest of our development will therefore be for the model in Figure 2.2.

In typical operation of the boost converter under what may be called constant-frequency PWM control, the transistor in Figure 2.1 is turned on every T seconds, and turned off $d_k T$ seconds later in the kth cycle, $0 < d_k < 1$, so d_k represents the duty ratio

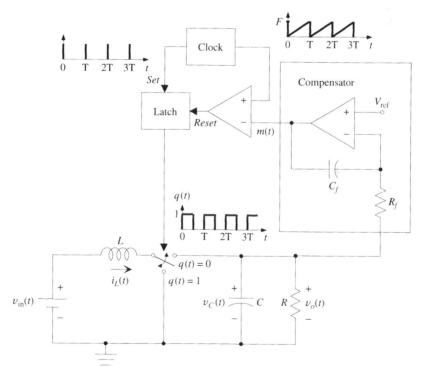

Figure 2.2 Controlling the boost converter. The operation of the switch is controlled by the latch. The switch is moved down every T seconds by a set pulse from the clock to the latch ($q(t) = 1$). The clock simultaneously initiates a ramp input of slope F/T to one terminal of the comparator. The modulating signal $m(t)$ is applied to the other terminal of the comparator. At the instant t_k in the kth cycle when the ramp crosses the level $m(t_k)$, the comparator output goes high, the latch resets ($q(t) = 0$), and the switch is moved up. The resulting duty ratio d_k in the kth cycle is $m(t_k)/F$.

in the kth cycle. If we maintain a positive inductor current, $i_L(t) > 0$, then when the transistor is on, the diode is off, and vice versa. This is referred to as the *continuous conduction mode*. In the *discontinuous conduction mode*, on the other hand, the inductor current drops all the way to zero some time after the transistor is turned off, and then remains at zero, with the transistor and diode *both* off, until the transistor is turned on again. Limiting our attention here to the case of continuous conduction, the action of the transistor/diode pair in Figure 2.1 can be represented in idealized form via the double-throw switch in Figure 2.2.

We will mark the position of the switch in Figure 2.2 using a *switching function* $q(t)$. When $q(t) = 1$, the switch is down; when $q(t) = 0$, the switch is up. The switching function $q(t)$ may be thought of as (proportional to) the signal that has to be applied to the base drive of the transistor in Figure 2.1 to turn it on and off as desired. Under the constant-frequency PWM switching discipline described above, $q(t)$ jumps to 1 at the start of each cycle, every T seconds, and falls to 0 an interval $d_k T$ later in its kth cycle. The average value of $q(t)$ over the kth cycle is therefore d_k; if the duty ratio is constant at the value $d_k = D$, then $q(t)$ is periodic, with average value D.

In Figure 2.2, $q(t)$ corresponds to the signal at the output of the latch. This signal is set to "1" every T seconds when the clock output goes high, and is reset to "0" later in the cycle when the comparator output goes high. The two input signals of the comparator are cleverly arranged so as to reset the latch at a time determined by the desired duty ratio. Specifically, the input to the "+" terminal of the comparator is a sawtooth waveform of period T that starts from 0 at the beginning of every cycle, and ramps up linearly to F by the end of the cycle. At some instant t_k in the kth cycle, this ramp crosses the level of the *modulating signal* $m(t)$ at the "−" terminal of the comparator, and the output of the comparator switches from low to high, thereby resetting the latch. The duty ratio thus ends up being $d_k = m(t_k)/F$ in the corresponding switching cycle. By varying $m(t)$ from cycle to cycle, the duty ratio can be varied.

Note that the *samples* $m(t_k)$ of $m(t)$ are what determine the duty ratios. We would therefore obtain the same sequence of duty ratios even if we added to $m(t)$ any signal that stayed negative in the first part of each cycle and crossed up through 0 in the kth cycle at the instant t_k. This fact corresponds to the familiar *aliasing* effect associated with sampling. Our assumption for the averaged models below will be that $m(t)$ is not allowed to change significantly *within* a single cycle (i.e., that $m(t)$ is restricted to vary considerably more slowly than half the switching frequency). As a result, $m(t) \approx m(t_k)$ in the kth cycle, so $m(t)/F$ at any time yields the prevailing duty ratio (provided also that $0 \le m(t) \le F$, of course—outside this range, the duty ratio is 0 or 1).

The modulating signal $m(t)$ is generated by a feedback scheme. For the particular case of output feedback shown in Figure 2.2, the output voltage of the converter is compared with a reference voltage, and the difference is applied to a compensator, which produces $m(t)$. Note that the ramp level F can also be varied in order to modulate the duty ratio, and this mechanism is often exploited to implement certain *feedforward* schemes that compensate for variations in the input voltage V_{in}.

Switched State-Space Model for the Boost Converter

We assume the reader is familiar with the notion of a state-space model, and with how to obtain such a model for an electrical circuit; an introduction may be found in

[2]. Choosing the inductor current and capacitor voltage as natural state variables, picking the resistor voltage as the output, and using the notation in Figure 2.2, it is easy to see that the following state-space model describes the idealized boost converter in that figure:

$$\frac{di_L(t)}{dt} = \frac{1}{L}[(q(t) - 1)v_C(t) + v_{in}(t)]$$

$$\frac{dv_C(t)}{dt} = \frac{1}{C}\left[(1 - q(t))i_L(t) - \frac{v_C(t)}{R}\right] \tag{2.2}$$

$$v_o(t) = v_C(t)$$

Denoting the state vector by $\mathbf{x}(t) = [i_L(t) \; v_C(t)]'$ (where the prime indicates the transpose), we can rewrite the above equations as

$$\frac{d\mathbf{x}(t)}{dt} = [(1 - q(t))\mathbf{A}_0 + q(t)\mathbf{A}_1]\mathbf{x}(t) + \mathbf{b}v_{in}(t)$$

$$v_o(t) = \mathbf{c}\mathbf{x}(t) \tag{2.3}$$

where the definitions of the various matrices and vectors are obvious from (2.2). We refer to this model as the switched or instantaneous model, to distinguish it from the averaged and sampled-data models developed in later paragraphs.

If our compensator were to directly determine $q(t)$ itself, rather than determining the modulating signal $m(t)$ in Figure 2.2, then the above bilinear and time-invariant model would be the one of interest. It is indeed possible to develop control schemes directly in the setting of the switched model (2.3). In [3], for instance, a switching curve in the two-dimensional state space is used to determine when to switch $q(t)$ between its two possible values, so as to recover from a transient with a minimum number of switch transitions, eventually arriving at a periodic steady state. Drawbacks include the need for full state measurement and accurate knowledge of system parameters.

Various sliding mode schemes have also been proposed on the basis of switched models such as (2.3); see, for instance, [4,5,6], and references in these papers. Sliding mode designs again specify a surface across which $q(t)$ switches, but now the (sliding) motion occurs on the surface itself, and is analyzed under the assumption of infinite-frequency switching. The requisite models are thus averaged models in effect, of the type developed below. Any practical implementation of a sliding control must limit the switching frequency to an acceptable level, and this is often done via hysteretic control, where the switch is moved one way when the feedback signal exceeds a particular threshold, and is moved back when the signal drops below another (slightly lower) threshold. Constant-frequency implementations similar to the one in Figure 2.2 may also be used to get reasonable approximations to sliding mode behavior.

As far as the design of the compensator in Figure 2.2 is concerned, we require a model describing the converter's response to the modulating signal $m(t)$ or the duty ratio $m(t)/F$, rather than the response to the switching function $q(t)$. Augmenting the model (2.3) to represent the relation between $q(t)$ and $m(t)$ would introduce time-varying behavior and additional nonlinearity, leading to a model that is hard to work with. The averaged and sampled-data models considered below are developed in response to this difficulty.

Averaged Model for the Boost Converter

To *design* the analog control scheme in Figure 2.2, we seek a tractable model that relates the modulating signal $m(t)$ or the duty ratio $m(t)/F$ to the output voltage. In fact, since the ripple in the instantaneous output voltage is made small by design, and since the details of this small-output ripple are not of interest anyway, what we really seek is a continuous-time dynamic model that relates $m(t)$ or $m(t)/F$ to the *local average* of the output voltage (where this average is computed over the switching period). Also recall that $m(t)/F$, the duty ratio, is the local average value of $q(t)$ in the corresponding switching cycle. These facts suggest that we should look for a dynamic model that relates the local average of the switching function $q(t)$ to that of the output voltage $v_o(t)$.

Specifically, let us define the local average of $q(t)$ to be the lagged running average

$$d(t) = \frac{1}{T}\int_{t-T}^{t} q(\tau)d\tau \tag{2.4}$$

and call it the *continuous duty ratio* $d(t)$. Note that $d(kT) = d_k$, the actual duty ratio in the kth cycle (defined as extending from $kT - T$ to kT). If $q(t)$ is periodic with period T, then $d(t) = D$, the steady-state duty ratio. Our objective is to relate $d(t)$ in (2.4) to the local average of the output voltage, defined similarly by

$$\bar{v}_o(t) = \frac{1}{T}\int_{t-T}^{t} v_o(\tau)\,d\tau \tag{2.5}$$

A natural approach to obtaining a model relating these averages is to take the local average of the state-space description in (2.2). The local average of the derivative of a signal equals the derivative of its local average, because of the linear, time-invariant (LTI) nature of the local averaging operation we have defined. The result of averaging the model (2.2) is therefore the following set of equations:

$$\frac{d\bar{i}_L(t)}{dt} = \frac{1}{L}[\overline{qv_C}(t) - \bar{v}_C(t) + \bar{v}_{\text{in}}(t)]$$

$$\frac{d\bar{v}_C(t)}{dt} = \frac{1}{C}\left[\bar{i}_L(t) - \overline{qi_L}(t) - \frac{\bar{v}_C(t)}{R}\right] \tag{2.6}$$

$$\bar{v}_o(t) = \bar{v}_C(t)$$

where the overbars again denote local averages.

The terms that prevent the above description from being a state-space model are $\overline{qv_C}(t)$ and $\overline{qi_L}(t)$; the average of a product is generally *not* the product of the averages. Under reasonable assumptions, however, we can write

$$\overline{qv_C}(t) \approx \bar{q}(t)\bar{v}_C(t) = d(t)\bar{v}_C(t)$$

$$\overline{qi_L}(t) \approx \bar{q}(t)\bar{i}_L(t) = d(t)\bar{i}_L(t) \tag{2.7}$$

One set of assumptions leading to the above simplification requires $v_C(\cdot)$ and $i_L(\cdot)$ over the averaging interval $[t - T, t]$ to not deviate significantly from $\bar{v}_C(t)$ and $\bar{i}_L(t)$ respectively. This condition is reasonable for a high-frequency switching converter operating with low ripple in the state variables. There are alternative assumptions that lead to the same approximations. For instance, if $i_L(\cdot)$ is essentially piecewise linear and has a slowly varying average, then the approximation in the second equation of (2.7) is reasonable even if the ripple in $i_L(\cdot)$ is large; this situation is often encountered.

With the approximations in (2.7), the description (2.6) becomes

$$
\frac{d\bar{\imath}_L(t)}{dt} = \frac{1}{L}[(d(t) - 1)\bar{v}_C(t) + \bar{v}_{\text{in}}]
$$
$$
\frac{d\bar{v}_C(t)}{dt} = \frac{1}{C}\left[(1 - d(t))\bar{\imath}_L(t) - \frac{\bar{v}_C(t)}{R}\right] \tag{2.8}
$$
$$
\bar{v}_o(t) = \bar{v}_C(t)
$$

What has happened, in effect, is that *all* the variables in the switched state-space model (2.2) have been replaced by their average values. In terms of the matrix notation in (2.3), and with $\bar{\mathbf{x}}(t)$ defined as the local average of $\mathbf{x}(t)$, we have

$$
\frac{d\bar{\mathbf{x}}(t)}{dt} = [(1 - d(t))\mathbf{A}_0 + d(t)\mathbf{A}_1]\bar{\mathbf{x}}(t) + \mathbf{b}\bar{v}_{\text{in}}(t)
$$
$$
\bar{v}_o(t) = \mathbf{c}\bar{\mathbf{x}}(t) \tag{2.9}
$$

This is a nonlinear but *time-invariant* continuous-time state-space model, often referred to as the *state-space averaged* model, [7]. The model is driven by the continuous-time control input $d(t)$—with the constraint $0 \leq d(t) \leq 1$—and by the exogenous input $\bar{v}_{\text{in}}(t)$. Note that, under our assumption of a slowly varying $m(t)$, we can take $d(t) \approx m(t)/F$; with this substitution (2.9) becomes an averaged model whose control input is the modulating signal $m(t)$, as desired.

The averaged model (2.9) leads to much more efficient simulations of converter behavior than those obtained using the switched model (2.3), provided only local averages of variables are of interest. This averaged model also forms a convenient starting point for various nonlinear control design approaches; see, for instance, the development of energy-based control in Section 8.3, and also [8,9,10], and references in these papers. The implementation of such nonlinear schemes would involve an arrangement similar to that in Figure 2.2, although the modulating signal $m(t)$ would be produced by some nonlinear controller rather than the simple integrator shown in the figure. More traditional small-signal control design to regulate operation in the neighborhood of a fixed operating point can be based on the corresponding LTI linearization of the averaged model.

It should also be noted that the averaged model, though derived here for the case of periodically switched converters, actually provides a good representation of the dynamics of local averages under a variety of other switching disciplines and regimes (e.g., hysteretic control and even chaotic operation).

Averaged Model for Current-Mode Control of the Boost Converter

In this subsection, we show how to modify the preceding averaged model so as to approximately represent the dynamics of a high-frequency PWM converter operated under so-called current-mode control [11]. The name comes from the fact that a fast inner loop regulates the inductor current to a reference value, while the slower outer loop adjusts the current reference to correct for deviations of the output voltage from its desired value. The current monitoring and limiting that are intrinsic to current-mode control are among its attractive features.

In constant-frequency peak-current-mode control, the transistor is turned on every T seconds, as before, but is turned off when the inductor current (or equivalently, the transistor current) reaches a specified reference or *peak* level, denoted by $i_P(t)$. The duty ratio, rather than being explicitly commanded via a modulating signal such as $m(t)$ in Figure 2.2, is now implicitly determined by the inductor current's relation to $i_P(t)$. Despite this modification, the averaged model (2.8) is still applicable in the case of the boost converter. (Instead of constant-frequency control, one could use hysteretic or other schemes to confine the inductor current to the vicinity of the reference current.)

A tractable and reasonably accurate continuous-time model for the dynamics of the outer loop is obtained by assuming that the average inductor current is approximately equal to the reference current:

$$\bar{i}_L(t) \approx i_P(t) \tag{2.10}$$

Making the substitution (2.10) in (2.8) and using the two equations there to eliminate $d(t)$, we are left with the following first-order model:

$$\frac{d\,\bar{v}_C^2(t)}{dt} + \frac{2}{RC}\,\bar{v}_C^2(t) = -\frac{2i_P(t)}{C}\left(L\frac{di_P(t)}{dt} - \bar{v}_{\text{in}}\right) \tag{2.11}$$

This equation is simple enough that one can use it to explore various nonlinear control possibilities for adjusting $i_P(t)$ to control $\bar{v}_C(t)$ or $\bar{v}_C^2(t)$. The equation shows that, for constant $i_P(t)$ (or periodic $i_P(t)$, as in the nominal operation of so-called unity-power-factor PWM rectifiers, where $i_P(t)$ is a rectified sinusoid that tracks the rectified sinusoidal input voltage), $\bar{v}_C^2(t)$ approaches its constant (respectively, periodic) steady state exponentially, with time constant $RC/2$. A linearized version of this equation can be used to design small-signal controllers for perturbations around a fixed operating point (even for operation in the chaotic regime, as in Section 8.6).

Sampled-Data Models for the Boost Converter

Sampled-data models are naturally matched to power electronic converters, first, because of the cyclic way in which power converters are operated and controlled, and second, because such models are well suited to the design of digital controllers, which are used increasingly in power electronics (particularly for machine drives). Like averaged models, sampled-data models allow us to focus on cycle-to-cycle behavior, ignoring details of the intracycle behavior. Sampled-data models play a central role in this book.

We illustrate how a sampled-data model may be obtained for our boost converter example. The state evolution of (2.2), (2.3) for each of the two possible values of $q(t)$ can be described very easily using the standard matrix exponential expressions for LTI systems, and the trajectories in each segment can then be pieced together by invoking the continuity of the state variables. Recall that the matrix exponential can be defined, just as in the scalar case, by the (very well behaved) infinite matrix series

$$e^{\mathbf{A}t} = \mathbf{I} + \mathbf{A}t + \mathbf{A}^2 t^2/2! + \cdots \tag{2.12}$$

from which it is evident that

$$\frac{d}{dt}e^{\mathbf{A}t} = \mathbf{A}e^{\mathbf{A}t} \tag{2.13}$$

Under the switching discipline of constant-frequency PWM, where $q(t) = 1$ for the initial fraction d_k of the kth switching cycle, and $q(t) = 0$ thereafter, and assuming the input voltage is constant at V_{in}, it can be shown (see [2], Chapter 12 for guidance on the derivation) that

$$\mathbf{x}(kT + T) = e^{(1-d_k)\mathbf{A}_0 T}\left(e^{d_k \mathbf{A}_1 T}\mathbf{x}(kT) + \Gamma_1 V_{\text{in}}\right) + \Gamma_0 V_{\text{in}} \tag{2.14}$$

where

$$\begin{aligned}
\Gamma_0 &= \int_0^{(1-d_k)T} e^{\mathbf{A}_0 t}\mathbf{b}\ dt \\
\Gamma_1 &= \int_0^{d_k T} e^{\mathbf{A}_1 t}\mathbf{b}\ dt
\end{aligned} \tag{2.15}$$

In situations where the sampling interval is understood, $\mathbf{x}(kT)$ is sometimes denoted $\mathbf{x}[k]$ or $\mathbf{x}_k$.

For a well designed high-frequency PWM dc/dc converter in continuous conduction, the state trajectories in each switch configuration are close to linear, because the switching frequency is much higher than the filter cutoff frequency. What this implies is that the matrix exponentials in (2.14) are well approximated by just the first two terms in their Taylor series expansions:

$$\begin{aligned}
e^{(1-d_k)\mathbf{A}_0 T} &\approx \mathbf{I} + (1 - d_k)\mathbf{A}_0 T \\
e^{d_k \mathbf{A}_1 T} &\approx \mathbf{I} + d_k \mathbf{A}_1 T
\end{aligned} \tag{2.16}$$

If we use these approximations in (2.14) and neglect terms in T^2, the result is the following approximate sampled-data model:

$$\mathbf{x}(kT + T) = \left(\mathbf{I} + (1 - d_k)\mathbf{A}_0 T + d_k \mathbf{A}_1 T\right)\mathbf{x}(kT) + \mathbf{b}T V_{\text{in}} \tag{2.17}$$

This model is easily recognized as the usual forward-Euler approximation of the continuous-time model in (2.9), obtained by replacing the derivative there by a forward difference. (Retaining the terms in T^2 leads to more refined, but still very simple, sampled-data models.) Alternative ways to approximate the derivative lead to other approximate sampled-data models [12].

The sampled-data models in (2.14) and (2.17) were derived from (2.2), (2.3), and therefore used samples of the natural state variables, $i_L(t)$ and $v_C(t)$, as state variables. However, other choices are certainly possible, and may be more appropriate for a particular implementation. For instance, we could replace $v_C(kT)$ by $\bar{v}_C(kT)$ (i.e., the sampled local average of the capacitor voltage).

2.1.4 Extensions

Generalized Averaging

It is often useful or necessary—for instance, in modeling the dynamic behavior of resonant converters—to study the *local fundamental* and *local harmonics*, [13], in addition to local averages of the form shown in (2.4) and (2.5). For a variable $x(t)$, the *local $\ell\omega_s$-component* or the ℓth *dynamic phasor* may be defined as the following lagged running average:

$$\langle x \rangle_\ell(t) = \frac{1}{T}\int_{t-T}^{t} x(\tau)e^{-j\ell\omega_s\tau}\,d\tau \qquad (2.18)$$

In this equation, ω_s is usually chosen as the switching frequency (i.e., $2\pi/T$), and ℓ is an integer. The local averages in (2.4) and (2.5) correspond to the choice $\ell = 0$ and are thus 0-phasors; the choice $|\ell| = 1$ yields the local fundamental or 1-phasor, while $|\ell| > 1$ yields the local $(\ell-1)$th harmonic or ℓ-phasor. A key property of the local $\ell\omega_s$-component or ℓ-phasor is that

$$\left\langle \frac{dx}{dt} \right\rangle_\ell = j\ell\omega_s\langle x \rangle_\ell + \frac{d}{dt}\langle x \rangle_\ell \qquad (2.19)$$

where we have omitted the time argument t to keep the notation simple. For $\ell = 0$, we recover the result that was used in obtaining (2.6) from (2.2), namely that the local average of the derivative equals the derivative of the local average. More generally, we could evaluate the local $\ell\omega_s$-component of both sides of a switched state-space model such as (2.3), for *several* values of ℓ. With suitable approximations, this leads to an augmented state-space model whose state vector comprises the local $\ell\omega_s$-components for *all* these values of ℓ. In the case of the boost converter, for instance, we could choose $\ell = +1$ and $\ell = -1$ in addition to the choice $\ell = 0$ that was used to get (2.9) from (2.3). A development along these lines may be found in [13] and [14], which show the significant improvements in accuracy that can be obtained this way, relative to the averaged model (2.9), while maintaining the basic simplicity and efficiency of averaged models relative to switched models. The former paper also shows how generalized averaging may be applied to the analysis of resonant converters. Extensions to 3-phase systems (i.e., to local dynamic symmetrical components) may be found in [15].

Generalized State-Space Models

A sampled-data model for a power converter will almost invariably involve a state-space description of the form

$$\mathbf{x}(kT + T) = \mathbf{f}\Big(\mathbf{x}(kT), \mathbf{u}_k, \mathbf{T}_k, k\Big) \qquad (2.20)$$

The vector $\mathbf{u}_k$ here comprises a set of parameters that govern the state evolution in the kth cycle (e.g., parameters that describe control choices and source variations during the kth cycle), and $\mathbf{T}_k$ is a vector of *switching times*, comprising the times at which switches in the converter open or close. The switching-time vector $\mathbf{T}_k$ will satisfy a set of constraints of the form

$$0 = \mathbf{c}\Big(\mathbf{x}(kT), \mathbf{u}_k, \mathbf{T}_k, k\Big) \tag{2.21}$$

If $\mathbf{T}_k$ can be solved for in (2.21), then the result can be substituted in (2.20) to obtain a standard sampled-data model in state-space form. However, there are many cases in which the constraint (2.21) is *not* explicitly solvable for $\mathbf{T}_k$, so one is forced to live with the pair (2.20), (2.21) as the sampled-data model. Such a pair, comprising a state evolution equation along with a side constraint, is what we refer to as a generalized state-space model.

REFERENCES

[1] G. C. Verghese, Dynamic modeling and control in power electronics, in *The Control Handbook* (W. S. Levine, ed.), Boca Raton, FL: CRC Press–IEEE Press, Sec. 17, Chap. 78.1, pp. 1413–1423, 1996.

[2] J. G. Kassakian, M. F. Schlecht, and G. C. Verghese, *Principles of Power Electronics.* Reading, MA: Addison-Wesley, 1991.

[3] W. W. Burns and T. G. Wilson, Analytic derivation and evaluation of a state trajectory control law for dc-dc converters, *IEEE Power Electronics Specialists' Conf. Rec.*, pp. 70–85, 1977.

[4] H. Sira-Ramírez, Sliding motions in bilinear switched networks, *IEEE Trans. on Circuits and Systems*, vol. 34, no. 8, pp. 919–933, 1987.

[5] S. R. Sanders, G. C. Verghese, and D. E. Cameron, Nonlinear control of switching power converters, *Control—Theory and Advanced Technology*. vol. 5, no. 4, pp. 601–627, 1989.

[6] L. Malesani, L. Rossetto, G. Spiazzi, and P. Tenti, Performance optimization of Ćuk converters by sliding-mode control, *IEEE Trans. on Power Electronics*, vol. 10, no. 3, pp. 302–309, 1995.

[7] R. D. Middlebrook and S. Ćuk, A general unified approach to modeling switching converter power stages, *IEEE Power Electronics Specialists' Conf. Rec.*, pp. 18–34, 1976.

[8] S. R. Sanders and G. C. Verghese, Lyapunov-based control for switched power converters, *IEEE Trans. on Power Electronics*, vol. 7, no. 1, pp. 17–24, 1992.

[9] H. Sira-Ramírez and M. T. Prada-Rizzo, Nonlinear feedback regulator design for the Ćuk converter, *IEEE Trans. on Automatic Control*, vol. 37, no. 8, pp. 1173–1180, 1992.

[10] N. Kawasaki, H. Nomura, and M. Masuhiro, The new control law of bilinear dc-dc converters developed by direct application of Lyapunov, *IEEE Trans. on Power Electronics*, vol. 10, no. 3, pp. 318–325, 1995.

[11] S. P. Hsu, A. Brown, L. Resnick, and R. D. Middlebrook, Modeling and analysis of switching dc-to-dc converters in constant–frequency current–programmed mode, in *IEEE Power Electronics Specialists' Conf. Rec.*, pp. 284–301, 1979.

[12] M. S. Santina, A. R. Stubberud, and G. H. Hostetter, Discrete-time equivalents to continuous-time systems, *The Control Handbook* (W. S. Levine, ed.), CRC Press–IEEE Press, Boca Raton, FL, Sec. 4, Chap. 13, pp. 281–300, 1996.

[13] S. R. Sanders, J. M. Noworolski, X. Z. Liu, and G. C. Verghese, Generalized averaging method for power conversion circuits, *IEEE Trans. on Power Electronics*, vol. 6, no. 2, pp. 251–259, 1991.

[14] V. A. Caliskan, G. C. Verghese, and A. M. Stanković, Multi–frequency averaging of dc/dc converters, *IEEE Trans. on Power Electronics*, vol. 14, no. 1, pp. 124-133, 1999.

[15] A. M. Stanković, S. R. Sanders, and T. Aydin, Analysis of unbalanced ac machines with dynamic phasors, *Naval Symposium on Electric Machines*, pp. 219–224, 1998.

2.2 A CLOSER LOOK AT SAMPLED-DATA MODELS FOR POWER CONVERTERS

Francesco Vasca
Mario di Bernardo
Gerard Olivar

2.2.1 Introduction

One of the most important steps in analytically investigating the occurrence of bifurcations and chaos in any physical dynamical system is deciding what model to use to describe its dynamics. Using an appropriate system description is essential to the success of the proposed study.

Power electronic circuits are most naturally modeled as piecewise-smooth systems of ordinary differential equations (ODEs). These models are in the continuous-time domain and can be used, as will be shown throughout this book, to obtain an analytical and numerical description of the dynamics of the physical systems they model. Nevertheless, when one's aim is to understand the nature of the nonlinear phenomena exhibited by a continuous-time dynamical sytem, the use of alternative, discrete-time models can be very useful.

A powerful method of discretization involves a so-called *Poincaré section*. A Poincaré section is a suitably chosen hypersurface in the state space that is crossed by the state trajectories as the system evolves. The states are then sampled every time the trajectory of interest crosses the chosen section, resulting in a new discrete-time system, or Poincaré map, that describes the dynamics from one crossing to the next.

Computing a Poincaré map is an established practice in the modern theory of nonlinear dynamics, and its origins go back to the end of the nineteenth century. This discretization tool first appeared in the pioneering work of the French scientist Henri de Poincaré [1]. His method, which has been improved since then and further developed over the years, is now widely used in many areas of applied science. In this chapter, we will detail the use of appropriate system discretizations or Poincaré maps to carry out the analysis of nonlinear phenomena in power electronic circuits.

The use of these discrete-time maps offers several advantages. For instance, these mappings usually have lower dimensionality than the original model of the physical system under investigation. Hence, several problems can be treated much more easily by dealing with the discrete-time maps rather than using the original system equations. For example, as will be shown later, the stability analysis of a periodic orbit of the original model can be reduced to that of the corresponding fixed point of some suitable Poincaré map.

2.2.2 Poincaré Maps for Smooth and Nonsmooth Dynamical Systems

The systems of concern for our study take the form

$$\dot{\mathbf{x}} = f(\mathbf{x}, t, \mu) \tag{2.22}$$

where $\mathbf{x}$ is the real, n-dimensional state vector, $\mathbf{x} \in I\!\!R^n$. Similarly $\mu \in I\!\!R^p$ is the parameter vector and f is a piecewise-smooth function. The state space of a general piecewise-smooth system can be divided into regions, in each of which the system has a different smooth functional form, where *smooth* denotes a function that is continuous and has continuous derivatives. At the boundaries of these regions f may be discontinuous, or may be continuous but have discontinuous first derivatives (nonsmooth). The system, then, switches between different configurations whenever the system flow (i.e., the state trajectories) crosses one of the state-space boundaries.

In what follows, we will briefly illustrate the derivation of an appropriate Poincaré map for (2.22). There is no general method for the construction of the Poincaré map associated with a generic system of ordinary differential equations. We will therefore detail only a few cases of relevance to this book.

First consider the case of an autonomous (i.e., time invariant) system, and with f being continuous. Denote by $\phi(\mathbf{x}_0, t; \mu)$ the trajectory (or flow) generated by system (2.22) starting from the initial condition $\mathbf{x}_0$, with μ fixed. The corresponding Poincaré map is a reduced-order, discrete-time model of system (2.22) of the form:

$$\mathbf{x}_{k+1} = P(\mathbf{x}_k, \mu) \qquad (2.23)$$

where $\mathbf{x}_k \in I\!\!R^{n-1}$ is given by sampling the system state $\mathbf{x}$ every time the system flow crosses a given hyperplane in the state space that is transversal to the system flow. This hyperplane, termed the *Poincaré section*, has to be carefully chosen, and usually its choice is guided by the particular geometry of the state space.

In the case of a three-dimensional system, for instance, a Poincaré section is a plane in the state space (see Figure 2.3). On this plane, fixed points will correspond to periodic solutions of the original system. Hence, the Poincaré map is an excellent tool to determine the stability of a given periodic orbit, since its stability properties correspond to those of the associated fixed point, and these are much simpler to investigate.

If the continuous-time system is nonautonomous (i.e., subject to some external forcing), a viable system discretization can be obtained instead by sampling the system state at regular intervals. The choice of the sampling time is crucial and varies according to the different system topologies. In the case of a continuous, periodically forced

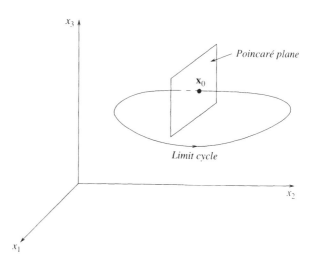

Figure 2.3 Poincaré section in 3-D space; $\mathbf{x}_0$ is the equilibrium point corresponding to the system limit cycles depicted in the figure.

dynamical system of ODEs, a Poincaré map, also called a *stroboscopic* map, is usually obtained by sampling the system states every T seconds, with T being the period of the external forcing term.

Finally, we consider the case of piecewise-smooth systems. These systems are widely used to model power electronic circuits and are therefore particularly relevant to our study. However, we begin with the following mechanical example.

The displacement $y(t)$ of a mechanical oscillator (Figure 2.4) with an obstacle at $y = s$ (an impact oscillator) can be described by the equation:

$$m\ddot{y} + c\dot{y} + ky = a\sin(\omega t) \qquad \text{for } y < s \qquad (2.24)$$

with $\dot{y} \mapsto -r\dot{y}, 0 < r < 1$ if $y(t) = s$ [2]. Thus, the state space for an impact oscillator is characterized by the presence of a boundary corresponding to the position, $y = s$, of the obstacle. On such a boundary, the functional form of the system is discontinuous since the velocity undergoes a jump (i.e., $\dot{y} \mapsto -r\dot{y}$).

An effective method to construct an appropriate Poincaré map for a piecewise-smooth system involves sampling the system state whenever the system flow crosses one of the boundaries between two regions in the state space. This produces a reduced-order discrete-time model, termed an *impact* or *switching* map, that can be associated with the original system dynamics. This map describes the system behavior from one switching to the next, and has been shown to be very useful in characterizing the occurrence of nonlinear phenomena in many circuits [3].

Notice that in the case of an externally forced piecewise-smooth system, both a stroboscopic and a switching map can be defined. The choice between the former and latter maps is suggested by the type of analysis that is to be carried out. These and other issues related to the use of discrete-time maps in modeling power electronic circuits will be the subject of the rest of this chapter.

2.2.3 Piecewise-Smooth Power Electronic Circuits

Most power electronic circuits can be considered to be made of suitably connected elementary components. These components can be roughly grouped in three main classes: sources (voltage and current sources), passive components (resistances, transformers, inductances, capacitors) and switching elements (diodes, thyristors, and controllable transistor switches such as the BJT, MOSFET or GTO).

In order to obtain simple circuit models that capture only those aspects of the system dynamics that are relevant to the intended analysis, some simplifying assumptions are usually introduced:

1. The switching elements in the circuit are considered to be ideal; i.e., each switch is considered as a short circuit when it is closed or on, and as an open circuit when it is open or off.

2. The commutations—or switch transitions between on and off states—are assumed to be instantaneous [4,5].

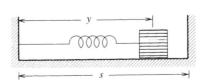

Figure 2.4 The impact oscillator.

3. The passive elements are assumed to be linear and time invariant.

Under these assumptions, the circuits can be considered to have a finite number of configurations or phases. Each phase is determined once the conducting conditions of the switching elements have been specified (i.e., once it is known which switches are conducting and which are not). The time sequence of the different phases is then determined by the logic that controls the switches. In particular, each phase can be modeled through a linear, time-invariant (LTI) system of ordinary differential equations. This system can be written as follows:

$$\dot{\mathbf{x}}(t) = A_i \mathbf{x}(t) + B_i \mathbf{u}(t), \tag{2.25}$$

where $\mathbf{x}$ is the state vector (usually the currents in the inductances and the voltages across the capacitors), $\mathbf{u}$ is the input vector, and the index $i = 1, \ldots N$ denotes the ith configuration among the N possible different converter configurations given by the on/off states of the switching elements. Typically, during standard operating regimes, the system cyclically switches among these configurations or phases while the inputs remain constant or periodic. As will be shown later, the possible periodicities of the phase sequence and of the input signals are important in determining the type of discrete-time map that is most appropriate in describing the dynamics of the converter.

In writing (2.25) we implicitly assumed that the state and input variables are the same for all the converter configurations. This is why the index i does not appear as a subscript of either the state vector $\mathbf{x}$ or the input vector $\mathbf{u}$. This assumption, although not strictly necessary, allows a simpler representation of the whole system. There are situations in which, for instance, the number and identity of the natural state variables changes as the converter moves from one phase to the next. However, it is generally still possible to find a representation of the form that we have assumed, with a fixed set of state variables. For instance, in a converter operating under discontinuous conduction mode, the current in an inductor, say x_r, goes to zero and cannot become negative because of the presence of a series-connected diode. When x_r drops to 0, the system topology changes, and x_r is identically zero during the new phase, so one might consider a lower-order state space representation for this new phase, eliminating x_r from the original state vector $\mathbf{x}$. However, as an alternative, the same choice of state variables as in the other phases can be maintained if we set $\dot{x}_r(t) = 0$ as the state equation corresponding to the evolution of the inductor current during the discontinuous conduction phase.

Finally, in what follows we will make the assumption—a very reasonable one in power electronics—that the state trajectory is continuous at the transitions between different phases. More general models can be used if this assumption is not satisfied [6].

We now present two different examples of converters that can be modeled as piecewise-smooth systems.

EXAMPLE: THE BUCK-BOOST CONVERTER

Let us consider the buck-boost dc/dc converter topology shown in Figure 2.5. Let us choose as state vector $\mathbf{x} = (i, v)^T$, where the superscript T denotes transposition of a matrix or vector, and assume continuous conduction mode. The converter presents two different configurations, depending on whether the controllable switch S is on (closed) or off (open). When S is on, the matrices in (2.25) are

$$A_1 = \begin{pmatrix} -R_l/L & 0 \\ 0 & -1/RC \end{pmatrix}, \quad B_1 = \begin{pmatrix} 1/L \\ 0 \end{pmatrix}$$

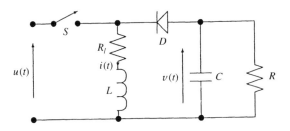

Figure 2.5 Buck-boost dc/dc converter topology; S represents a controllable electronic switching element and D a diode.

whereas, when S is off, we have

$$A_2 = \begin{pmatrix} -R_l/L & 1/L \\ -1/C & -1/RC \end{pmatrix}, \quad B_2 = \begin{pmatrix} 0 \\ 0 \end{pmatrix}$$

The *events* that determine the change of configuration are the turning on and the turning off of the switch S. This situation is schematically shown in Figure 2.6. This figure also considers the possibility of discontinuous conduction, where the current may go to zero while the switch is off, causing the converter to change its configuration from the second to the third one, thereby entering discontinuous conduction mode. The third configuration is characterized by the matrices:

$$A_3 = \begin{pmatrix} 0 & 0 \\ 0 & -1/RC \end{pmatrix}, \quad B_3 = \begin{pmatrix} 0 \\ 0 \end{pmatrix}$$

When the system is in the third configuration, the inductor current is identically zero and the only actual state variable during this phase is the capacitor voltage. As already mentioned, this situation can be analytically taken into account by assuming $\dot{x}_1(t) = 0$ and by maintaining a second-order representation (see the first row of A_3 and B_3 and recall that the state is assumed to be continuous at the transitions).

When the switch is turned on while the circuit is in the third phase, a transition to the first phase occurs and the phase sequence then starts again.

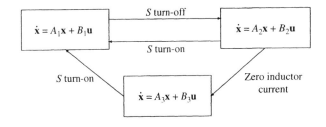

Figure 2.6 The hybrid system scheme that describes the variation of the buck-boost configurations under continuous and discontinuous conduction mode.

EXAMPLE: THE SINGLE-PHASE STATIC VAR SYSTEM

We now consider the single-phase static VAR system represented in Figure 2.7 (the word *phase* in *single-phase* is now used in the more traditional sense of power systems, to distinguish the system from *three-phase*). The converter is assumed to operate in discontinuous conduction mode.

Let us choose as state vector $\mathbf{x} = (i_s, v, i_r)^T$ and suppose that S_1 is on. Under this configuration the system matrices are:

$$A_1 = \begin{pmatrix} -R_s/L_s & -1/L_s & 0 \\ 1/C & 0 & -1/C \\ 0 & 1/L_r & -R_r/L_r \end{pmatrix}, \quad B_1 = \begin{pmatrix} 1/L_s \\ 0 \\ 0 \end{pmatrix}$$

When the current i_r goes to zero, the circuit configuration changes and the matrices of the new state space model are:

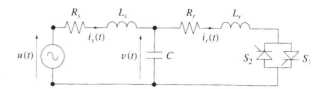

Figure 2.7 Single-phase static VAR system; S_1 and S_2 are thyristors.

$$A_2 = \begin{pmatrix} -R_s/L_s & -1/L_s & 0 \\ 1/C & 0 & -1/C \\ 0 & 0 & 0 \end{pmatrix}, \quad B_2 = \begin{pmatrix} 1/L_s \\ 0 \\ 0 \end{pmatrix}$$

Note that in the discontinuous-mode configuration the time derivative of the current i_r is identically zero. Thus, when the capacitor voltage becomes negative, the thyristor S_2 becomes forward biased and a turn-on signal can be sent to the gate of S_2 in order to switch it on. Following this event, the system switches back to the first configuration and the current i_r becomes negative. Again, the system remains in the first configuration until the current goes to zero, which determines the transition to the second configuration. After that, when the capacitor voltage is positive, the thyristor S_1 is forward biased and a turn-on signal can be sent to its gate. For further details on how the converter operates, see [7].

The circuit configurations and the transitions between them are shown in Figure 2.8.

2.2.4 Power Electronic Systems as Hybrid Systems

In the previous section we have shown how to model the different configurations of a power electronic circuit and how to describe the transitions among these on the basis of the circuit operating conditions. In order to obtain a discrete-time model of the system, we still need to analytically characterize the events that determine the transitions among the different configurations. Keep in mind that the system typically switches from one configuration to another when at least one of the following conditions is satisfied:

- The state or the input become equal to some given value.
- A directly controllable switch is turned on or turned off.
- A specific time instant is reached.

These events can all be modeled analytically by introducing appropriate *switching conditions* [4] that govern the commutation of a switch. These conditions can be written in the following form:

$$\sigma_{ij}(\mathbf{x}(\bar{t}), \mathbf{u}(\bar{t}), \mathbf{e}(\bar{t}), \bar{t}) = 0 \tag{2.26}$$

where $i = 1, \ldots N$, $j = 1, \ldots N$, and $\bar{t}$ is the time instant at which the circuit goes from the ith to the jth configuration, while $\mathbf{e}$ is a control vector whose elements are somehow linked to the variables that determine the commutations of directly controllable

Figure 2.8 The hybrid system scheme that describes the behavior of the static VAR system.

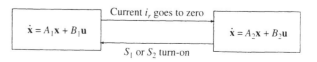

switches, for instance, the "turn on" and "turn off" commands in the previous examples. For a converter that exhibits some type of periodic behavior with period T, it can be useful to consider the normalized time variable $\tau = t$ mod T (where mod T denotes the remainder after division by period T) as a component of the external variable vector **e**. (When the switching condition depends explicitly on time, it may be useful to consider time as a further state variable with derivative identically equal to one, i.e., to consider $\dot{t} = 1$ as a further state equation.)

The explicit form of the switching condition (2.26) depends on the converter configuration and operating conditions. In what follows we reconsider the examples introduced above in order to show how the switching events described in Figures 2.6 and 2.8 can be described by (2.26).

EXAMPLE: SWITCHING CONDITIONS FOR THE BUCK-BOOST CONVERTER

In Figure 2.6 we described the switching between different configurations using a set of logical conditions ("S turn on," "S turn off," "current goes to zero"). Now, we aim to express these conditions analytically in the case of the converter operating under feedback control.

Assume that the converter is controlled through so-called voltage-mode control. Under this type of control (see Figure 2.9) a control voltage v_c is compared with a T-periodic sawtooth waveform, say v_r, and a commutation occurs whenever these two signals first become equal in any switching cycle. In other words, the transitions from the first to the second phase occur at time instants $\bar{t}$ such that $v_c(\bar{t}) = v_r(\bar{t})$. Note that, under open-loop operation, the control voltage and the ramp signal can both be considered as external variables, and the switching condition (2.26) can be rewritten as $e_1(\bar{t}) = 0$ where $e_1(t) = v_c(t) - v_r(t)$.

The turning on of the switch (which implies commutation from the second or the third phase to the first one—see Figure 2.6) occurs at time instants that are integer multiples of T. At these time instants the ramp v_r is discontinuous and the control signal v_c crosses it (see Figure 2.9). This

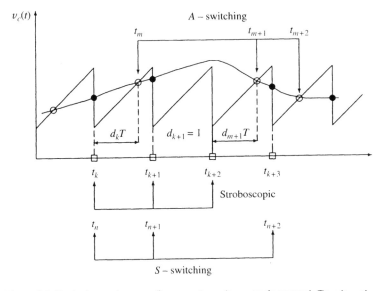

Figure 2.9 Typical v_r and v_c waveforms under voltage-mode control. Two iterations of A-switching, stroboscopic and S-switching maps (discussed in Sections 2.2.5 and 2.2.6) are represented.

condition can be analytically modeled by introducing the external variable $e_2(t) = (t \bmod T)/T$; in fact, $e_2 = 1$ when t is an integer multiple of T; hence the switch S is turned on, setting the system model to the first configuration.

Voltage-mode control may be implemented through a state feedback strategy that involves both the current and the voltage. In particular, the control signal is a linear combination of the current and the error in the output voltage (which is the difference between a reference voltage and the measured capacitor voltage); see Figure 2.10 for a block diagram of this scheme. In this case commutation occurs at $\bar{t}$ such that:

$$\mathbf{g}^T \mathbf{x}(\bar{t}) = v_r(\bar{t}) + g_v V_{\text{ref}} \tag{2.27}$$

where $\mathbf{g}^T = (g_i, g_v)$ is a suitable gain vector, and g_i, g_v are suitable constants (see [3] for details). Again (2.27) can be rewritten in the form of the switching condition (2.26) by defining $e_1(t) = v_r(t) + g_v V_{\text{ref}}$.

The transition from the second phase to the phase describing a discontinuous operating mode (the third phase in Figure 2.6) takes place at time instants $\bar{t}$ such that

$$x_1(\bar{t}) = 0 \tag{2.28}$$

The transition back to the first phase is finally achieved when a multiple of the modulating period is reached.

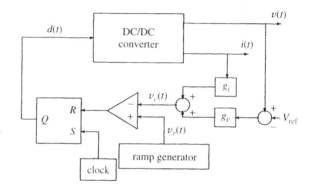

Figure 2.10 Scheme for implementation of voltage-mode-controlled dc/dc converter; the flip-flop may be eliminated, thus allowing multiple pulsing.

A schematic representation, equivalent to Figure 2.6 but incorporating the switching conditions in analytical form, is shown in Figure 2.11.

It is possible to show that the scheme in Figure 2.11 is also valid under current-mode control [3].

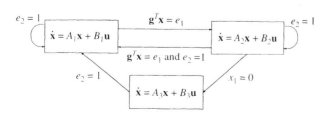

Figure 2.11 Operating scheme of the buck-boost converter under voltage-mode control. A maximum of one switch operation within each period (or two under discontinuous mode) is assumed. The event $e_2 = 1$ being in the first or the second configurations correspond to skipped cycles (see Section 2.2.6). The condition $\mathbf{g}^T \mathbf{x} = e_1$ in the commutation from the second to the first configuration is used, with abuse of notation, meaning that the control signal at a multiple of T is included into the discontinuity jump of the ramp (see Figure 2.9).

EXAMPLE: THE SWITCHING CONDITION FOR THE SINGLE-PHASE STATIC VAR SYSTEM

Following a similar procedure to that presented for the buck-boost converter, it is possible to represent the operating scheme of the static VAR system as shown in Figure 2.12. In particular, the external variable e represents the firing signal. This is equal to $+1$ when an impulse is sent to the gate of the thyristor S_1, equal to -1 when an impulse is sent to the gate of the thyristor S_2, and zero otherwise.

2.2.5 Stroboscopic Maps

Earlier we mentioned that the notion of a stroboscopic map—obtained by sampling the state at time instants that are integer multiples of a certain period—can be very useful in analyzing a periodically forced system. Notice that in this case the sampling time instants are not necessarily switching instants as well. If the sampling time instants of the stroboscopic map are also synchronous with switching events, we will then term the corresponding map a *stroboscopic switching map*.

As already shown, for the systems under investigation we can have two different types of periodicity, according to whether the system naturally goes through a periodic phase sequence or the periodicity is forced upon the system by an external periodic forcing input (as in the case of the static VAR system). In this section, we start by considering the case of systems exhibiting simple cyclic repetition of the system phases, with all other inputs constant; this covers, for example, the buck-boost dc/dc converter. We then analyze how the stroboscopic switching map changes in the presence of non-constant inputs. Finally, we describe a case of a stroboscopic map in which the sampling times do not correspond to switching instants.

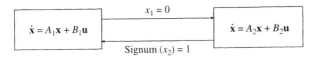

Figure 2.12 Operating scheme of the single-phase static VAR system.

EXAMPLE: STROBOSCOPIC SWITCHING MAP FOR THE BUCK-BOOST CONVERTER

Let us consider the buck-boost converter shown in Figure 2.5, but with a constant input voltage $u(t) = U$. If we assume the converter operates in continuous conduction mode, the system model configuration will alternate between only two phases. Let us indicate by $d_{k1}T$ and $d_{k2}T = (1 - d_{k1})T$ the respective time durations of the two phases during the kth period of the modulating waveform; here d_{k1} is the same as the duty ratio d_k shown in Figure 2.9, but the refined notation is helpful in this section. Consider the time instant $t_k = kT$ as the *first* sampling time for the map construction. The state at the succeeding switching instant $t_k + d_{k1}T$ can be obtained by solving (2.25) between t_k and $t_k + d_{k1}T$ given the initial condition $\mathbf{x}(t_k)$:

$$\mathbf{x}(t_k + d_{k1}T) = N_1(d_{k1})\mathbf{x}(t_k) + M_1(d_{k1}) \tag{2.29}$$

with

$$N_1(d_{k1}) = e^{A_1 d_{k1} T}, \quad M_1(d_{k1}) = \int_0^{d_{k1}T} e^{A_i \alpha} B_i U d\alpha = A_1^{-1}\left(e^{A_1 d_{k1} T} - I\right)B_1 U \tag{2.30}$$

where the matrix exponential e^{At} is as defined in (2.12), and the matrices A_1 and B_1 have been defined in the previous section.

By considering the second system phase and solving (2.25) between $t_k + d_{k1}T$ and $t_{k+1} = kT + T$, the state at the next stroboscopic switching time instant can be written as:

$$\mathbf{x}(t_{k+1}) = N_2(d_{k2})\mathbf{x}(t_k + d_{k1}T) + M_2(d_{k2}) \tag{2.31}$$

where N_2 and M_2 can be straightforwardly obtained from (2.30) by replacing the subscript 1 with 2.

By substituting (2.29) in (2.31), one obtains the following stroboscopic switching map for the buck-boost converter:

$$\mathbf{x}_{k+1} = N(d_{k1})\mathbf{x}_k + M(d_{k1}) \tag{2.32}$$

where $\mathbf{x}_k = \mathbf{x}(t_k)$, and

$$N(d_{k1}) = N_2(1 - d_{k1})N_1(d_{k1}), \quad M(d_{k1}) = N_2(1 - d_{k1})M_1(d_{k1}) + M_2(1 - d_{k1})$$

(where we have used the relation $d_{k2} = 1 - d_{k1}$).

The stroboscopic map changes if the converter operates in discontinuous conduction mode. The system then has three different phases, as already shown in Figure 2.6. The first phase corresponds to the switch being on, the second phase to the switch being off and the current being nonzero, and the third phase to the switch being off and current being zero. Let us denote by $d_{k1}T$, $d_{k2}T$ and $d_{k3}T$ the time durations of these three phases, respectively, during the kth period. Following a procedure similar to that presented above, the state at $t_k + d_{k1}T$ can be obtained as a function of $\mathbf{x}(t_k)$, and so on, proceeding switching by switching and phase by phase. Thus, simple substitutions allow one to write the state $\mathbf{x}(t_{k+1})$ as a function of the state $\mathbf{x}(t_k)$. In particular, the stroboscopic map for the buck-boost converter under discontinuous conduction mode takes the following form:

$$\mathbf{x}_{k+1} = N(\Delta_k)\mathbf{x}_k + M(\Delta_k) \tag{2.33}$$

where $\Delta_k = (d_{k1}, d_{k2}, d_{k3})^T$ and the matrices are given by:

$$N(\Delta_k) = N_3(d_{k3})N_2(d_{k2})N_1(d_{k1}),$$

$$M(\Delta_k) = N_3(d_{k3})N_2(d_{k2})M_1(d_{k1}) + N_3(d_{k3})M_2(d_{k2}) + M_3(d_{k3})$$

Note that the three components of the duty cycle vector Δ_k are not independent. In particular, since the second phase ends when the inductor current goes to zero, the following condition must be satisfied:

$$\mathbf{c}^T[N_2(d_{k2})N_1(d_{k1})\mathbf{x}_k + N_2(d_{k2})M_1(d_{k1}) + M_2(d_{k2})] = 0 \tag{2.34}$$

where $\mathbf{c}^T = (1, 0)$ is the output vector that selects the first component of the state vector, namely the inductor current. Equation (2.34) implies that d_{k2} depends on d_{k1} and $\mathbf{x}_k$. Moreover, since the sum of the duty ratios during each period must be equal to one, we can write: $d_{k3} = 1 - d_{k1} - d_{k2}$. In other words, as should have been expected, the only independent duty ratio is d_{k1}.

The preceding procedure for the construction of the stroboscopic switching map of the buck-boost converter can be generalized in an obvious way to the case of a piece-wise-smooth system whose configuration changes among N different phases according to a fixed and cyclically repeated sequence, rather than just the two or three phases treated in the buck-boost case; see [4, Sections 12.5-7 and 13.4]. For this kind of system it is important, in order to simplify the derivation of the stroboscopic map, to choose carefully the *first* sampling time. The best option is usually to choose a switching instant as the initial sampling time (i.e., a time instant at which a commutation between two system phases has just occurred).

Periodic Phase Sequence and Time-Varying Inputs

If the input $u(t)$ is not constant, one can follow a procedure similar to that previously presented in order to arrive at a model of the form

$$\mathbf{x}_{k+1} = N(\Delta_k)\mathbf{x}_k + M\big(\Delta_k, u_{k,k+1}\big) \qquad (2.35)$$

where $u_{k,k+1}$ indicates the input signal $u(t)$ for $t \in (kT, (k+1)T)$. The matrix $N(\Delta_k)$ is of the same form as before, but $M(\Delta_{k,u_k,k+1})$ is given by a more complicated expression (though still involving convolution integrals of matrix exponentials with the input). The dependence of the map on u introduces some difficulties in the analytical manipulations, even though the input u is assumed completely known. For a possible modeling solution to this problem, see [4, Section 12-5].

One of the interesting cases is when u is periodic with period T_u. In this situation, a question naturally arises: Is the corresponding system dynamical behavior periodic with period T_u or with period T? If T_u is an integer multiple of T (as often happens in cases of interest), say $T_u = rT$ with r integer, the stroboscopic map can be defined as the map (2.35), or as the map obtained by iterating r times the *elementary* map (2.35). In the first case, the continuous-time periodic evolution of the state with period T_u will correspond to a sequence of r different intersections with the Poincaré section, whereas in the second case the same elementary periodic behavior will correspond to a fixed point of the overall mapping.

A Stroboscopic Nonswitching Map

Although many power electronic circuits typically evolve according to a periodic phase sequence, it is convenient for certain systems to consider an alternative stroboscopic map obtained by sampling the state at periodic time instants that may or may not correspond to switching events. An example of this type of system is the previously described static VAR converter. A detailed description of how the stroboscopic map for this circuit can be obtained is reported in [7] and in Chapter 6 of this book.

The static VAR converter is an example of a circuit in which the periodic phase sequence has a period T that is different from the period T_u of the external forcing input. A detailed discussion of the advantages and disadvantages of considering maps obtained by sampling the state with period T_u or with period T can be found in [7]. In particular, looking at Figure 2.12, the map with period T_u is obtained by considering the sequence $(1, 2, 1, 2)$ as the elementary phase sequence; whereas the map with sampling period T will describe the elementary phase sequence $(1, 2)$. Obviously, relevant differences between these two maps become apparent only when the circuit behavior is not time symmetric with respect to the sequence repetition $(1, 2)$.

Closed-Loop Maps

In the previous section, in order to obtain the stroboscopic switching map, we have assumed that the variables d and u are directly manipulable. More generally, the duty ratios are used as control signals and are defined through suitable control actions that typically depend on the system states. Therefore, as is evident by considering (2.33) and

(2.35), the closed-loop map becomes nonlinear, even in the case of linear control schemes.

For example, consider again the voltage-mode-controlled buck-boost converter. As shown above, both under continuous and discontinuous mode, the stroboscopic map depends only on the duty ratio d_{k1}. By using (2.29) and by assuming $v_r(t) = \alpha + \beta(t \bmod T)$, the switching condition (2.27) can be rewritten as follows:

$$\mathbf{g}^T(N_1(d_{k1})\mathbf{x}_k + M_1(d_{k1})) = \alpha + \beta d_{k1}T + g_v V_{ref} \qquad (2.36)$$

or, in a more general form, as:

$$\sigma(d_{k1}, \mathbf{x}_k) = 0 \qquad (2.37)$$

Note that the switching condition (2.37) depends upon both the duty ratio d_{k1} and the state $\mathbf{x}_k$. Whether it is possible to solve (2.37) for the duty ratio depends on the specific structure of the matrices describing the system under investigation. If a solution is available, then the explicit closed-loop map can be simply obtained by substituting the solution of (2.37) in (2.33). Otherwise, if an explicit expression for d_{k1} as a function of the state $\mathbf{x}_k$ cannot be derived from (2.37), then the closed-loop map will take the following general form:

$$\mathbf{x}_{k+1} = f(\mathbf{x}_k, \Delta_k) \qquad (2.38)$$

$$\Sigma(\Delta_k, \mathbf{x}_k) = 0 \qquad (2.39)$$

where f is defined through (2.33) and we have generalized the model to the case of a duty ratio vector.

2.2.6 Switching Maps

In the previous section we considered the case in which the state is sampled periodically at switching time instants that are integer multiples of the period of the phase sequence or of the system input. However, it is possible (and sometimes indeed extremely useful) to construct other types of switching maps. Here, the emphasis is not placed on the periodic behavior of the system under ordinary operating conditions, but rather on the location of the commutations between different system configurations.

In what follows we briefly describe how the *S-switching* and *A-switching* maps can be obtained for dc/dc converters operating under voltage-mode control [8].

S-Switching Maps for DC/DC Converters

When considering a dc/dc converter under voltage-mode control, it may happen that, under large transients, the control voltage v_c exceeds the ramp signal v_r for time durations that are longer than the ramp period. This leads to a *skipped cycle* (or several such cycles); i.e., a cycle with no switchings, see Figure 2.9. The stroboscopic map that was previously described, which samples at integer multiples of T, does not discriminate between cycles in which a switching has occurred (the discrete-time instants indexed by n in Figure 2.9) and skipped cycles. In order to take into account this aspect, a different discrete-time map—the *S-switching map*—can be used.

The S-switching map is obtained by sampling the state vector only at selected integer multiples of T, say t_n, namely those that correspond to cycles in which a switching event has occurred. Among the time instants kT, those at which a skipped cycle has taken place will not be included in the S-switching map. Taking the case shown in Figure 2.9, the stroboscopic time instant t_{k+2} will not be considered as a S-switching time instant because no switching has occurred in the cycle that ends at t_{k+2}. By cascading the solutions of (2.25) for the two phases of the system model, the S-switching map can be written as

$$\mathbf{x}_{n+1} = N_2(1 - d_n + s_{n2})N_1(d_n + s_{n1})\mathbf{x}_n$$
$$+ N_2(1 - d_n + s_{n2})M_1(d_n + s_{n1}) + M_2(1 - d_n + s_{n2}) \qquad (2.40)$$

where $\mathbf{x}_n = \mathbf{x}(t_n)$, s_{n1} and s_{n2} are the numbers of skipped cycles in the corresponding phases between the time instants t_n and t_{n+1}. For instance, for the situation shown in Figure 2.9, $s_{n1} = 0$, $s_{n2} = 0$, $s_{n+1,1} = 1$, $s_{n+1,2} = 0$ and so on. A straightforward comparison of (2.40) and (2.32) shows that the stroboscopic map can be obtained from (2.40) by assuming $s_{n1} = 0$ and $s_{n2} = 0$ for all n.

A-Switching Maps for DC/DC Converters

The A-switching map is obtained when the state vector is asynchronously sampled at switching times internal to the modulating period, which are denoted by t_m in Figure 2.9. As in the previous analysis, by assuming that after an A-switching the converter enters in phase 2, by cascading the solutions of (2.25) for the second and the first phase of the system model, the A-switching map can be written as follows:

$$\mathbf{x}_{m+1} = N_2(d_{m+1,1})N_1(1 - d_{m1})\mathbf{x}_m$$
$$+ N_1(d_{m+1,1})M_2(1 - d_{m1}) + M_1(d_{m+1,1}) \qquad (2.41)$$

where $\mathbf{x}_m = \mathbf{x}(t_m)$, the subscript m has been used instead of k in order to specify that the state is not sampled at the multiples of T previously denoted by t_k and, for notational simplicity, we do not consider the case of skipped cycles.

The A-switching map can be very useful for the construction of a closed-loop map. For this, (2.41) must be rewritten by highlighting the dependence of the variables $d_{m+1,1}$ and d_{m1} on the state samples $\mathbf{x}_m$ and $\mathbf{x}_{m+1}$. In particular, under voltage-mode control, using (2.36), we can write the following A-switching conditions:

$$\mathbf{g}^T \mathbf{x}_m = \alpha + \beta d_{m1} T + g_v V_{\text{ref}} \qquad (2.42)$$

$$\mathbf{g}^T \mathbf{x}_{m+1} = \alpha + \beta d_{m+1,1} T + g_v V_{\text{ref}} \qquad (2.43)$$

Note that, in contrast with what is observed for the stroboscopic switching map, this mapping depends on the duty ratios of two successive periods, with the switching conditions being linear in the state vector. It is now possible to obtain the closed-loop map. Let us assume $\beta \neq 0$. By computing d_{m1} from (2.42) and $d_{m+1,1}$ from (2.43) and substituting in (2.41), we obtain the following closed-loop A-switching map:

$$\mathbf{x}_{m+1} = N_1(\gamma^T \mathbf{x}_{m+1} - a)N_2(1 - \gamma^T \mathbf{x}_m + a)\mathbf{x}_m$$
$$+ N_1(\gamma^T \mathbf{x}_{m+1} - a)M_2(1 - \gamma^T \mathbf{x}_m + a) + M_1(\gamma^T \mathbf{x}_{m+1} - a) \qquad (2.44)$$

where $\gamma^T = g^T/(\beta T)$ and $a = (\alpha + g_v V_{\text{ref}})/(\beta T)$. Note that (2.44) has a closed form, although it is an implicit map.

Similar considerations allow one to obtain the closed-loop A-switching map in the case $\beta = 0$ (see [3] for details).

The switching conditions (2.42) and (2.43) show that the duty ratios are affinely (i.e., linearly plus a constant) related to the current and voltage (the state vector) at the A-switching times. This suggests the possibility of replacing the current or the voltage in the state vector with the duty ratio. Such a replacement can yield some interesting results, as reported in [8] and elsewhere in the literature dealing with converters operating in discontinuous conduction mode (for an extensive bibliography see [3]).

Stroboscopic maps, stroboscopic switching maps, and A-switching maps are almost equivalent for the analysis of the periodic behavior of a circuit. Things may be different if the system does not operate periodically. Under this situation the sampling times for the stroboscopic map are still periodic, but those corresponding to the A-switching map are not. In fact, the A-switching sampling times will be determined by the satisfaction of the switching conditions whenever they occur (periodically or not).

2.2.7 Simplified Maps

The introduction of some realistic hypotheses on the converter operating conditions allows the derivation of some useful *simplified* maps.

A typical assumption, usually satisfied by dc/dc power electronic converters, is that the time duration of each phase of the converter is much less than the corresponding time constants. This means that the matrix exponential in (2.30) can be approximated by the first few terms of its series expansion, for example,

$$N_1(d) \approx I + A_1 dT + \frac{1}{2} A_1^2 (dT)^2 \qquad (2.45)$$

In [3] it is shown how the simplifying assumption (2.45) in the case of dc/dc converters allows one to write explicit closed-loop maps, both under voltage-mode and current-mode control. The availability of an *explicit* map is important for analysis, and can also be important in obtaining numerical results through simpler codes.

Another typical simplifying assumption introduced for the analysis of dc/dc and other types of power converters, one that is often verified under realistic operating conditions, is that some capacitors are large enough for the corresponding voltage to be essentially constant. Under this assumption the converter can be modeled by a reduced-order system whose state does not include as state variable the voltage across these capacitors.

2.2.8 Conclusions

The definition of a discrete-time map is the first fundamental step for analytical and numerical investigations of the dynamic behavior of power converters. In this section we have presented a general procedure for the derivation of discrete-time maps for some typical power electronic systems, both under open-loop and closed-loop control schemes. Each map is obtained by considering an appropriate criterion to select the sampling time instants.

Once a specific trajectory (or dynamical scenario) exhibited by the converter has been identified, these maps can be used as an essential tool to carry out their analysis. For instance, the definition of the Jacobian (i.e., first-derivative matrix) of a map corresponding to a given periodic orbit can enormously facilitate the task of determining analytical conditions for its stability and possible bifurcations [3]. Moreover, by introducing suitable simplifying assumptions (based on realistic hypotheses related to the system operating conditions), it is often possible to obtain simplified maps. These maps can have explicit form with respect to the state vector and hence allow the definition of faster numerical codes for simulations and simpler analytical models.

Discrete-time mappings of different types will be extensively used in the rest of the book, where they will be used to investigate nonlinear phenomena in a variety of power electronic converters.

REFERENCES

[1] H. Poincaré, *Les Methodes Nouvelle de la Mecanique Celeste*. Paris: Gauthier-Villars, 1899.

[2] M. Oestreich, N. Hinrichs, K. Popp, and C. J. Budd, Analytical and experimental investigation of an impact oscillator, *Proc. ASME 16th Biennial Conf. on Mech. Vibrations and Noise*, 1996.

[3] M. di Bernardo and F. Vasca, On discrete time maps for the analysis of bifurcations and chaos in dc/dc converters, *IEEE Trans. on Circuits and Systems—I*, vol. 47, no. 2, pp. 130–143, 2000.

[4] J. Kassakian, M. Schlecht, and G. Verghese, *Principles of Power Electronics*. Reading, MA: Addison-Wesley, 1991.

[5] N. Mohan, T. Undeland, and W. Robbins, *Power Electronics*. 2nd ed. New York: Wiley, 1995.

[6] IEEE, *Special Issue on Hybrid Systems*, May 1998. *IEEE Trans. on Automatic Control*.

[7] S. Jalali, I. Dobson, R. H. Lasseter, and G. Venkataramanan, Switching time bifurcations in a thyristor controlled reactor, *IEEE Trans. on Circuits and Systems—I*, vol. 43, no. 3, pp. 209–218, 1996.

[8] M. di Bernardo, F. Garofalo, L. Glielmo, and F. Vasca, Switchings, bifurcations and chaos in dc/dc converters, *IEEE Trans. on Circuits and Systems—I: Fundamental Theory*, vol. 45, no. 2, pp. 133–141, 1998.

Chapter 3

BASICS OF BIFURCATION AND CHAOS THEORY

3.1 INTRODUCTION TO NONLINEAR DYNAMICS AND CHAOS

Soumitro Banerjee

3.1.1 System State, and State-Space Models

Matter exists in motion. There is nothing static or unchangeable in this material world. Things that may appear to be static, like the mighty mountains or the stars in the sky, are also changing.

Ever since humankind recognized this fact, the study of dynamics has been a major pursuit. Initially the investigations were piecemeal: Newtonian scientists were studying the dynamics of moving bodies; chemists were studying the changes of chemical properties of matter; biologists were probing the changes in living organisms. Slowly it came to be recognized that though the objects of study in various disciplines are different, there are common elements in all changes. A body of knowledge gradually emerged, which deals with dynamical systems in general.

Any system whose status changes with time is called a *dynamical system*. This does not refer only to moving systems; any sort of change in a system comes under the purview of dynamics. The change in chemical composition of a solution is also a problem of dynamics; electronic circuits with no moving parts are also dynamical systems.

In studying a dynamical system, one identifies a minimum number of variables that uniquely specify the state of the system. These are called the *state variables*. The study of dynamics is essentially an investigation of how these state variables change with time. Mathematically this is done by relating the rate of change of these state variables to their current values, via a system of first-order differential equations. Thus, if $\{x_i, i = 1 \text{ to } n\}$ are the state variables, one expresses the dynamics as an nth-order state-space model of the form

$$\dot{x}_i = f_i(x_1, x_2, \ldots x_n) \tag{3.1}$$

or in vector form,

$$\dot{\mathbf{x}} = f(\mathbf{x}) \tag{3.2}$$

The ability to express $\dot{\mathbf{x}}$ purely as a function of $\mathbf{x}$ is what identifies x_i as state variables (and we shall shortly allow dependence on forcing functions and/or time).

Some systems change discretely. Such a situation may arise when a system is actually changing continuously but is observed only at certain intervals. Most power electronic circuits can be conveniently modeled this way. There can also be inherently discrete systems as in digital electronic systems or populations of various species. In such cases the state variables at the $(n+1)$th instant are expressed as a function of those at the nth instant :

$$\mathbf{x}_{n+1} = f(\mathbf{x}_n) \tag{3.3}$$

(The subscript on boldface $\mathbf{x}$ denotes the time instant, not a conponent of the vector.)

Equations of forms (3.2) or (3.3) with a given set of initial conditions can be solved either analytically or numerically, and the solution gives the future states of the system as functions of time.

Geometrically, the dynamics can be visualized by constructing a space with the state variables as the coordinates. This is called the *state space* or *phase space*. The state of the system at any instant is represented by a point in this space. Starting from any given initial position, the state-point moves in the state space, and this movement is completely determined by the state equations. The path of the state-point is called the *orbit* or the *trajectory* of the system that starts from the given initial condition. These trajectories are obtained as the solutions of the differential equations (3.2) or iterates of the map (3.3).

In one-dimensional discrete systems there is a very simple way of working out the orbit originating from a given initial condition. This is illustrated in Figure 3.1. In this figure, the function $x_{n+1} = f(x_n)$ is graphed and the 45° line is also shown. The intersection of the graph of the function and the 45° line is the fixed point $x_{n+1} = x_n = x^*$. In order to iterate the function, start from any initial condition x_0. Draw a vertical line to meet the graph of the map. The vertical height to this point is x_1. In order to get this value on the horizontal axis, draw a horizontal line to meet the 45° line and come down to the horizontal axis. This mapping from x_0 to x_1 is shown by a curved arrow on the horizontal axis. The iterations $x_1 \rightarrow x_2 \rightarrow x_3 \rightarrow x_4 \rightarrow x_5$ are done in the same way, as shown. This procedure of working out the orbit gives the picture a cobweb-like appearance. Hence it is called a *cobweb diagram*.

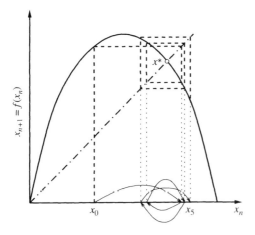

Figure 3.1 The graphical method of working out the iterates of a one-dimensional map, also called *cobweb diagram*. Note that in actually implementing the iteration, you only need to follow the dashed line, never coming down to the horizontal axis on the dotted line.

3.1.2 Autonomous Systems and Nonautonomous Systems

If the system equations do not have any externally applied time-varying input or other time variations, the system is said to be *autonomous*. In such systems the right-hand side of (3.2) does not contain any time-dependent term. A typical example is the simplified model of atmospheric convection, known as the Lorenz system:

$$\frac{dx}{dt} = -3(x - y)$$

$$\frac{dy}{dt} = -xz + rx - y \qquad (3.4)$$

$$\frac{dz}{dt} = xy - z$$

where r is a parameter.

Systems with external inputs or forcing functions or time variations in their definition are called *nonautonomous* systems. In such systems, the right-hand side of (3.2) contains time-dependent terms. As a typical example, one can consider a pendulum with an oscillating support, the equations of which are

$$\frac{dx}{dt} = y$$

$$\frac{dy}{dt} = -y - g \sin x + F \cos \omega t \qquad (3.5)$$

Likewise, power electronic circuits with clock-driven control logic are nonautonomous systems.

3.1.3 Vector Fields of Linear, Linearized, and Nonlinear Systems

In studying the dynamical behavior of a given system, one has to compute the trajectory starting from a given initial condition. However, it is generally not necessary to compute all possible trajectories in order to study a given system. It may be noted that the left-hand side of (3.2) gives the rate of change of the state variables. This is a vector, which is expressed as a function of the state variables. The equation (3.2) thus defines a vector at every point of the state space. The properties of a system can be studied by studying this *vector field*.

If the system is linear, this procedure becomes quite straightforward. In that case it is customary to express (3.2) in the form

$$\dot{\mathbf{x}} = \mathbf{A}\mathbf{x} + \mathbf{B}\mathbf{u} \qquad (3.6)$$

where $\mathbf{A}, \mathbf{B}$ are time-independent matrices, and the components of the vector $\mathbf{u}$ are the externally imposed inputs of the system.

The points where the $\dot{\mathbf{x}}$ vector has zero magnitude (i.e., where $\dot{\mathbf{x}} = \mathbf{A}\mathbf{x} + \mathbf{B}\mathbf{u} = 0$) are called the *equilibrium points*. From (3.6) it is evident that the term $\mathbf{B}\mathbf{u}$ has the effect of shifting the equilibrium point, while stability of the equilibrium point is given by the matrix $\mathbf{A}$. Therefore, while studying the stability of the equilibrium point, one considers the unforced system—where $\mathbf{A}$ operates on the vector $\mathbf{x}$ to give the vector $\dot{\mathbf{x}}$. In general, the vector $\mathbf{x}$ and the vector $\dot{\mathbf{x}}$ lie along different directions. But there may be some special directions in the state space such that if the vector **x** is in that direction, the resultant vector $\dot{x}$ also lies along the same direction. Any vector along these special directions is called an *eigenvector*. The factor by which any eigenvector expands or contracts when it is operated on by the matrix **A** is called the *eigenvalue*. The eigenvalues are obtained by solving the equation $|\mathbf{A} - \lambda\mathbf{I}| = 0$, and the eigenvectors are obtained, for each real eigenvalue, from the equation $(\mathbf{A} - \lambda\mathbf{I})\mathbf{x} = 0$. One can then construct the solution of the differential equations as a linear combination of the solutions along the eigenvectors $\mathbf{v}_i$ expressed in the form $e^{\lambda_i t}\mathbf{v}_i$.

From this, it follows that if the real parts of the eigenvalues are negative, the system is stable in the sense that any perturbation from an equilibrium point eventually dies out and the system settles back to the equilibrium point. Such a stable equilibrium point is called a *node*. If the real parts of the eigenvalues are positive, the system is unstable. If the imaginary parts of all eigenvalues are zero, the system does not oscillate; otherwise it does. If some eigenvalues are real and negative while others are real and positive, the system is stable in the subspace spanned by the eigenvectors associated with the negative eigenvalues, and is unstable away from this. Such an equilibrium point is called a *saddle*, and a system with a saddle equilibrium point is globally unstable. If the eigenvalues are purely imaginary, the response of the system is oscillatory without any damping, and such an equilibrium point is called a *center*. Some typical vector fields of linear systems are shown in Figure 3.2.

In linear systems there can be only one equilibrium point, and the gross structure of the vector field over the whole state space is determined by the eigenvalues and

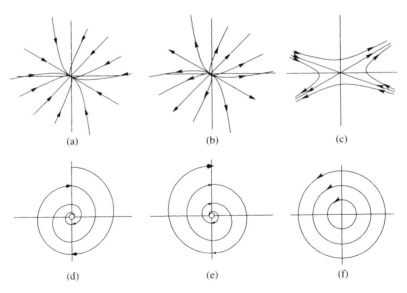

Figure 3.2 Vector field in the state space of two-dimensional linear systems in the vicinity of an equilibrium point. (a) An attracting node: both eigenvalues negative and real. (b) A repelling node: both eigenvalues positive and real. (c) A saddle: one eigenvalue positive and the other negative, both real. (d) A spiralling attractor: complex conjugate eigenvalues with negative real part. (e) A spiralling repeller: complex conjugate eigenvalues with positive real part. (f) A center: purely imaginary eigenvalues.

eigenvectors of the matrix **A**. Therefore linear systems are simple to analyze and easy to handle, and a lot of research attention has been directed toward such systems. Today almost all the methodologies for control system design depend on linear system theory.

Unfortunately, one rarely finds truly linear systems in nature or in engineering. In nonlinear systems the behavior of the vector field may be different for different parts of the state space, and there can be more than one equilibrium point. In such cases one can study the *local* properties of the state space by *linearizing* the system around the equilibrium points, given by $\dot{\mathbf{x}} = f(\mathbf{x}) = 0$.

This is done by using the Jacobian matrix of the functional form $f(\mathbf{x})$ at an equilibrium point. For example, if the state space is two-dimensional, given by

$$\dot{x} = f_1(x, y)$$

$$\dot{y} = f_2(x, y)$$

then the local linearization at an equilibrium point (x^*, y^*) is given by

$$\begin{pmatrix} \delta\dot{x} \\ \delta\dot{y} \end{pmatrix} = \begin{pmatrix} \frac{\partial f_1}{\partial x} & \frac{\partial f_1}{\partial y} \\ \frac{\partial f_2}{\partial x} & \frac{\partial f_2}{\partial y} \end{pmatrix} \begin{pmatrix} \delta x \\ \delta y \end{pmatrix} \tag{3.7}$$

where $\delta x = x - x^*$, $\delta y = y - y^*$. The matrix containing the partial derivatives is called the *Jacobian matrix* and the numerical values of the partial derivatives are calculated at the equilibrium point. This is really just a (multivariate) Taylor series expanded to first order.

The properties of the vector field in a close neighborhood of a given equilibrium point can be obtained by studying the eigenvalues and eigenvectors of the Jacobian matrix. This representation is widely used in engineering because, in general, the nominal operating point of any system is located at an equilibrium point, and if perturbations are small then the linear approximation gives a simple workable model of the dynamical system.

In nonlinear systems one might expect to study the properties of the whole vector field by studying it part by part, breaking it up into linear regions. Unfortunately, the whole is not just a sum of the parts. There are global properties that cannot be found by working piecemeal.

In some simple cases of nonlinear systems, the reductionist approach works. In such cases one linearizes the system at the equilibrium points, obtains the characteristics of the vector fields in the neighborhood of these points, and joins the field lines to work out the approximate structure of the vector field over the whole state space. Figure 3.3 shows an example of a vector field obtained by this method.

3.1.4 Attractors in Nonlinear Systems

To illustrate some typical features of nonlinear systems, we take the system given by $\ddot{x} - \mu(1 - x^2)\dot{x} + x = 0$, known as the *Van der Pol* equation. Figure 3.4 shows the vector field of this system with state variables x and $y = \dot{x}$. If the parameter μ is varied from a negative value to a positive value, a fundamental change in the property of the vector field occurs. The stable equilibrium point becomes unstable and the field lines spiral outward. But it does not become globally unstable as the field lines at a distance from the equilibrium point still point inward. Where the two types of field lines meet,

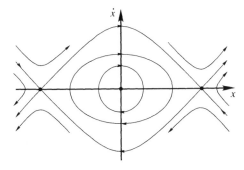

Figure 3.3 Vector field of the two-dimensional nonlinear system given by $\ddot{x} + x - x^3 = 0$. The equilibrium point at $(0, 0)$ is a center (imaginary eigenvalues) and the equilibrium points at $(-1, 0)$ and $(1, 0)$ are saddles. One can construct a good approximation of the vector field of the whole state space from the local vector fields in the neighborhood of the equilibrium points.

there develops a stable periodic behavior. This is called a *limit cycle* (Figure 3.4(b)). It is a *global* behavior whose existence can never be predicted from linear system theory.

Note that there is a fundamental difference between the periodic behaviors in Figure 3.2(f) (i.e., a linear system with purely imaginary eigenvalues) and Figure 3.4(b). In the first case a different periodic orbit (though of constant period) is attained for initial conditions at different radii, while in the case of the limit cycle, trajectories starting from different initial conditions converge on to the same periodic behavior. The limit cycle appears to attract points of the state space. This is an example of an *attractor*.

Thus in a two-dimensional nonlinear system one can come across periodic attractors as in Figure 3.4(b). If the state space is of higher dimension, say three, the orbit has the freedom to loop around without intersecting itself. Therefore in such systems one can have more intricate attractors.

Let us explain this point with reference to Figure 3.5. Suppose a third-order dynamical system is going through oscillations and when we plot one of the variables against time, it has a periodic waveform as shown in Figure 3.5(a), corresponding to a state-space trajectory that shows a single loop. When some parameter is varied, the

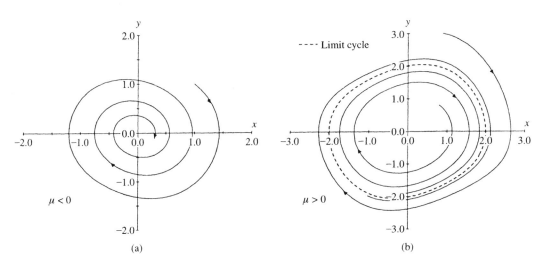

Figure 3.4 The vector fields for $\ddot{x} - \mu(1 - x^2)\dot{x} + x = 0$, (a) for $\mu < 0$ and (b) for $\mu > 0$. The dashed line shows the limit cycle.

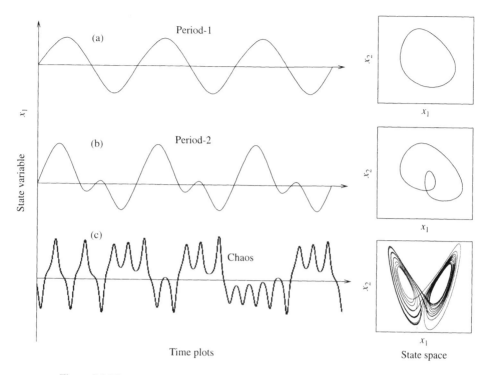

Figure 3.5 The appearance of period-1, period-2, and chaotic waveforms in the time
domain and in state space, for a system of order ≥ 3.

waveform can change to the type shown in Figure 3.5(b), which has twice the period of
the earlier periodic waveform. In order for such orbits to exist, the state space must
have three or higher dimensions. (Note that the figure actually shows a projection of a
3-D state space onto two dimensions—a real state-space orbit cannot cross itself
because there is a unique velocity vector $\dot{x}$ associated with every point in the state
space.)

Sometimes there is one periodicity superimposed on another, and we have a torus-
shaped attractor in the state space. This is called a *quasiperiodic attractor*. For a graphic
illustration, please refer to Figure 4.5 in Chapter 4.

One interesting possibility opens up in systems of order 3 or greater: bounded
aperiodic orbits, as shown in Figure 3.5(c). In such a case the system state remains
bounded—within a definite volume in the state space, but the same state never repeats.
In every loop through the state space the state traverses a new trajectory. This situation
is called *chaos* and the resulting attractor is called a *strange attractor*. When such a
situation occurs in an electronic circuit, the system undergoes apparently random
oscillations.

3.1.5 Chaos

What are the characteristics of a chaotic system? The first person to chance upon a
strange attractor was Lorenz [1], who was studying the dynamics of weather systems
through the mathematical model given by (3.4). Lorenz noticed that for certain ranges

of values of r, the trajectories starting from very close initial conditions diverge fast and lead to entirely different future states. This is called *sensitive dependence on initial conditions*, which is a hallmark of chaos.

This sensitivity has a very important implication. Much of the study of dynamical systems is propelled by the necessity of predicting future states of systems. But the sensitive dependence on initial conditions in chaotic systems renders prediction impossible beyond a short time frame. This is because initial conditions cannot be measured or specified with infinite accuracy. Minuscule errors in defining the initial condition do not matter in stable nonchaotic systems, for orbits starting from slightly different initial conditions do not diverge exponentially. But in a chaotic attractor, arbitrarily close states diverge exponentially and prediction becomes impossible unless the initial condition is known with infinite information and specified with infinite accuracy, which is impossible.

There is another aspect of this phenomenon. Suppose we take a set of initial conditions enclosed within a ball in the state space, and investigate how this *set* evolves. We would find that it expands in one direction—initial conditions get further from each other in this direction. But there is a theorem due to Liouville, which states that in conservative systems (systems with no dissipation, i.e., no resistances or friction) the *volume* of the set of initial states is preserved. In dissipative systems the volume contracts. Therefore, if the shape of the set of points expands in one direction, it must contract in another.

Something must be done to the expanding direction to keep it within bounds. This is what happens in chaotic systems: the expanding direction *folds*. It is the continuous stretching and folding in the state space that ensures that the state remains bounded while nearby initial conditions diverge.

Smale [2] showed that this stretching-and-folding property of chaotic systems can be captured by a transformation known as *horseshoe mapping*. To visualize this, consider a two-dimensional system. Take a square area in the state space (Figure 3.6). Stretch it in one direction and squeeze it in another. Then fold the expanding direction to shape it like a horseshoe. Now take the area that contains this horseshoe shape and

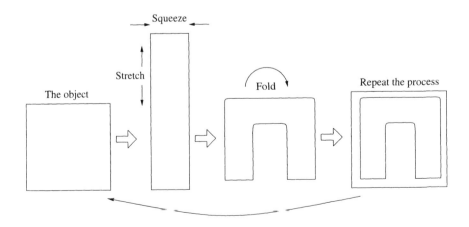

Figure 3.6 One iteration of Smale's horseshoe mapping.

repeat the process. Repeated application of this mapping leads to a fine layered structure that is found in the strange attractors of chaotic systems.

3.1.6 Poincaré Map

In many situations it is more convenient to analyze a system if it can be expressed as a discrete mapping rather than a continuous-time system. This is achieved by a method invented by Henri Poincaré. In this method, one places a surface, called the *Poincaré section*, at a suitable place in the state space. The *Poincaré map* is then the mapping of a point of intersection of a trajectory with the surface onto the subsequent intersection from the same side. In this way continuous-time evolution in state space is reduced to a map of the form (3.3) in a lower-dimensional space, as illustrated in Figure 3.7(a).

In explicitly time-dependent cases (nonautonomous systems), where the forcing function has period T, the natural choice for a Poincaré map is the mapping of the variables at time t onto those at $t + T$ (Figure 3.7(b)). This has the effect of placing a stroboscope in the state space to illuminate the state at intervals of T. The method of sampled-data modeling of power electronic circuits, as explained in Chapters 1 and 2, follows this technique.

If the continuous-time orbit is periodic, there are a finite number of points on the Poincaré section. The discretized system thus reveals the periodicity of the orbit of the underlying continuous-time system via the periodicity of the corresponding sampled state. For this to hold true, the Poincaré section has to be placed such that it experiences a maximum number of intersections with a given orbit. In nonautonomous systems, the periodicity is given by the repetitiveness of the orbit as integer multiples of the period of the external input, and not by the number of loops in the continuous-time state-space trajectory. Thus, if the continuous-time orbit shows one loop in the state-space, but there are two cycles of the forcing function within that period, the orbit will be said to be period-2.

In the case of quasiperiodic orbits, since the two frequencies are incommensurate, points in the sampled model will not fall on each other and will be arranged in a closed loop (see Figure 4.5 in Chapter 4 for illustration). For a chaotic orbit the asymptotic behavior in discrete time shows an infinite number of points, contained within a finite

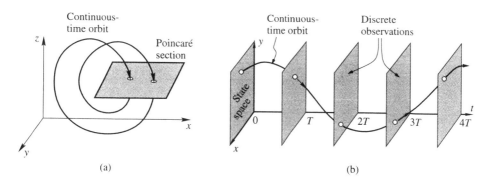

(a)

(b)

Figure 3.7 Obtaining discrete model of a continuous-time dynamical system: (a) for an autonomous system, and (b) for a nonautonomous system where the external input has period T.

volume, and distributed over a region of very intricate structure (see, for example, Figure 1.6). This is the strange attractor in the discrete domain.

3.1.7 Dynamics of Discrete-Time Systems

In analyzing discrete-time nonlinear systems expressed in the form

$$\mathbf{x}_{n+1} = f(\mathbf{x}_n)$$

one first finds the fixed points $\mathbf{x}_{n+1} = \mathbf{x}_n = \mathbf{x}^*$. Then one can locally linearize the discrete system in the neighborhood of a fixed point by obtaining the Jacobian matrix. The eigenvalues of the Jacobian matrix indicate the stability of the fixed point.

But there is a subtle difference between what the eigenvalues indicate in a continuous-time system and what they indicate in a discrete-time system. In a continuous-time system the Jacobian matrix, when operated on a state-vector, gives the velocity vector corresponding to that state. But in a discrete-time system the operation of the Jacobian matrix on a state-vector gives the state-vector at the next iterate.

To illustrate, imagine a two-dimensional discrete-time system

$$\begin{bmatrix} x_{n+1} \\ y_{n+1} \end{bmatrix} = \begin{bmatrix} J_{11} & J_{12} \\ J_{21} & J_{22} \end{bmatrix} \begin{bmatrix} x_n \\ y_n \end{bmatrix}$$

linearized around the fixed point $(0, 0)$. Let λ_1 and λ_2 be the eigenvalues of the Jacobian matrix. Imagine an initial condition placed on the eigenvector associated with a real eigenvalue λ_1. Now if λ_1 has magnitude less than unity, the next iterate will fall on another point of the eigenvector closer to the fixed point, and in subsequent iterates the state will converge onto the fixed point along the eigenvector (Figure 3.8(a)). If both the eigenvalues are real and have magnitude less than unity, any initial condition will asymptotically move toward an eigenvector and will converge onto the fixed point. Thus the system is stable and the fixed point is an *attractor*.

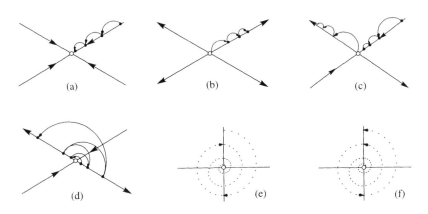

Figure 3.8 Examples of fixed points in a linearized two-dimensional discrete system. (a) An attractor: eigenvalues real, $0 < \lambda_1, \lambda_2 < 1$. (b) A repeller: eigenvalues real, $\lambda_1, \lambda_2 > 1$. (c) A regular saddle: eigenvalues real, $0 < \lambda_1 < 1, \lambda_2 > 1$. (d) A flip saddle: eigenvalues real, $0 < \lambda_1 < 1$, $\lambda_2 < -1$. (e) A spiral attractor: eigenvalues complex, $|\lambda_1|, |\lambda_2| < 1$. (f) A spiral repeller: eigenvalues complex, $|\lambda_1|, |\lambda_2| > 1$.

On the other hand, if these eigenvalues have magnitudes greater than unity, the system is unstable and the fixed point is a *repeller* (Figure 3.8(b)). If these eigenvalues are real, with λ_1 of magnitude less than unity and λ_2 of magnitude greater than unity, the system is stable in one direction (the eigenvector associated λ_1) and unstable in the other. Such a fixed point is called a *saddle*. If one of the eigenvalues is negative, the orbit flips between two sides of the fixed point, which is called a *flip saddle* (Figure 3.8(d)). If both the eigenvalues are positive, a state on one side of the fixed point remains on the same side on subsequent iterates, and the fixed point is called a *regular saddle* (Figure 3.8(c)).

If λ_1 and λ_2 are complex conjugate and have magnitude less than unity, any initial condition moves in a spiral fashion and converges onto the fixed point. Such a fixed point is said to be *spirally attracting* (Figure 3.8(e)). Likewise, if the eigenvalues are complex conjugate and with magnitude greater than unity, the fixed point is *spirally repelling* (Figure 3.8(f)). These ideas can easily be extended to higher-dimensional systems, though one rarely comes across maps of dimension greater than two in power electronic circuits.

It is important to note that in a discrete system a fixed point is stable if all the eigenvalues of the Jacobian matrix have magnitude less than unity, while in a continuous-time system an equilibrium point is stable if the eigenvalues have negative real part.

Outside the small neighborhood of a fixed point, the above description of linearized system behavior no longer remains valid. For example, if the fixed point is a saddle, the eigenvectors in the small neighborhood have the property that if an initial condition is placed on the eigenvector, subsequent iterates remain on the eigenvector. Outside the small neighborhood the lines with this property no longer remain straight lines. One can therefore identify *curved lines* passing through the fixed point with the property that any initial condition placed on the line forever remains on it under iteration of the map. Such curves are called *invariant manifolds* (Figure 3.9). If iterates of an initial condition approach the fixed point along a manifold, then it is said to be *stable*; if iterates move away from the fixed point (or move toward it under the application of the inverse map $\mathbf{x}_n = f(\mathbf{x}_{n+1})$) then the manifold is said to be *unstable*. It is clear that the eigenvectors in the linearized model are locally tangent to the stable and unstable manifolds of the fixed point.

If an initial condition is not placed on one of the invariant manifolds, on iteration of the map the state moves away from the stable manifold and closer to the unstable manifold (Figure 3.10). The unstable manifold *attracts* points in the state space, and therefore if there is a saddle fixed point in a system, all attractors must lie on the unstable manifold of the saddle. The stable manifold, on the other hand, *repels* points in the state space. Therefore, if there are multiple attractors in a system, the stable manifold of a saddle fixed point acts as the boundary separating the basins of attraction. Thus the manifolds play an important role in determining the dynamics.

In nonlinear systems these manifolds can curve and bend and wander around the state space in a quite complicated way. The nature, structure, and intersections of the manifolds play an important role in defining the dynamics of nonlinear systems. This will be explained in a greater detail in Section 3.6.

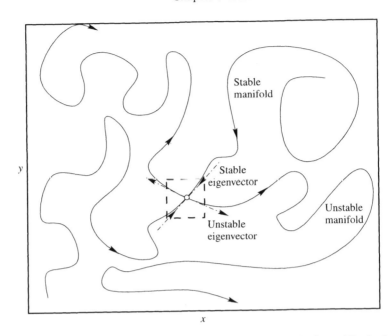

Figure 3.9 Schematic view of the stable and unstable manifold of a saddle fixed point. The locally linearized neighborhood of the fixed point is also shown.

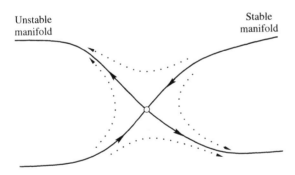

Figure 3.10 The unstable manifold of a regular saddle attracts points in the state space while the stable manifold repels.

3.1.8 Fractal Geometry

We have seen that the chaotic attractor in discrete domain is a collection of an infinite number of points, distributed over a very complicated structure—a practical example of which was shown in Figure 1.6. The mathematical description of such a chaotic attractor requires the concept of *fractals*.

Fractal geometry was invented by Mandelbrot [3] to describe the geometrical properties of natural objects. In contrast with the Euclidean objects that one comes across in school textbooks (like the triangle, sphere, parabola, rectangular parallelopiped, etc.), natural objects have some distinguishable properties.

Mandelbrot made his point with his famous question: "How long is the coastline of England?" A little thought makes it evident that the measured length depends on the yardstick of measurement. If the yardstick is long, much of the detail of the coastline geometry will be missed. As smaller and smaller yardsticks are used, the creeks and bends come in view and the measured length increases. And in the limit, the measured length of the coastline becomes infinitely large. It follows that the coastline is a curve of infinite length enclosing a finite area. The same is true for all 3-D natural objects enclosed by natural surfaces. Furthermore, natural surfaces are not differentiable anywhere. At higher and higher levels of magnification (over a range of several orders of magnitude) the surface does not flatten into a plane, but retains its complexity.

These objects, therefore, need a new mathematical tool for characterization. This is done by defining a *fractal dimension*.

It is necessary at this stage to distinguish between the dimension of an object and that of the embedding space. The dimension of the embedding space is given by the degrees of freedom, or the number of variables used to describe the object, which obviously is an integer. In contrast, the dimension of an object is defined according to the way it fills the space defined by these variables.

To illustrate, let us consider a known two-dimensional object—the square. To see how it fills space, we cover it with a grid of length ϵ on each side, and successively reduce ϵ. The number N of grid elements (or boxes) needed to cover the object increases as the square of the reciprocal of grid length: $N(\epsilon) \approx \left(\frac{1}{\epsilon}\right)^2$ asymptotically, for small ϵ.

If, instead, we take a right-angled triangle as our object, as ϵ is reduced, the count of covering boxes approaches $N(\epsilon) \approx \frac{1}{2}\left(\frac{1}{\epsilon}\right)^2$. And for a circle we have $N(\epsilon) \approx \frac{\pi}{4}\left(\frac{1}{\epsilon}\right)^2$. In general we can write

$$N(\epsilon) \approx K\left(\frac{1}{\epsilon}\right)^2$$

for small ϵ, where K is a constant.

We can now extract the dimension (2 in this case) from $N(\epsilon)$ as follows:

$$\ln N(\epsilon) \approx \ln K + 2\ln\frac{1}{\epsilon}$$

$$2 \approx \frac{\ln N(\epsilon)}{\ln\frac{1}{\epsilon}} - \frac{\ln K}{\ln\frac{1}{\epsilon}}$$

The second term vanishes as $\epsilon \to 0$. Thus the dimension of the object is given by

$$D = \lim_{\epsilon \to 0}\frac{\ln N(\epsilon)}{\ln\frac{1}{\epsilon}} \tag{3.8}$$

If this method is applied to natural objects, the value of D typically turns out to be well defined, but to be a fraction. Indeed, this is the most important property that distinguishes natural objects from idealized Euclidean geometrical objects. Geometrical objects with fractional dimensions are called *fractals*.

In dynamical systems it has been found that chaotic attractors are fractal objects. The determination of the fractal dimension is thus one of the methods of characterizing a chaotic attractor.

The technique for estimating the fractal dimension of an attractor is to divide the embedding space into volume elements or *boxes* with sides of length ϵ. If $N(\epsilon)$ is the number of volume elements needed to cover the attractor, and ϵ is successively made smaller, the fractal dimension (D_F) is given by (3.8). It is also called the *capacity dimension* or *box-counting* dimension.

3.1.9 Lyapunov Exponent

Another way to characterize a chaotic system is to quantify the rates of stretching and squeezing in the state space. This is done by means of *Lyapunov exponents*. In the direction of stretching, two nearby trajectories diverge, while in the directions of squeezing, nearby trajectories converge. If we approximate this divergence and convergence by exponential functions, the rates of stretching and squeezing would be quantified by the exponents. These are the Lyapunov exponents. Since the exponents vary over the state space, one has to take the long time average exponential rates of divergence (or convergence) of nearby orbits.

The total number of Lyapunov exponents is equal to the degrees of freedom of the system. If system trajectories have at least one positive Lyapunov exponent, then those trajectories are either unstable or chaotic. If the trajectories are bounded and have positive Lyapunov exponent, the system definitely includes chaotic behavior. The larger the positive exponent, the shorter the time scale of system predictability. The estimation of the largest exponent therefore assumes a special importance. Computational methods of determining the maximal Lyapunov exponent are discussed in Section 4.2.

3.1.10 Bifurcation

A qualitative change in the dynamics which occurs as a system parameter is changed is called a *bifurcation*. In power electronics, converters are generally designed to work at a specific operating condition that gives specific output voltage ripple, spectral characteristics, and so on. However, the operating mode can change substantially and qualitatively when a parameter like input voltage or load changes. Therefore the study of bifurcations assumes great importance in such systems.

A convenient way of studying bifurcations is through *bifurcation diagrams*. In such a graphical representation, a parameter is varied and plotted along the x-axis. On the y-axis, the asymptotic behavior of a *sampled* state variable is plotted as discrete points. If the system is operating in period-1 (i.e., period equal to the sampling interval) for some parameter value, there will be only one point corresponding to that parameter value. If it is in period-2, there will be two points. If the system behaves chaotically for some other parameter value, there will be a large (theoretically infinite) number of points corresponding to that parameter value. Such a bifurcation diagram summarizes the change in system behavior in response to the variation of a parameter. Figure 3.11 shows a representative bifurcation diagram for the dynamical system given by $x_{n+1} = \mu x_n(1 - x_n)$.

In this section we have presented the basic issues in nonlinear dynamics briefly in a rather informal language. More detailed treatment can be found in [4,5,6,7]. The following sections of this chapter will present the mathematical framework in which bifurcations in general are understood.

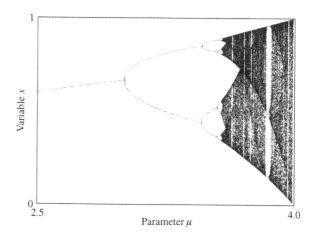

Figure 3.11 Bifurcation diagram for the logistic map $x_{n+1} = \mu x_n(1 - x_n)$ when the parameter μ is varied from 2.5 to 4.0.

REFERENCES

[1] E. N. Lorenz, Deterministic nonperiodic flow, *J. Atmospheric Sciences*, vol. 20, no. 2, pp. 130–141, 1963.

[2] S. Smale, Differentiable dynamical systems, *Bulletin of the American Mathematical Society*, vol. 73, p. 747, 1967.

[3] B. B. Mandelbrot, *The Fractal Geometry of Nature*. San Francisco: Freeman, 1982.

[4] T. S. Parker and L. O. Chua, Chaos: A tutorial for engineers, *Proc. IEEE*, vol. 75, no. 8, pp. 982–1008, 1987.

[5] K. T. Alligood, T. D. Sauer, and J. A. Yorke, *Chaos: An Introduction to Dynamical Systems*. New York: Springer-Verlag, 1996.

[6] S. H. Strogatz, *Nonlinear Dynamics and Chaos*. Reading, MA: Addison-Wesley, 1994.

[7] E. Ott, *Chaos in Dynamical Systems*. Cambridge University Press, Cambridge: 1993.

3.2 BIFURCATIONS OF SMOOTH MAPS

Jonathan H. B. Deane

We now illustrate by examples the four main types of bifurcation that take place in smooth maps. The table below gives the name of four bifurcations and an example map that displays the bifurcation.

Name of Bifurcation	Map, f
Pitchfork	$x \mapsto (1 + \mu)x - x^3$
Saddle-node	$x \mapsto \mu + x - x^2$
Period doubling or flip	$x \mapsto -(1 + \mu)x + x^3$
Neimark	$(x, y) \mapsto (y, \mu/2 + (\mu + 1)(y - x - 2xy))$

Each of these four maps is a simple model for the bifurcations indicated. We do not discuss tests for finding if a given map undergoes any of the above bifurcations; the details concerning this are to be found in [1].

The maps are to be understood to generate a sequence $x_0, x_1, \ldots$ by $x_{n+1} = f(x_n)$ (or the two-dimensional equivalent). In what follows, we drop the subscript n for clarity.

In order to illustrate the characteristics of the bifurcations we need to assume that

1. x and y are real.
2. x is, where necessary, within δx of a fixed point, with δx small enough that we can neglect terms $O(\delta x^2)$.
3. $|\mu| \ll 1$, so that we can neglect terms $O(\mu^2)$.

We now deal with each map in turn.

3.2.1 The Pitchfork Bifurcation

Given a map of the form

$$x \mapsto f(x) = (1 + \mu)x - x^3 \tag{3.9}$$

we first find the fixed points $\bar{x} = f(\bar{x})$, and then determine their stability. The fixed points are given by $\mu\bar{x} = \bar{x}^3$ which has solutions (I) $\bar{x} = 0$ and (II) $\bar{x} = \pm\sqrt{\mu}$. To determine the stability, we need to consider small displacements from these fixed points. Performing a Taylor expansion around a fixed point gives

$$f(\bar{x} + \delta x) \approx \bar{x} + \delta x f'(\bar{x}) + O(\delta x^2)$$

Case I: $\bar{x} = 0$ In this case, $f(\bar{x} + \delta x) \approx \delta x(1 + \mu)$. Hence, the fixed point $\bar{x} = 0$ is stable for $\mu < 0$ and unstable for $\mu > 0$.

Case II: $\bar{x} = \pm\sqrt{\mu}$ This pair of fixed points is real only for $\mu > 0$. A Taylor series approximation gives $f(\bar{x} + \delta x) \approx \pm\sqrt{\mu} + \delta x(1 - 2\mu)$. Hence, both fixed points are stable for $\mu > 0$.

A bifurcation is said to take place at $\mu = 0$, that is, a sudden qualitative change in the behavior of solutions to equation (3.9) as μ goes from negative to positive. In this case, the stable fixed point at $x = 0$ for negative μ splits into a pair of stable fixed points at $x = \pm\sqrt{\mu}$ for $\mu > 0$, while the fixed point at $x = 0$ becomes unstable. Initial conditions determine which of the fixed points for $\mu > 0$ is observed. Figure 3.12, which was calculated for equation (3.9), illustrates this. There is a background vector field shown in this and other figures in this section. The direction of the field is the direction in which successive iterates move, and the magnitude is $|f(f(x)) - f(x)|$, which is an indication of how rapidly they move in the direction indicated.

3.2.2 The Saddle-Node Bifurcation

Proceeding in the same way with a map of the form

$$x \mapsto f(x) = \mu + x - x^2 \tag{3.10}$$

the fixed points are the solutions to $\bar{x} = \mu + \bar{x} - \bar{x}^2$, and hence $\bar{x} = \pm\sqrt{\mu}$. The series approximation around the fixed points gives

$$f(\bar{x} + \delta x) \approx \pm\sqrt{\mu} + \delta x(1 \mp 2\sqrt{\mu})$$

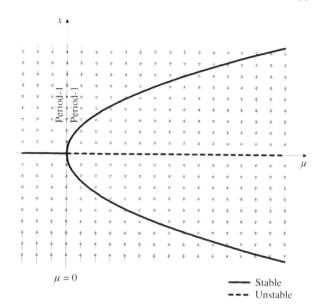

Figure 3.12 A bifurcation diagram for equation (3.9), showing the typical characteristics of a pitchfork bifurcation.

$\mu = 0$

—— Stable
--- Unstable

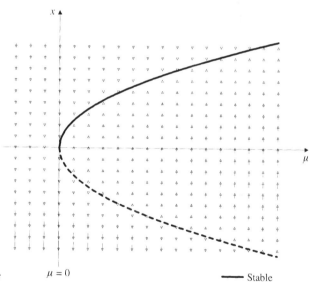

Figure 3.13 The saddle-node bifurcation for equation (3.10).

$\mu = 0$

—— Stable
--- Untable

indicating that for $\mu > 0$ the fixed point at $+\sqrt{\mu}$ is stable and the one at $-\sqrt{\mu}$ is unstable. There are no fixed points for $\mu < 0$. This phenomenon is called saddle-node bifurcation and is illustrated in Figure 3.13. In some literature it is also called *fold bifurcation*.

3.2.3 The Period-Doubling Bifurcation

To illustrate the period-doubling bifurcation, we start from the map

$$x \mapsto f(x) = -(1 + \mu)x + x^3 \qquad (3.11)$$

for which the fixed points are (I) $\bar{x} = 0$ and (II) $\bar{x} = \pm\sqrt{2 + \mu}$.

Case I: $\bar{x} = 0$ This case is easily treated. The series expansion gives

$$f(\bar{x} + \delta x) \approx -\delta x(1 + \mu)$$

so the fixed point at $x = 0$ is clearly stable for $\mu < 0$ and unstable for $\mu > 0$.

Case II: $\bar{x} = \pm\sqrt{2 + \mu}$ Proceeding as before, we find

$$f(\bar{x} + \delta x) \approx \pm\sqrt{2 + \mu} + \delta x(5 + 2\mu)$$

and both of these fixed points are unstable. Let us now consider the second iterate of $f(x)$. We then have

$$f(f(x)) = -(1 + \mu)\left[-(1 + \mu)x + x^3\right] + x^3\left[-(1 + \mu) + x^2\right]^3$$

$$\approx (1 + 2\mu)x - 2(1 + 2\mu)x^3 + O(x^5)$$

expanding to first order in μ. The fixed points $\bar{\bar{x}}$ of $f(f(x))$ are defined as $f(f(\bar{\bar{x}})) = \bar{\bar{x}}$. Hence

$$\bar{\bar{x}} \approx (1 + 2\mu)\bar{\bar{x}} - 2(1 + 2\mu)\bar{\bar{x}}^3$$

which has solutions $\bar{\bar{x}} = 0 (= \bar{x})$, stable for $\mu < 0$, and $\bar{\bar{x}} \approx \pm\sqrt{\mu(1 - 2\mu)}$. The stability of the latter pair of fixed points is determined in the usual way:

$$f(f(\bar{\bar{x}} + \delta x)) = \bar{\bar{x}} + \delta x\left[f(f(\bar{\bar{x}}))\right]' + O(\delta x^2)$$

$$\approx \pm\sqrt{\mu(1 - 2\mu)} + \delta x(1 - 4\mu)$$

Hence, we conclude that the second iterate of f has stable fixed points at $x = \pm\sqrt{\mu(1 - 2\mu)}$ for $\mu > 0$. The interpretation of this is as follows. For $\mu > 0$ the fixed point at $x = 0$ becomes unstable, and is replaced by a pair of fixed points, $\sqrt{\mu(1 - 2\mu)} = x_a$ say, and $-\sqrt{\mu(1 - 2\mu)} = x_b$. Neither of these is stable, but they have the property that $f(x_a) = x_b$ and $f(x_b) = x_a$, and this *is* stable. The phenomenon of a period-1 solution giving way to a period-2 solution is known as *period doubling*. Since the bifurcated orbit flips between two points, it is also known as the *flip bifurcation*. This is illustrated in Figure 3.14. Further illustrations of the period-doubling bifurcation can be found in Figures 3.41 and 3.42 in Section 3.5, which show the graph of a map and its second iterate as a period doubling occurs. Repeated period doublings, known as a *period-doubling cascade*, can result in aperiodic (or chaotic) orbit.

3.2.4 The Neimark Bifurcation

This is substantially different from the other three bifurcations. The major differences are that (i) the map must be at least two-dimensional, and (ii) the nonlinear terms cannot be assumed to be small (assumption 2 at the beginning of this section), so we cannot make certain simplifying assumptions. We study the following map:

$$(x, y) \mapsto f(x, y) = (y, \mu/2 + (\mu + 1)(y - x - 2xy)) \tag{3.12}$$

Proceeding as before, we first find the fixed points. These are $(\bar{x}, \bar{y}) = (-1/2, -1/2)$ and $(\mu/2(\mu + 1), \mu/2(\mu + 1))$. Only the second of these undergoes a Neimark bifurcation. To establish stability we need to calculate the eigenvalues of J, the Jacobian of f, given by

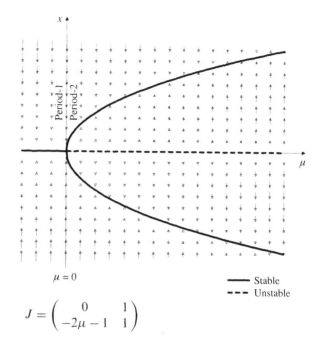

Figure 3.14 The period-doubling bifurcation for $f(f(x))$ defined in equation (3.11). The period-1 solution for $\mu < 0$ gives way to a period-2 solution for $\mu > 0$.

$$J = \begin{pmatrix} 0 & 1 \\ -2\mu - 1 & 1 \end{pmatrix}$$

which has eigenvalues $\lambda_\pm = [1 \pm j\sqrt{3 + 8\mu}]/2 \approx (1 + \mu)\exp \pm j(\pi/3 + \mu/\sqrt{3}) + O(\mu^2)$. This shows that the fixed point $(0,0)$ is stable for $\mu < 0$, since $|\lambda_\pm| < 1$. For $\mu > 0$ we can state that:

- This fixed point becomes unstable.
- We expect that, close to $(0,0)$, $x_n \approx x(n) \exp jn\pi/3$, where $x(n)$ is some function of n.

The first point is illustrated in Figure 3.15, which is a bifurcation diagram for x as μ varies. The second is at least suggested by Figure 3.16, in which x, y phase plane plots

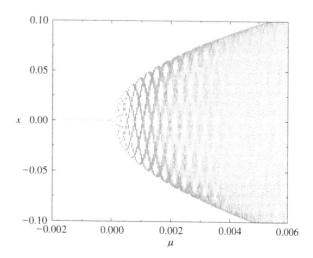

Figure 3.15 A bifurcation diagram illustrating the Neimark bifurcation in relation to solutions to equation (3.12).

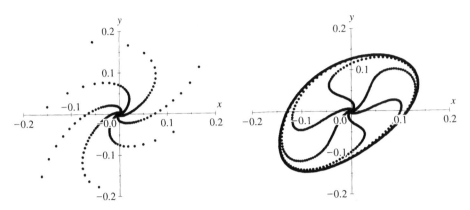

Figure 3.16 Phase plane diagrams for equation (3.12). Left: $\mu < 0$. The fixed point at $(0,0)$ is stable and iterates spiral in to it. Right: $\mu > 0$. The fixed point at $(0,0)$ has become unstable so that iterates spiral away from it and onto a limit cycle instead.

are drawn for $\mu < 0$ and $\mu > 0$. The left-hand figure shows what happens for $\mu < 0$: iterations from an initial condition spiral in to the fixed point, the 6-fold rotational symmetry being a result of the factor $\exp jn\pi/3$. The right-hand figure results when $\mu > 0$, for which the fixed point is unstable. Iterates spiral out to an invariant curve. The shape of this curve, and the fact that it is stable, are both a result of the nonlinear term in $f(x, y)$, and cannot be accounted for by linearizing around the fixed point. This is because the excursions in x and y are *not* small.

Since maps arise out of making a Poincaré section in the state space of continuous-time systems, it is interesting to note what happens in the continuous-time system when a Neimark bifurcation occurs. Before the bifurcation, the Poincaré section shows a stable fixed point—therefore in the continuous-time system it is a stable limit cycle. After the bifurcation, the Poincaré section shows a closed loop, which implies a quasi-periodic orbit (see Figure 4.5). Therefore, in continuous time, a Neimark bifurcation marks a transition from a periodic orbit to a quasiperiodic orbit.

How is a limit cycle born in a continuous-time system? This question, obviously, cannot be probed by considering discrete maps—because there is no intersection with the Poincaré section before the birth of a limit cycle. Therefore one has to consider the change in the orbit of a nonlinear system represented by the set of differential equations $\dot{\mathbf{x}} = f(\mathbf{x})$. In this case the equilibrium point is given by $f(\mathbf{x}) = \mathbf{0}$, and one can probe the stability of the equilibrium point through the Jacobian matrix.

Consider, for example, the second-order system given by $\ddot{x} - \mu(1 - x^2)\dot{x} + x = 0$, which can be written as a set of first-order equations:

$$\dot{x} = y$$

$$\dot{y} = \mu(1 - x^2)y - x$$

Its fixed point is $(0,0)$ and the Jacobian calculated at this fixed point is

$$\begin{bmatrix} 0 & 1 \\ -1 & \mu \end{bmatrix}$$

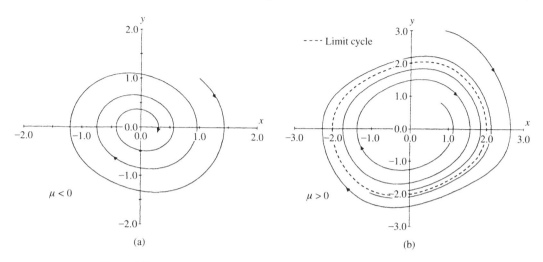

Figure 3.17 The birth of a limit cycle in a continuous-time system represented by $\ddot{x} - \mu(1 - x^2)\dot{x} + x = 0$, through a Hopf bifurcation. (a) $\mu < 0$ and (b) $\mu > 0$.

This has eigenvalues

$$\lambda = \frac{\mu \pm \sqrt{\mu^2 - 4}}{2}$$

By standard linear control theory, when the real part of the eigenvalues is negative (for negative μ), the orbit is stable. When μ increases through zero, a pair of complex conjugate eigenvalues crosses the imaginary axis. If such a situation occurs in a linear system, it becomes unstable. But in a nonlinear system a limit cycle is born, as shown in Figure 3.17. This is called a *Hopf bifurcation*.

REFERENCES

[1]   J. Guckenheimer and P. Holmes, *Nonlinear Oscillations, Dynamical Systems, and Bifurcations of Vector Fields*, New York: Springer-Verlag, 1983.

[2]   D. K. Arrowsmith and C. M. Place, *An Introduction to Dynamical Systems*, Cambridge, UK: Cambridge University Press, 1990.

3.3 BIFURCATIONS IN PIECEWISE-SMOOTH MAPS

Soumitro Banerjee
Celso Grebogi

Many past investigations on the dynamics of power electronic circuits have revealed that such systems exhibit bifurcation phenomena which cannot be explained in terms of the theory developed for smooth (everywhere differentiable) systems. It has also been shown that discrete models of many systems obtained by observing the state vector at every clock instant yield *piecewise-smooth maps* [1,2,3]; some systems yield *discontinuous maps.* We first illustrate, with two examples, why some power electronic circuits are modeled by piecewise-smooth maps.

Figure 3.18 shows a peak-current-controlled boost converter. In this system, the inductor current increases during the *on* period and when it reaches a reference current I_{ref}, the switch is turned off. The switch is turned on by a free-running clock: if the switch is off then the next clock pulse turns it on and if the switch is on then the clock pulses are ignored. A standard waveform resulting from this control logic is shown in Figure 3.19.

Now if we consider the observations at each clock instant, it is clear that there can be two distinct *types* of evolution between two clock instants: one, when the inductor current reaches I_{ref} and the switch turns off before the next observation instant, and the other when the inductor current doesn't reach I_{ref} and the switch remains on throughout the clock period. Obviously these two courses of evolution would be represented by two different functions in the resulting map. The value of the inductor current at a clock instant for which i reaches I_{ref} exactly at the next clock instant represents the borderline case I_{border}.

Now consider two values of the inductor current at a clock instant—one slightly above I_{border} and the other slightly below I_{border} (see Figure 3.20). It is clear that small deviations from I_{border} at one clock instant result in small deviations at the next clock instant. This gives a clue to two other properties of such a map—first, it is continuous, and second, the partial derivatives are finite. The map is formally derived in Section 5.1.

Next consider the buck converter with duty cycle controlled by voltage feedback as shown in Figure 3.21. In this controller, a constant reference voltage V_{ref} is subtracted from the output voltage and the error is amplified with gain A to form a control signal $v_{\text{con}} = A(v - V_{\text{ref}})$. The switching signal is generated by comparing the control signal with a periodic sawtooth (ramp) waveform (v_{ramp}). S turns on whenever v_{con} goes below v_{ramp} and a latch allows it to switch off only at the end of the ramp cycle.

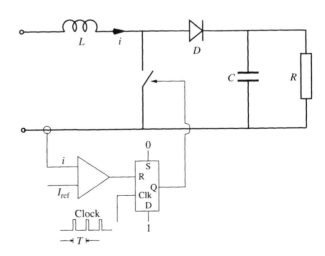

Figure 3.18 Schematic circuit diagram of the current-mode-controlled boost converter.

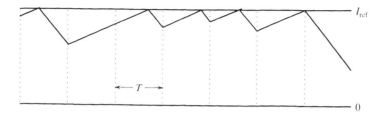

Figure 3.19 Waveform of inductor current of the boost converter.

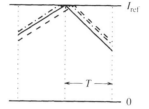

Figure 3.20 Small perturbations of the inductor current above and below the borderline value causes small deviation in the next sampling instant.

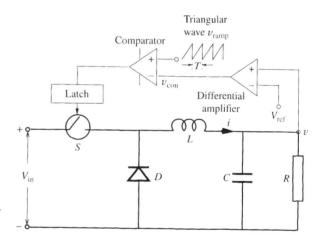

Figure 3.21 The schematic circuit diagram of the voltage-controlled buck converter.

There are three ways in which the state can move from one observation point to the next: (1) the control voltage is throughout above the ramp waveform and the switch remains off; (2) the cycle involves an *off* period and an *on* period; (3) the control voltage is throughout below the ramp waveform and the switch remains on. The three cases are shown in Figure 3.22. These are represented by three different expressions of the map (though in this case the map cannot be derived in closed form). The borderlines are given by the condition where the control voltage grazes the top or bottom of the ramp waveform. Therefore there are three compartments in the phase space, separated by two borderlines, and we have a piecewise-smooth map. It is also easy to appreciate that the map would also be continuous and the derivatives at the two sides are finite (see Figure 3.23).

Likewise it can be shown that the stroboscopic sampling of many other converter topologies and control schemes also yield piecewise-smooth maps. It is therefore

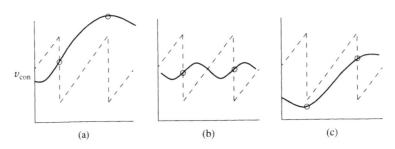

v_{con}

(a)   (b)   (c)

Figure 3.22 There are three possible courses of evolution from the beginning of one ramp cycle to the next. In the stroboscopic map these three possibilities would be marked by three regions in the state space, separated by borderlines representing grazing conditions.

necessary to have a theory of bifurcations in piecewise-smooth maps in order to analyze the bifurcation phenomena in a large class of power electronic circuits.

Mathematically, such systems can be described as follows. Consider a general two-dimensional piecewise-smooth map $g(\hat{x}, \hat{y}; \rho)$ which depends on a single parameter ρ (Figure 3.24). We represent the state variables with a hat because we reserve the unhatted variables for another representation which we'll come to soon. Let Γ_ρ, given by $h(\hat{x}, \hat{y}, \rho) = 0$ denote a smooth curve that divides the phase plane into two regions R_A and R_B. The map is given by

$$g(\hat{x}, \hat{y}; \rho) = \begin{cases} g_1(\hat{x}, \hat{y}; \rho) \text{ for } \hat{x}, \hat{y} \text{ in } R_A, \\ g_2(\hat{x}, \hat{y}; \rho) \text{ for } \hat{x}, \hat{y} \text{ in } R_B \end{cases} \qquad (3.13)$$

It is assumed that the functions g_1 and g_2 are both continuous and have continuous derivatives. The map g is continuous but its derivative is discontinuous at the line Γ_ρ, called the *border*. It is further assumed that the one-sided partial derivatives at the border are finite. The question is, What kinds of bifurcations can such a system exhibit as the parameter ρ is varied?

If a bifurcation occurs when the fixed point of the map is in one of the smooth regions R_A or R_B, it is one of the generic types, namely, period doubling, saddle-node or Neimark bifurcation (Figure 3.25). Such bifurcations have been presented in the last section. But if a fixed point collides with the borderline, there is a discontinuous jump in the eigenvalues of the Jacobian matrix. In such a case an eigenvalue may not *cross* the unit circle in a smooth way, but rather *jumps* over it as a parameter is continuously varied. Such a discontinuous change in the eigenvalues (or characteristic multipliers) of

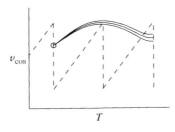

v_{con}

T

Figure 3.23 Small perturbations at the two sides of the grazing condition cause only small change in state variables at the next observation instant. This implies that the map is continuous at the border. Moreover, the partial derivatives at the two sides of the border are finite.

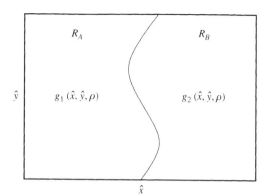

Figure 3.24 Schematic representation of a two-dimensional piecewise-smooth map.

Figure 3.25 In smooth maps the classification of bifurcations depends on where an eigenvalue crosses the unit circle. In this illustration, as a parameter changes, the eigenvalues of a fixed point move in the complex plane in the direction shown. (a) A period doubling bifurcation: eigenvalue crosses the unit circle on the negative real line. (b) A saddle-node or fold bifurcation: an eigenvalue touches the unit circle on the positive real line. (c) A Neimark bifurcation: a complex conjugate pair of eigenvalues cross the unit circle. In piecewise-smooth maps the eigenvalues can discontinuously jump across the unit circle.

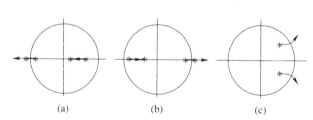

a fixed point has been observed in the boost converter [4] and the buck converter [5]. One cannot classify the bifurcations arising from such border collisions as those occurring for smooth systems where the eigenvalues cross the unit circle smoothly. This phenomenon has been called *border collision bifurcation* [6,7,8].

3.3.1 The Normal Form

Since the local structure of border collision bifurcations depends only on the local properties of the map in the neighborhood of the border, we can study the border collision bifurcations with the help of *normal forms*—the piecewise-affine approximations of g in the neighborhood of the border.

The easiest way of understanding this is in the context of the one-dimensional map as shown in Figure 3.26. If we are interested in studying the bifurcations that may occur when a fixed point moves across the break-point (the one-dimensional equivalent of the border) in this map, it will suffice to use the simpler piecewise-linear form

$$G_1(x; \mu) = \begin{cases} a\,x + \mu & \text{for } x \leq 0, \\ b\,x + \mu & \text{for } x > 0 \end{cases} \tag{3.14}$$

where the border has been moved to $x = 0$ with a simple coordinate transformation, and a and b are the slopes of the tangents at the two sides of the border.

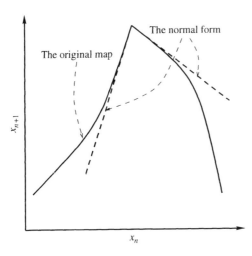

Figure 3.26 A one-dimensional piecewise-smooth map and its normal form.

In the case of the more general two-dimensional piecewise-smooth maps we use the same method. However, the derivation of the normal form is much more involved. Here we'll present the normal form without the derivation which is available in [6,9,10].

Through a series of coordinate transformations, the normal form of the two-dimensional piecewise-smooth map in the neighborhood of the border is obtained as:

$$G_2(x, y; \mu) = \begin{cases} \begin{pmatrix} \tau_L & 1 \\ -\delta_L & 0 \end{pmatrix} \begin{pmatrix} x \\ y \end{pmatrix} + \mu \begin{pmatrix} 1 \\ 0 \end{pmatrix}, & \text{for } x \leq 0, \\[2ex] \begin{pmatrix} \tau_R & 1 \\ -\delta_R & 0 \end{pmatrix} \begin{pmatrix} x \\ y \end{pmatrix} + \mu \begin{pmatrix} 1 \\ 0 \end{pmatrix}, & \text{for } x > 0 \end{cases} \tag{3.15}$$

The structure of the state space as represented by the normal form is shown in Figure 3.27. In this representation, τ_L and δ_L are simply the trace and the determinant of the Jacobian matrix of the fixed point P_ρ on the R_A side of the border Γ. Let P_ρ denote a fixed point of the original map $g(\hat{x}, \hat{y}; \rho)$ defined on $\rho_0 - \epsilon < \rho < \rho_0 + \epsilon$ for some small $\epsilon > 0$; that is, P_ρ depends continuously on ρ. Assume that P_ρ is in region R_A

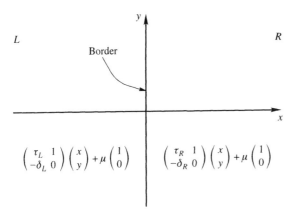

Figure 3.27 The structure of the state space in the two-dimensional normal form.

when $\rho < \rho_0$ and in region R_B when $\rho > \rho_0$, and that P_ρ is on Γ when $\rho = \rho_0$. For $\rho < \rho_0$, the eigenvalues of the Jacobian matrix of the fixed point P_ρ are denoted as λ_1 and λ_2. Since the trace and the determinant of the Jacobian are invariant under the transformation of coordinates, we can obtain the values of τ_L and δ_L as

$$\tau_L = \lim_{\rho \to \rho_0^-} (\lambda_1 + \lambda_2)$$

$$\delta_L = \lim_{\rho \to \rho_0^-} (\lambda_1 \lambda_2) \tag{3.16}$$

The values of τ_R and δ_R can be calculated in a similar way for $\rho > \rho_0$ (see illustration in Figure 3.28). This property is very important in numerical computations. For a border-crossing periodic orbit with higher period, we examine the pth (if the period is p) iterate of the map. The matrices in (3.15) then correspond to the pth iterate rather than the first iterate of the map.

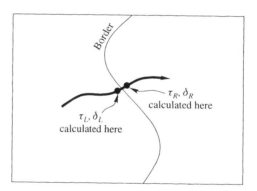

Figure 3.28 Computation of the eigenvalues before and after border collision gives τ_L, δ_L, τ_R, and δ_R.

3.3.2 Bifurcations in the One-Dimensional Normal Form

Various combinations of the values of a and b exhibit different kinds of bifurcation behaviors as μ is varied. In Figure 3.29, we break up the a-b parameter space into regions with the same qualitative bifurcation phenomena. If (a, b) is inside a region, then the original piecewise-smooth map and G_1 will have the same types of bifurcations. If it is on a boundary, then higher-order terms need to be studied to determine the bifurcations of the original map. Note that the map G_1 is invariant under the transformation: $x \to -x$, $\mu \to -\mu$, $a \rightleftharpoons b$. Thus any bifurcation that occurs as μ increases through zero also occurs as μ decreases through zero if we interchange the values of a and b. Hence the line $a = b$ in Figure 3.29 is a line of symmetry and it suffices to consider only the region $a \geq b$ in the following discussions. Essentially the same types of bifurcations take place in $a < b$ regions. Here we enumerate only the results which can easily be checked using cobweb diagrams.

The fixed points at the two sides of the border (if they exist) are given by

$$x_L^* = \frac{\mu}{1-a} \quad \text{and} \quad x_R^* = \frac{\mu}{1-b}, \quad \text{where } x_L^* < 0 < x_R^* \tag{3.17}$$

There are two cases in which no bifurcation occurs at $\mu = 0$:

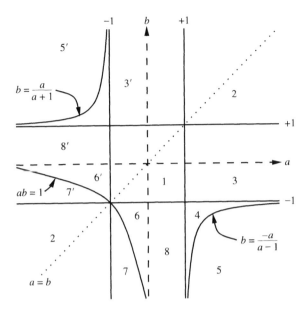

Figure 3.29 The partitioning of the parameter space into regions with the same qualitative bifurcation phenomena. Numbering of the cases is as discussed in the text. (1) Period-1 to period-1. (2) No attractor to no attractor. (3) No fixed point to period-1. (4) No fixed point to chaotic attractor. (5) No fixed point to unstable chaotic orbit (no attractor). (6) Period-1 to period-2. (7) Period-1 to no attractor. (8) Period-1 to periodic or chaotic attractor. The regions shown in primed numbers have the same bifurcation behavior as the unprimed ones when μ is varied in the opposite direction.

Case 1: If $-1 < b \leq a < 1$, then there exists one stable fixed point for $\mu < 0$ and one stable fixed point for $\mu > 0$.

Case 2: If (a) $1 < b \leq a$ or (b) $b \leq a < -1$, then there is one unstable fixed point for both positive and negative values of μ. No attractors exist.

For (a), trajectories to the left of the unstable fixed point go monotonically to $-\infty$ and those to the right of the unstable fixed point go monotonically to $+\infty$. For (b), trajectories oscillate between L and R and its magnitude approaches ∞.

Border Collision Pair Bifurcation

If $b < 1 < a$, then, for $\mu < 0$, we have $G_1(x; \mu) < x$ for all x, and all orbits approach $x = -\infty$. Thus there is no fixed point for $\mu < 0$. For $\mu > 0$, there are two fixed points, one on each side of $x = 0$. Since the two fixed points are created at the border, we call such bifurcation a *border collision pair bifurcation*.

Note that the border collision pair bifurcation is analogous to a tangent (or saddle-node) bifurcation occurring in 1-D smooth maps. The difference is that, for smooth maps, one of the two fixed points created at a tangent bifurcation is stable; for the normal form G_1, one fixed point may be stable or both may be unstable.

Depending on the values of a and b, there can be three possible outcomes of a border collision pair bifurcation:

Case 3: If $-1 < b < 1 < a$, there is a bifurcation from no attractor to a fixed point attractor at $\mu = 0$.

Case 4: If $a > 1$ and $-\frac{a}{a-1} < b < -1$, then there is a bifurcation from no attractor to a chaotic attractor at $\mu = 0$. The basin of this chaotic attractor is the interval $[x_L^*, (x_L^* - \mu)/b]$.

Case 5: If $a > 1$ and $b < -\frac{a}{a-1}$, then there is a bifurcation from no fixed point to an unstable chaotic orbit. There is no attractor for $\mu > 0$.

Border-Crossing Bifurcations

If $a > -1$ and $b < -1$, the fixed point crosses the border as μ varies through zero. We call this a *border-crossing bifurcation*.

Case 6: If $b < -1 < a < 0$ and $ab < 1$, as μ is varied through 0, there is a bifurcation from a period-1 attractor to a period-2 attractor. Note that the shape of the bifurcation diagram for a standard period-doubling bifurcation is different from that of a border collision period-doubling bifurcation of a piecewise-smooth map. In the former case, the period-2 points diverge perpendicularly from the μ axis near the critical parameter value; in the latter case, they may diverge at an angle that is less than 90^0 from the μ axis (see Figure 3.30).

Case 7: If $b < -1 < a < 0$ and $ab > 1$, then there is a period-1 attractor for $\mu < 0$ and no attractor for $\mu > 0$.

Case 8: If $0 < a < 1$ and $b < -1$, then a fixed point attractor can bifurcate into a periodic attractor or a chaotic attractor as μ is varied from less than 0 to greater than 0.

This parameter region has been extensively studied by Takens [11], Nusse and Yorke [12] and Maistrenko et al. [13]. They have shown that, for $\mu > 0$, all trajectories are bounded, and various periodic attractors as well as chaotic attractors can exist. Figure 3.31 gives a schematic diagram of the complicated phenomenology occurring in this regime. Note that for $\mu > 0$ there can be a period-adding cascade, with chaotic windows sandwiched between periodic windows, as the magnitudes of a or b (or both) are increased.

It may be noted that for $\mu > 0$, the behavior is chaotic for the whole region of Case 4 and a significant portion of Case 8. A chaotic attractor is said to be *robust* [14] if, for its parameter values, there exists a neighborhood in the parameter space with no periodic attractor and the chaotic attractor is unique in that neighborhood. The chaotic attractor of the one-dimensional normal form is robust in the contiguous region of the parameter space where no periodic windows exist.

The results outlined above give a complete description of the bifurcations of the normal form G_1 as μ is varied. They also describe the local bifurcations of the original piecewise-smooth map in the neighborhood of the border collisions.

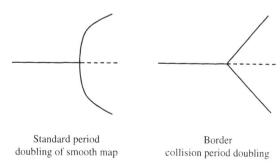

Figure 3.30 Representative bifurcation diagrams for a standard period-doubling bifurcation of a smooth map and that of a border collision period-doubling bifurcation of a piecewise-smooth map. The solid lines indicate attracting orbits while the dashed lines indicate repellers.

Standard period doubling of smooth map

Border collision period doubling

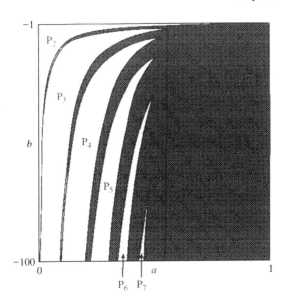

Figure 3.31 Schematic drawing of the parameter region $0 < a < 1$, $b < -1$ (Case 8) showing the type of attractor for $\mu > 0$. The white regions correspond to period-n attractors and shaded regions have chaotic attractors.

3.3.3 Bifurcations in the Two-Dimensional Normal Form

Let us first look at some properties of the normal form (3.15) which will be made use of in the subsequent analysis.

The fixed points of the system at both sides of the boundary are given by

$$L^* = \left(\frac{\mu}{1 - \tau_L + \delta_L}, \frac{-\delta_L \mu}{1 - \tau_L + \delta_L} \right)$$

$$R^* = \left(\frac{\mu}{1 - \tau_R + \delta_R}, \frac{-\delta_R \mu}{1 - \tau_R + \delta_R} \right)$$

and the stability of each of them is determined by the eigenvalues $\lambda_{1,2} = \frac{1}{2}\left(\tau \pm \sqrt{\tau^2 - 4\delta} \right)$. Here τ and δ refer both to τ_L and τ_R, and to δ_L and δ_R respectively. If the eigenvalues are real, the slopes of the corresponding eigenvectors are given by $-(\delta/\lambda_1)$ and $-(\delta/\lambda_2)$, respectively. Since in engineering we encounter only dissipative systems (systems with resistance or friction), we assume $|\delta_L| < 1$ and $|\delta_R| < 1$. For a positive determinant, there can be four basic types of fixed points.

1. When $\delta > \tau^2/4$, both eigenvalues of the Jacobian are complex, indicating that the fixed point is spirally attracting. If $\tau > 0$, it is a clockwise spiral, and if $\tau < 0$ the spiraling motion is counterclockwise.
2. When $\delta < \tau^2/4$, both eigenvalues are real. If $2\sqrt{\delta} < \tau < (1 + \delta)$ then the eigenvalues are positive (and less than 1 in modulus). Hence the fixed point is a regular attractor. If $-2\sqrt{\delta} > \tau > -(1 + \delta)$ then the eigenvalues are negative and it is a flip attractor.
3. If $\tau > (1 + \delta)$, $0 < \lambda_2 < 1$ and $\lambda_1 > 1$. The fixed point is a regular saddle.
4. If $\tau < -(1 + \delta)$, then $-1 < \lambda_1 < 0$ and $\lambda_2 < -1$. The fixed point is a flip saddle.

If the determinant is negative, the eigenvalues are always real and spiraling orbits cannot exist. Moreover, the fixed point cannot be regular attractor or regular saddle. Thus there can be only two types of fixed points:

1. For $-(1+\delta) < \tau < (1+\delta)$, one eigenvalue is positive and the other negative, and both are less than 1 in modulus—which means that the fixed point is a flip attractor.

2. For $\tau > (1+\delta)$, $\lambda_1 > 1$ and $-1 < \lambda_2 < 0$, i.e., the fixed point is a flip saddle. If $\tau < -(1+\delta)$, then $\lambda_2 < -1$ and $0 < \lambda_1 < 1$. The fixed point is again a flip saddle.

When referring to sides L and R, these quantities have the appropriate subscripts (i.e., $\lambda_{1L}, \lambda_{2L}$ are the eigenvalues in side L and $\lambda_{1R}, \lambda_{2R}$ are the eigenvalues in side R). As a fixed point collides with the border, its character can change from any one of the above types to any other. This provides a way of classifying border collision bifurcations.

It may be noted that in some portions of the parameter space there may be no fixed point in half of the phase space. For example, the location of L^* may turn out to be in side R. In such cases, the dynamics in L is determined by the character of the *virtual* fixed point. If the eigenvalues are real, invariant manifolds of these virtual fixed points still exist and play an important role in deciding the system dynamics.

A special feature of the normal form (3.15) is that the unstable manifolds fold at every intersection with the x-axis, and the image of every fold point is a fold point. The stable manifolds fold at every intersection with the y-axis and the pre-image of every fold point is a fold point. The argument is as follows. Forward iterate of points on the unstable manifold remains on the same manifold. In the normal form, points on the y-axis map to points on the x-axis. As an unstable manifold crosses the y-axis, one linear map changes to another linear map. Therefore the slope of the unstable manifold in the two sides of the x-axis cannot be the same unless the parameters of the normal form in the two sides of the border are the same (implying a smooth map). In the case of the stable manifold, the same argument applies for the inverse map.

Since the system is linear in each side, period-2 (or higher period) fixed points cannot exist with all points in L or all points in R. However, in some parts of the parameter space, a period-2 fixed point may exist with one point in L and one point in R. The eigenvalues of the second iterate are

$$\frac{1}{2}\left(\tau_L\tau_R - \delta_R - \delta_L \pm \sqrt{\tau_L^2\tau_R^2 - 2\tau_L\tau_R\delta_R - 2\tau_L\delta_L\tau_R + \delta_R^2 - 2\delta_R\delta_L + \delta_L^2}\right)$$

From this, the condition of stability of the period-2 orbit is obtained as follows:

1. If the eigenvalues are real,

$$1 - \tau_L\tau_R + \delta_L + \delta_R + \delta_L\delta_R > 0 \quad \text{for } \lambda_1 < +1 \qquad (3.18)$$

$$1 + \tau_L\tau_R - \delta_L - \delta_R + \delta_L\delta_R > 0 \quad \text{for } \lambda_2 > -1 \qquad (3.19)$$

2. If the eigenvalues are complex, i.e., if $(\tau_L\tau_R - \delta_L - \delta_R)^2 < 4\delta_L\delta_R$, then the condition for stability is $\delta_L\delta_R < 1$, which is always satisfied in the parameter space under consideration. Therefore for the normal form G_2, an eigenvalue of the period-2 fixed point can go out of the unit circle only on the real line.

The conditions of stability of fixed points of higher periodicities can be calculated in a similar manner. However, the expressions become unwieldy and are excluded in the present analysis.

3.3.4 Classification of Border Collision Bifurcations

Various combinations of the values of τ_L, τ_R, δ_L, and δ_R exhibit different kinds of bifurcation behaviors as μ is varied through zero. To present a complete picture, we break up the four-dimensional parameter space into regions with the same qualitative bifurcation phenomena.

As in case of the one-dimensional map, if a certain kind of bifurcation occurs when μ is increased through zero, the same kind of bifurcation would also occur when μ is decreased through zero if the parameters in L and R are interchanged. Therefore, there exists a symmetry in the parameter space and it suffices to describe the bifurcations in half the parameter space. In the following discussion we will present the various types of border collision bifurcations that may occur in the 2-D normal form. Since the reasoning involves mathematical logic which cannot be presented in full details in the present book, we refer the reader to [10,15,16] for more comprehensive presentations.

Border Collision Pair Bifurcation

$$\text{If} \quad \tau_L > (1 + \delta_L) \quad \text{and} \quad \tau_R < (1 + \delta_R) \qquad (3.20)$$

then there is no fixed point for $\mu < 0$ and there are two fixed points, one each in L and R, for $\mu > 0$. The two fixed points are born on the border at $\mu = 0$. This is therefore a *border collision pair* bifurcation. An analogous situation occurs if $\tau_L < (1 + \delta_L)$ and $\tau_R > (1 + \delta_R)$ as μ is reduced through zero. Due to the symmetry of the two cases, we consider only the parameter region (3.20).

The parameter region (3.20) can be subdivided into a few compartments with different bifurcation behaviors.

$$\text{If} \quad (1 + \delta_R) > \tau_R > -(1 + \delta_R) \qquad (3.21)$$

then R^* is an attracting fixed point while L^* is a saddle. Therefore it is like a saddle-node bifurcation, where a periodic attractor appears at $\mu = 0$. The distinctive feature is that the fixed points are born on the border and move away from it as μ is increased.

$$\text{If} \quad \tau_L > (1 + \delta_L) \quad \text{and} \quad \tau_R < -(1 + \delta_R) \qquad (3.22)$$

then there are three possibilities depending on the position in the parameter space:

1. No fixed point to period-2
2. No fixed point to chaotic attractor
3. No fixed point to unstable chaotic orbit

When both the determinants are positive, there is a robust chaotic attractor for $\mu > 0$ if

$$\delta_L \tau_R \lambda_{1L} - \delta_R \lambda_{1L} \lambda_{2L} + \delta_R \lambda_{2L} - \delta_L \tau_R + \tau_L \delta_L - \delta_L^2 - \lambda_{2L} \delta_L > 0 \qquad (3.23)$$

Otherwise, there is an unstable chaotic orbit for $\mu > 0$.

For $-1 < \delta_R < 0$, if the condition (3.19) is satisfied, the attractor for $\mu > 0$ is period-2. If

$$\frac{\lambda_{1L} - 1}{\tau_L - 1 - \delta_L} > \frac{\lambda_{2R} - 1}{\tau_R - 1 - \delta_R} \tag{3.24}$$

then (3.23) still gives the parameter range for existence of a robust chaotic attractor. But if (3.24) is not satisfied, the condition of existence of the chaotic attractor changes to

$$\frac{\lambda_{2R} - 1}{\tau_R - 1 - \delta_R} < \frac{\delta_L(\tau_L - \delta_L - \lambda_{2L})}{(\tau_L - 1 - \delta_L)(\delta_R\lambda_{2L} - \delta_L\tau_R)} \tag{3.25}$$

For $\delta_L < 0$ and $\delta_R < 0$, the same partitioning as above applies. But if $\delta_L < 0$ and $\delta_R > 0$, there is no analytic relationship defining the boundary between the conditions of existence of the chaotic attractor and unstable chaotic orbit and it has to be determined numerically.

Note that these bifurcations may also occur in the higher iterates of a piecewise-smooth map. If a chaotic attractor is born through a border collision pair bifurcation while a periodic attractor exists at a lower iterate, the chaotic attractor will not be robust. The condition for robustness stated above only ensures the absence of multiple attractors at iterates higher than that at which the border collision pair bifurcation occurs. The absence of periodic windows in some neighborhood of the parameter space is, however, guaranteed.

All three types of border collision pair bifurcations have been observed in switching circuits, notably the voltage-controlled buck converter.

Border-Crossing Bifurcations

In all regions of the parameter space except (3.20), a fixed point crosses the border as μ is varied through zero. The resulting bifurcations are called *border-crossing* bifurcations. In the following discussions we consider the bifurcations caused by the change of the character of a fixed point as μ varies from a negative value to a positive value.

Regular attractor to flip saddle: There is a bifurcation from a period-1 attractor to a chaotic attractor as μ is increased through zero. This chaotic attractor is robust.

Regular attractor to spiral attractor, regular attractor to regular attractor, flip attractor to flip attractor, regular attractor to flip attractor and spiral attractor to spiral attractor having the same sense of rotation: In all these cases, there is a unique period-1 attractor for both $\mu < 0$ and $\mu > 0$; only the path of the orbit changes at $\mu = 0$.

Spiral attractor to flip attractor and spiral attractor to spiral attractor with opposite sense of rotation: A period-1 attractor exists at both sides of $\mu = 0$. There may also be coexisting attractors in the higher iterates.

Flip attractor to flip saddle: For $\mu < 0$, a unique period-1 attractor exists. For $\mu > 0$, if (3.18) and (3.19) are satisfied, a period-2 orbit exists. There are no coexisting attractors.

This is like a period-doubling bifurcation occurring on the borderline. In contrast with standard period-doubling bifurcation, the distinctive feature of the border collision period doubling is that as μ is varied through zero, the bifurcated orbit does not emerge orthogonally from the orbit before the bifurcation.

If (3.18) is not satisfied, no attractor can exist. If (3.19) is not satisfied, there is a chaotic orbit.

Spiral attractor to flip saddle: For $\mu < 0$, L^* is an attracting fixed point. There may be coexisting attractors in the higher iterates.

For $\mu > 0$, R^* is unstable. If conditions (3.18) and (3.19) are satisfied, then there is a stable period-2 orbit with points in L and R. If (3.18) is not satisfied, the eigenvalue of the twice-iterated map is greater than unity. In that case any initial

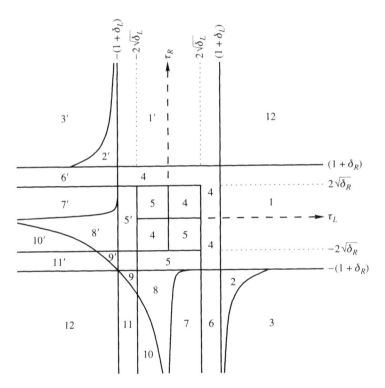

Figure 3.32 Schematic diagram of the parameter space partitioning for $1 > \delta_L > 0$ and $1 > \delta_R > 0$ into regions with the same qualitative bifurcation phenomena. (1) No fixed point to period-1. (2) No fixed point to chaotic attractor. (3) No fixed point to unstable chaotic orbit, no attractor. (4) Period-1 to period-1. (5) Period-1 plus coexisting attractors to period-1 plus different coexisting attractors. (6) Period-1 to chaos. (7) Period-1 plus coexisting attractors to chaos or high periodic orbit plus coexisting attractors. (8) Period-1 plus coexisting attractors to period-2 plus coexisting attractors. (9) Period-1 to period-2. (10) Period-1 plus coexisting attractors to no attractor. (11) Period-1 to no attractor. (12) No attractor to no attractor. The regions shown in primed numbers have the same bifurcation behavior as the unprimed ones when μ is varied in the opposite direction.

condition will diverge along the unstable eigenvector and no attractor can exist. If (3.19) is not satisfied, the period-2 attractor becomes unstable. In that condition, a chaotic attractor or a periodic attractor of higher periodicity can exist. There can also be coexisting attractors.

Flip saddle to flip saddle and regular saddle to regular saddle: There is no attractor for both positive and negative values of μ.

This gives a complete description of the bifurcations that can occur at various regions of the parameter space of the normal form (3.15). The resulting partitioning of the parameter space for the positive determinant case is shown in Figure 3.32 and that for negative determinant case is shown in Figure 3.33.

Most of these bifurcations have been observed in power electronic converters, and the theory presented here helps us to understand these atypical nonlinear phenomena that are very common in such converters.

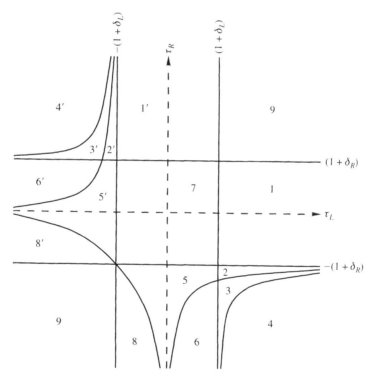

Figure 3.33 Schematic diagram of the parameter space partitioning for $-1 < \delta_L < 0$ and $-1 < \delta_R < 0$ into regions with the same qualitative bifurcation phenomena. (1) No fixed point to period-1. (2) No fixed point to period 2. (3) No fixed point to chaos. (4) No fixed point to unstable chaotic orbit, no attractor. (5) Period-1 to period-2. (6) Period-1 to chaos. (7) Period-1 to period-1. (8) Period-1 to no attractor. (9) No attractor to no attractor. The regions shown in primed numbers have the same bifurcation behavior as the unprimed ones when μ is varied in the opposite direction.

REFERENCES

[1] S. Banerjee and K. Chakrabarty, Nonlinear modeling and bifurcations in the boost converter, *IEEE Trans. on Power Electronics*, vol. 13, no. 2, pp. 252–260, 1998.

[2] S. Banerjee, E. Ott, J. A. Yorke, and G. H. Yuan, Anomalous bifurcations in dc-dc converters: Borderline collisions in piecewise smooth maps, *Power Electronics Specialists' Conference*, pp. 1337–1344, 1997.

[3] G. H. Yuan, S. Banerjee, E. Ott, and J. A. Yorke, Border collision bifurcations in the buck converter, *IEEE Trans. on Circuits and Systems—I*, vol. 45, no. 7, pp. 707–716, 1998.

[4] W. C. Y. Chan and C. K. Tse, Study of bifurcations in current programmed dc/dc boost converters: From quasiperiodicity to period doubling, *IEEE Trans. on Circuits and Systems—I*, vol. 44, no. 12, pp. 1129–1142, 1997.

[5] G. Olivar, Chaos in the buck converter. PhD diss., Technical University of Catalonia, Barcelona, Spain, 1997. Servei de Publicacions de la UPC (3640).

[6] H. E. Nusse and J. A. Yorke, Border-collision bifurcations including "period two to period three" for piecewise smooth maps, *Physica D*, vol. 57, pp. 39–57, 1992.

[7] H. E. Nusse, E. Ott, and J. A. Yorke, Border-collision bifurcations: An explanation for observed bifurcation phenomena, *Physical Review E*, vol. 49, pp. 1073–1076, 1994.

[8] W. Chin, E. Ott, H. E. Nusse, and C. Grebogi, Universal behavior of impact oscillators near grazing incidence, *Physics Letters A*, vol. 201, pp. 197–204, 1995.

[9] G. H. Yuan, Shipboard crane control, simulated data generation and border collision bifurcations. PhD diss., University of Maryland, College Park, USA, 1997.

[10] S. Banerjee and C. Grebogi, Border collision bifurcations in two-dimensional piecewise smooth maps, *Physical Review E*, vol. 59, no. 4, pp. 4052–4061, 1999.

[11] F. Takens, Transition from periodic to strange attractors in constrained equations, in *Dynamical Systems and Bifurcation Theory* (M. Camacho, M. Pacifico, and F. Takens, eds.), vol. 160 of *Pitman Research Notes in Mathematics Series*, pp. 399–421, Longman Scientific and Technical, 1987.

[12] H. E. Nusse and J. A. Yorke, Border-collision bifurcations for piecewise smooth one dimensional maps, *Int. J. Bifurcation and Chaos*, vol. 5, no. 1, pp. 189–207, 1995.

[13] Y. L. Maistrenko, V. L. Maistrenko, and L. O. Chua, Cycles of chaotic intervals in a time-delayed Chua's circuit, *Int. J. Bifurcation and Chaos*, vol. 3, pp. 1573–1579, 1993.

[14] S. Banerjee, J. A. Yorke, and C. Grebogi, Robust chaos, *Physical Review Letters*, vol. 80, pp. 3049–3052, 1998.

[15] S. Banerjee, P. Ranjan, and C. Grebogi, Bifurcations in two-dimensional piecewise smooth maps: Theory and applications in switching circuits. *IEEE Trans. on Circuits and Systems—I*, vol. 47, no. 5, pp. 633–643, 2000.

[16] M. di Bernardo, M. I. Feigin, S. J. Hogan, and M. E. Homer, Local analysis of C-bifurcations in *n*-dimensional piecewise smooth dynamical systems, *Chaos, Solitons & Fractals*, vol. 10, no. 11, 1999.

3.4 NONSTANDARD BIFURCATIONS IN DISCONTINUOUS MAPS

Ian Dobson
Soumitro Banerjee

Not all power electronic circuits yield continuous, smooth, or piecewise-smooth Poincaré maps and so the bifurcation theory developed in Sections 3.3 and 3.4 is not universally applicable. Indeed, there is an important class of power electronic systems that yield *discontinuous maps* under discrete modeling, and nonstandard bifurcation phenomena can occur in such systems.

In this section, we will illustrate some bifurcation phenomena peculiar to discontinuous maps and then illustrate switching time bifurcations. To facilitate simple graphical illustration, we will base our discussion on one-dimensional maps. More complicated phenomena may occur in higher dimensions.

The first point to note is that a discontinuity can give rise to multiple fixed points of different character. Multiple fixed points can occur in continuous systems also, but there are some important differences between the two classes.

The case where two *stable* fixed points exist at the two sides of the discontinuity is illustrated in Figure 3.34(a). Figure 3.34(b) shows the corresponding situation in a continuous map. Notice that for the continuous map there must be an unstable fixed point C between the stable fixed points while for the discontinuous map there is no such requirement. For the continuous map, C acts as the boundary of the two basins of attraction, while in the discontinuous case the point of discontinuity separates the two basins.

Now suppose the graph of the map changes with the variation of a parameter. Figure 3.35 illustrates such a situation in a discontinuous map and a continuous map.

In Figure 3.35(a), for some parameter value there are two stable fixed points in the two disconnected segments of the map. As the parameter is varied, one of the fixed points approaches the point of discontinuity. Beyond a critical parameter value, one of the stable fixed points *disappears* as now only one map segment intersects the 45° line. If the system state was at the stable fixed point B for the parameter below the critical value, the state undergoes a transient and then converges to the fixed point A as the parameter is varied through the critical value.

The corresponding situation for continuous maps is illustrated in Figure 3.35(b). In this case, as the parameter is varied, a stable fixed point and an unstable fixed point

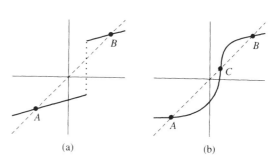

Figure 3.34 Illustration of maps with two stable fixed points: (a) discontinuous map and (b) continuous map.

(a)   (b)

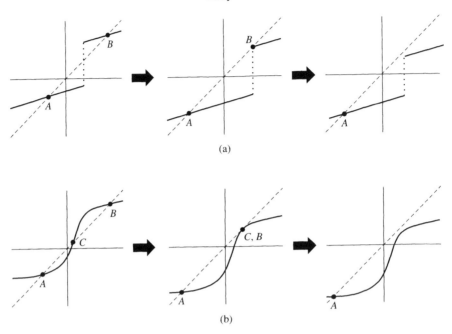

Figure 3.35 Discontinuous bifurcation due to the change of a parameter: (a) discontinuous map and (b) continuous map.

approach each other, and at a critical parameter value they coalesce and subsequently disappear. This is the standard saddle-node bifurcation.

The distinctive feature of the discontinuous map is that the disappearance of a fixed point does not follow the merging of a stable and an unstable fixed point. In continuous maps fixed points can appear or disappear only in pairs, while in discontinuous maps single fixed points may appear or disappear. Moreover, unlike the continuous and smooth maps, the eigenvalues (in case of 1-D maps, the slope) of a fixed point do not signal the onset of a bifurcation.

Where there is a fixed point on each side of a discontinuity, all stability combinations of the fixed points are possible. Figure 3.36 illustrates the case where one fixed point is stable and the other unstable, and the case where both the fixed points are unstable.

It may be noted that a map like Figure 3.36(a) will exhibit stable periodic behavior with basin of attraction $(-\infty, B]$. A map of type Figure 3.36(b) may exhibit chaotic behavior (since neither fixed point is stable, but any initial condition in $[A, B]$ will remain bounded in that region). In the latter case, if a fixed point disappears with the variation of a parameter, the chaotic orbit can no longer exist and all initial conditions go to infinity.

We thus find that systems represented by discontinuous maps may exhibit sudden changes in the system behavior as a parameter is changed. These *jump* phenomena are not due to crises (as explained in Section 3.6), but are caused by discontinuities in the discrete models.

In two-dimensional discontinuous Poincaré maps, there can also be two fixed points at the two sides of the line of discontinuity. In two-dimensional smooth maps,

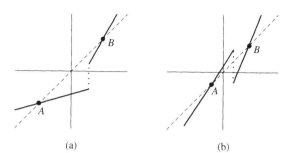

Figure 3.36 Other possible stability combinations in one-dimensional discontinuous maps: (a) one fixed point stable and the other unstable, (b) both unstable.

(a)

(b)

two attracting fixed points are separated by a basin boundary which can be the stable manifold of a saddle-type fixed point or an unstable periodic orbit. In contrast, for discontinuous maps the basin boundary can be formed from points of discontinuity. We'll see a practical example of this phenomenon in Chapter 6.

After a brief discussion of nonlinearity in diode and thyristor circuits we illustrate switching time bifurcations in a simple ac thyristor circuit. Switching time bifurcations in thyristor circuits can cause Poincaré maps to be discontinuous.

Diode switchings are uncontrolled: an ideal diode switches on when its voltage starts to become positive and switches off when its current starts to become negative. Thus the times of the diode switchings are determined by the circuit currents or voltages. The dependence of the diode switching times on the circuit currents or voltages is a cause of circuit nonlinearity, even if the individual circuits obtained with the diode either on or off are linear. (In contrast, a succession of linear circuits switching at fixed, predetermined times is linear.)

Thyristor switchings are uncontrolled at turn-off in the same way as diodes: an ideal thyristor switches off when its current starts to become negative. This uncontrolled switch-off is similarly a cause of circuit nonlinearity. Thyristor switch-on requires both positive voltage and the presence of a firing pulse. In particular, thyristor switch-on is inhibited when the firing pulse is absent. This detail of the thyristor switch-on can have particularly strong and nonstandard effects on the dynamics, and the main phenomenon underlying these effects is called a *switching time bifurcation*.

To illustrate switching time bifurcations in a simple context, consider the circuit shown in Figure 3.37. The source voltage has time dependence,

$$u(t) = \cos 2\pi t + p \cos 6\pi t$$

The parameter p controls the amount of third harmonic distortion in the voltage. This example considers very large amounts of third harmonic distortion. The thyristor is

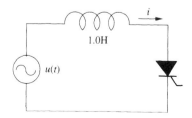

Figure 3.37 Simple thyristor circuit which shows switching time bifurcations.

fired at times 0, 1, 2, ... The thyristor firing pulse is assumed to be very short so that the thyristor can turn on only at the firing times.

First assume that the circuit is in steady state with regular thyristor switchings and that $0 \leq p < 3$. Then the current $i(t)$ is given by integrating the source voltage in the first half cycle, and is zero in the second half cycle:

$$i(t) = \begin{cases} \frac{1}{2\pi}(\sin 2\pi t + \frac{p}{3}\sin 6\pi t) & ; \ 0 \leq t \leq 0.5 \\ 0 & ; \ 0.5 < t \leq 1 \end{cases}$$

Observe that the steady state is stable for $0 \leq p < 3$; any small perturbation in the current in the first half cycle is damped to zero after the (perturbed) thyristor switch-off; subsequent cycles are exactly at the steady state. Indeed this is a simple example of the thyristor switch-off damping the perturbation (see Section 6.6). There is no resistive damping in the circuit.

For steady-state operation with $0 \leq p < 3$, the thyristor switches off when the current reaches zero at time 0.5 as shown in Figure 3.38(a). However, as the parameter p is slowly increased through 3, the third harmonic distortion becomes so severe that the current dip near time 0.25 actually hits zero, and there is a sudden jump or bifurcation in the switching time: for p exceeding 3 by a small amount, the thyristor turns off near time 0.25 as shown in Figure 3.38(b). As p increases through 3, the stable steady state suddenly disappears and another, new stable steady state applies. In this simple example, there is no transient before the new steady state is achieved. This disappearance of the earlier stable steady state in the switching time bifurcation is not a conventional loss of stability, and it cannot be analyzed using conventional stability measures (the Poincaré map eigenvalue is always zero in this case).

Let's now consider the effect of an initial current $i(0)$ on the transient behavior of the circuit of Figure 3.37. It is convenient to assume $p = 0$ so that there is no third harmonic distortion. If $i(0)$ is less than $\frac{1}{2\pi}$, then the thyristor switches off before time 1 and then turns on at time 1. The current following time 1 is then the steady-state behavior. On the other hand, if the current $i(0)$ is larger than $\frac{1}{2\pi}$, then the thyristor never switches off and there is a different steady-state periodic orbit. These two behaviors are shown in Figure 3.39 and are separated by a switching time bifurcation in which the lower fold of current in Figure 3.39(a) hits zero.

We can now examine the Poincaré map. The stroboscopic Poincaré map P advances the current i by one unit of time and we choose to examine the currents at

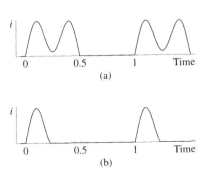

Figure 3.38 Thyristor current near switching time bifurcation: (a) $p < 3$, (b) $p > 3$.

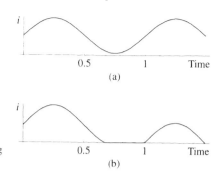

Figure 3.39 Thyristor current near switching time bifurcation: (a) $i(0) > \frac{1}{2\pi}$, (b) $i(0) < \frac{1}{2\pi}$.

times 0, 1, 2, ... For example, $P(i(0)) = i(1)$. If the current $i(0)$ is less than $\frac{1}{2\pi}$, then $P(i(0)) = 0$. On the other hand, if the current $i(0)$ is larger than $\frac{1}{2\pi}$, then the thyristor never switches off and $P(i(0)) = i(0)$. Thus the Poincaré map is as shown in Figure 3.40.

The switching time bifurcation of Figure 3.39 causes the Poincaré map to be discontinuous at $i(0) = \frac{1}{2\pi}$. (Note that the switching time bifurcation of Figure 3.39 is not detectable in the Poincaré map; this appears to be an exceptional case because the circuit example is one-dimensional.) As explained above and in contrast with the case of smooth or piecewise-smooth maps, there is little mathematical structure which constrains what generically happens in discontinuous maps. However, the underlying switching time bifurcations can be understood by examining fold structures in the thyristor current waveform, as explained in more detail in Chapter 6. Thus bifurcation of thyristor switching times has a profound effect on the Poincaré map. It is also apparent that the Poincaré map in Figure 3.40 has zero gradient and is noninvertible between zero and $\frac{1}{2\pi}$.

Another consideration in diode and thyristor circuits concerns the changes in the dynamical system equations when diodes or thyristors switch off or on. The state space dimension (the number of states or *order*) of the dynamical system changes. In particular, when a diode is off, its current is constrained to zero. If an alternative current path to a series inductor is available during the diode switch-off period (as in continuous-current dc/dc converters), the state space dimension remains unaltered. However, in many circuits this zero-current constraint reduces the dimension of the state space by one while the diode is off. The state space dimension increases by one when the diode turns on and the constraint no longer applies.

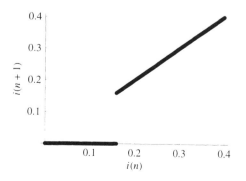

Figure 3.40 Poincaré map.

In general, the changing constraints when diodes or thyristors switch on or off can be represented by corresponding changes in the state space dimension. Stability is studied by examining how perturbations to a steady state grow or decay. It is important to note that not only the diode current but also the *perturbation* of a diode current is constrained to zero when the diode is off. These modeling issues are further explained in Chapter 6 using examples of a static VAR control circuit and a dc/dc converter in discontinuous conduction mode.

3.5 THE METHOD OF SCHWARZIAN DERIVATIVES

Chi K. Tse

3.5.1 Background

The essential operation that characterizes most power electronics circuits is the cyclic switching of the circuit configuration from one linear system to another. Such periodic switching operation naturally permits discrete-time modeling of power electronics systems in the form of iterative functions. In this section we review iterative functions of the first order and in particular discuss a technique for deciding if a given iterative function can period-double to chaos. The main tool that we use for this study is the *Schwarzian derivative,* which was defined by the German mathematician Hermann Schwarz in 1869 for studying complex value functions. Moreover, our purpose here is to try to make use of certain important properties of the Schwarzian derivative for constructing a necessary condition for the occurrence of period-doubling cascades in first-order iterative functions.

As we will see, this technique finds application in establishing the possibility of a route to chaos via period doubling in dc/dc converters that operate in discontinuous mode.

3.5.2 Problem Description

Let x be a variable which generates itself through an iterative function f. Also, for consistency with most power electronics system descriptions, we introduce a control parameter μ which affects f and hence the outcome of the iteration. The general form of first-order iterative function under study is

$$x_{n+1} = f(x_n, \mu) \qquad (3.26)$$

The above system is said to be *open loop* if μ is independent of x. Moreover, we may construct a *closed-loop* system by defining μ as a function of x. This defining function is called a *feedback control function* $g(.)$:

$$\mu_n = g(x_n) \qquad (3.27)$$

i.e., the value of the control parameter at the nth iterate is $g(x_n)$.

Since our study is motivated by phenomena observed in power electronics systems which invariably contain a feedback loop, we further focus our attention on the closed-loop system where μ is controlled via (3.27). In particular, it is of interest to know the condition on the form of $g(.)$ such that the closed-loop system can be driven to chaos via a period-doubling cascade. In the following we discuss an analytical approach to derive the required condition, based on the *Schwarzian derivative*.

3.5.3 Mechanism of Period Doubling

Before embarking on the formal use of the Schwarzian derivative for studying period-doubling bifurcations, it is helpful to review the basic process through which a period doubling emerges.

Consider the iterative function $x_{n+1} = f(x_n, \mu)$, where μ denotes a general parameter. In power electronics context, μ can be a control parameter as in (3.26). For the purpose of illustrating the essential process involved in a period doubling, we keep μ as a general bifurcation parameter, changing which may alter the qualitative behavior of the iterative function. Now suppose f has a fixed point at $x = x_s$, and $f'(x_s)$ is greater than, equal to, and less than -1, respectively, for μ slightly less than, equal to, and larger than μ_c. This is a familiar scenario in which the fixed point x_s loses stability at $\mu = \mu_c$. Moreover, if we wish to know whether a period doubling would occur, we need to examine the second iterate f^2 and the formation of new fixed points whose stability implies that a period doubling has occurred.

This simple view of the typical period-doubling process is best exemplified by the logistic map shown in Figure 3.41. To see the emergence of two period-2 fixed points, we consider the second iterate, $f^2(x)$, which is shown in Figure 3.42. Note that the fixed point of f (e.g., $x = 0.6667$ for $\mu = \mu_c$) is also a fixed point of f^2, as shown in the figure.

Taking a closer look at the vicinity of the fixed point, we observe that the emergence of period doubling relies on a special geometrical property of the f^2 map in the proximity of the bifurcation point, which corresponds to the characteristic behavior shown in Figure 3.43. This special geometrical property is guaranteed if the following is satisfied:

$$\frac{f'''(x)}{f'(x)} - \frac{3}{2}\left[\frac{f''(x)}{f'(x)}\right]^2 < 0 \tag{3.28}$$

Furthermore, in order for f^2 (and in general f^n) to possess the same property, $Sf^n < 0$ is a necessary condition. This was first recognized by the American mathematician David Singer back in 1978 [1].

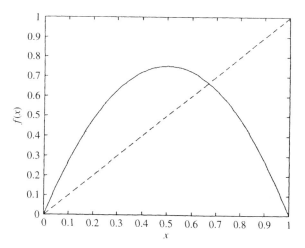

Figure 3.41 Logistic map $f(x) = \mu x(1 - x)$, where $\mu = 3$.

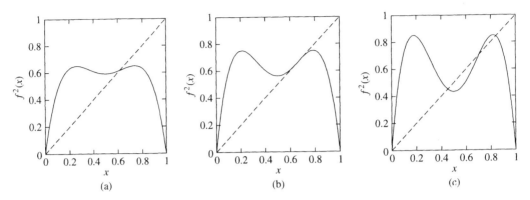

Figure 3.42 Illustration of period doubling in the logistic map $x_{n+1} = \mu x_n(1 - x_n)$: (a) at $\mu = 2.4$, stable period-1 orbit exists; (b) at $\mu = \mu_c = 3$, a period doubling occurs; (c) at $\mu = 3.4$, new stable period-2 orbit exists.

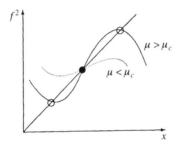

Figure 3.43 Geometrical manifestation of $Sf < 0$.

3.5.4 Schwarzian Derivative and Period Doublings
ad infinitum

We now formally define the Schwarzian derivative Sf of an iterative map f as follows.

Definition 3.1 Let f be defined on the interval I, and assume that the third derivative f''' is continuous on I. The Schwarzian derivative Sf is defined by

$$(Sf)(x) = \frac{f'''(x)}{f'(x)} - \frac{3}{2}\left[\frac{f''(x)}{f'(x)}\right]^2$$

In the previous subsection we have seen how a negative Schwarzian derivative is manifested geometrically. In fact, the possession of a negative Schwarzian derivative is a necessary condition for period doubling to occur. For more rigorous treatments of this property, the readers may refer to Argyris et al. [2] and Gulick [3].

At this point, we may proceed to consider the condition for repeated period doublings or *period doublings ad infinitum*. To do this, we need two further properties of the Schwarzian derivative, which are summarized below. For brevity, we denote the composite function $f(h(.))$ by $f \circ h$.

Property 3.1: Suppose $Sf < 0$ and $Sh < 0$. Then $S(f \circ h) < 0$.

Sketch of Proof: Using the chain rule, one computes that $(f \circ h)''(x) = f''(h(x))(h'(x))^2 + f'(h(x))h''(x)$ and $(f \circ h)'''(x) = f'''(h(x))(h'(x))^3 + 3f''(h(x))h''(x)h'(x) + f'(h(x))h'''(x)$. It follows that $S(f \circ h)(x) = Sf(h(x))(h'(x))^2 + Sh(x) < 0$, *q.e.d.*

Property 3.2: Suppose that $Sf < 0$. Then $Sf^n < 0$ for any positive integer n.

Sketch of Proof: Since $Sf^{(n)} < 0$ implies $Sf^{(n+1)} < 0$, and $Sf < 0$, the result follows from induction, *q.e.d.*

As we will see, Property 3.2 is instrumental to proving the following theorem, which addresses the condition for repeated period doublings in a system described by an iterative function $x_{n+1} = f(x_n)$.

Theorem on Period Doubling   A necessary condition for the occurrence of period-doubling cascades for the iterative function f is $Sf < 0$ in the proximity of the bifurcation points [2].

Sketch of Proof: First we recognize that the emergence of a period doubling in f requires $Sf < 0$ as a necessary condition. In general, a period-p orbit bifurcates to a period-2p orbit only if $Sf^p < 0$. Thus, we need $Sf^n < 0$ for all $n \geq 1$, for the system to period-double repeatedly. From Property 3.2, $Sf < 0$ implies $Sf^n < 0$ for all $n > 1$. Hence, $Sf < 0$ is a necessary condition for the iterative system to repeatedly double its period to chaos.

3.5.5 Application to Power Electronics

The application of the foregoing technique requires the system under study to be modeled adequately by a first-order iterative map. This remains a major limitation of the technique, but there are fortunately systems in power electronics which lend themselves to such analysis.

Specifically, simple dc/dc converters, when operating in discontinuous mode, are effectively first-order systems. This is because the inductor current, which assumes a zero value at the start of every period, disqualifies itself as a storage element. The remaining dynamic element is the output capacitor. Thus, we may choose the output voltage as the only state variable, and attempt to derive a first-order iterative map for the system.

Let x be the output voltage, E be the input voltage, T be the switching period, and d the duty cycle. The derivation of the describing iterative function essentially involves successive substitution of solutions of the involving linear subsystems. Specifically, since the circuit is linear in each subinterval of time, analytical solution can be obtained that effectively expresses the value of x at the end of a subinterval in terms of that at the beginning of the subinterval. Then, by stacking up solutions of consecutive subintervals, we may express x at $t = (n+1)T$ in terms of that at $t = nT$, where n is a non-negative integer. Mathematical details of the derivation can be found in Section 5.5 and also in Tse [4,5]. In particular, the required iterative function for a boost converter operating in discontinuous mode has been found as

$$x_{n+1} = \alpha x_n + \frac{\beta g(x_n)^2 E^2}{x_n - E} \tag{3.29}$$

where α and β are circuit parameters defined by

$$\alpha = 1 - \frac{T}{CR} + \frac{T^2}{2C^2 R^2}, \quad \beta = \frac{T^2}{2LC}$$

Likewise, for the buck converter operating in discontinuous mode, the describing iterative function is

$$x_{n+1} = \alpha x_n + \frac{\beta g(x_n)^2 E(E - x_n)}{x_n} \tag{3.30}$$

As said before, our objective is to find the condition on $g(.)$ such that the system can exhibit a period-doubling route to chaos. We will illustrate, using the boost converter as an example, the use of the Schwarzian derivative to derive the required result.

3.5.6 Illustrative Example: The Boost Converter

Before we proceed to work on the Schwarzian derivative for the iterative function (3.29), some important properties of the function should be noted which will simplify drastically the subsequent derivations and will eliminate unnecessarily lengthy algebra. First of all, the feedback control function $g(.)$ is always equal to the duty cycle in the steady state. Denoting the steady-state output voltage and duty cycle by X and D respectively, we have

$$g(X) = D \tag{3.31}$$

Also, when the boost converter operates in discontinuous mode, the steady-state duty cycle D can be found by putting $x_{n+1} = x_n = X$ in (3.29), i.e.,

$$g(X) = \sqrt{\frac{(1 - \alpha)(1 - M)}{\beta M^2}} \tag{3.32}$$

where M is the input-to-output voltage ratio, E/X. We further assume that the closed-loop system has a negative characteristic multipler (i.e., $f'(x) < 0$). This condition remains necessary for the exhibition of period-doubling routes to chaos, since no period doubling would ever occur otherwise. Writing $f'(x)$ explicitly, we have

$$f'(x) = \alpha - \frac{\beta E^2 g(x)^2}{(x - E)^2} + \frac{2\beta E^2 g(x)g'(x)}{x - E} \tag{3.33}$$

In the neighborhood of the steady-state operating point (3.33) can be written as

$$f'(X) = \alpha - \frac{\beta E^2 D^2}{(X - E)^2} + \frac{2\beta E^2 D g'(X)}{X - E} < 0$$

$$\text{or } g'(X) < \frac{X - E}{2\beta E^2 D}\left(\frac{\beta E^2 D^2}{(X - E)^2} - \alpha\right) \tag{3.34}$$

By substituting the value of D from (3.32) in (3.34), we can state the following:

Consider the boost converter operating in discontinuous mode described by $x_{n+1} = f(x_n, g(x_n))$, where $f(.)$ is defined in Eq. (3.29). $f'(.) < 0$ if, and only if,

$$g'(.) < -h_u$$

where

$$h_o = \frac{\alpha(2 - M) - 1}{2E\sqrt{\beta(1 - \alpha)(1 - M)}}$$

The above results enable the condition on the characteristic multiplier to be placed on the control function, and hence permit the subsequent analysis to be performed in terms of the control function. Note that the value of h_o is positive for all practical component values and parameters of the discontinuous-mode boost converter.

In order to produce an infinite cascade of period doublings, the Schwarzian derivative of f must be negative (i.e., $Sf < 0$). From the definition, we can write $Sf < 0$ as

$$\frac{3}{2}(f'')^2 - f'''f' > 0 \tag{3.35}$$

which can be expanded, using (3.29), to give

$$\frac{6\alpha\beta E^2 g(x)^2}{(x - E)^4} - \frac{12\alpha\beta E^2 g(x)g'(x)}{(x - E)^3} + \frac{6\alpha\beta E^2 g'(x)^2}{(x - E)^2} + \frac{6\beta^2 E^4 g(x)^2 g'(x)^2}{(x - E)^4}$$

$$- \frac{12\beta^2 E^4 g(x)g'(x)^3}{(x - E)^3} + \frac{6\beta^2 E^4 g'(x)^4}{(x - E)^2} + \frac{6\alpha\beta E^2 g(x)g''(x)}{(x - E)^2} + \frac{6\beta^2 E^4 g(x)^3 g''(x)}{(x - E)^4}$$

$$- \frac{6\alpha\beta E^2 g'(x)g''(x)}{x - E} - \frac{6\beta^2 E^4 g(x)^2 g'(x)g''(x)}{(x - E)^3} + \frac{6\beta^2 E^4 g(x)^2 g''(x)^2}{(x - E)^2}$$

$$- \frac{2\alpha\beta E^2 g(x)g'''(x)}{x - E} + \frac{2\beta^2 E^4 g(x)^3 g'''(x)}{(x - E)^3} - \frac{4\beta^2 E^4 g(x)^2 g'''(x)g(x)}{(x - E)^2}$$

$$> 0 \tag{3.36}$$

Based on (3.36), we may derive the main result that applies to the boost switching regulator operating in discontinuous mode regarding the condition for possible exhibition of period-doubling cascades as follows.

First of all, we observe that if $g' < -h_o < 0$ and $g'' \geq 0$, the sum of all terms but the last three in LHS of (3.36) is positive, i.e.,

$$-T_g - (f')^2 Sf > 0 \tag{3.37}$$

where T_g denotes the sum of the last three terms in LHS of (3.36) and is given by

$$T_g = -\frac{2\alpha\beta E^2 g(x)g'''(x)}{x - E} + \frac{2\beta^2 E^4 g(x)^3 g'''(x)}{(x - E)^3} - \frac{4\beta^2 E^4 g(x)^2 g'(x)g'''(x)}{(x - E)^2} \tag{3.38}$$

Equation (3.38) can be factorized as

$$T_g = -\frac{2E^2 g(x)g'''(x)\beta f'(x)}{x - E} \tag{3.39}$$

Finally, since $g' < -h_o$ forces $f'(x) < 0$ as noted earlier, $g'''(x) \geq 0$ leads to $T_g \geq 0$, and hence, from (3.37), we arrive at the following important result.

Main Result: The boost regulator described by Eq. (3.29) has a negative Schwarzian derivative if, for the range of operation,

1. $g'(x) < -h_o$,
2. $g''(x) \geq 0$, and
3. $g'''(x) \geq 0$.

Let us now apply the above result to a particular control scenario in which the discontinuous-mode boost converter is controlled by a linear or parabolic function (i.e., $g(x) = a + bx$ or $g(x) = a + bx + cx^2$), where a, b, and c are constant. Here, we can easily show that a sufficient condition for the system to have a negative Schwarzian derivative is

$$g'(x) \leq -h_o, \tag{3.40}$$

over the control range. Moreover, for higher-order control functions, additional sufficient conditions are $g''(x) \geq 0$ and $g'''(x) \geq 0$.

3.5.7 Interpretation and Application of the Result

The importance of the above result can be appreciated as it provides an indication of the possibility of a period-doubling route to chaos. To probe further into the result, we note that g' corresponds to the small-signal gain of the feedback control since $g' = \partial d/\partial x$. The condition on g', as stated in the previous subsection, is equivalent to a sufficiently large small-signal gain. In other words, *when the feedback gain is sufficiently large, possibility exists for a period-doubling path to chaos.* This conclusion is in perfect agreement with the result reported in Tse [4] (see also Chapter 5), as we now illustrate.

Consider a boost converter having circuit parameters as follows: $\alpha = 0.8872$, $\beta = 1.2$, $E = 16\text{V}$, $X = 25\text{V}$, and $h_o = 0.03$. We assume that the dc/dc converter is controlled by a simple proportional feedback of the form:

$$g(x) = D + \kappa(x - X) \tag{3.41}$$

The range of $g'(x)$ is between -0.07 and -0.16. Hence, $g' < -h_o$ and $g'' = g''' = 0$. The system thus has a negative Schwarzian derivative.

3.5.8 Remarks and Summary

Applying the same technique to the buck converter, we obtain a similar set of conditions for the occurrence of period-doubling routes to chaos [6]. In short, the essential requirement is that the magnitude of the small-signal feedback gain be large enough. In Chapter 5 we will re-examine period-doubling bifurcation in this type of converter circuits using both computer simulations and laboratory experiments.

REFERENCES

[1] D. Singer, Stable orbits and bifurcation of maps of the interval, *SIAM J. Appl. Math.*, vol. 35, pp. 260–267, 1978.

[2] J. Argyris, G. Faust, and M. Haase, *An Exploration of Chaos*. Amsterdam: Elsevier Science BV, 1994.

[3] D. Gulick, *Encounters with Chaos*. New York: McGraw Hill, 1992.

[4] C. K. Tse, Flip bifurcation and chaos in three-state boost switching regulators, *IEEE Trans. Circ. & Syst. Part I*, vol. 41, no. 1, pp. 16–23, Jan. 1994.

[5] C. K. Tse, Chaos from a buck switching regulator operating in discontinuous mode, *Int. J. Circ. Theory Appl.*, vol. 22, no. 4, pp. 263–278, Jul.–Aug. 1994.

[6] W. C. Y. Chan and C. K. Tse, On the form of feedback function that can lead to chaos in discontinuous-mode dc/dc converters," *IEEE Power Electron. Spec. Conf. Rec.*, pp. 1317–1322, June 1997.

3.6 COEXISTING ATTRACTORS, BASINS OF ATTRACTION, AND CRISES

Enric Fossas
Gerard Olivar

In the previous section, we saw how the Poincaré map can be a very useful tool for studying some types of dynamical systems. This also applies to time-dependent cases when the forcing function is T-periodic, as in many dc/dc converter schemes. The construction of a Poincaré map is especially useful when the stability of a periodic orbit must be studied, since the problem is then reduced to the stability character of a fixed point of the corresponding Poincaré map.

3.6.1 Characteristic (Floquet) Multipliers

The stability of a periodic solution is determined by its characteristic multipliers, also called Floquet multipliers. Characteristic multipliers are a generalization of the eigenvalues at an equilibrium point.

Consider a fixed point $\mathbf{x}^* = (x_1^*, x_2^*)$ of a map P, in a two-dimensional space. The local behavior of the map near $\mathbf{x}^*$ is determined by linearizing the map at $\mathbf{x}^*$. In particular, the linear map

$$\delta \mathbf{x}_{k+1} = DP(\mathbf{x}^*)\delta \mathbf{x}_k$$

where $DP(\mathbf{x}^*)$ is the Jacobian of the map at $\mathbf{x}^*$, governs the evolution of a perturbation $\delta \mathbf{x}_0$ in a neighborhood of the fixed point.

In general, let p be the dimension of the Poincaré section. Let the eigenvalues of $DP(\mathbf{x}^*)$ be $m_i \in C$, with corresponding eigenvectors $\eta_i \in C^n$ for $i = 1, \ldots p$. Assuming that the eigenvectors are distinct, the orbit of P with initial condition $\mathbf{x}^* + \delta \mathbf{x}_0$ is, to first order,

$$\mathbf{x}_k = \mathbf{x}^* + \delta \mathbf{x}_k = \mathbf{x}^* + (DP(\mathbf{x}^*))^k \delta \mathbf{x}_0 = \mathbf{x}^* + c_1 m_1^k \eta_1 + \cdots + c_p m_p^k \eta_p$$

where $c_i \in C$ are constants obtained from the initial condition.

The eigenvalues m_i are called the *characteristic multipliers* of the periodic solution. Like eigenvalues at an equilibrium point, the characteristic multipliers' position in the complex plane determines the stability of the fixed point. If m_i is real, then η_i and c_i are also real, and it is clear that the characteristic multiplier is the amount of contraction (if $|m|_i < 1$) or expansion (if $|m|_i > 1$) near $\mathbf{x}^*$ in the direction of η_i for one iteration of the map. In the case of complex eigenvalues, the magnitude of m_i again gives the amount of contraction (if $|m_i| < 1$) or expansion (if $|m_i| > 1$) for one iteration of the map; the angle of the characteristic multiplier is the frequency of rotation.

3.6.2 Invariant Sets and Invariant Manifolds

Let $S \subseteq R^n$ be a set. Then,

1. *Continuous time:* S is said to be invariant under the vector field $\dot{\mathbf{x}} = f(\mathbf{x})$ if for any $\mathbf{x}_0 \in S$ we have that the solution passing through $\mathbf{x}_0$, $\phi_t(\mathbf{x}_0) \in S$ for all $t \in R$.
2. *Discrete time:* S is said to be invariant under the map $\mathbf{x} \mapsto g(\mathbf{x})$ if for any $\mathbf{x}_0 \in S$, we have $g^n(\mathbf{x}_0) \in S$ for all n.

In applications, manifolds are most often met as q-dimensional surfaces embedded in the state space. Roughly speaking, an *invariant manifold* is a surface contained in the state space of a dynamical system whose property is that orbits starting on the surface remain on the surface throughout the course of their dynamical evolution.

Additionally, the sets of orbits which approach or recede from a fixed point $\mathcal{M}$ asymptotically in time under certain conditions are also invariant manifolds, which are called the *stable* and *unstable manifolds*, respectively, of $\mathcal{M}$ (see Figure 3.44).

Knowledge of the invariant manifolds of a dynamical system, as well as the intersection of their respective stable and unstable manifolds, is absolutely crucial in order to obtain a complete understanding of the global dynamics.

Suppose that we have

$$\dot{\mathbf{x}} = f(\mathbf{x}) \qquad \mathbf{x} \in R^n \tag{3.42}$$

where $f : R^n \longrightarrow R^n$. We make the following assumptions about a point $\mathbf{x}_0$ of the state space:

1. $f(\mathbf{x}_0) = 0$.
2. $Df(\mathbf{x}_0)$ has $n - k$ eigenvalues having positive real parts and k eigenvalues having negative real parts.

Then, $\mathbf{x}_0$ is called a *hyperbolic fixed point* for $\dot{\mathbf{x}} = f(\mathbf{x})$ if $Df(\mathbf{x}_0)$ has no eigenvalues with zero real part.

Linearizing the system ($\delta\dot{\mathbf{x}} = Df(\mathbf{x}_0)\delta\mathbf{x}$), we can denote by

$$\mathbf{v}_1, \dots \mathbf{v}_{n-k}$$

the generalized eigenvectors corresponding to the eigenvalues having positive real parts, and

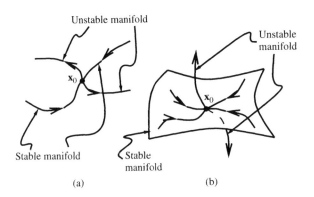

Unstable manifold

Unstable manifold

$\mathbf{x}_0$

$\mathbf{x}_0$

Stable manifold

Stable manifold

(a)

(b)

Figure 3.44 (a) Invariant manifolds in a two-dimensional state space. Each one is one-dimensional. (b) Invariant manifolds in a three-dimensional state space. The stable manifold is two-dimensional.

$$\mathbf{v}_{n-k+1}, \ldots \mathbf{v}_n$$

the generalized eigenvectors corresponding to the eigenvalues having negative real parts. Then, the linear subspaces of R^n defined as

$$E^u = span\{\mathbf{v}_1, \ldots \mathbf{v}_{n-k}\} \qquad E^S = span\{\mathbf{v}_{n-k+1}, \ldots \mathbf{v}_n\}$$

are invariant manifolds for the linear system, which are known as the unstable and stable subspaces, respectively.

The stable manifold theorem for hyperbolic fixed points [1] tells us that in a neighborhood U of the fixed point $\mathbf{x}_0$, there exist a $(n - k)$-dimensional surface $W^u(\mathbf{x}_0)$ tangent to E^u at $\mathbf{x}_0$ and a k-dimensional surface $W^s(\mathbf{x}_0)$ tangent to E^s at $\mathbf{x}_0$, with the properties that orbits of points on $W^u(\mathbf{x}_0)$ approach $\mathbf{x}_0$ in reverse time (i.e., as $t \to -\infty$) and orbits of points on $W^s(\mathbf{x}_0)$ approach $\mathbf{x}_0$ in positive time (i.e., as $t \to \infty$). $W^u(\mathbf{x}_0)$ and $W^s(\mathbf{x}_0)$ are known as the *local unstable* and *local stable manifolds*, respectively, of $\mathbf{x}_0$.

All of this can be defined when a map $P : R^n \longrightarrow R^n$ is considered. Intuitively, the stable manifold of a fixed point $\mathbf{x}_0$ of P will be the set of all points $\mathbf{x}$ such that $P^k(\mathbf{x})$ approaches $\mathbf{x}_0$ as $k \to \infty$, and the unstable manifold will be defined as the set of all points $\mathbf{x}$ such that $P^k(\mathbf{x})$ approaches $\mathbf{x}_0$ as $k \to -\infty$ (and thus, $P^k(\mathbf{x}_0)$ moves away from $\mathbf{x}_0$ as $k \to +\infty$).

One important point to take into account is that the invariant manifolds are as smooth as the orbits. In some models for power converters, the orbits are continuous but only piecewise differentiable. Then the manifolds would also be expected to be piecewise differentiable.

Homoclinic and Heteroclinic Orbits

In continuous-time nonlinear systems there are special trajectories which globally organize the structure of the state space. These trajectories are called *homoclinic* and *heteroclinic orbits*.

Let $\mathbf{x}_0$ and $\mathbf{y}_0$ be two different equilibrium points for a dynamical system. An orbit in the state space that lies in $W^s(\mathbf{x}_0) \cap W^u(\mathbf{x}_0)$ and which is nonconstant is called a homoclinic orbit (see Figure 3.45 (a)). An orbit in the state space that lies in $W^s(\mathbf{x}_0) \cap W^u(\mathbf{y}_0)$ is called a heteroclinic orbit. Note that a homoclinic orbit approaches $\mathbf{x}_0$ both in forward and backward time, while a heteroclinic orbit approaches one of the equilibrium points in forward time and the other one in backward time. Note also that the orbits approach the equilibrium points in infinite time, since otherwise they would violate the uniqueness theorem for orbits.

One can define homoclinic and heteroclinic orbits in a similar way for maps. Let $\mathbf{x}_0$ be a fixed point with stable and unstable manifolds $W^s(\mathbf{x}_0)$ and $W^u(\mathbf{x}_0)$ respectively. Assume that the manifolds intersect at a different point $\mathbf{x}_0^*$. Then the orbit from $\mathbf{x}_0^*$ in forward and backward time is called a *homoclinic orbit*, and $\mathbf{x}_0^*$ is called a *homoclinic point*.

The existence of a homoclinic point leads to an interesting situation. Since $\mathbf{x}_0^*$ lies in the stable manifold of $\mathbf{x}_0$, therefore all future iterates of $\mathbf{x}_0^*$ must also lie on the stable manifold. At the same time since $\mathbf{x}_0^*$ lies also on the unstable manifold, all iterates starting from that point must also lie on the unstable manifold. It follows that all the points on the homoclinic orbit are the points on intersection of the stable manifold and the unstable manifold. Not only that, all backward iterates (the orbit that leads to $\mathbf{x}_0^*$)

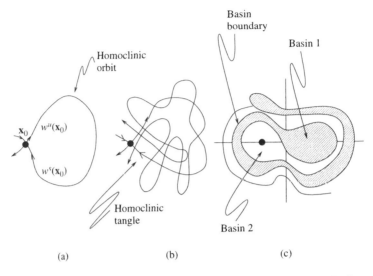

Figure 3.45 (a) Homoclinic orbit; (b) homoclinic tangle; (c) different basins of attraction.

are also the points of intersection of the two manifolds. Thus, the existence of a single homoclinic intersection indicates an infinity of such intersections and hence an infinite complexity of the manifold structures (see Figure 3.45(b)).

The two manifolds may or may not intersect transversally. If they do not intersect transversally, then they must be identical, and the structure is called a *homoclinic connection*. Almost any perturbation in the system definition will destroy the connection. If the manifolds intersect transversally (manifold tangle), the resulting stretching and folding actions give the map an embedded horseshoe structure (as explained in Section 3.1), which can lead to chaotic dynamics. Moreover, this is a robust structure, and it is not destroyed by generic perturbations.

As a parameter is varied, the stable and the unstable manifolds of a fixed point can also change qualitatively from intersection at the equilibrium point x_0 to infinite intersection points. This leads to qualitative changes in the state space.

3.6.3 Coexisting Attractors

It is worth noting that in a linear system only one attractor can exist in the state space. This is not true for nonlinear systems; in fact, it is one of the most important differences between linear and nonlinear systems. As will be seen later, it has serious consequences for applications. Thus, in power electronics, it is common to find coexisting attractors in the state space, and this will be extensively shown in Chapter 5. The existence of multiple attractors brings with it a natural partition of the state space into several regions, so-called basins or domains of attraction, each one associated with a different attractor (one of the attractors can be at infinity). The closure of the set of initial conditions whose stationary state is a given attractor is called the *basin of attraction of the attractor*. These regions are limited by the basin boundaries (or separatrix), which also have a role in the organization of the orbits (see Figure 3.45(c)).

The existence of basins of attraction causes the stationary behavior of the system to be different, depending on in which basin the initial conditions are located. In applications this is quite important, since if initial conditions are placed near a basin boundary, a minor change in any of the circuit parameters can produce a change in the boundary, and thus a different stationary state will be reached. Due to this fact, the study of the changes of the basins and their boundaries as a parameter is varied (basin metamorphoses) is of great interest in applications. A critical situation occurs when one of the basin boundaries is not smooth but has fractional dimension. Near the boundary, one cannot be sure of placing initial conditions in a particular basin to obtain a prescribed stationary behavior, which implies a significant lack of predictability.

3.6.4 The Role of Invariant Manifolds and Basins of Attraction

Stable and unstable manifolds of the saddles play a fundamental role in basin organization, and when obtained systematically with a varying bifurcation parameter, they permit a thorough geometrical understanding of the structures of the attractors and their basins of attraction. Therefore, to get a global view of the dynamics of a given system, invariant manifolds and basins of attraction have to be computed and their dependency on the system parameters has to be studied.

Indeed, the main global bifurcations are associated with homoclinic and heteroclinic tangencies between the stable and unstable manifolds of a given m-periodic regular (R^m) or flip (F^m) saddle, and two different regular or flip saddles, respectively. The closure of the stable manifold of a regular saddle coincides with the border between two basins of attraction of different solutions of a given system. Likewise, the closure of the stable manifold of a flip saddle coincides with the boundary between the two distinct subdomains identifiable in the basin of a given solution under a mapping P^{2m}.

When varying the bifurcation parameter, occurrence of manifold tangling is a sufficient condition for fractal basin boundary (and thus for unpredictability of response) and a necessary condition for the onset of a chaotic attractor in one basin. If it exists, this attractor is contained within the closure of one of the branches of the unstable manifold of the regular saddle located on the relevant basin boundary. If the chaotic attractor originates through repeated period-doubling bifurcations, then the attractor also occurs within the closures of the unstable manifolds of the flip saddles resulting from those period doublings. The sequence of homoclinic and heteroclinic tangencies, and intersections of stable and unstable manifolds corresponding to coexisting unstable periodic solutions, govern the sequence of bifurcations and sudden changes in system behavior.

To illustrate the above with an example, the basins of attraction are computed for certain values of the parameters of a model buck converter introduced in Chapter 1 (please refer to Section 5.3 for details of this converter). Although the concept of basin boundary is a simple one, the large time-consumption often poses computational difficulties in a numerical simulation. There exist specific algorithms for constructing the basins efficiently: one way is to compute the invariant manifolds [2]; the other way is via cell mapping algorithms [3,4]. In our example, the second method was adopted.

At the parameter value $V_{in} = 29.0$V, there are two attractors: the main $2T$-periodic stable orbit ($2T$ signifies a periodicity twice that of the clock) and a $6T$-periodic attractor. It has been shown in Section 3.2 that such attractors come into being through saddle-node bifurcations, where a regular saddle is created simultaneously with the attractor. By computing the stable manifolds of the regular saddle (which originated along with the main $2T$-periodic attractor at a saddle-node bifurcation), one finds that the closure of the stable manifold is the boundary of the basins of the competing attractors.

When the main attractor bifurcated from period-1 to period-2 at $V_{in} = 24.516$V, it resulted in a flip-saddle fixed point, which also exists in the state space at the parameter value $V_{in} = 29.0$V. The stable and unstable manifolds of this fixed point are shown in Figure 3.46(a). It can be seen that the two manifolds do not intersect each other. When the parameter is changed to $V_{in} = 30.0$V (Figure 3.46(b)), the two manifolds intersect. After the homoclinic tangency, fractal basin boundaries are expected.

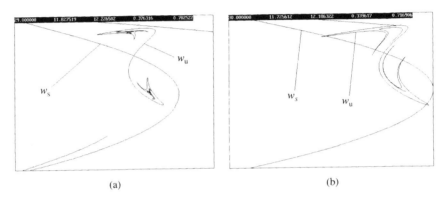

(a)   (b)

Figure 3.46 (After [8]) (a) Invariant manifolds for the main flip saddle just before the homoclinic tangency. $V_{in} = 29.0$V. v range is (11.70,13.13); i range is (0.38,0.70). (b) Invariant manifolds for the main flip saddle just after the homoclinic tangency. Fractal boundary basins are expected. $V_{in} = 30.0$V. v range is (11.67,13.08); i range is (0.34,0.72).

Figure 3.47(a) shows the basins of attraction of the two attractors after the homoclinic tangency. It can be seen that over a large portion of the state space, the two basins of attraction get mixed in an arbitrarily fine scale—which becomes clearer when we zoom onto a smaller portion of the state space (Figure 3.47(b)). If the dimension of the basin boundary is calculated using the procedure outlined in Section 3.1, it turns out to be a fractal.

3.6.5 Crises

Crises are collisions between a chaotic attractor and a coexisting unstable fixed point or periodic orbit (or its stable manifold). Grebogi, Ott, and Yorke were the first to observe that such collisions lead to sudden changes in the chaotic attractor [5]. A simple example occurs in the period-3 window of the one-dimensional quadratic map $x_{n+1} = a - x_n^2$, where three stable and three unstable fixed points are generated at a tangent (or saddle-node) bifurcation. The bifurcation diagram is shown in Figure 3.48, and the above bifurcation occurs at the onset of the period-3 window seen in the figure. As the parameter value is increased, the period-3 attractor undergoes repeated period-doubling bifurcations and becomes chaotic. Then suddenly the attractor seems to expand. This happens when the unstable period-3 fixed point touches the chaotic attractor. These unstable fixed points then repel the trajectory out of the subbands in

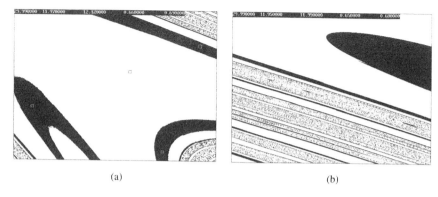

(a)   (b)

Figure 3.47 (After [8]) (a) Basins of attraction for $V_{in} = 29.990$V. Black corresponds
to the $6T$ basin, white to the $2T$ basin. Points in the attractors are
marked with rectangles. v range is (11.97,12.12); i range is (0.66,0.69).
(b) Detail of the basins for $V_{in} = 29.990$V. v range is (11.95,11.99);
i range is (0.65,0.68).

such a way that the regions between the bands are also filled chaotically—resulting in
an expansion of the attractor. This is called an *interior crisis*.

As the crisis point is approached, one often finds transient chaos (i.e., sudden
bursts of chaotic behavior amid regular periodic operation). Almost all sudden changes
in chaotic attractors are due to crises.

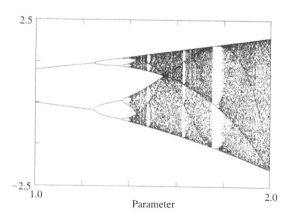

Figure 3.48 Bifurcation diagram for the
quadratic map. A period-3 window is clearly
visible. Afterward, a crisis occurs and the
attractor is enlarged.

Interior Crises

There can be two types of interior crises. At an interior crisis of *merging type*
(or an interior crisis of the first kind, called IC1), due to the collision with a n-
periodic flip saddle F_j^n, the $2n$ pieces of a chaotic attractor merge two by two, giving
rise to a n-piece chaotic attractor (see Figure 3.49(a)). A crisis of this kind is also
seen in the bifurcation diagram in Figure 3.48. A detailed explanation of this crisis
is found in [6].

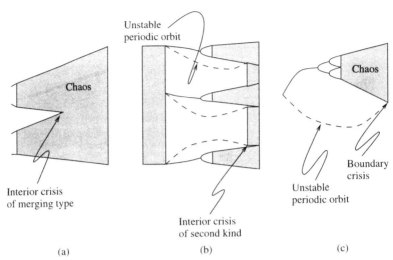

Figure 3.49 Schematic diagrams illustrating the different types of crises: (a) interior crisis of merging type; (b) interior crisis of second kind; (c) boundary crisis.

At an interior crisis of the second kind (called IC2), the chaotic attractor collides with an unstable periodic orbit which is in the interior of its basin of attraction, and the attractor suffers a sudden enlargement in its size, while the basin of attraction does not change (see Figure 3.49(b)). Just after the crisis, the orbit on the attractor spends a long time in the region where the smaller attractor existed before the crisis, and intermittently jumps from this region to the other regions of the new large attractor created after the crisis. This dynamical behavior is called *crisis-induced intermittency*. It can be found, for example, in the Ikeda map [7].

All the above types of crisis have been observed in the buck converter [8].

Boundary Crises

When multiple attractors occur in a system, there is always a saddle fixed point in between—whose stable manifold separates the basins of attraction. Now imagine that a chaotic attractor (and its basin) exist for a value $\lambda < \lambda_{bc}$ and let $\lambda \to \lambda_{bc}$. At a boundary crisis ($\lambda = \lambda_{bc}$) the chaotic attractor collides with the saddle fixed point that lies on the basin boundary. Subsequently this orbit no longer remains stable (see Figure 3.49(c)). After the boundary crisis the motion can be regular or chaotic, depending on the type of the other attractor, which now has a larger basin of attraction.

Just after the crisis, the attractor that existed for $\lambda < \lambda_{bc}$ becomes an unstable chaotic orbit. Trajectories in this region are transiently chaotic, and finally converge on the other attractor. It can be said that the attractor develops a region from which orbits can escape to another zone of the state space. For some cases, the transient has a lifetime of the order $(\lambda - \lambda_{bc})^{-\delta}$, δ being the critical exponent of the crisis.

Often, coexisting attractors can come into existence through saddle-node bifurcation and can go out of existence, at a different parameter value, through a boundary crisis. This type of sequence has also been observed in the buck converter [8] (see Figure 5.18 in Section 5.3).

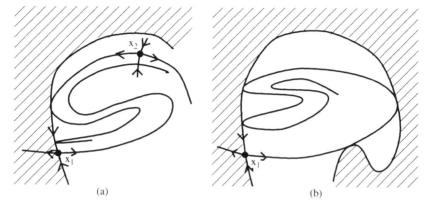

Figure 3.50 (a) Heteroclinic tangency; (b) homoclinic tangency.

Two-Dimensional Maps

Some results concerning two-dimensional maps have been given by Grebogi et al. [7,9], where the crisis phenomenon is due to a tangency of the stable manifold of a periodic orbit on the basin boundary with the unstable manifold of an unstable periodic orbit on the attractor. It is conjectured that these types of crises are the only ones which can occur in strictly dissipative two-dimensional maps. For these systems one of the following takes place (see Figure 3.50):

1. *Heteroclinic tangency crisis:* the stable manifold of an unstable periodic orbit x_1 becomes tangent with the unstable manifold of an unstable periodic orbit x_2. During the crisis process, x_1 is on the boundary and x_2 is on the attractor.
2. *Homoclinic tangency crisis:* the stable and unstable manifolds of an unstable periodic orbit x_1 are tangent.

In both cases, during the crisis process, the basin boundary is the closure of the stable manifold of x_1. At the crisis parameter value, the chaotic attractor is the closure of the unstable manifold of x_1. In the heteroclinic tangency crisis, before the crisis parameter value, the attractor is also the closure of the unstable manifold of x_2.

REFERENCES

[1] S. Wiggins, *Introduction to Applied Nonlinear Dynamical Systems and Chaos.* New York: Springer-Verlag, 1990.

[2] T. S. Parker and L.O. Chua, *Practical Numerical Algorithms for Chaotic Systems.* New York: Springer-Verlag, 1989.

[3] C. S. Hsu, *Cell-to-Cell Mapping.* New York: Springer-Verlag, 1987.

[4] H. E. Nusse and J. A. Yorke, *Dynamics: Numerical Explorations.* New York: Springer-Verlag, 1994.

[5] E. Ott, *Chaos in Dynamical Systems.* Cambridge: Cambridge University Press, 1993.

[6] C. Grebogi, E. Ott, and J. A. Yorke, Basin boundary metamorphoses: Changes in accessible boundary orbits, *Physica D*, vol. 24, pp. 243–262, 1987.

[7] C. Grebogi, E. Ott, F. Romeiras, and J. A. Yorke, Critical exponents for crisis-induced intermittency, *Phys. Rev. A*, vol. 36, 5365, 1987.

[8] M. Di Bernardo, E. Fossas, G. Olivar, and F. Vasca, Secondary bifurcations and high periodic orbits in voltage controlled buck converter, *Intern. J. Bif. Chaos*, vol. 7, no. 12, pp. 2755–2771, 1997.

[9] C. Grebogi, E. Ott, and J. A. Yorke, Critical exponents of chaotic transients in nonlinear dynamical systems, *Phys. Rev. Lett.*, vol. 57, 1284, 1986.

Chapter 4

EXPERIMENTAL AND COMPUTATIONAL TECHNIQUES FOR INVESTIGATION OF NONLINEAR PHENOMENA

4.1 TECHNIQUES OF EXPERIMENTAL INVESTIGATION

Chi K. Tse

4.1.1 Introduction

Sensitive dependence on initial conditions and lack of long-term predictability are key features of chaotic systems, which have profound implications on the approaches taken to study such systems. From the computational standpoint, exact trajectories cannot be sought for a chaotic system, no matter how accurate the numerical simulations and the models used in the simulations are. Any computed trajectory will "eventually be wrong." This is particularly true with modern digital computers which introduce roundoff errors, and depending on the algorithms used, the errors can accumulate and render any solution eventually inaccurate. This leads to the question of how much we can trust our analysis and simulation. Putting it a different way, how accurate should the analytical model be, and how do we tell if the analytical result is reflecting the true behavior of the system under study?

An equally important mode of investigation of nonlinear phenomena in power electronics is to begin with experimentation. Certain phenomena may be observed unintentionally while developing a practical power electronics system. The quest for an explanation for the observed unusual behavior motivates in-depth analysis of the underlying mechanism. This finally calls for appropriate analytical models which fit the observed phenomena and provide adequate analytical basis to predict the occurrence of similar phenomena.

4.1.2 Overview of Simulation Study and Verification

Very often, for the purpose of studying chaotic systems, analytical models need not be very accurate since exact trajectories are never wanted. What is needed is perhaps a simple model that contains adequate salient nonlinear features of the system under study. After all, sensitive dependence will strike and render the model useless for generating exact trajectories. In fact, what we want the model to predict is really qualitative behavior, such as bifurcations and exhibition of chaotic attractors. If simulations are performed to verify the predicted qualitative behavior or to study certain behavior, they must be viable ones in order to reflect the true behavior of the system. Thus, using the same analytical model to simulate the system can only be regarded as part of the analysis (which is done numerically) and should not be claimed as a verification or simulation study. For electronic circuits, any viable verification or computer simulation study should be performed using real circuit models. Some existing packages such as PSPICE may help in this respect.

Computer simulation alone, however, is not completely convincing as a verification or investigation tool since numerical procedures are always subject to roundoff errors, however small, and the model used for simulation may not fully describe the system. What we see in the computer-simulated waveforms may sometimes contain artifacts due to numerical errors or flaws in the simulating model. Hence, laboratory experiments remain an indispensable form of verification. Furthermore, as mentioned earlier, experimentation can sometimes be well ahead of any analysis and simulation, particularly for power electronics circuits, whose popularity in practical use often precedes any detailed analysis.

In summary, rigorous analysis, viable simulations, and laboratory experiments are all indispensable, and they complement one another [1]. In this chapter we focus our attention on experimental investigation, and specifically on some essential laboratory techniques for capturing Poincaré sections and bifurcation diagrams.

4.1.3 Experimental Investigation

From what has been said, experimental study plays the dual role of *verifying* and *establishing* certain nonlinear phenomena in physical systems. It thus becomes obvious that experiments should be designed to focus the kinds of investigation that would be used in analysis and/or simulations. Usually, we examine nonlinear phenomena in one or more of the following aspects:

1. Time-domain waveforms
2. Phase portraits
3. Frequency spectra
4. Poincaré sections
5. Bifurcation diagrams.

While time-domain waveforms, phase portraits and frequency spectra are familiar to most electronics engineers, the way to obtain Poincaré sections and bifurcation diagrams on the oscilloscope may appear nontrivial. Nonetheless, we will briefly review the commonly used instruments for capturing time-domain waveforms, phase portraits, and frequency spectra, and will then go into details of displaying Poincaré sections and bifurcation diagrams on the oscilloscope.

4.1.4 Displaying Time-Domain Waveforms, Attractors, and Spectra

It should be straightforward enough for most engineers to capture periodic waveforms using an analog oscilloscope. For aperiodic waveforms such as those of quasi-periodicity or chaos, the waveforms appear to be shaking—which is generally a signature of these nonlinear phenomena. If a digital storage oscilloscope (DSO) is used, one can *freeze* the waveforms at a certain instant and then the irregular behavior of a chaotic waveform becomes apparent. To display phase portraits, one can simply use the X-Y mode of the oscilloscope instead of a sweeping time base [2].

The phase portraits provide a handy tool to identify chaotic behavior. Chaos is characterized by phase portraits that cover a well-defined bounded region, and can be easily distinguished from random noise which shows fuzzy edges on phase portraits. Furthermore, chaotic signals are wideband signals, and hence can be easily distinguished from periodic signals by inspecting their frequency spectra. This can be done with a spectrum analyzer. Moreover, some DSOs actually provide spectral analysis by performing the *fast Fourier transform* (FFT) on the signal being measured. Thus, we may use a DSO to obtain frequency spectra for periodic and non-periodic waveforms.

As an example, we consider the Ćuk converter operating under fixed-frequency current-mode control [3]. Figure 4.1 shows the experimental converter circuit which can be constructed and tested in the laboratory without much difficulty. The operation of the circuit can be briefly described as follows. The essential control variable is the sum of the two inductor currents, which is picked up by the 1Ω sensing resistor. The voltage across this sensing resistor is then compared with an adjustable threshold voltage which serves as a bifurcation parameter. The on-off status of the power switch (5N06) is determined by the output of the comparator (LM311). Essentially, when the power switch is on, the voltage across the sensing resistor ramps up, and as it reaches the threshold voltage, the RS flip-flop (actually a pair of NOR gates) is reset and the power switch is turned off. Then, the control

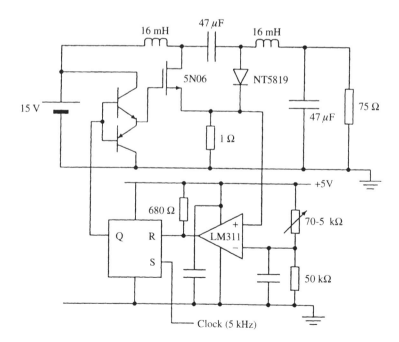

Figure 4.1 Experimental Ćuk converter circuit under fixed-frequency current-mode control. The RS flip-flop block is constructed from a pair of NOR gates [2].

variable ramps down, until the clock pulse sets the RS flip-flop again and turns the switch back on. The cycle repeats at 5kHz. Analysis has shown that changing the parameter values affects the qualitative behavior of the system. Here, we include in Figures 4.2 and 4.3 some typical waveforms, phase portraits, and frequency spectra obtained from this circuit.

4.1.5 Displaying Poincaré Sections

One of the approaches to studying nonlinear systems is to examine a Poincaré section of a trajectory. To keep our discussion simple, we initially consider *third-order autonomous circuits*. For simplicity we define a Poincaré section as a two-dimensional (2-D) plane that intersects the trajectory.

By examining the way the steady-state trajectory (sometimes referred to as the *attractor*) intersects the Poincaré section, one can tell if the steady-state motion is periodic, quasi-periodic, or chaotic [4]. The following is what we will typically see on a Poincaré section. First, if the motion is periodic, we will see a finite number of points on the Poincaré section. If the motion is quasi-periodic (torus), we will see a closed loop on the Poincaré section. And if the motion is chaotic, we will see a large number of irregularly and densely located points on the Poincaré section.

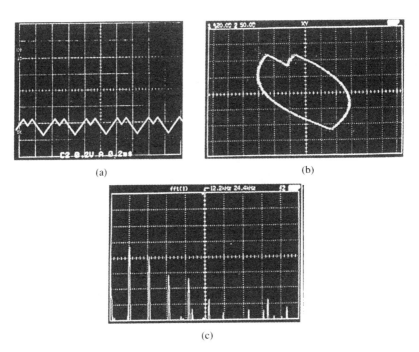

(a)   (b)

(c)

Figure 4.2 Experimental waveform, phase portrait, and frequency spectrum from oscilloscope for Ćuk converter operating under current-mode control showing period-2 operation. Reference for $i_1 + i_2$ set at 0.49A. (a) Inductor current (1 × 0.2A/div, 0.2ms/div, lowest horizontal grid line is 0A); (b) phase portrait of inductor current against a capacitor voltage, (c) FFT of inductor current [2].

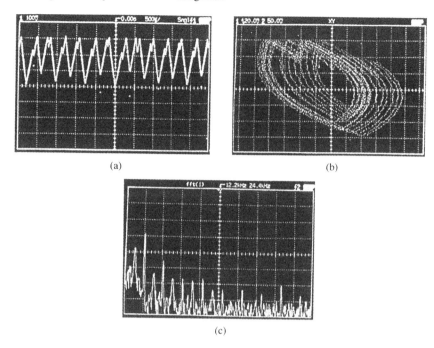

(a)

(b)

(c)

Figure 4.3 Experimental waveform, phase portrait, and frequency spectrum from oscilloscope for Ćuk converter operating under current-mode control showing chaotic operation. Reference for $i_1 + i_2$ set at 0.74A. (a) Inductor current (1×1V/div, 500μs/div, lowest horizontal grid line is 0A); (b) phase portrait of inductor current against a capacitor voltage; (c) FFT of inductor current [2].

Principle of Poincaré Section Measurement

Obviously, since the oscilloscope can only display 2-D phase portraits, we can at best view a projection of an attractor. Using the X-Y mode of the oscilloscope, we can display a 2-D projection (effectively a phase portrait) from any two given signals. This is adequate as long as the 2-D projection clearly reflects the kind of attractor. For most cases, we are still able to confidently tell, from a 2-D projection, if it is a periodic orbit. However, for a torus or chaotic attractor, we usually cannot make a definite conclusion unless we know what its Poincaré section looks like. Fortunately, it is not difficult to show a Poincaré section on the oscilloscope along with the 2-D projection of the attractor. What we need to do is to highlight the attractor when it cuts through a certain 2-D plane which has been chosen as the Poincaré section.

Suppose the system's variables are x, y, and z, and the oscilloscope is now plotting x against y using the X-Y mode [2]. Thus, the oscilloscope is showing the projection of the attractor on the x-y plane. We may define a Poincaré surface of section as $z = k$, where k is a suitable constant. Imagine that the attractor is traversing in 3-D space and is cutting through the plane $z = k$ in both upward and downward directions, as shown in Figure 4.4. Further suppose that we have a means to highlight the intersecting points on the projection. (We will explain how to do it later.) If the motion is periodic, such as

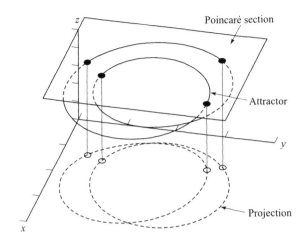

Figure 4.4 An attractor and Poincaré section.

the one shown in the figure, the projection should adequately reflect the periodicity of the motion. In this case, we see stationary points on the projection being highlighted.

It should be noted that, by definition, the Poincaré section captures only one direction of crossing so that the period, if finite, can be correctly found. In a period-2 orbit as shown in Figure 4.5(a), for example, the Poincaré section should correctly show only two crossing points instead of four. Moreover, if the motion is quasi-periodic, we should see a closed loop on the projection, as shown in Figure 4.5(b), and likewise for chaotic motion.

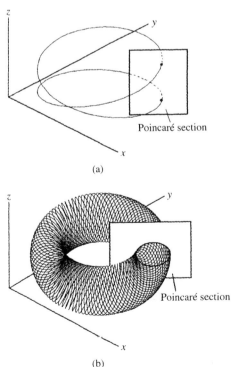

Figure 4.5 Poincaré section of (a) a period-2 orbit; and (b) a quasi-periodic orbit.

Clearly, we need a comparator circuit to determine when the attractor is hitting the plane $z = k$. This can be easily done using the circuit shown in Figure 4.6. The function of this circuit is to produce a pulse whenever the signal z is equal to the value k which is set by a potentiometer. The display of the Poincaré section is then left to the oscilloscope. The idea is to make use of the Z-axis modulation function of the oscilloscope, which momentarily brightens the trace when its Z-input receives a pulse. Thus, if the output from the circuit described above is applied to the Z-input of the oscilloscope, the trace will momentarily brighten whenever the attractor intersects the plane $z = k$. This technique was also used by Deane and Hamill [5] in their experimental study of chaos in power electronics.

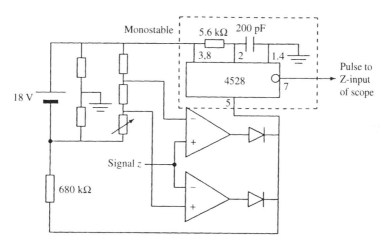

Figure 4.6 Circuit for detecting intersection of attractor and Poincaré section.

Example: Free-Running Ćuk Converter

As an example, we consider a third-order autonomous Ćuk converter. The experimental circuit is shown in Figure 4.7. This circuit operates under a *free-running current-mode control*, which is effectively a bang-bang type of control. The sum of the inductor currents, sampled by a 0.1Ω sensing resistor, is compared with a reference signal which is derived continuously from the output voltage via a feedback circuit. The comparison is actually done by a Schmitt trigger circuit, which also provides adjustment for the width of the hysteretic band. Referring to the circuit diagram of Figure 4.7, the feedback voltage gain is adjusted by R_μ and the inductor dc current level is adjusted by R_K. The $1\text{M}\Omega$ variable resistor sets the width of the hysteretic band and hence the switching frequency.

As will be shown in Chapter 5, analysis of the dynamics of this converter reveals the possibility of a Hopf bifurcation, and computer simulation consistently reveals the characteristic sequence of changes in qualitative behavior starting from fixed point, via limit cycles and quasi-periodic orbits, to chaos. Experimental study would inevitably require examining Poincaré sections since quasi-periodic and chaotic attractors can be

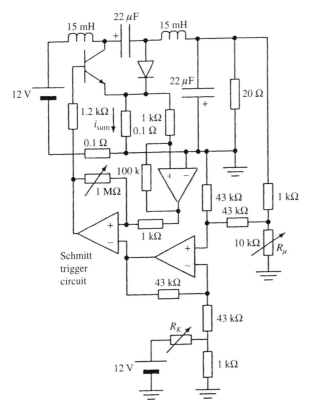

Figure 4.7 Experimental circuit of free-running autonomous Ćuk converter (refer to Chapter 5 for analysis).

distinguished only from the appearance of their Poincaré sections. Figure 4.8 shows the sequence of phase portraits starting from fixed point, through limit cycle and quasi-periodic orbit, to chaotic orbit.

4.1.6 Poincaré Sections for Nonautonomous Circuits

Power converters controlled with a fixed-frequency clock are nonautonomous. For this kind of system, Poincaré sections can be obtained in a similar manner with the Z-axis modulation set to sample at the clock frequency of the converter under study. The resulting display contains bright dots along with the attractor, and the number of bright dots indicates the period of repetition in the case of periodic and subharmonic motion. Specifically, N bright dots means that the system is attracted to a subharmonic orbit whose period is N times the switching period. A large number of irregularly and densely located points may indicate chaos.

4.1.7 Displaying Bifurcation Diagrams

Bifurcation diagrams are frequently used for identifying the way in which a system's qualitative behavior changes as some chosen parameters are varied. To display a bifurcation diagram, we need to construct a circuit which generates the necessary signals to the oscilloscope for displaying a bifurcation diagram [6]. We will begin with basic operational requirements and then discuss the details of the implementation. For brevity, we will refer to the power converter or dynamical circuit being studied as *system under test* (SUT).

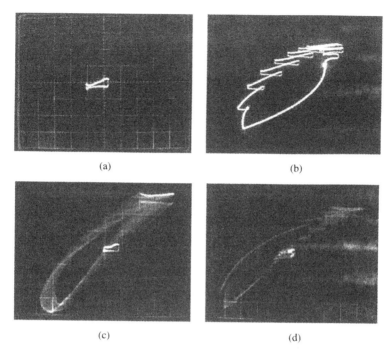

Figure 4.8 Phase portraits from autonomous Ćuk converter showing (a) fixed point;
(b) limit cycle; (c) quasi-periodic orbit; (d) chaotic orbit. The Poincaré
sections are highlighted in (b), (c) and (d). The output voltage across the
20Ω load is used as input to the Poincaré section detector circuit of
Figure 4.6.

Basic Operational Requirements

We first examine what a bifurcation diagram contains. A typical bifurcation diagram, as shown in various sections of this book, has its horizontal axis corresponding to a variation of a parameter and its vertical axis corresponding to the sampled steady-state value of a variable from the SUT. Obviously, we can make use of the X-Y mode of the oscilloscope to display a bifurcation diagram provided the necessary signals are applied to the X and Y input channels. In order to generate these signals, we need to perform two basic processes:

1. Vary a given parameter of the SUT according to a slowly swept sawtooth voltage that is applied to the X-input of the oscilloscope.

2. Sample a given signal from the SUT and send the sampled data to the Y-input of the oscilloscope.

Moreover, these two functions must be performed in a well-coordinated manner. Firstly, the sawtooth must sweep relatively slowly, and the value of the bifurcation parameter is set according to the sawtooth voltage in a stepwise manner. Then, for each value of the bifurcation parameter, the SUT is sampled to give enough data to the Y-input channel. Figure 4.9 shows the functional block diagram of the measurement system.

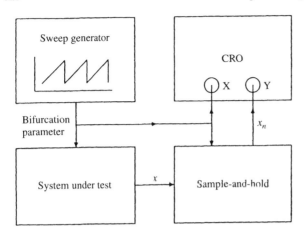

Figure 4.9 Block diagram of the system for displaying bifurcation diagrams. x denotes the variable to be sampled from the system under test (SUT). The CRO can be replaced by a computer which acquires the data from the sample-and-hold and the sawtooth generator, and plots/prints the bifurcation diagram.

Digital Implementation and Related Issues

We will consider a digital implementation of the required measurement system. The sawtooth voltage can be generated by a D/A converter which reads the output from one or more digital counters. The horizontal resolution of the bifurcation diagram is determined by the number of bits of the D/A converter. A 12-bit D/A converter, for instance, will offer 4096 steps, and hence will give 4096 points along the horizontal axis of the bifurcation diagram to be displayed on the scope. Figure 4.10 shows the block diagram of a possible implementation of the sawtooth generator. The next question is how fast we should drive the counter (i.e., how fast should the sawtooth sweep?).

The value of the sawtooth voltage controls the value of the bifurcation parameter used in the SUT. At each step of the sawtooth voltage, we have to ensure that enough

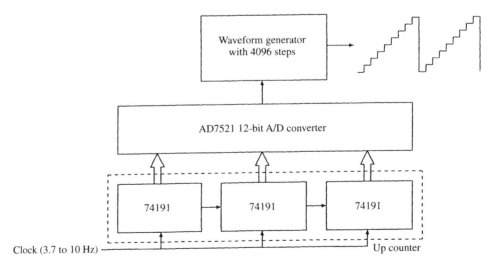

Figure 4.10 Block diagram of sawtooth generator. Output serves as voltage analog of bifurcation parameter to be sent to X-input channel of oscilloscope and the system under test (SUT).

time is given to sample enough data points from the SUT which are to be sent to the Y-input of the oscilloscope. If the sampling is done at a frequency f_s Hz, and N data points are to be displayed for each value of the bifurcation parameter, then the sawtooth must sweep as slowly as N/f_s second per step. Thus, if a 12-bit D/A converter is used, the sweep rate of the sawtooth should be lower than $f_s/(4096N)$ Hz.

Finally, the vertical resolution is controlled by the amount of sampled data displayed during each step of the slowly swept sawtooth. Usually 500 samples are adequate. This value, denoted by N above, will affect the sweep rate of the sawtooth.

Other Methods, Problems, and Practical Issues

If the bifurcation parameter is a signal variable (e.g., the reference current), the sawtooth sweep method can be used. But if one intends to study the bifurcations in response to the variation of a power variable (e.g., the input voltage) or a physical parameter (e.g., the load resistance), other methods have to be used. There is a simpler way to display a bifurcation diagram on the oscillocope. The idea is to use the Z-axis modulation to implicitly sample the required variable. This will eliminate the sample-and-hold circuit described above. If the clock pulse of the PWM in the converter is available, we may simply use it to drive the Z-input of the scope and hence eliminate the need for constructing a separate driving circuit as mentioned in the previous subsection. It is worth noting that the use of Z-axis modulation for obtaining bifurcation diagrams is simpler, but is less flexible compared to the use of an extra sample-and-hold circuit which allows the use of a computer for plotting, storing, and further manipulating the data obtained from the SUT.

Furthermore, it is possible to generate the sweeping voltage manually with a voltage supply. If we can do it steadily and slowly, we can still get a reasonably good bifurcation diagram. The capturing of the diagram can be done by a DSO, or by a camera using a long exposure time if an analog scope is used.

Finally, there is an important criterion for displaying a bifurcation diagram on the oscilloscope. The bifurcation parameter has to be a voltage or represented by a voltage. In the case where the bifurcation parameter is a current or value of a component (e.g., a resistance), we need to devise a way to make a voltage analog of the bifurcation parameter. This would vary from case to case. For instance, if the load resistance is the bifurcation parameter, we need to produce a voltage proportional to the resistance value, sweep it through a suitable range, and feed it to the X-input of the oscilloscope. A handy way to do this is to use a two-limb rheostat with a common jockey. A portion of one limb is connected as the load, and the other limb connected to a separate voltage source. The voltage across the same portion of the second limb is fed to the X-input channel of the oscilloscope. Thus, the variation (i.e., manual sweep) of the load resistance is proportional to the voltage fed to the X-input.

Example: Boost Converter Under Current-Mode Control

As an example, we consider a current-mode-controlled boost converter. The bifurcation parameter is the reference current I_{ref}, which sets the peak value of the inductor current. Figure 4.11 shows the schematic of the converter under study. Our aim is to

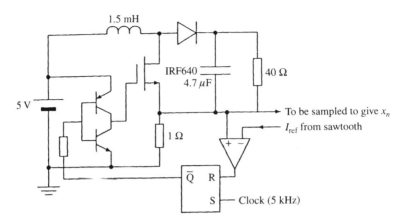

Figure 4.11 Schematic of experimental current-mode-controlled boost converter, I_{ref} being the bifurcation parameter supplied by the sawtooth generator. The RS flip-flop block consists of a pair of NOR gates.

display the bifurcation diagram, with I_{ref} as the bifurcation parameter (horizontal axis) and the inductor current as the sampled data (vertical axis).

The operation of the circuit can be described briefly as follows. A 5kHz clock periodically turns on the power switch. While the switch is on, the inductor current climbs up linearly until its value is equal to I_{ref}, which is the bifurcation parameter. When the inductor current reaches (just exceeds) I_{ref}, the comparator goes high, resetting the RS flip-flop. This turns off the power switch. Once the switch is turned off, the inductor current ramps down until the next clock pulse sets the RS flip-flop again and turns the switch back on. The clock repeats periodically at 5kHz.

The sampling is to be done at the switching frequency of the boost converter (i.e., 5kHz). The variable to be sampled is the inductor current which is picked up by the 1Ω sensing resistor. The slowly swept sawtooth effectively defines I_{ref}, and is also sent to the X-input of the oscilloscope. Five hundred samples of inductor current are displayed at each step of the bifurcation parameter.

Figure 4.12 shows a photograph of the oscilloscope display as the trace sweeps horizontally from left to right, corresponding to I_{ref} swept from 0 to about 1A. A 12-bit

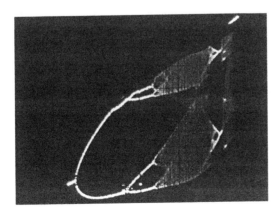

Figure 4.12 Bifurcation diagram from oscilloscope for the current-mode-controlled boost converter, inductor current being the variable (vertical axis) and peak inductor current I_{ref} being the bifurcation parameter (horizontal axis).

A/D converter is used for the sawtooth generator (i.e., a maximum of 4096 horizontal steps can be recorded). At each step 500 samples are displayed.

Note on Sampling. The inductor current in a switching converter typically exhibits a piecewise linear waveform, with ringings (fast oscillatory pulses) sandwiched between linear segments due to the presence of parasitic inductance and capacitance. When sampling the inductor current, care should be taken to avoid sampling at the ringings. We can either apply suitable filtering or deliberately delay the sampling instant. In our experimental circuit, sampling is synchronized with the turn-on instants of the power switch, but with a small delay to avoid the ringing pulses.

ACKNOWLEDGMENTS

The bifurcation measurement circuit described in this chapter was first constructed with the help of Mr. Bruce Tang, during his final-year study at Hong Kong Polytechnic University. The efforts of Dr. Y. M. Lai and Mr. Philip Li in refining the bifurcation circuits and performing measurements of Poincaré sections are gratefully acknowledged.

REFERENCES

[1] M. Hasler, Electrical circuits with chaotic behavior, *Proc. IEEE,* vol. 75, no. 8, pp. 1009–1021, August 1987.

[2] S. Prentiss, *The Complete Book of Oscilloscopes.* New York: McGraw-Hill, 1992.

[3] C. K. Tse, S. C. Fung, and M. W. Kwan, Experimental confirmation of chaos in a current-programmed Ćuk converter, *IEEE Trans. on Circ. Syst. I,* vol. 43, no. 7, pp. 605–607, July 1996.

[4] T. S. Parker and L. O. Chua, *Practical Numerical Algorithms for Chaotic Systems.* New York: Springer-Verlag, 1989.

[5] J. H. B. Deane, and D. C. Hamill, Instability, subharmonics, and chaos in power electronics systems, *IEEE Trans. on Power Electron.,* vol. 5, no. 3, pp. 260–268, 1990.

[6] C. K. Tse, and W. C. Y. Chan, Experimental verification of bifurcations in current-programmed dc/dc converters: From quasi-periodicity to period-doubling, *European Conf. Circ. Theory & Design,* Budapest, pp. 1274–1279, September 1997.

4.2 TECHNIQUES OF NUMERICAL INVESTIGATION

Soumitro Banerjee
David C. Hamill

4.2.1 Simulation of Power Electronic Circuits

As with other nonlinear systems, computer simulation has a major role in investigations of power converters. However, the characteristics of switching circuits give rise to some distinctive problems [1]. To follow a trajectory numerically, the system of ordinary differential equations (ODEs) is solved by performing an approximate integration. For the general system $dx/dt = \mathbf{f}(\mathbf{x}, t)$, $\mathbf{x}(t = 0) = \mathbf{x}_0$, the trajectory is found by repeated application of

$$\mathbf{x}(t + h) = \mathbf{x}(t) + \int_t^{t+h} \mathbf{f}(\mathbf{x}, t)dt \qquad (4.1)$$

where h is the time step for some domain $t \in [0, t_{\text{end}}]$. Equation (4.1) can also be formulated as a Taylor series:

$$\mathbf{x}(t + h) = \mathbf{x}(t) + h\mathbf{f}[\mathbf{x}(t), t] + \frac{h^2}{2!}\mathbf{f}'[\mathbf{x}(t), t] + \frac{h^3}{3!}\mathbf{f}''[\mathbf{x}(t), t] + \dots \qquad (4.2)$$

Thus numerical integration of the ODE is equivalent to summing an infinite series. Two assumptions are usually made: (1) the solution $\mathbf{x}(t)$ is smooth (of class C^∞ over the domain $[0, t_{\text{end}}]$), so all the terms of the series exist; and (2) by choosing h sufficiently small, the series may be made to converge rapidly, so a few terms are sufficient for accuracy. Both assumptions are routinely violated by power electronics circuits.

Problems Arising from Varying Topology

As exemplified by the buck converter equations presented in Chapter 1, it is clear that the ODEs of ideal power converters have discontinuous right-hand sides (i.e., $\mathbf{x}(t)$ is of class C^0: the derivative exists, but contains jump discontinuities at the switching instants). Therefore, the first assumption of numerical integration is contravened: the derivatives in (4.2) do not exist for all $t \in [0, t_{\text{end}}]$. Because $\mathbf{f}$ is undefined at the switching instants, integrating across such a discontinuity (e.g., by using a fixed step size) is likely to incur a large error, even with a small value of h.

To circumvent this difficulty, power electronics simulators can take one of two actions. Switched-circuit simulators determine the switching instant t_s accurately, then integrate up to t_s^-, apply the new value of $\mathbf{f}$ at t_s^+, set $\mathbf{x}(t_s^+) = \mathbf{x}(t_s^-)$, and integrate onward. Unfortunately, there are still problems when one switching event leads to another: for example, in the buck converter, when S opens, D immediately starts conducting. Yet S and D must never conduct simultaneously, or infinite current would flow; nor must they block simultaneously or infinite voltage would be generated. Dealing with such situations requires *a priori* knowledge of circuit operation to be incorporated in the program [2].

The alternative is to replace the ideal switches with nonideal ones. For example, PSpice [3], a commercial development of the public-domain circuit simulator SPICE, provides a switch model that has a nonzero on-resistance and a finite off-resistance, and which must transfer between on and off in nonzero time. The justification is that real switching devices behave in a similar way. The drawbacks to this approach are twofold: first, small time-constants are introduced, necessitating a stiff ODE solver (SPICE uses the trapezoidal method as standard), which, though stable, can introduce high frequency artifacts into the solutions; second, $\partial\mathbf{f}/\partial t$ is very large during switching transitions, causing very small values of h, and possible nonconvergence of the inner iterations of the implicit ODE solver (Newton-Raphson in SPICE).

Problems Arising from Incompatible Boundary Conditions

Unfortunately, matters are sometimes even worse. In certain circuits with ideal switches, $\mathbf{x}(t)$ is itself discontinuous! This can happen, for example, at the closing of a switch across a capacitance—perhaps the inherent capacitance of a switching device. If the capacitance C has an initial voltage $v \neq 0$, then an infinite current flows at the switching instant, dissipating energy $\frac{1}{2}Cv^2$. To reduce such losses, a major class of power converters is designed so the switches close only when there is no voltage across them (zero voltage switching converters). Although this desirable condition may be obtained in the steady state, it may not extend to startup and transient conditions.

Discontinuous left-hand sides can be handled by switched-circuit simulators if provision is made within the ODE solver to reset the state variables to their appropriate values: $\mathbf{x}(t_s^+) \neq \mathbf{x}(t_s^-)$. In SPICE-like simulators, a small time step must again be used to ensure accuracy during the transition. The price is that simulations take a long time if a slow transient is to be observed; run times of several hours are typical.

Computer simulation is a powerful tool for investigating nonlinear systems, but unfortunately the switched nature of power electronics causes some inherent numerical problems that cannot easily be sidestepped. Perhaps further development of the mathematics of discontinuous systems could help in this respect.

We now present three computation techniques widely used in analyzing nonlinear dynamical systems:

1. Bifurcation diagrams
2. Basins of attraction
3. The maximal Lyapunov exponent

More detailed accounts of these and many other numerical techniques are available in [4,5].

4.2.2 Obtaining Bifurcation Diagrams

Bifurcation diagrams are the basic tool of studying the change in system behavior in response to the variation of system parameters. Generally one of the system parameters is varied while the others are kept fixed. Multiple-parameter bifurcation diagrams are possible, but that requires 3-D or color graphics. Presently we will confine our discussion to one-parameter bifurcation diagrams, which will be used extensively in this book.

Bifurcation diagrams use discrete models of systems. If the system is autonomous, the discrete model is obtained by the method of Poincaré surface of section. If the system is nonautonomous—like power electronics systems having clock input—then the discrete model is obtained by observing the state variables at every clock instant. One can also use the *impact map*, where observations are made at every switch-on instant of a controllable switch. Such techniques have been discussed in detail in Chapter 2.

One has to choose *one* of the state variables for the purpose of plotting bifurcation diagrams. This choice is arbitrary and one generally chooses that state variable which makes the bifurcation phenomena more apparent.

The parameter is incremented in steps and is plotted in the x-coordinate. The y-coordinate should show the asymptotic behavior of the sampled variable for each parameter value. To obtain this, one starts the iteration of the discrete map from an initial condition for the first parameter value. The first few (about 1000) iterates (generally called *preiterates*) are not plotted in order to eliminate the transient, and then subsequent iterates are plotted. It generally suffices to plot 100–500 points, depending on the desired density of the diagram. If the system is period-1 for a parameter value, all the points will fall at the same location—thus showing just one point. If the system is period-2 for another parameter value, there will be two points on the y-coordinate for that value of the x-coordinate. Likewise, if the system is chaotic then the points will fall at different locations on the vertical line corresponding to that parameter value.

After the points corresponding to a parameter value are plotted, the parameter is incremented to the next step and the final state for the last parameter value is taken as the initial condition for the current parameter value. Again some 100–500 points are plotted for this parameter value. This procedure is repeated until the end of the parameter range to complete the bifurcation diagram.

If there are multiple attractors for some ranges of the parameter values, and if one intends to observe the evolution of all these attractors, one has to take a slightly different procedure. In that case, for each parameter value, a *set* of initial conditions is taken so that at least one initial condition falls in the basin of attraction of each of the attractors. For each initial condition, a good number of preiterates are eliminated and some 10–50 points are plotted. Then the next initial condition is taken and the procedure is repeated. When the points corresponding to all the initial conditions are plotted, the parameter value is incremented to the next step. This procedure takes much more computation time, which scales as the number of initial conditions used for each parameter value.

In order to place the initial conditions, one has to find out the range of values of the state variables. Since the attractors must remain within this range of state variables, the initial conditions can be placed within a rectangle formed by the minimum and maximum values of the state variables. Experience has shown that instead of placing the initial conditions evenly over this area, it is a good idea to place the initial conditions on one of the diagonals of the rectangle. That way one can use a relatively smaller number of initial conditions, reducing the computation time.

It is difficult to specify the number of preiterates required to be eliminated at each computation step. Sometimes there are unstable chaotic orbits in a system and if an initial condition falls in such an unstable chaotic orbit, it may take a large number of iterates before it settles into the asymptotic stable orbit. If an insufficient number of preiterates are eliminated, the behavior for that parameter value may appear to be chaotic (points scattered along a vertical line for that parameter value). It has been found that such chaotic transients may sometimes last for tens of thousands of iterates. When the occurrence of such unstable chaotic orbits is anticipated in a system, it is a good idea to plot the bifurcation diagram a few times, using different numbers of preiterates. One should take that number beyond which any increase of the number of preiterates does not make any difference in the resulting bifurcation diagram.

4.2.3 Plotting Basins of Attraction in Systems with Multiple Attractors

When there are multiple attractors in a system, each point in the state space is in the basin of attraction of one of the attractors. If the system is two-dimensional, one can draw such basins of attraction on the computer screen and print on paper.

To obtain such a plot, one starts with a rectangular area with the x-coordinate representing one state variable and the y-coordinate representing the other state variable. The sides of this rectangular area are adjusted to accommodate the range of state variables containing the attractors and related areas of interest.

One then divides this area into boxes (say 100×100). If a high resolution is desired, each box may be one pixel in size. Then iterations of the map are started with initial conditions placed in the midpoint of the boxes.

The basin of each attractor is assigned a specific color. Suppose there is a system with two attractors (one of them may be at infinity) and their basins are to be colored red and blue respectively. Starting from an initial condition, if the state goes to the red attractor, the color of that box is changed to red. Otherwise, its color is changed to blue. This way the whole range of the state space is scanned and each point is assigned a color—red or blue. It is not really necessary to do the computation with each box as initial condition because if an orbit converges to the red attractor, all the points (boxes) in its path are also in the basin of the red attractor.

Deciding whether an attractor has been reached is relatively simple for periodic attractors—one only has to check if the same state is repeated after some iterates. In the case of chaotic attractors, the same state never repeats, but the attractor remains confined to a finite number of boxes. This has to be checked over a preassigned number of iterates. In the case of an attractor at infinity, one cannot rigorously show that the state is really going toward infinity. In that case, the box is assigned a color corresponding to the attractor at infinity when the state goes out of a given area.

4.2.4 Computing the Maximal Lyapunov Exponent

Lyapunov exponents can be calculated both from the system model—continuous or discrete—and from a time series. The time series can be obtained either from simulation or from experiment. Here we discuss a basic algorithm to compute the largest or *maximal* Lyapunov exponent [6]. For specific systems one may have to make small changes to suit the characteristics of the system.

We assume that the time series of all the state variables of a system are available for a sufficiently long period of time. This is often possible in experiment using multi-channel data acquisition systems. In case the time series of only one state variable $x(t)$ is available, one can still reconstruct the m-dimensional state space using the delay coordinate method.

In this approach, a point on the attractor is constructed as $\{x(t), x(t + \tau), ... x(t + [m - 1]\tau)\}$, where τ is a chosen delay time. A convention is to pick τ to be roughly one-third the number of points in the mean period of motion. In discrete-time systems τ would be unity. The dimension m of the reconstructed space should not be so small that the reconstruction is topologically incorrect. In case of power electronics systems, however, the actual dimensionality of the state space is known and the choice of m is not guesswork.

We take a point $y(t_o)$ on the attractor at time t_o (Figure 4.13). The evolution of the state starting from this initial condition is calculated from the model. This is called the *fiducial trajectory*. In the case of a time series, the fiducial trajectory is the time series itself starting from $y(t_o)$.

We then take another initial condition $z_o(t_o)$ at a small distance from $y(t_o)$. In the case of a time series, one has to look for another point z in the attractor within a neighborhood radius of ϵ. We then assume that both y and z are states of the system

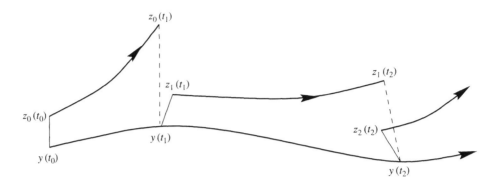

Figure 4.13 Schematic representation of the algorithm to compute λ_1, the maximal Lyapunov exponent.

at instant t_o. We want to observe how the trajectories with these two points as initial conditions evolve.

The two initial conditions are evolved by the model to obtain the trajectories. For a time series, their paths of evolution are available from the time series. The trajectories would diverge for a chaotic system. We observe the divergence for a time span Δt. The rate of divergence at this portion of the state space can be calculated from this data.

The divergence is, however, accompanied by folding. If we rely on Euclidean distance between two points, the apparent divergence cannot continue forever. It would thus be necessary to keep aligned along the direction of stretching. This is achieved by renormalizing the separation at regular intervals. If at any step t_1 the separation L'_o between them exceeds some chosen value ϵ', z is changed to some other point closer to y, but along the same direction. The evolved point $y(t_1)$ is retained, and a new neighbor $z_1(t_1)$ is sought such that the distance $L_1 = \|y(t_1) - z_1(t_1)\|$ is less than ϵ and such that $z_1(t_1)$ lies as close as possible to $y(t_1)$ in the same direction from $y(t_1)$ to $z_o(t_1)$. This is trivial if the model is used. But if only a time series is available it may be impossible to locate a point from the time series exactly in this direction. It is therefore necessary to specify an angle (say 30°) within which the point z_1 should be located.

We then observe the divergence of the trajectories y and z_1. The procedure continues until the fiducial trajectory y has been followed to the end of the time series. The largest Lyapunov exponent of the attractor is estimated as

$$\lambda_1 = \frac{1}{N \Delta t} \sum_{i=0}^{M-1} \ln \frac{L'_i}{L_i} \qquad (4.3)$$

where M is the number of replacement steps and N is the total number of time steps.

This algorithm is relatively insensitive to reasonable changes in the search radius ϵ and the evolution time step Δt. Its data requirements are also modest; only a few thousand attractor points are needed to estimate λ_1 to within 10% of the true value.

Though this method is quite standard for systems defined by continuous differential equations, it has been found that it gives erroneous results for power electronics systems where two (or more) *sets* of differential equations define the dynamics during various switching phases. In such situations there are two possible remedies. The first is to use a sampled data model instead of differential equations. If the discretized model yields a continuous map (which is true in many cases, see Section 3.3), the method yields an acceptable result. The other is to use the continuous-time model, taking certain corrective steps. This method has been recently developed and is available in [7].

REFERENCES

[1] D. C. Hamill, Time-domain simulation of resonant and other dc-dc converters, in *IEEE Workshop on Computers in Power Electronics* (Lewisburg, PA), pp. 93–107, August 1990.

[2] D. Skowronn, D. Li, and R. Tymerski, Simulation of networks with ideal switches, *Int. J. Electronics*, vol. 77, no. 5, pp. 715–730, 1994.

[3] P. W. Tuinenga, *SPICE: A Guide to Circuit Simulation and Analysis Using PSpice*. Englewood Cliffs, NJ: Prentice-Hall, 1988.

[4] H. E. Nusse and J. A. Yorke, *Dynamics: Numerical Explorations*. New York: Springer-Verlag, 1997.

[5] T. S. Parker and L. O. Chua, *Practical Numerical Algorithms for Chaotic Systems*. New York: Springer-Verlag, 1989.

[6] M. Wolf, J. B. Swift, H. I. Swinney, and A. Vastano, Determining Lyapunov exponents from time series, *Physica D*, vol. 16, no. 3, pp. 285–317, 1985.

[7] Y. H. Lim and D. C. Hamill, Problems of computing Lyapunov exponents in power electronics, *Proc. 1999 IEEE Int. Symp. on Circuits and Systems*, vol.5, pp. 297-301, May 30–June 2, 1999, Orlando, FL.

4.3 COMPUTATION OF AVERAGES UNDER CHAOS

José Luis Rodríguez Marrero
George C. Verghese
Roberto Santos Bueno
Steven H. Isabelle

4.3.1 Introduction

In dc/dc converters under current-mode control, the controller specifies a peak inductor current in each cycle, rather than the duty ratio. For constant-frequency operation, a switch is turned on every T seconds but is turned off when the inductor current reaches a specified reference level, I_{ref}. This reference level is now the primary control variable; the duty ratio D becomes an indirectly controlled auxiliary variable. Steady-state operation with period T and with $D > 0.5$ (approximately) is impossible when I_{ref} is held constant, because this periodic solution is unstable [1]. The waveforms for $D > 0.5$ under the condition that I_{ref} is constant assume complicated forms, corresponding either to periodic operation at some multiple of T (subharmonic operation) or to chaotic variation from cycle to cycle; see for instance [2,3,4,5]. A stabilizing ramp is normally introduced in order to prevent these instabilities and extend the range for stable periodic operation beyond $D > 0.5$.

The spectral modifications associated with chaotic operation of dc/dc converters provide an important motivation for actual operation in this regime [6,7]. Conventional dc/dc converters generate electromagnetic interference (EMI) consisting of the switching frequency and many harmonics. Such interference gives rise to important electromagnetic compatibility (EMC) problems, which can be reduced by filtering (for conducted interference) and screening (for radiated interference). Other strategies for interference reduction are spectral shaping based on randomized modulation [8], and chaotic operation, which flattens the switching spectrum (at the expense of a corresponding broadening). Exploiting the random nature of the chaotic operation of dc/dc converters could provide a way of obtaining a prescribed power spectrum [9].

In this section we show how to compute the input-output gain, and various other averages of interest, for the chaotic regime of dc/dc converters under current-mode control and in continuous-conduction mode. Our approach involves recognizing and exploiting the *ergodicity* of the sampled inductor current in a simplified first-order model of the converter. What this means is that the evolution of the inductor current samples is governed by a unique "probability" density. Time averages (whose direct determination would require tedious, costly, and possibly unreliable time-domain simulations) can now be replaced by ensemble averages computed with respect to this density. We demonstrate that very good results are obtained even if the density is approximated. In contrast, traditional computations based on the nominal (period-T, unstable) periodic solution can be considerably in error in the chaotic regime. Our treatment provides convenient analytical expressions to support design for operation in the chaotic regime, and thereby enables more serious evaluation of the potential advantages of chaotic operation. Results are presented for the boost, buck-boost, buck, and Ćuk converters.

4.3.2 Chaotic Operation of DC/DC Converters Under Current-Mode Control

Describing Chaotic Behavior via Densities

Figure 4.14 shows a typical segment of the inductor current i_L of a dc/dc converter under current-mode control, in the chaotic regime. It is assumed that the converters analyzed in this section are operating in continuous-conduction mode. The switch is controlled by clock pulses that are spaced T seconds apart. When the switch is closed,

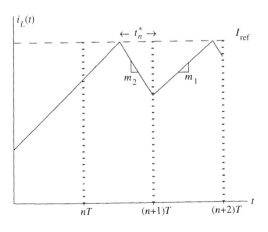

Figure 4.14 Typical segment of inductor current for chaotic regime, $\alpha > 1$.

the inductor current increases until it reaches the specified reference value, I_{ref}, at which point the switch opens. Any clock pulse that arrives while the switch is closed is ignored. Once the switch has opened, the next clock pulse causes it to close again. Under the assumption that the inductor current is essentially piecewise linear, the dynamics of the controlled current is described by the following map:

$$i_{n+1} = \begin{cases} i_n + m_1 T & \text{if} \quad i_n \le I_{\text{ref}} - m_1 T \\ I_{\text{ref}} - m_2 t_n^* & \text{if} \quad i_n > I_{\text{ref}} - m_1 T \end{cases} \tag{4.4}$$

where $i_n = i_L(nT)$ is the value of the inductor current at the clock instant nT; m_1 and m_2 are respectively the magnitudes of the slopes on the increasing and decreasing segments of i_L; and the time t_n^* is the duration of the off-time in the cycle between nT and $nT + T$, if the switch turns off in this cycle, and satisfies the following equation:

$$i_n + m_1(T - t_n^*) = I_{\text{ref}} \tag{4.5}$$

Under steady-state operation in periodic or chaotic mode, with a constant input voltage V_{in} and a low-ripple output voltage of constant average value V_{out}, m_1 and m_2 in (4.4) will be constants that can be expressed in terms of V_{in}, V_{out} and the circuit parameters. We restrict our analysis to such quasi-steady-state operation.

Since the values of i_n satisfy the constraint $I_{\text{ref}} - m_2 T \le i_n \le I_{\text{ref}}$, it is possible to *normalize* the inductor current according to the linear relationship

$$z(t) = a i_L(t) + b \tag{4.6}$$

so that $0 \le z(t) \le 1$. Choosing

$$a = \frac{1}{m_2 T} \qquad b = 1 - \frac{I_{\text{ref}}}{m_2 T} \tag{4.7}$$

the map for the *normalized current samples* $z_n = z(nT)$ becomes

$$z_{n+1} = \begin{cases} z_n + \frac{1}{\alpha} & \text{if} \quad 0 \le z_n \le 1 - \frac{1}{\alpha} \\ \alpha(1 - z_n) & \text{if} \quad 1 - \frac{1}{\alpha} < z_n \le 1 \end{cases} \tag{4.8}$$

where α is the ratio of the slope magnitudes:

$$\alpha = \frac{m_2}{m_1}$$

This map is plotted in Figure 4.15 for the case $\alpha > 1$. Since this is a 1-D map, its behavior can be analyzed using the method introduced in Section 3.1. The fixed point is the intersection of the graph of the map with the 45° line, and the stability of the fixed point is given by the slope of the map at that point. It is easy to see that if the magnitude of the slope (equal to α) is less than unity, the fixed point is stable—implying a regular periodic behavior. If $\alpha > 1$, then the fixed point is unstable, and no high-periodic orbit can be stable. For $\alpha > 1$, or $D > 0.5$, the equation has no stable equilibrium point or stable periodic solution, so stable operation of the converter with period T or any multiple of T is not possible (at least to the extent that the simplified model in (4.4) actually describes the circuit behavior). These facts are well known in the power electronics literature, see [1].

A qualitative, empirical analysis of the dynamics of the map (4.8) can be done as follows. Pick an arbitrary initial state $z_0 \in [0, 1]$. Iterate equation (4.8) to obtain a sequence of states, $z_1, z_2, z_3, \ldots$, called the *trajectory* or *orbit* emanating from or passing

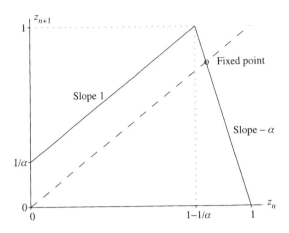

Figure 4.15 Map for (4.8).

through z_0. Plotting these points (omitting the first few if the transient is not of interest) for a range of α values results in an empirical bifurcation diagram of the type shown in Figure 4.16.

The figure shows that for $\alpha < 1$ the trajectory converges to a fixed state that is independent of the chosen initial state. This is consistent with the periodic behavior of the converter in this range of α: the inductor current samples at the switching times are always the same in the periodic steady state. However, for $\alpha > 1$ and for most initial states, the trajectory does not converge to anything regular; it is erratic or chaotic. Moreover, the trajectory is significantly altered by a slight change in the initial state. It can be shown that all states are visited upon iteration for $\alpha \geq \alpha_0$, where $\alpha_0 = (1 + \sqrt{5})/2 \approx 1.618$ is the golden mean.

Although any particular trajectory for $\alpha > 1$ is significantly altered by even a slight change in the initial state z_0, the general appearance of the bifurcation diagram in Figure 4.16 is hardly affected by picking different initial states. In other words, the distribution of points on the trajectory seems not to be much affected by the initial state. This fact suggests that the *distribution* of points visited on a trajectory may

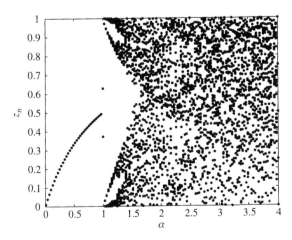

Figure 4.16 Bifurcation diagram for (4.8).

constitute a more useful representation of the behavior of the system than knowledge of the specific points. This is the idea that is elaborated on and exploited in the remainder of this section.

Suppose, for a given α, we pick the initial state z_0 *randomly* according to an arbitrary probability density function $f_\alpha^{(0)}(z)$. Then the next state z_1 will be random too, with a probability density function $f_\alpha^{(1)}(z)$, and similarly at future time steps. It is possible to track the evolution of probability densities from the initial density forward, using what is termed the *Frobenius-Perron operator* associated with the map (4.8); this operator is described in more detail in the appendix to this section. It turns out (see [10] and [11]) that the properties of the map (4.8) ensure the densities will converge to a unique *invariant density* $f_\alpha(z)$. The reason for the label *invariant* is that if z_0 is distributed or picked according to this density, then all future z_n are also distributed according to this density.

The key result for us is a celebrated theorem of Birkhoff—his *ergodic* theorem—which shows that the invariant density can be used to replace time averages of functions of z_n by ensemble (i.e., probabilistic) averages of these functions, computed with respect to the invariant density [12]. The quantities of typical interest in power electronics are indeed time averages, and what the ergodic theorem allows us to do is evaluate these averages more directly once the invariant density is known.

Numerical evaluation of the invariant density can certainly be carried out, starting from some arbitrary initial density (e.g., a uniform density) and then tracking the evolution of the density through its convergence to the invariant density. However, it turns out that for certain special values of α the map (4.8) is what is known as a *Markov map*, for which a more refined computation of the invariant density is possible—through determination of the dominant eigenvector associated with a particular matrix that is simply derived from the map [10]; an example is provided in the appendix.

The map (4.8) is Markov for *integer* $\alpha \geq 2$. The corresponding density turns out to be a piecewise-constant function [13] given by:

$$f_\alpha(z) = \frac{2k}{1+\alpha} \quad \text{for} \quad \frac{k-1}{\alpha} \leq z < \frac{k}{\alpha} \qquad (4.9)$$

where $k = 1, 2, \ldots \alpha$. The form of this density is consistent with the visual appearance of the bifurcation diagram for sections taken at $\alpha = 2, 3$, and 4: there are noticeable breaks in the shading at precisely the expected points, and the density—indicated by the intensity of the shading—proceeds steadily in piecewise-constant steps, from low density in the smallest band of z_n values to high density in the largest band of z_n values.

Another family of values of α for which the map (4.8) is Markov is the set of solutions of

$$\alpha^2 - R\alpha - N = 0 \qquad (4.10)$$

where $R = 1, 2, \ldots$ and $N = 1, 2, \ldots R$. For the special case defined by $N = R$, i.e., for roots of the equation

$$\alpha^2 - R\alpha - R = 0 \qquad (4.11)$$

explicit expressions are possible [13]. For this set of values of α we divide the unit interval into $R + 1$ subintervals:

$$I_2 = \left[0, \frac{1}{\alpha}\right)$$

$$I_2 = \left[\frac{1}{\alpha}, \frac{2}{\alpha}\right)$$

$$I_{R-1} = \left[\frac{R-2}{\alpha}, \frac{R-1}{\alpha}\right)$$

$$I_R = \left[\frac{R-1}{\alpha}, \frac{R}{\alpha}\right)$$

$$I_{R+1} = \left[\frac{R}{\alpha}, 1\right]$$

The invariant density $f_\alpha(z)$ is then the piecewise-constant function given by:

$$f_\alpha(z) = \begin{cases} \dfrac{2\alpha k}{(R+3)R} & \text{for } z \in I_k \quad k = 1, 2, \ldots R \\[2ex] \dfrac{2\alpha^2}{(R+3)R} & \text{for } z \in I_{R+1} \end{cases} \tag{4.12}$$

For example, the density for $\alpha = \alpha_0$ (the golden mean) is obtained from (4.12) with $R = 1$, and is given by

$$f_{\alpha_0}(z) = \begin{cases} \dfrac{\alpha_0}{2} & \text{for} \quad 0 \le z < 1/\alpha_0 \\[2ex] \dfrac{(\alpha_0)^2}{2} & \text{for} \quad 1/\alpha_0 \le z \le 1 \end{cases} \tag{4.13}$$

Again, the appearance of the bifurcation diagram in a section taken at $\alpha = \alpha_0 \approx 1.618$ is entirely consistent with this computed density.

Calculation of the Time-Average of the Inductor Current

In this section we will use the results of the previous section to calculate the time-average $\langle i_L \rangle$ of the inductor current i_L in the chaotic regime, see Figure 4.14. The result will then be used in the next section to characterize the operation of various dc/dc converter circuits under current-mode control.

In order to calculate $\langle i_L \rangle$, we define a new discrete-time variable ξ_n to be the time-average of $z(t)$ over the nth switching period, which is the one that commences at nT and ends T seconds later:

$$\xi_n = \frac{1}{T} \int_{nT}^{(n+1)T} z(t)dt \tag{4.14}$$

Computing the integral using (4.6), we obtain:

$$\xi_n = \begin{cases} z_n + 1/2\alpha & \text{for} \quad 0 \le z_n \le 1 - \frac{1}{\alpha} \\[2ex] \left[1/2 - \alpha(z_n^2 - 1)/2 - \alpha^2(z_n - 1)^2/2\right] & \text{for} \quad 1 - \frac{1}{\alpha} < z_n \le 1 \end{cases} \tag{4.15}$$

Thus the one-cycle time average ξ_n depends on $z_n = z(nT)$, which is the normalized inductor current sample at the switching instant that initiates the cycle. We can make this dependence explicit in our notation by writing the one-cycle average as $\xi(z)$, the specific functional form of which is displayed in (4.15).

Note now from the definition (4.14) that the discrete-time average of ξ_n equals the continuous-time average of $z(t)$. The discrete-time average can be computed via ensemble averaging, invoking Birkhoff's theorem and our knowledge of the invariant density of z_n. The continuous-time average of $z(t)$ and thereby of $i_L(t)$ can then be determined. This calculation is developed next.

By Birkhoff's theorem, the discrete-time average $\langle \xi \rangle$ of ξ_n can be computed by ensemble averaging as follows:

$$\langle \xi \rangle = \int_0^1 \xi(z) f_\alpha(z) dz \tag{4.16}$$

In the case of integer α ($\alpha \geq 2$), the density $f_\alpha(z)$ is piecewise constant, so the required calculation is easy. For this case, evaluation of the above integral using (4.9) and (4.15) gives

$$\langle \xi \rangle = \frac{2}{3} \qquad \text{for } \alpha = 2, 3, \ldots \tag{4.17}$$

For arbitrary values of α, it is not possible in general to evaluate (4.16) other than numerically. However, it can be shown [13] that for $\alpha > 1$ the following bounds apply:

$$\frac{1}{2} \leq \langle \xi \rangle \leq \frac{3}{4} \leq \frac{2\alpha + 1}{2\alpha + 2} \tag{4.18}$$

The larger upper bound corresponds to periodic (although unstable) operation of the converter.

The time-average inductor current $\langle i_L \rangle$ can now be obtained from (4.6):

$$\langle i_L \rangle = I_{\text{ref}} - m_2 T (1 - \langle \xi \rangle) \tag{4.19}$$

This expression is valid for all the converters that will be analyzed in the next section.

Analysis of DC/DC Converters

In this section we analyze the boost, buck-boost, buck, and Ćuk converters operating under current-mode control in the chaotic regime. We use power balance arguments to establish a relationship between $\langle i_L \rangle$ and V_{out}.

Boost Converter. The boost converter analyzed in this section is shown in Figure 4.17. Under the assumptions stated earlier, the inductor current waveform has the form suggested in Figure 4.14, and the dynamics of the converter are described by the map in equation (4.8), with α given by

$$\alpha = \frac{(V_{\text{out}} - V_{\text{in}})/L}{V_{\text{in}}/L} = \frac{V_{\text{out}} - V_{\text{in}}}{V_{\text{in}}} \tag{4.20}$$

Using the fact that (for our idealized lossless converter model) the time-averaged power input of the converter must equal the time-averaged power output, we find that

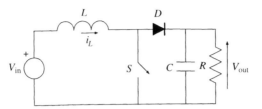

Figure 4.17 Boost converter circuit.

$$V_{in}\langle i_L\rangle = \frac{V_{out}^2}{R} \tag{4.21}$$

Substitution of $\langle i_L\rangle$ obtained from this equation into (4.19) gives

$$I_{ref} = \frac{(1+\alpha)^2 V_{in}}{R} + \frac{\alpha V_{in} T}{L}(1 - \langle\xi\rangle) \tag{4.22}$$

For α integer, substitution of (4.17) in (4.22) gives the relation

$$I_{ref,int} = \frac{(1+\alpha)^2 V_{in}}{R} + \frac{\alpha V_{in} T}{3L} \tag{4.23}$$

For most noninteger values of α, the determination of $f_\alpha(z)$ for the evaluation of (4.16) and (4.22) becomes considerably more complicated. However, it turns out that using the constraint in (4.23) as an approximation for all $\alpha > 1$ yields good results. For instance, when $\alpha = \alpha_0$ (the golden mean), we can easily evaluate (4.16) using (4.13) to obtain

$$\langle\xi\rangle = \frac{2\alpha_0 + 1}{4\alpha_0} \tag{4.24}$$

Inserting this result in (4.22) gives the relation

$$I_{ref} = \frac{(1+\alpha_0)^2 V_{in}}{R} + \frac{(2\alpha_0 - 1)V_{in} T}{4L} \tag{4.25}$$

which is in good agreement with the approximate solution (4.23) for $\alpha = \alpha_0$, because $(\alpha_0/3) \approx (2\alpha_0 - 1)/4$.

Using (4.18), it can be established for general $\alpha > 1$ that

$$I_{ref,per} \le I_o + \frac{\alpha V_{in} T}{4L} \le I_{ref} \le I_o + \frac{\alpha V_{in} T}{2L} \tag{4.26}$$

where $I_o = (1+\alpha)^2 V_{in}/R$ and

$$I_{ref,per} = \frac{(1+\alpha)^2 V_{in}}{R} + \frac{\alpha V_{in} T}{(1+\alpha)2L} \tag{4.27}$$

is the value of I_{ref} that would be needed to obtain a given V_{out} with the (unstable) solution of period T.

To compute V_{out} for a given I_{ref}, we first determine α from (4.23) and then substitute in (4.20) to get V_{out}. Figure 4.18 shows a plot (middle solid line) of V_{out} in volts as a function of I_{ref} in amps, computed in this way for a particular example. The circuit parameters are $V_{in} = 10$ volts, $R = 20\Omega$, $T = 100\mu s$, $L = 1mH$, and $C = 200\mu F$. We also show a plot (dotted line) of the average output voltage obtained from simulations

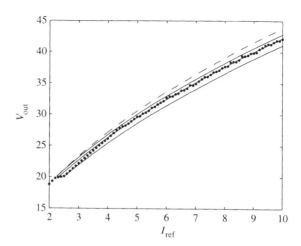

Figure 4.18 V_{out} in volts as a function of I_{ref} in amps, for the boost converter parameters given in the text.

of the circuit using an accurate second-order sampled-data model implemented in Simulink (MathWorks Inc.). For the most part, the match is very good. The major discrepancy occurs for $V_{out} = 20$ volts. Since $V_{in} = 10$ volts, this point corresponds to $\alpha = 1$, which is the onset of instability. For values of $\alpha > 1.4$, our approximate results are in excellent agreement with those obtained through the simulation. Notice that the results from the simulation are relatively far from those that would be obtained (dashed line) using the expression (4.27) associated with the periodic solution, except for $V_{out} \leq 20$ volts, where the circuit does indeed have a stable periodic solution. The top and bottom solid lines in the figure correspond to the bounds in (4.26).

For a more detailed look at representative results from Figure 4.18, we present in Table 4.1 a listing of V_{out} for selected values of I_{ref} (in amps). The second column is the value of α computed from (4.23), and the third column gives the value of V_{out} (in volts) obtained by using this value of α in (4.20). The fourth column—labeled $V_{out,sim}$—shows the average output voltage obtained from simulations of the circuit using the second-order sampled-data model implemented in Simulink (essentially identical results are obtained in Spice simulations of the second-order continuous-time circuit as well). Comparing the third and fourth columns, it is evident that our approximate analysis performs well.

The last column in Table 4.1, labeled $I_{ref,per}$, shows what I_{ref} would be needed in order to obtain the indicated value of α (and hence V_{out}) if the inductor current waveform were *periodic* with period T. Although the periodic solution is *unstable* for the range of α's shown in Table 4.1, our intent is to see what sorts of results would be obtained if calculations that are routinely done for $D < 0.5$ are blindly extended to

TABLE 4.1 Boost converter comparisons.

I_{ref}	α, (4.23)	V_{out},(4.20)	$V_{out,sim}$	$I_{ref,per}$
4.0	1.629	26.29	26.2	3.766
5.0	1.95	29.5	29.6	4.682
6.0	2.24	32.41	32.5	5.594
7.0	2.51	35.11	35.1	6.518

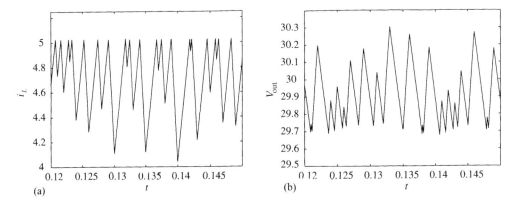

Figure 4.19 Waveforms of (a) inductor current and (b) output voltage, obtained by simulation.

$D > 0.5$. It is evident from Table 4.1 that $I_{\text{ref,per}}$ is a poor approximation to the true I_{ref}. This fact indicates the need for a direct analysis (even if approximate) for the chaotic regime, and provides some justification for our efforts.

Typical waveforms from the Simulink simulation are shown in Figure 4.19. Note that, despite the erratic—chaotic—appearance of the waveforms, the converter is in continuous conduction and the output voltage is essentially constant, with only a small ripple.

Buck-Boost Converter. The buck-boost converter is shown in Figure 4.20 (note the reference polarity we have chosen for V_{out}). The inductor current waveform again has the appearance of Figure 4.14, and the dynamics of the converter are still described by the map in equation (4.8), but with α now given by

$$\alpha = \frac{m_2}{m_1} = \frac{V_{\text{out}}/L}{V_{\text{in}}/L} = \frac{V_{\text{out}}}{V_{\text{in}}} \tag{4.28}$$

The time-averaged power input of the converter must equal the time-averaged power output, and the time-averaged current through the capacitor must be zero, so

$$V_{\text{in}}\left(\langle i_L \rangle - \frac{V_{\text{out}}}{R} \right) = \frac{V_{\text{out}}^2}{R} \tag{4.29}$$

from which

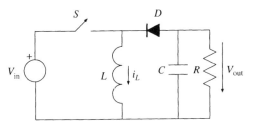

Figure 4.20 Buck-boost converter circuit.

$$\langle i_L \rangle = (1 + \alpha) \frac{V_{\text{out}}}{R} \tag{4.30}$$

Substituting (4.17), (4.19) and (4.28) into this relation, we obtain the following relation in the case of integer $\alpha > 1$:

$$I_{\text{ref,int}} = \frac{\alpha(1 + \alpha)V_{\text{in}}}{R} + \frac{\alpha V_{\text{in}} T}{3L} \tag{4.31}$$

We shall use the result in (4.31) as an approximation for all $\alpha > 1$. Given I_{ref}, we can compute α from (4.31) and V_{out} from (4.28). As a comparison, note that an exact solution of this problem for $\alpha = \alpha_0$, where the density is given by (4.13), yields the relation

$$I_{\text{ref}} = \frac{(1 + 2\alpha_0)V_{\text{in}}}{R} + \frac{(2\alpha_0 - 1)V_{\text{in}} T}{4L} \tag{4.32}$$

This relation is in very good agreement with the approximate constraint (4.31) for $\alpha = \alpha_0$.

Figure 4.21 shows a plot (solid line) of V_{out} in volts as a function of I_{ref} in amps, computed from (4.31). The parameters of the buck-boost circuit used for this example are $V_{\text{in}} = 10$ volts, $R = 20\Omega$, $L = 1\text{mH}$, $C = 200\mu\text{F}$, and $T = 100\mu\text{s}$. Figure 4.21 also shows a plot (dotted line) of the average output voltage obtained from simulations of the circuit using Simulink. Our approximate results are in excellent agreement with those obtained through the simulation of the circuit. The dashed line in the figure plots the relationship between V_{out} and I_{ref} that would be obtained using the expressions that apply for the periodic solution; it is evident that our approximate analysis of the chaotic case does much better.

Buck Converter. The buck converter circuit is shown in Figure 4.22. The inductor current waveform again has the appearance in Figure 4.14, and the dynamics of the converter are still described by the map in equation (4.8), but with α now given by

$$\alpha = \frac{m_2}{m_1} = \frac{V_{\text{out}}/L}{(V_{\text{in}} - V_{\text{out}})/L} = \frac{V_{\text{out}}}{V_{\text{in}} - V_{\text{out}}} \tag{4.33}$$

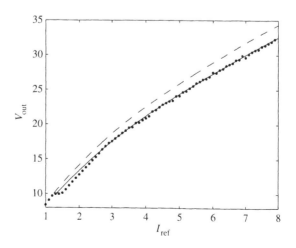

Figure 4.21 V_{out} in volts as a function of I_{ref} in amps, for the buck-boost converter parameters given in the text.

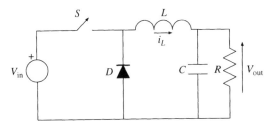

Figure 4.22 Buck converter circuit.

Noting that the average capacitor current must be zero, we can directly write

$$V_{\text{out}} = \langle i_L \rangle R \tag{4.34}$$

Substituting (4.17), (4.19) and (4.33) in this, we obtain the following relation in the case of integer $\alpha > 1$:

$$I_{\text{ref,int}} = \frac{\alpha V_{\text{in}}}{(1+\alpha)R} + \frac{\alpha V_{\text{in}} T}{(1+\alpha)3L} \tag{4.35}$$

The relation again turns out to be a very good approximation for noninteger α as well. For instance, an exact solution of this problem for $\alpha = \alpha_0$ using the density in (4.13) yields

$$I_{\text{ref}} = \frac{V_{\text{in}}}{\alpha_0 R} + \frac{(3-\alpha_0)V_{\text{in}} T}{\alpha_0 4L} \tag{4.36}$$

which is in good agreement with the approximate solution (4.35) for $\alpha = \alpha_0$.

Figure 4.23 shows a plot (solid line) of V_{out} in volts as a function of I_{ref} in amps, computed from (4.35). The parameters of the buck circuit used for this example are $V_{\text{in}} = 20$ volts, $R = 10\Omega$, $L = 10$mH, $C = 200\mu$F, and $T = 100\mu$s. The figure also shows (dotted line) the average output voltage obtained from simulations of the circuit using Simulink. Our approximate results are once more in very good agreement with those obtained through the simulation of the circuit. The dashed line in the figure plots the relationship between V_{out} and I_{ref} that would be obtained using the expressions that

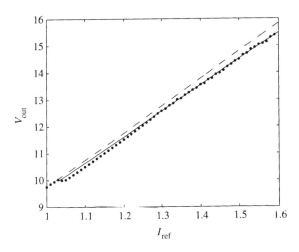

Figure 4.23 V_{out} in volts as a function of I_{ref} in amps, for the buck converter parameters given in the text.

apply for the periodic solution; it is evident that our approximate analysis of the chaotic case does much better.

Ćuk Converter. The Ćuk converter analyzed in this section is shown in Figure 4.24. The current in L_1 is the controlled current, and is treated as i_L was in the converters presented thus far. The inductor current waveform i_{L_1} has the appearance of i_L in Figure 4.14. The dynamics of the converter are still described by the map in equation (4.8), but with α now given by

$$\alpha = \frac{m_2}{m_1} = \frac{V_{\text{out}}/L_1}{V_{\text{in}}/L_1} = \frac{V_{\text{out}}}{V_{\text{in}}} \tag{4.37}$$

To calculate V_{out} for $\alpha > 1$ we again invoke the fact that the time-averaged power input of the converter must equal the time-averaged power output:

$$V_{\text{in}}\langle i_{L_1}\rangle = \frac{V_{\text{out}}^2}{R} \tag{4.38}$$

from which, using (4.37), we get

$$\langle i_L \rangle = \frac{\alpha^2 V_{\text{in}}}{R} \tag{4.39}$$

Combining (4.17), (4.19), (4.37), and (4.39) we obtain the following relation in the case of integer $\alpha > 1$:

$$I_{\text{ref,int}} = \frac{\alpha^2 V_{\text{in}}}{R} + \frac{\alpha V_{\text{in}} T}{3 L_1} \tag{4.40}$$

As before, this relation is a very good approximation for noninteger α as well.

Figure 4.25 shows a plot (solid line) of V_{out} in volts as a function of I_{ref} in amps, computed from (4.40). The parameters of the Ćuk circuit used for this example are $V_{\text{in}} = 10$ volts, $R = 20\Omega$, $L_1 = L_2 = 1\text{mH}$, $C = 200\mu\text{F}$, $C_1 = 500\mu\text{F}$, and $T = 100\mu\text{s}$. The figure also shows (dotted line) the average output voltage obtained from simulations of the circuit using Simulink. Our approximate results are again in good agreement with those obtained through the simulation. The dashed line in the figure plots the relationship between V_{out} and I_{ref} that would be obtained using the expressions that apply for the periodic solution; it is evident that our approximate analysis of the chaotic case does much better.

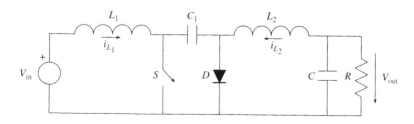

Figure 4.24 Ćuk converter circuit.

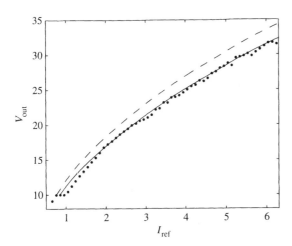

Figure 4.25 V_{out} in volts as a function of I_{ref} in amps, for the Ćuk converter parameters given in the text.

Average Switching Frequency and Average Duty Ratio

A quantity of interest in the analysis of dc/dc converters is the *average switching frequency*, $\langle s \rangle$, defined as the average number of switch openings per sampling period. For periodic operation $\langle s \rangle = 1$. However, for chaotic operation $\langle s \rangle < 1$, since there are sampling periods in which the switch does not open. In fact, the development that led up to (4.8) shows that the switch opens in the interval $[nT, (n + 1)T]$ precisely when

$$1 - 1/\alpha < z_n \leq 1$$

Invoking ergodicity, we can then write

$$\langle s \rangle = \text{Probability} \left\{ 1 - 1/\alpha < z \leq 1 \right\} \tag{4.41}$$

For integer $\alpha > 1$, this probability evaluates to $2/(\alpha + 1)$; see [13]. For large α, therefore, the average switching frequency becomes small. This result may provide another motivation for chaotic operation of the converter, since switching losses—a primary factor in converter efficiency—are thereby reduced. However, it should be noted that the switching ripple will correspondingly increase; the trade-off needs to be examined, and compared to the analogous trade-off under periodic operation (where indeed the switching frequency can be reduced, but again with a concomitant increase in ripple).

The duty ratio is perhaps less significant for chaotic operation than it is for periodic operation. The duty ratio varies from cycle to cycle, so it is the average duty ratio that is of interest, and there are two types of average duty ratio that can be defined. The first is the average of the *local* duty ratio, where the latter is defined as the fraction of time the switch is on in a full switching cycle (whose length is some integer number of sampling or clock intervals); the local duty ratio in general varies from one full switching cycle to the next (unlike for periodic operation). The second definition is a *global* average, defined as the limiting ratio of total on-time to total time, as the averaging window becomes infinitely large.

It can be shown that the global average duty ratio is the same as the duty ratio $D_{per} = \alpha/(1 + \alpha)$ corresponding to the (unstable) periodic solution; see [13]. The average local duty ratio, on the other hand, can be easily computed using the invariant density. For integer $\alpha > 1$, we get

$$\langle D \rangle = 1 - \frac{1}{2\alpha} \sum_{k=1}^{\alpha} \frac{1}{k} \tag{4.42}$$

This result has been verified using simulations, and also compared with D_{per}. For example, when $\alpha = 2$, computation of the duty ratio for the (unstable) periodic solution yields $D_{\text{per}} = 2/3 = 0.667$, while our calculations with the invariant density yield $\langle D \rangle = 0.625$. For $\alpha = \alpha_0$, computation of the duty ratio for the (unstable) periodic solution gives $D_{\text{per}} = 0.618$, while our calculations yield $\langle D \rangle = 0.602$. More generally, it turns out that $\langle D \rangle < D_{\text{per}}$ throughout the chaotic regime. In each case, detailed simulations of the circuit—for the converter parameters above as well as for other choices of parameters—have confirmed our results computed from the invariant density.

4.3.3 Experimental Results

We now explore experimentally the application of the foregoing theory to an actual boost converter, shown schematically in Figure 4.17, and with nominal parameter values of $R = 195\Omega$, $L = 3.2\text{mH}$, $C = 100\mu\text{F}$ and $T = 50\mu\text{s}$. The converter operates under current-mode control. The reference current I_{ref} can be set at different levels.

Figure 4.26 displays the empirical *stroboscopic* map $z_{n+1} = F(z_n)$ obtained from sampled and normalized measurements of the inductor current. It is evident that the results are of the form shown in Figure 4.15. Specifically, the results in the figure here are obtained for $V_{\text{in}} = 4\text{V}$, and the measured $V_{\text{out}} = 14.25\text{V}$. The corresponding value of α determined by (4.20) is $\alpha = 2.56$.

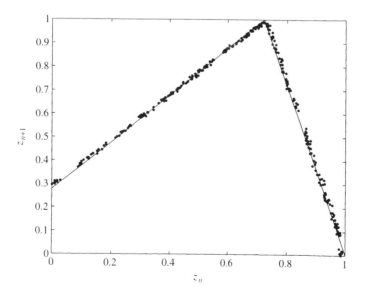

Figure 4.26 Empirical map for the boost converter in the chaotic regime.

The full line in Figure 4.26 represents the best fit to a map of the form in Figure 4.15, but corresponds to $\alpha = 3.6$. The discrepancy in the value of α is attributed to the nonidealities of the experimental circuit, most importantly the voltage drops across the inductor series resistance and the switches. These voltage drops can be roughly captured by taking the effective V_{in} to be smaller than the measured value. The best-fit value of $\alpha = 3.6$ corresponds—again using equation (4.20)—to an effective $V_{in,eff} = 3.1V$. With this correction, the fit of theory to experiment is evidently excellent, verifying the validity of the assumptions that underlie the derivation of the map in Figure 4.15. Note that the inductor current samples in Figure 4.26 are distributed throughout the allowed range, as expected from the fact that the invariant density associated with the chaotic operation for this α is nonzero throughout the interval [0,1].

The output voltage has been measured for different values of I_{ref} and with $V_{in} = 4V$. The results are summarized in Table 4.2. The first two columns are the measured output voltage V_{out} and the corresponding measured value of I_{ref}. In order to approximately take into account the voltage drops across the inductor series resistance and the switches, we follow the approach described in the paragraph above. Sampled measurements of the inductor current were taken for each operating point (i.e., for each I_{ref}) to construct an experimental portrait of the attractor, as in Figure 4.26, from which the corresponding value of α was computed; these values are listed in the third column of Table 4.2. The fourth column then gives the associated I_{ref} computed by determining the $V_{in,eff}$ that goes with the experimental α, and then using the approximate relation (4.23). The last column of the table gives (using standard computations [1]) the I_{ref} value that would be needed to obtain the measured V_{out} using the (unstable) periodic solution; note that the value of V_{in} is not needed for this computation.

It is evident from Table 4.2 that our analysis of behavior in the chaotic mode is in very good agreement with the experimental results, and significantly better than the predictions that would be obtained using formulas for the periodic case. Although our good results were obtained at the expense of experimentally determining the effective α and corresponding $V_{in,eff}$, results that are notably better than the periodic calculations are also obtained by picking a single $V_{in,eff}$, say 3.3V, to use in equation (4.23) for all the operating conditions in the table. Note that $I_{ref,per}$ in the last column is consistently

TABLE 4.2 Experimental and theoretical results.

V_{out} (V)	I_{ref} (mA)	α	$I_{ref,int}$ (mA)	$I_{ref,per}$ (mA)
10.0	174.5	1.9	182.8	166.4
10.5	204.3	2.1	204.0	184.8
11.0	213.2	2.2	219.9	199.0
11.5	248.8	2.5	249.2	224.7
12.0	263.1	2.6	266.7	240.3
12.5	287.5	2.8	291.6	262.5
13.0	314.2	3.0	317.5	285.7
13.5	340.1	3.2	344.3	309.9
14.0	376.2	3.4	372.2	335.1
14.5	400.9	3.6	401.2	361.3

smaller than the other I_{ref} values for each listed V_{out}, which is consistent with the results in Figure 4.18.

4.3.4 Conclusions

In this section we have shown how to analyze dc/dc converters under current-mode control in the chaotic regime, using a state-densities approach. We have also presented a simplification that yields very good approximations and tractable analytical expressions. Finally, it has been established that simpleminded extension of traditional computations from the stable periodic regime into the chaotic regime produces results that can be significantly in error.

The waveforms in Figure 4.19 make clear that the chaotic regime is not necessarily one to be avoided; although stable period-T operation is lost, the waveforms are still well-behaved, and the output voltage ripple is small. A potential advantage of chaotic operation is that the switching spectrum is reduced in intensity (although at the expense of a corresponding broadening); see, for example [5,7]. As noted in [8], which deals with actively randomized modulation, this spectral shaping may be desirable in some situations. Another potential advantage of chaotic operation is that the average switching frequency is lower than for the periodic case.

ACKNOWLEDGMENTS

The authors would like to acknowledge the important roles played by Sashi Venkataraman and José María Font at the early stages of this study; their ideas, persistence, and hard work got us off the ground. The appendix is based on notes by Soumitro Banerjee. José Luis Rodríguez Marrero is grateful for partial support by a grant from Comisión Interministerial de Ciencia y Tecnología of Spain (TIC97-0370).

APPENDIX: THE FROBENIUS-PERRON OPERATOR

In this appendix we present the statistical properties of maps. Imagine that a map is iterated a large number of times starting from an arbitrary initial condition, and that we keep track of the locations where the iterates fall. If the map is chaotic, the iterates will typically be spread over a large part (or maybe the whole) of the state space. We'll find that some regions are more densely populated by the iterates, meaning that these regions were visited more frequently, and that some regions will be sparsely populated. If we (indirectly) define the density $\rho(x)$ of iterates at various values of the state variable x via

$$\int_{\Delta x} \rho(x)dx = \frac{\text{Number of iterates in the range } \Delta x}{\text{Total number of iterates}}$$

and plot $\rho(x)$ against x, we obtain a curve such as that in Figure 4.27. This *density function* is an important indicator of a dynamical system, especially if we are interested in the time-averaged value of a variable that depends on x.

There is another way of looking at the problem. Suppose we pick an initial condition according to some initial probability density. Now applying the map once will result in the next state being distributed according to some new probability density. If

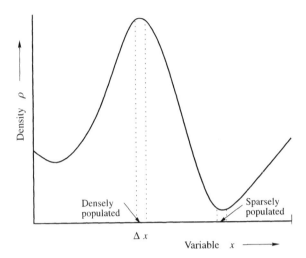

Figure 4.27 Schematic representation of the density of iterates $\rho(x)$.

we apply the map once more, the density function will change further. The *operator* that transforms these density functions under the action of the map is called the *Frobenius-Perron* (F-P) operator associated with the map. If we apply this operator to an initial density, the density will converge (subject to some conditions, see [12]) to an *invariant density*, one that does not change under application of the operator. The invariant density is thus the *fixed point* of the Frobenius-Perron operator.

The content of the Birkhoff ergodic theorem [12] is essentially the equivalence of the iterate density and the probability density viewpoints of the action of the map; the iterate density relates to taking time averages, whereas the probability density relates to taking ensemble averages.

Let us now obtain the Frobenius-Perron operator for any given map. Let the density function at the nth iteration of the F-P operator be $\rho_n(x)$. Figure 4.28 shows that upon one further iteration of the map, the states in the range $[x_1, x_1 + \Delta x]$ become spread over a range of $[x_1, x_1 + \Delta x]|\frac{dy}{dx}|_{x_1}$ and thus their contribution to the density in that range becomes $\rho_n(x)/|\frac{dy}{dx}|_{x_1}$. For the particular map shown in Figure 4.28, the points in the range $[y, y + \Delta y]$ come from three ranges in x. Therefore, at the

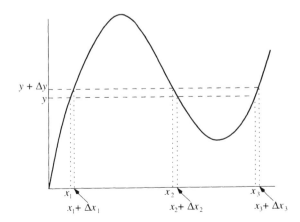

Figure 4.28 Illustration for the derivation of the Frobenius-Perron operator for a map $y = f(x)$.

$(n + 1)$th iteration of the F-P operator the density in the range $[y, y + \Delta y]$ is given in terms of the densities in the nth iterate by

$$\rho_{n+1}(y, y + \Delta y) = \frac{\rho_n(x_1, x_1 + \Delta x_1)}{\left|\frac{dy}{dx}\right|_{x_1}} + \frac{\rho_n(x_2, x_2 + \Delta x_2)}{\left|\frac{dy}{dx}\right|_{x_2}} + \frac{\rho_n(x_3, x_3 + \Delta x_3)}{\left|\frac{dy}{dx}\right|_{x_3}} \quad (4.43)$$

Thus the density at any point y can be written in terms of the summation

$$\rho_{n+1}(y) = \sum_m \frac{\rho_n(x_m)}{|f'(x_m)|} = \sum_m \frac{\rho_n(f^{-1}(y))}{|f'(f^{-1}(y))|} \quad (4.44)$$

where the x_m are the solutions of $x = f^{-1}(y)$. This transformation of ρ_n to ρ_{n+1} defines the Frobenius-Perron operator associated with the map f.

EXAMPLE

We consider the class of maps given by

$$x_{n+1} = \begin{cases} x_n + \alpha^{-1} & \text{for} \quad 0 \leq x_n \leq 1 - \alpha^{-1} \\ \alpha - \alpha x_n & \text{for} \quad 1 - \alpha^{-1} \leq x_n \leq 1 \end{cases} \quad (4.45)$$

where α is an integer greater than 1. The map is illustrated in Figure 4.29, and coincides with the map (4.8). Note that if we partition the [0,1] range of the state variable into α partitions with breakpoints $1/\alpha, 2/\alpha, 3/\alpha, \ldots 1$ then breakpoints map to breakpoints. In this case the map is a *Markov map* or, more specifically, an eventually expanding, piecewise-linear, Markov map [10]. Expanding piecewise-linear Markov maps are interesting because their probability density functions turn out to be constant in each partition interval and can be found exactly, as follows.

Using $\rho_n(a, b)$ to denote a value of $\rho_n(x)$ that is constant in the interval $x \in (a, b)$, and noting that partitions map to partitions in our example, we can define the Frobenius-Perron operator P via

$$P\rho(x) : \begin{cases} \rho_{n+1}(0, \alpha^{-1}) & = & \rho_n(1 - \alpha^{-1}, 1)/\alpha \\ \rho_{n+1}(\alpha^{-1}, 2\alpha^{-1}) & = & \rho_n(0, \alpha^{-1}) + \rho_n(1 - \alpha^{-1}, 1)/\alpha \\ \rho_{n+1}(2\alpha^{-1}, 3\alpha^{-1}) & = & \rho_n(\alpha^{-1}, 2\alpha^{-1}) + \rho_n(1 - \alpha^{-1}, 1)/\alpha \\ \vdots \\ \rho_{n+1}(1 - \alpha^{-1}, 1) & = & \rho_n(1 - 2\alpha^{-1}, 1 - \alpha^{-1}) + \rho_n(1 - \alpha^{-1}, 1)/\alpha \end{cases} \quad (4.46)$$

For example, if we take $\alpha = 3$ we will have three partitions: $I_1 = (0, 1/3)$, $I_2 = (1/3, 2/3)$, and $I_3 = (2/3, 1)$. Then (4.46) can be written as

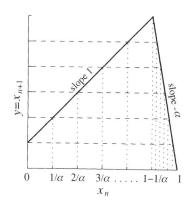

Figure 4.29 The graph of the map (4.45) for $\alpha = 6$.

$$\begin{bmatrix} \rho_{n+1}(I_1) \\ \rho_{n+1}(I_2) \\ \rho_{n+1}(I_3) \end{bmatrix} = \begin{bmatrix} 0 & 0 & 1/3 \\ 1 & 0 & 1/3 \\ 0 & 1 & 1/3 \end{bmatrix} \begin{bmatrix} \rho_n(I_1) \\ \rho_n(I_2) \\ \rho_n(I_3) \end{bmatrix} \tag{4.47}$$

Now in order to calculate the invariant density function $\rho^*(x)$, we set $P\rho^*(x) = \rho^*(x)$:

$$\begin{bmatrix} \rho^*(I_1) \\ \rho^*(I_2) \\ \rho^*(I_3) \end{bmatrix} = \begin{bmatrix} 0 & 0 & 1/3 \\ 1 & 0 & 1/3 \\ 0 & 1 & 1/3 \end{bmatrix} \begin{bmatrix} \rho^*(I_1) \\ \rho^*(I_2) \\ \rho^*(I_3) \end{bmatrix} \tag{4.48}$$

That is, the invariant density function $\rho^*(x)$ is determined by the eigenvector of the Frobenius-Perron operator with eigenvalue equal to unity. Solving, we obtain

$$\rho^*(I_3) = 3\rho^*(I_1)$$
$$\rho^*(I_2) = 2\rho^*(I_1) \tag{4.49}$$

Recalling also that the area under the density function should be unity, we require

$$\frac{1}{3}[\rho^*(I_1) + \rho^*(I_2) + \rho^*(I_3)] = 1 \tag{4.50}$$

Solving these equations, we get

$$\rho^*(I_1) = \tfrac{1}{2}$$
$$\rho^*(I_2) = 1$$
$$\rho^*(I_3) = \tfrac{3}{2} \tag{4.51}$$

For the general case of α partitions we get

$$\rho^*(x) = \frac{2k}{1+\alpha} \quad \text{for} \quad \frac{k-1}{\alpha} \le x < \frac{k}{\alpha} \tag{4.52}$$

where $k = 1, 2, 3 \ldots \alpha$.

REFERENCES

[1] J. G. Kassakian, M. F. Schlecht, and G. C. Verghese, *Principles of Power Electronics*. Reading, MA: Addison-Wesley, 1991.

[2] J. H. B. Deane and D. C. Hamill, Chaotic behavior in a current-mode controlled dc-dc converter, *Electronics Letters*, vol. 27, pp. 1172–1173, 1991.

[3] J. H. B. Deane, Chaos in a current-mode controlled boost dc-dc converter, *IEEE Trans. on Circuits and Systems*, vol. CAS-32, pp. 680–683, August 1992.

[4] C. K. Tse and W. C. Y. Chan, Instability and chaos in a current-mode controlled Ćuk converter, *IEEE Power Electronics Specialists Conference* (Atlanta, GA), pp. 608–613, 1995.

[5] I. Zafrany and S. Ben-Yaakov, A chaos model of subharmonic oscillations in current mode PWM boost converters, *IEEE Power Electronics Specialists Conference* (Atlanta, GA), pp. 1111–1117, 1995.

[6] J. L. R. Marrero, J. M. Font, and G. C. Verghese, Analysis of the chaotic regime for dc-dc converters under current-mode control, *IEEE Power Electronics Specialists Conference* (Baveno, Italy), pp. 1477–1483, 1996.

[7] J. H. B. Deane and D. C. Hamill, Improvement of power supply EMC by chaos, *Electronics Letters*, vol. 32, p. 1045, 1996.

[8] A. M. Stankovic, G. C. Verghese, and D. J. Perreault, Analysis and synthesis of randomized modulation schemes for power converters, *IEEE Trans. on Power Electronics*, vol. 10, no. 6, pp. 680–693, 1995.

[9] L. O. Chua, Y. Yao, and Q. Yang, Generating randomness from chaos and constructing chaos with desired randomness, *Int. J. Circuit Theory and Applications*, vol. 18, pp. 215–240, 1990.

[10] S. H. Isabelle, A Signal Processing Framework for the Analysis and Applications of Chaotic Systems. PhD diss., EECS Department, MIT, February 1995.

[11] A. L. Baranovski, A. Mögel, W. Schwarz, and O. Woywode, Statistical analysis of a dc-dc converter, *Proc. Nonlinear Dynamics of Electronics Systems (NDES 99)*. (Rønne, Bornholm, Denmark), July 15–17, 1999.

[12] A. Lasota and M. C. Mackey, *Chaos, Fractals and Noise: Stochastic Aspects of Dynamics*, 2nd ed., New York: Springer-Verlag, 1994.

[13] R. Santos Bueno, Análisis y Control de Convertidores dc-dc en Régimen Caótico. PhD diss., ETS Ingeniería ICAI, Universidad Pontificia Comillas, 2000.

4.4 CALCULATION OF SPECTRAL PEAKS IN A CHAOTIC DC/DC CONVERTER

Jonathan H. B. Deane

4.4.1 Characterization of Spectral Properties

One possible application of chaos in power electronics involves shaping the spectral properties of waveforms occurring in a switch-mode power supply. The spectral properties of a *periodic* waveform $p(t)$, such as would be associated with conventional steady-state operation of a power converter, follow from its Fourier series representation,

$$p(t) = \sum_{m=-\infty}^{\infty} A_m e^{jm\omega_c t} \tag{4.53}$$

It is easy to show that the squared magnitudes of the Fourier coefficients, $\tilde{A}_m = |A_m|^2$, give the *time-averaged power*[1] at the corresponding frequency $m\omega_c$ associated with the fundamental ($|m| = 1$), harmonics ($|m| > 1$), or dc ($m = 0$). In what follows, we shall generally drop the word *average* when referring to average power, unless needed for clarity. Note that the distribution of (average) power among the fundamental and its harmonics depends only on the magnitudes of the A_m. Thus phase shifts in the individual frequency components don't affect the power in these components; so there is actually an infinite family of signals with period $2\pi/\omega_c$ that has the same distribution of power over frequency.

The standard way of describing the distribution of signal power as a function of frequency is through the *power spectral density* (PSD) or *power density spectrum* (PDS) of the signal. The PDS is defined to be that function of frequency whose *integral* in any particular frequency range yields the (average) power associated with the frequency components of the waveform in that frequency range. Hence, for a periodic waveform

---

[1] The word *power* is conventionally used for the square of a voltage or current, and more generally for the square of any signal of interest.

$p(t)$, the PDS $\tilde{P}(\omega)$ consists of δ-functions or *impulses* at the fundamental and harmonics, with areas (or strengths or intensities) equal to the squared magnitudes of the Fourier coefficients:

$$\tilde{P}(\omega) = \sum_{m=-\infty}^{\infty} |A_m|^2 \delta(\omega - m\omega_c) = \sum_{m=-\infty}^{\infty} \tilde{A}_m \delta(\omega - m\omega_c) \tag{4.54}$$

Clearly, integrating this over any particular frequency range yields precisely the power of the frequency components in that range.

An alternative approach to the PDS in the periodic case—and an approach that turns out to generalize naturally to the nonperiodic case—is the following. First compute the time-averaged *autocorrelation* function of the waveform,

$$R_p(\tau) = \lim_{T \to \infty} \frac{1}{2T} \int_{-T}^{T} p(t)p(t + \tau)\,dt \tag{4.55}$$

which simplifies to

$$R_p(\tau) = \sum_{m=-\infty}^{\infty} |A_m|^2 e^{-m\omega_c \tau}$$

as is easily verified on noting that the only term in the product $p(t)p(t + \tau)$ which in the limit survives the integration and division by $2T$ is the term that involves no exponentials in t. Taking the Fourier *transform* of $R_p(\tau)$ yields $\tilde{P}(\omega)$. Thus, the PDS is given by the *Fourier transform of the autocorrelation function*. Note that variations in the phases of the individual frequency components of $p(t)$ do not affect $R_p(\tau)$ because the autocorrelation depends only on $|A_m|^2$.

Now consider a possibly *nonperiodic* signal $i(t)$ but one that is *persistent* (i.e., a signal with nonzero average power). Examples of interest include a sum of sinusoids at nonrationally related frequencies (a *quasiperiodic* signal); a waveform associated with a power converter in the chaotic regime; or a (sample function of a) stationary random process. The definition of the PDS $\tilde{I}(\omega)$ for such a signal is, in principle, the same as in the periodic case: the PDS is that function of frequency whose integral in any particular frequency range yields the average power associated with the frequency components of the signal in that frequency range. However, making precise the notion of average power associated with a frequency range is more subtle for a nonperiodic signal than for a periodic one, and we refer the reader to [1, Chapter 3] for a more elaborate treatment.

For our purposes, it suffices to note that the PDS $\tilde{I}(\omega)$ in the nonperiodic case can still in general be computed by the procedure illustrated for the periodic case: first compute the autocorrelation function $R_i(\tau)$ of the signal, where the autocorrelation is defined exactly as in (4.55) with $i(t)$ instead of $p(t)$; then $\tilde{I}(\omega)$ is the Fourier transform of $R_i(\tau)$. The proof of this claim constitutes the Wiener-Khintchine or Einstein-Wiener-Khintchine theorem [1]. Just as in the periodic case, it turns out that variations in the phase shifts associated with the frequency components of $i(t)$ do not affect $R_i(\tau)$ and hence do not affect $\tilde{I}(\omega)$. A given $R_i(\tau)$ or $\tilde{I}(\omega)$ will therefore characterize a whole family of signals, rather than just a single one—perhaps an entire stationary random process rather than just an individual sample path, or an entire family of chaotic waveforms corresponding to different initial conditions in a given converter.

For the nonperiodic (and nonquasiperiodic) case, the PDS will typically contain a continuous part that reflects the frequency distribution of average power in the nonperiodic (and nonquasiperiodic) part of the signal, plus an impulsive part referred to as the *line spectrum*, which reflects the presence of any sinusoidal components at the corresponding frequencies.

4.4.2 Motivation and Outline

The motivation for our study of spectral properties is twofold:

1. In [2] we gave experimental evidence, in the form of analog spectrum analyzer measurements, for the idea that chaos might be used to improve the electromagnetic compatibility (EMC) of switch-mode power supplies; see Figure 4.30. This possibility was also mentioned in [3].
2. Given that the electromagnetic interference (EMI) requirement for switch-mode power supplies is, in essence, that the input current PDS should be less than a specified frequency-dependent level, it is important to be able to calculate certain spectral features, for instance, the height of spectral peaks.

This section describes, for a chaotically operated current-mode-controlled boost converter, how to calculate (in more-or-less closed form, under certain assumptions) the PDS of the input current at the clock frequency and its harmonics (i.e., the line spectrum of the input current). The treatment is based on [4]. Computation of the continuous part of the PDS is also of interest, and is discussed in [5]; see also [6] for computation of the PDS in converters with randomized modulation.

The boost converter is a two-dimensional system described by a two-dimensional mapping, but calculations are simplified if the underlying mapping is one-dimensional. In practice, the requirement for low output voltage ripple in real designs leads to a circuit that *is* well described by a one-dimensional mapping, and we derive this first. This mapping has the properties required for it to display *robust chaos*— see later in this section—so we can guarantee chaotic behavior of the circuit over certain parameter ranges.

We calculate the invariant density of the one-dimensional mapping (see Section 4.3 as well as the references cited below for the notion of an invariant density), and from this determine the line spectrum associated with the periodic components of the inductor current. We then provide experimental and numerical evidence of the validity of the

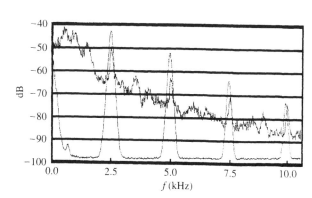

Figure 4.30 Experimental spectra from a boost converter in periodic (peaks) and chaotic (broadband) operation. (Reproduced by kind permission of *Electronics Letters.*)

model and our PDS calculation. Evidence is also presented to show that chaos can be exploited to reduce spectral peaks, significantly in some cases, which leads us to consider *EMC compliance by chaos.*

Figure 4.31 gives an overview of the calculation; its contents can be briefly summarized as follows. Starting from the circuit, a one-dimensional mapping can be derived (Section 4.4.3) that describes its behavior remarkably well. The mapping turns out to depend on one parameter, α, and the relationship between α and the circuit parameters is derived in Section 4.4.4. Using the appropriate definition of power density spectrum, the inductor current's PDS at the clock frequency, ω_c, and its harmonics can be computed (1) by brute force, and (2) directly from the invariant density using *Birkhoff's ergodic theorem*; this development is in Section 4.4.5. (Birkhoff's theorem has already been invoked in Section 4.3.) An algorithm is described in Section 4.4.6 for calculating the required invariant density. Computational and experimental results—including comparisons of the brute force and invariant density approaches—are presented in Sections 4.4.7 and 4.4.8.

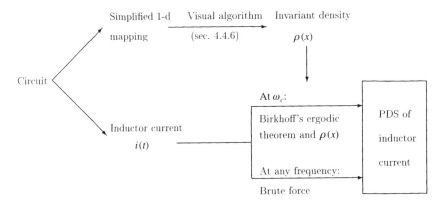

Figure 4.31 An overview of the spectral calculation.

4.4.3 The Simplified Mapping

A two-dimensional mapping that describes the boost converter in Figure 4.32 was presented in [7]. For the ranges of parameter values used in real designs, this mapping can be simplified to a one-dimensional version that nonetheless captures the features of interest in the behavior of a real circuit. In order to derive this one-dimensional mapping, we make the following assumptions:

1. The capacitor voltage v_c is sufficiently close to a constant, $\overline{V}o$, so that the inductor current waveform can be approximated by a linear ramp at all times. This requires that $CR \gg T_c$, where T_c is the clock period. In practical circuits, C is sufficiently large for this to be valid.

2. All components are ideal.

3. The clock pulses are of infinitesimal duration and period T_c.

With these assumptions, the power circuit can be modeled in the simplified form shown in Figure 4.33. The switch S is controlled by feedback. It closes when a clock pulse arrives and opens at the instant when $i(t)$ reaches the reference current I_{ref}.

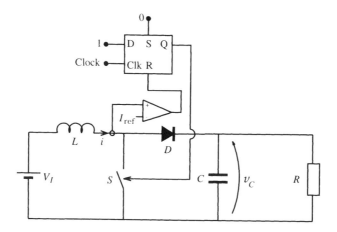

Figure 4.32 The peak current-mode-controlled boost converter.

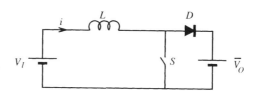

Figure 4.33 Simplified model of the power circuit of the boost converter.

The inductor current $i(t)$ is sketched in Figure 4.34. While S is closed, $i(t)$ satisfies

$$\frac{di}{dt} = \frac{V_I}{L}$$

Defining $t = 0$ at the instant of S closing, then $i(t)$ is

$$i(t) = i_n + \frac{V_I}{L}t \tag{4.56}$$

valid until time t_n at which $i(t_n) = I_{\text{ref}}$. Using equation (4.56)

$$t_n = \frac{(I_{\text{ref}} - i_n)L}{V_I} \tag{4.57}$$

Switch S now opens and $i(t)$ is given by

$$\frac{di}{dt} = \frac{V_I - \overline{V}_O}{L}$$

with initial condition $i(0) = I_{\text{ref}}$. (We reset the time origin for convenience.) Hence,

$$i(t) = I_{\text{ref}} + \frac{V_I - \overline{V}_O}{L}t \tag{4.58}$$

until the next clock pulse arrives. This happens at a time

$$t'_n = T_c - T_c \left(\frac{t_n}{T_c}\right) \text{ mod } 1 \tag{4.59}$$

after S last opened.

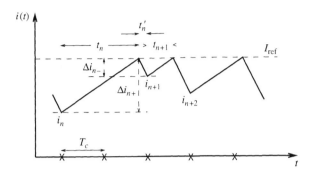

Figure 4.34 The current waveform $i(t)$ in the simplified boost converter.

The mapping $i_n \mapsto i_{n+1}$ can now be derived. By definition, $i_{n+1} = i(t'_n)$, which according to equations (4.57), (4.58), and (4.59), is

$$i_{n+1} = I_{\text{ref}} + \frac{(V_I - \overline{V}_O)T_c}{L}\left[1 - \left(\frac{(I_{\text{ref}} - i_n)L}{V_I T_c}\right) \text{mod } 1\right] \quad (4.60)$$

Introducing a dimensionless time x_n, defined as

$$x_n = \frac{t_n}{T_c} = \frac{(I_{\text{ref}} - i_n)L}{V_I T_c}$$

and parameter α, defined as

$$\alpha = \frac{\overline{V}_O}{V_I} - 1$$

the mapping (4.60) can be written in the simplified form

$$x_{n+1} = F(x_n) = \alpha(1 - x_n \text{ mod } 1) \quad (4.61)$$

Since $\overline{V}_O > V_I$ for a boost converter, $\alpha > 0$. (This is an *S-switching* mapping, in the language of Section 2.2, and can be contrasted with the stroboscopic one-dimensional mapping of the boost converter that was derived in Section 4.3.)

Note that the switching times can be recovered from (4.61):

$$t_n = T_c x_n \quad \text{and} \quad t'_n = T_c(1 - x_n \text{ mod } 1) \quad (4.62)$$

When $\alpha > 1$, the sequence $\{x_0, x_1, \ldots x_n \ldots\}$ will be chaotic, with each x_n lying between 0 and α. The mapping (4.61) and scaled versions of it have been studied in several contexts, most notably by Rényi [8] and Lasota and Mackey [9, Section 6.2]. The fact that F is the Rényi transformation means that several technical conditions are satisfied. Specifically, the Frobenius-Perron operator P, defined in the appendix of Section 4.3 and in Section 4.4.6, has a piecewise-constant invariant density $\rho(x)$ that governs the distribution over the interval $[0, \alpha)$ of the points in the sequence $\{x_0, x_1, \ldots x_n \ldots\}$. This density is calculated in Section 4.4.6.

4.4.4 Approximation of the Mean State Variables

In order to use the mapping (4.61), we need to know the value of α, and hence there remains the problem of relating $\overline{V}_O$, the mean output voltage, to known quantities in the circuit. We obtain an estimate of $\overline{V}_O$ by assuming that S is closed on average for a time $\overline{D}T_c$ per clock cycle, and $\overline{V}_O$ is constant. Here, $\overline{D}$ is to be interpreted as the mean duty factor of S. On average,

$$\overline{V}_O = R(1 - \overline{D})\left(I_{\text{ref}} - \frac{\overline{\Delta i}}{2}\right) \qquad (4.63)$$

where R is the load resistance shown in Figure 4.32. This assumes that the current through the diode is either zero (for a fraction $\overline{D}$ of the time) or $I_{\text{ref}} - \overline{\Delta i}/2$ (for a fraction $1 - \overline{D}$ of the time—see Figure 4.34). When S is closed, i rises at a rate of V_I/L for a time $\overline{D}T_c$ on average; when S is open, i falls at a rate $(\overline{V}_O - V_I)/L$ for a time $(1 - \overline{D})T_c$. In terms of the mean current rise, $\overline{\Delta i}_+$, and fall, $\overline{\Delta i}_-$,

$$\overline{\Delta i}_+ = \frac{V_I}{L}\overline{D}T_c, \quad \overline{\Delta i}_- = \frac{\overline{V}_O - V_I}{L}(1 - \overline{D})T_c \quad \text{and} \quad \overline{\Delta i}_+ = \overline{\Delta i}_- = \overline{\Delta i} \qquad (4.64)$$

Eliminating $\overline{\Delta i}_+$, $\overline{\Delta i}_-$, $\overline{\Delta i}$, and $\overline{D}$ between equations 4.63 and 4.64, gives

$$\overline{V}_O^3 + \overline{V}_O(V_I T_c/2L - I_{\text{ref}})RV_I - RT_c V_I^3/2L = 0 \qquad (4.65)$$

which can be solved for $\overline{V}_O$ by selecting the real root $\overline{V}_O > V_I$. Hence, α can be found.

The preceding calculation implicitly used assumptions that apply to the periodic regime rather than the chaotic regime, but the estimate produced by (4.65) is sufficiently accurate for our purposes here. A relation analogous to (4.65), but derived specifically for the chaotic case, was presented in equation (4.23) in Section 4.3, and this could be used instead of (4.65) if further accuracy is desired.

4.4.5 The Power Density Spectrum of the Inductor Current

We now calculate the PDS of the inductor current $i(t)$ from the invariant density of the mapping F. To this end, we first introduce the autocorrelation function of the current, $R_i(\tau)$, and calculate its Fourier transform (invoking certain assumptions). We then apply Birkhoff's ergodic theorem [10] to calculate the periodic component of the PDS, and compare results for periodic and chaotic operation of the converter.

Recall that the PDS, which we will write as $\tilde{I}(\omega)$, is the Fourier transform of the autocorrelation function

$$R_i(\tau) = \lim_{T \to \infty} \frac{1}{2T} \int_{-T}^{T} i(t)i(t + \tau)dt$$

Now define

$$A(\omega) = \lim_{T \to \infty} \frac{1}{2T} \int_{-T}^{T} i(t)e^{-j\omega t}dt \qquad (4.66)$$

and write $A_m = A(m\omega_c)$ for arbitrary integer m, where $\omega_c = 2\pi/T_c$ is the clock angular frequency. Note that if $i(t)$ were periodic with period T_c, then A_m would be just the Fourier coefficient associated with the harmonic component of $i(t)$ at $m\omega_c$. Of course, in such a periodic case we could extract A_m by simply averaging $i(t)e^{-jm\omega_c t}$ over an interval of length T_c rather than over the infinite interval $[-T, T]$, $T \to \infty$. However, the virtue of the more general expression in equation (4.66) is that it extracts this Fourier coefficient even when the periodic signal has some nonperiodic part added to it. Accordingly, returning to the case of a general nonperiodic $i(t)$, let us write

$$p(t) = \sum_{m=-\infty}^{\infty} A_m e^{jm\omega_c t} \quad \text{and} \quad c(t) = i(t) - p(t)$$

so that the current $i(t)$ is split into a component p that is periodic with the clock period T_c, and a residual component c. With this decomposition, we directly obtain

$$R_i(\tau) = R_{c+p}(\tau) = R_c(\tau) + R_p(\tau) + \lim_{T \to \infty} \frac{1}{2T} \int_{-T}^{T} c(t)[p(t+\tau) + p(t-\tau)]dt \quad (4.67)$$

We now assume that $c(t)$ has a continuous PDS $\tilde{I}_c(\omega)$, with no line spectrum (which implies $A(\omega) = 0$ for all $\omega \neq m\omega_c$, integer m). It can be shown that this implies the integral in equation (4.67) above vanishes; we say that $p(t)$ and $c(t)$ are *uncorrelated* in this case. Hence, under our assumption,

$$R_i(\tau) = R_c(\tau) + R_p(\tau)$$

Taking transforms, we find the PDS of the current to be

$$\tilde{I}(\omega) = \int_{-\infty}^{\infty} R_c(\tau)e^{-j\omega\tau}d\tau + \int_{-\infty}^{\infty} R_p(\tau)e^{-j\omega\tau}d\tau$$

$$= \tilde{I}_c(\omega) + \sum_{m=-\infty}^{\infty} \tilde{A}_m \delta(\omega - m\omega_c)$$

where, by [11, Theorems 11.9 and 11.10] we have

$$\tilde{A}_m = |A_m|^2 \quad (4.68)$$

Our assumption on $c(t)$ thus leads us to conclude that $\tilde{I}(\omega)$ is a sum of two types of functions, namely a continuous function $\tilde{I}_c(\omega)$ that corresponds to broadband noise, and a set of δ-functions or impulses at the clock frequency and its harmonics.

Under assumption (1) of Section 4.4.3, d^2i/dt^2 is a sequence of Dirac δ-functions of alternating sign and area $\overline{V}_O/L$. Using the dimensionless time $x = t/T_c$, the situation is then as illustrated in Figure 4.35.

The integration property of Fourier transforms

$$g(t) \rightleftharpoons G(\omega) \Longrightarrow \int_{-\infty}^{t} g(u)du \rightleftharpoons \frac{1}{j\omega} G(\omega)$$

also applies to the transformation in equation (4.66), provided $\lim_{t \to \pm\infty} i(t)$ is finite. Using the time shift property of Fourier transforms, $g(t) \rightleftharpoons G(\omega) \Longrightarrow g(t - \tau) \rightleftharpoons e^{-j\omega\tau}G(\omega)$, and the fact that $\delta(t) \rightleftharpoons 1$, equation (4.66) becomes

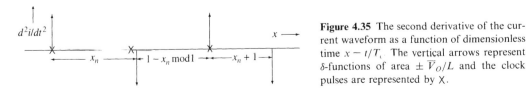

Figure 4.35 The second derivative of the current waveform as a function of dimensionless time $x - t/T_c$. The vertical arrows represent δ-functions of area $\pm \overline{V}_O/L$ and the clock pulses are represented by X.

$$A(\omega) = -\frac{\overline{V}_O}{\omega^2 L} \lim_{N \to \infty} \frac{1}{T_N} [\{1 - \exp(-j\omega T_c x_1)\}$$

$$+ \exp(-j\omega T_c [1 + \lfloor x_1 \rfloor])\{1 - \exp(-j\omega T_c x_2)\} + \dots \tag{4.69}$$

$$+ \exp(-j\omega T_c [N - 1 + \lfloor x_1 \rfloor + \dots + \lfloor x_{N-1} \rfloor])\{1 - \exp(-j\omega T_c x_N)\}]$$

where $\lfloor x \rfloor$ is the integer part of x; the factor $-1/\omega^2$ represents integrating twice; and the factor $\overline{V}_O/L$, which is the difference between di/dt when the switch is closed and when it is open, sets the vertical scale correctly. The time T_N is the total time taken for N on-and-off switchings of S, and is therefore defined as

$$T_N = T_c \sum_{n=1}^{N} 1 + \lfloor x_n \rfloor$$

Defining the integers J_n as

$$J_n = \begin{cases} 0 & n = 1 \\ \sum_{k=1}^{n-1} 1 + \lfloor x_k \rfloor = n - 1 + \sum_{k=1}^{n-1} \lfloor x_k \rfloor & n > 1 \end{cases}$$

equation (4.69) can be rewritten

$$A(\omega) = -\frac{\overline{V}_O}{\omega^2 L} \lim_{N \to \infty} \frac{1}{T_N} \sum_{n=1}^{N} e^{-j\omega T_c J_n} \{1 - e^{-j\omega T_c x_n}\} \tag{4.70}$$

This expression simplifies when $\omega = m\omega_c$, corresponding to the clock frequency or its harmonics; in this case, which is the case of interest to us here, equation (4.70) becomes

$$A_m = -\frac{\overline{V}_O}{m^2 \omega_c^2 L} \lim_{N \to \infty} \frac{1}{T_N} \sum_{n=1}^{N} 1 - e^{-2j\pi m x_n} \tag{4.71}$$

The preceding equation can be evaluated by *brute force* simulation. We follow a more refined approach (which also has the merit of demonstrating that the limit in equation (4.71) does not generally depend on x_1). There are two tasks involved in the more refined approach: (1) calculation of the asymptotic behavior of T_N and (2) evaluation of the sum in (4.71). Both of these can be addressed by using the invariant density ρ of the mapping F, in conjunction with Birkhoff's ergodic theorem [10].

Given an expanding mapping F with invariant density $\rho(x)$ on $[0, \alpha)$, Birkhoff's theorem states that

$$\lim_{N \to \infty} \frac{1}{N} \sum_{n=1}^{N} \phi\left(F^{[n-1]}(x)\right) = \int_0^\alpha \phi(y)\rho(y)dy \tag{4.72}$$

for almost all initial conditions x and for any integrable function ϕ that maps $[0, \alpha)$ to $\mathcal{R}$. Here, $F^{[i]}(x)$ is the ith iterate of F. (See also the discussion and use of this result in Section 4.3.) Thus, in order to apply the theorem we need to calculate the invariant density of F, namely $\rho(x)$. This is the normalized density of iterates $\{x_0, x_1, \dots\}$ of the mapping over the interval $[0, \alpha)$ and its calculation is described in Section 4.4.6.

The asymptotic behavior of T_N is such that

$$\lim_{N \to \infty} \frac{T_N}{N} = \langle T_0 \rangle = T_c(1 + \langle \lfloor x \rfloor \rangle)$$

and $\langle T_0 \rangle$, the mean time between successive off-on transitions of S, can be calculated from (4.72):

$$\frac{\langle T_0 \rangle}{T_c} = 1 + \int_0^\alpha \lfloor x \rfloor \rho(x) dx \qquad (4.73)$$

Since $T_N \sim N\langle T_0 \rangle$ we can evaluate the sum in (4.71) in the same way:

$$A_m = \frac{\overline{V}_O}{m^2 \omega_c^2 L \langle T_0 \rangle} \int_0^\alpha \left(1 - e^{-2j\pi mx}\right) \rho(x) dx$$

$$= \frac{\overline{V}_O}{m^2 \omega_c^2 L \langle T_0 \rangle} \left[\int_0^\alpha e^{-2j\pi mx} \rho(x) dx - 1 \right]$$

where the second equality follows from the fact that $\rho(x)$ is normalized. Hence

$$\tilde{A}_m = |A_m|^2 = \left[\frac{\overline{V}_O}{m^2 \omega_c^2 L \langle T_0 \rangle} \right]^2 \left[\left(\int_0^\alpha \rho(x) \cos 2\pi mx \, dx - 1 \right)^2 \right.$$

$$\left. + \left(\int_0^\alpha \rho(x) \sin 2\pi mx \, dx \right)^2 \right] \qquad (4.74)$$

which allows us to calculate the PDS at the mth harmonic of the clock frequency.

It is now interesting to compare chaotic and periodic operation of the boost converter under the same conditions (i.e., with the same $\overline{V}_O$). Experimentally, this can be achieved by removing the feedback loop in Figure 4.32 and instead driving the switch periodically, using a pulse generator with duty ratio $(\overline{D})$ set so that the mean output voltage is the same as in the chaotic case. Starting from equation (4.71), this is equivalent to setting $x_n = \overline{D}, \forall n$, and $T_N = NT_c$. Hence

$$A_m = -\frac{\overline{V}_O}{m^2 \omega_c^2 L} \lim_{N \to \infty} \frac{1}{NTc} \sum_{n=1}^N 1 - e^{-2j\pi m\overline{D}} \qquad (4.75)$$

Eliminating $\overline{\Delta i}$ between (4.63) and (4.64) gives $\overline{D} = \alpha/(1 + \alpha)$ and so the PDS for periodic operation is

$$\tilde{A}_m = |A_m|^2 = 2 \left[\frac{\overline{V}_O}{m^2 \omega_c^2 L T_c} \right]^2 \left(1 - \cos \frac{2\pi m\alpha}{1 + \alpha} \right) \qquad (4.76)$$

4.4.6 The Invariant Density Algorithm

The Frobenius-Perron operator P related to the mapping F (4.61), is defined by

$$P\{g(x)\} = \sum_{y=F^{-1}(x)} \frac{1}{|F'(y)|} g(y)$$

(see the appendix in Section 4.3 for an outline of the derivation of this operator). Its invariant density $\rho(x)$ can be calculated as the limit $\lim_{n \to \infty} \rho_n(x)$ where $\rho_n(x) = P(\rho_{n-1})(x)$, with $\rho_0(x)$ any nonnegative function with integral unity (e.g., constant at the value $1/\alpha$). This is illustrated in Figure 4.36, in which $\alpha = 2.3$, but the scheme converges for any $\alpha > 1$.

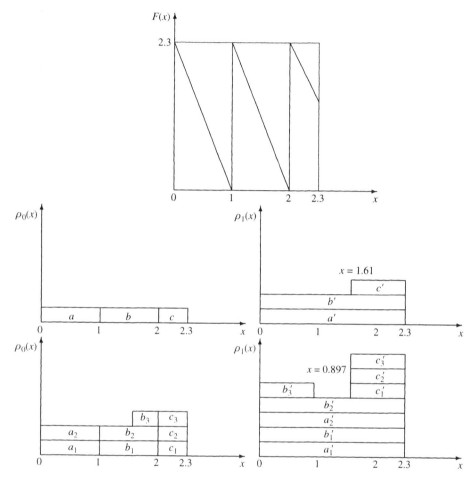

Figure 4.36 The mapping $F(x)$ and the algorithm for calculating the invariant density $\rho(x)$ by successive approximations $\rho_0, \rho_1, \rho_2, \ldots$, with $\alpha = 2.3$.

Take the obvious choice for $\rho_0(x)$, namely

$$\rho_0(x) = \begin{cases} \frac{1}{\alpha} & 0 \le x < \alpha \\ 0 & \text{otherwise} \end{cases}$$

which is clearly normalized. The x-axis is split into three intervals, labeled $a = [0, 1)$; $b = [1, 2)$; and $c = [2, \alpha]$. The densities on a and b are mapped by F uniformly and with equal weighting (because $|dF/dx| = \alpha$ everywhere except at $x = 1, 2, \ldots$) onto a' and b', where $a' = b' = [0, \alpha]$, but the density on c is mapped onto $c' = [\alpha, F(\alpha)] = [2.3, 1.61]$. The first approximation to ρ, denoted by $\rho_1(x)$, then comprises the sum of these three densities. To calculate ρ_2 from ρ_1, the same procedure is used, except that this time there are eight subintervals to be considered, a_1, a_2, $b_1 \ldots b_3$, and $c_1 \ldots c_3$. All the intervals except $b_3 = [1.61, 2)$ have been treated before, and are dealt with in the same way again; b_3 is mapped to $b_3' = [0, F^{[2]}(\alpha)] = [0, 0.897]$. The sum of these eight densities gives ρ_2. This algorithm has been implemented by computer.

Note that this algorithm can also be modified to calculate the invariant density of any piecewise-linear mapping, provided that the densities on each interval are mapped with a weighting factor equal to the modulus of the reciprocal of the gradient of the mapping in each interval. For special $F(x)$, more specialized algorithms can be developed; specifically, if α is such that $F(x)$ has a finite number of Markov partitions [12], then the eigenvector computation illustrated in the appendix of Section 4.3 can be used.

4.4.7 Practical Results

The approximate $\rho(x)$, using 20 iterations of the scheme described above, for $\alpha = 2.65$, is shown in Figure 4.37.

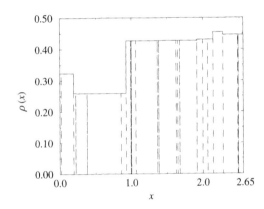

Figure 4.37 The invariant density for $\alpha = 2.65$. The dotted lines show the partitions.

Equation (4.73) can now be used to calculate $\langle T_0 \rangle$ as a function of α. A plot of $\langle T_0 \rangle / T_c$ against α is shown in Figure 4.38; note that $\langle T_0 \rangle = 2.00538 T_c$ at $\alpha = 2.65$.

Using the invariant density and equation (4.74) we can calculate the PDS at the clock frequency and its harmonics, as explained in the previous section. It is also possible, although slower, to estimate this by brute force—equation (4.71)—and for the purposes of comparison, this was also carried out. In Figure 4.39, these two calculations are compared with each other and also with the results from an experimental $i(t)$ waveform. (The vertical scale on all plots whose vertical axis is labeled *PDS (dB)* is $10 \log_{10} \tilde{A}_m$.) The experimental details are given in the next section.

It is interesting to note how the PDS at a given harmonic of the clock frequency varies with α, and this is displayed in Figure 4.40 for the fundamental and the second, fifth, and tenth harmonics. For a comparison of chaotic with periodic operation, see Figure 4.41. This shows how the calculated spectral peaks at ω_c, $2\omega_c$, $5\omega_c$, and $10\omega_c$ vary with α, when the converter is allowed to operate chaotically, and when it is forced to operate periodically, $\overline{V}_O$ being the same in each case. Purely experimental results on this were published in [2]. It can be seen from Figure 4.41 that for the fundamental and the second harmonic, the PDS is always less in the chaotic case, and that this is usually but not always so for the other harmonics considered.

4.4.8 Experimental Results

The peak-current-controlled boost converter shown in Figure 4.32 was built using $R = 293\Omega$, $L = 104\text{mH}$, $C = 220\mu\text{F}$, $V_I = 10.45\text{V}$, $\overline{V}_O = 32.3\text{V}$, reference current $I_{\text{ref}} = 0.5\text{A}$, and clock period $T_c = 400\mu\text{s}$. The total inductor series resistance $r_L = (1.0 + 2.33)\Omega$, the 1Ω being a current-sensing resistor. With these parameter values the circuit behaved chaotically. The output voltage had a chaotic ripple of peak-to-peak

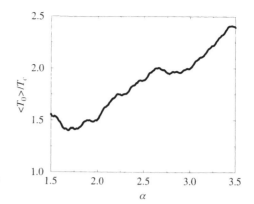

Figure 4.38 The variation of $\langle T_0 \rangle / T_c$ with α. This was calculated from equation (4.73), using $\rho(x)$ approximated by 50 iterations of the scheme outlined in Section 4.4.6.

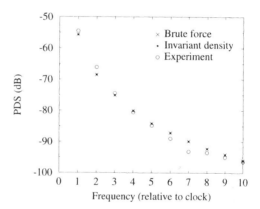

Figure 4.39 Three-way comparison of the inductor current PDS at the clock frequency and its harmonics, by brute force (4.71), the invariant density method, and experiment. The value of α is 2.65.

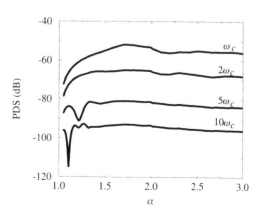

Figure 4.40 The PDS at ω_c, $2\omega_c$, $5\omega_c$, and $10\omega_c$ as a function of α, calculated from the invariant density.

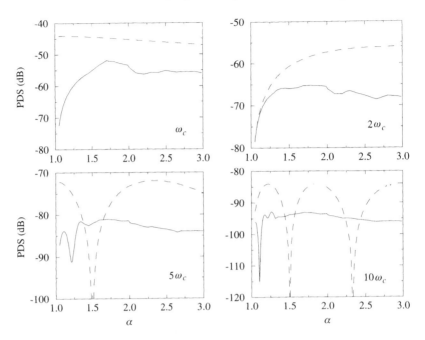

Figure 4.41 A comparison of chaotic and periodic operation. The PDS at ω_c (top left), $2\omega_c$ (top right), $5\omega_c$ (bottom left), and $10\omega_c$ (bottom right) is shown, as a function of α. Chaotic: continuous line; periodic operation with the same $\bar{V}_O$: dashed line.

amplitude ≈ 0.6V. Assumption (1) of Section 4.4.3 requires that $CR/T_c \gg 1$, which is satisfied by our values, for which $CR/T_c \approx 160$.

An analog-to-digital converter was used to monitor the inductor current $i(t)$. A section of the experimental current waveform, sampled at 40kHz, is shown in Figure 4.42. An experimental version of the mapping given in (4.61) was reconstructed from the turning points of a rather longer portion of this waveform, and is shown in Figure 4.43. The shape is exactly that expected from the theoretical mapping (4.61), confirming that the one-dimensional approximation is appropriate.

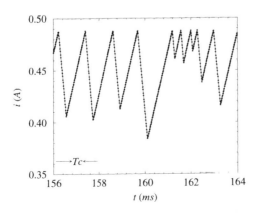

Figure 4.42 Experimental chaotic current waveform.

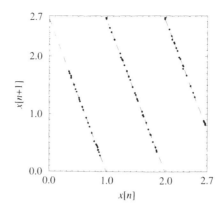

Figure 4.43 Experimental mapping deduced from the measured waveform; cf. equation 4.61. The dotted lines are a least-squares fit to each of the linear portions of the mapping.

The value of α that applies to this converter can be estimated in several ways (see similar calculations in Section 4.3). These include:

1. By definition, $\alpha =$ effective $(\overline{V}_O/V_I) - 1$. The actual value of V_I was 10.45 ± 0.06V. The mean voltage drop across r_L was measured as 1.4 ± 0.06V, and hence the effective $V_I = 9.05 \pm 0.12$V. The measured value of $\overline{V}_O$ was 32.3 ± 0.26V and the drop across the diode was 0.8 ± 0.1V; hence the effective $\overline{V}_O = 33.1 \pm 0.36$V. These results give $\alpha = 2.66 \pm 0.09$.

2. The gradients, with standard errors, of the three negative slope portions of the experimental mapping in Figure 4.43 were found by least-squares fit, giving a second experimental estimate of $\alpha = 2.66 \pm 0.02$.

3. Equation (4.65), derived from an approximate model of the circuit neglecting parasitic resistances (and with simplifications equivalent to assuming periodic operation), has the three roots $\overline{V}_O = -35.4$, -0.32, and 35.7V, the first two of which can be rejected since they are negative. This gives $\alpha = 2.94$. (A similar approach to fitting α—but using (4.23) instead of (4.65) in order to account more accurately for the converter being in the chaotic rather than periodic regime—is described in Section 4.3.)

4.4.9 Discussion

The simplified one-dimensional mapping is adequate for practical versions of a chaotically operating boost converter. From this, the power density spectrum of the input current can be calculated. Measurements verify the accuracy of our calculations, which may therefore be used with confidence as a design tool in the specification of boost converter circuits where the EMC targets are tightly controlled. One important application is in high-power-factor single-phase rectifiers for main power supplies.

Reference [13] sets out the conditions for the existence of *robust chaos*, which is characterized by the absence of periodic windows and of coexisting attractors in some region of parameter space. Note that the mapping F has a single chaotic attractor for $\alpha > 1$, and hence the boost converter described here is robustly chaotic.

Although the full circuit description results in a two-dimensional mapping, the practical benefits of extending the analysis to this case are minimal, particularly in view of the extra theoretical and computational difficulties this would raise.

These results may, of course, be applied to other physical systems that can be described by a piecewise-linear one-dimensional mapping.

We have shown how to calculate spectral peaks of the input current in a boost converter in chaotic operation. The predictions are in satisfactory agreement with measurements on a practical circuit. The calculation method may therefore be adopted by engineers for design purposes.

REFERENCES

[1] D. Middleton, *An Introduction to Statistical Communication Theory*. New York: McGraw-Hill, 1960; reprinted by IEEE Press in 1996.

[2] J. H. B. Deane and D. C. Hamill, Improvement of power supply EMC by chaos, *Electronics Letters*, vol. 32 no. 12, p. 1045, June 1996.

[3] J. L. R. Marrero, J. M. Font, and G. C. Verghese, Analysis of the chaotic regime for dc-dc converters under current-mode control, *IEEE Power Electronics Specialists Conf.* (Baveno, Italy), pp. 1477–1484, 1996.

[4] J. H. B. Deane, P. Ashwin, D. C. Hamill, and D. J. Jefferies, Calculation of the periodic spectral components in a chaotic dc–dc converter, *IEEE Trans. on Circuits and Systems*, vol. 46, no. 11, pp. 1313–1319, November 1999.

[5] A. L. Baranovski, A. Mögel, W. Schwarz, and O. Woywode, Statistical analysis of a dc-dc converter, *Proc. Nonlinear Dynamics of Electronic Systems (NDES 99)* (Rønne, Bornhom, Denmark), July 1999.

[6] A. M. Stankovic, G. C. Verghese, and D. J. Perreault, Analysis and synthesis of randomized modulation schemes for power converters, *IEEE Trans. on Power Electronics*, vol. 10, no. 6, pp. 680–693, 1995.

[7] J. H. B. Deane, Chaos in a current-mode controlled boost dc-dc converter, *IEEE Trans. on Circuits and Systems*, vol. 39, no. 8, pp. 680–683, August 1992.

[8] A. Rényi, Representation for real numbers and their ergodic properties, *Acta Math. Acad. Sci. Hung.*, vol. 8, pp. 477–493, 1957.

[9] A. Lasota and M. Mackey, *Chaos, Fractals and Noise: Stochastic Aspects of Dynamics*. Applied Mathematical Sciences series, vol. 97. New York: Springer-Verlag, 1994.

[10] J-P. Eckmann and D. Ruelle, Ergodic theory of chaos and strange attractors, *Reviews of Modern Physics*, vol. 57, no. 3, part 1, pp. 617–656, 1985.

[11] D. C. Champeney, *A Handbook of Fourier Theorems*. Cambridge, UK: Cambridge University Press, 1987.

[12] S. H. Isabelle, A signal processing framework for the analysis and application of chaotic systems, Ph.D. diss., MIT, 1995.

[13] S. Banerjee, J. A. Yorke, and C. Grebogi, Robust chaos, *Physical Review Letters*, vol. 80, pp. 3049–3052, 1998.

4.5 COMPUTER METHODS TO ANALYZE STABILITY AND BIFURCATION PHENOMENA

Yasuaki Kuroe

4.5.1 Introduction

In the usual steady-state operating modes of power electronic circuits and systems, the state trajectories or orbits or waveforms are periodic, with period equal to that of the (periodic) switching operations of the power semiconductor devices. When such periodic modes become unstable, the system behavior changes to subharmonic oscillations or chaos through various types of bifurcations discussed in Chapter 3. The focus of this section is to present approaches to computer-aided analysis of stability and bifurcations of periodic solutions.

The methods discussed here are applicable to discrete-time systems that are obtained from underlying continuous-time systems by suitable sampling. Various ways of obtaining such Poincaré maps in power electronic circuits have been discussed in Chapter 2. This section presents a method for computing the Jacobian matrix of the Poincaré map (i.e., the linearization of the discrete-time model obtained by sampling), and describes a procedure for testing stability. The results are developed in terms of general nonlinear state-space models, both time varying (nonautonomous) and time invariant (autonomous). The implementation of the analysis using transient simulators for power electronic circuits and systems is also described, as is a computer method to determine bifurcation values (i.e., values of system parameters at which bifurcations occur). Applications of the methods of this section to a power electronics system [11] can be found in Section 7.5.

4.5.2 Nonlinear Systems and Stability of Periodic Solutions

Consider a nonlinear dynamical system described by a system of differential equations in state-space form:

$$\frac{dx}{dt} = \dot{x} = f(t, x) \tag{4.77}$$

where x and f are n-dimensional vectors, and $f(t, x)$ is piecewise continuous in t and has a continuous first derivative with respect to x. We assume that $f(t, x)$ is periodic in t of period T, so $f(t, \cdot) = f(t + T, \cdot)$. Also assume that (4.77) has a periodic solution $\varphi(t)$ of period T, so $\varphi(t) = \varphi(t + T)$. Our concern here is stability of the periodic solution or orbit. Let $\phi(t, t_0, x_0)$ be the general solution of (4.77) at time t with the initial condition $x(t_0) = x_0$:

$$\phi(t, t_0, x_0) = \int_{t_0}^{t} f(\tau, \phi(\tau, t_0, x_0))d\tau + x_0 \tag{4.78}$$

An important tool for investigating the stability of the periodic orbit is the Poincaré map, which replaces a continuous-time system by an appropriately chosen discrete-time system. Define the map $P : R^n \to R^n$ by

$$P(x_0) = \phi(t_0 + T, t_0, x_0) \tag{4.79}$$

The map thus defined corresponds to the stroboscopic Poincaré map for the nonauto-nomous system (4.77). It maps the initial values of the state to those one period T later, and thereby defines a discrete-time dynamical system in state-space form:

$$x_0^{k+1} = P(x_0^k) \tag{4.80}$$

where x_0^k denotes the state of the discrete-time system at the kth strobe (or sampling) instant. Note that if we choose a point $x_0 = \varphi(t_0) = \varphi_0$ on the periodic solution $\varphi(t)$ as the initial state, then

$$\varphi_0 = P(\varphi_0) \tag{4.81}$$

because $\varphi(t) = \varphi(t + T)$. This implies that a periodic solution of the system (4.77) corresponds to a fixed point of the Poincaré map (4.79) (i.e., to an equilibrium solution of the sampled model). Thus the problem of stability analysis of a periodic solution of the continuous-time system is reduced to stability analysis of the corresponding fixed point of the discrete-time system (4.80). The stability of the fixed point is in turn investigated via the linearized model of the discrete-time system (4.80) at the fixed point, which is given by

$$\xi^{k+1} = [DP(\varphi_0)]\xi^k \tag{4.82}$$

where by definition $\xi = x_0 - \varphi_0$ and $DP(\varphi_0)$ is the Jacobian matrix of P with respect to x_0 (i.e., the matrix of partial derivatives of P with respect to the various components of x_0) evaluated at $x_0 = \varphi_0$. The fixed point φ_0 is (locally) asymptotically stable when all the eigenvalues of the matrix $DP(\varphi_0)$ are inside the unit circle on the complex plane. To summarize, the stability of the periodic solution of the continuous-time system (4.77) can be evaluated as specified in the following theorem.

Theorem 1: Let $P : R^n \to R^n$ be a map defined by (4.79) and let DP be the Jacobian matrix of P with respect to x_0. The periodic solution $\varphi(t)$ of (4.77) is asymp-totically stable if all the eigenvalues of the Jacobian matrix $DP(\varphi_0)$, where $\varphi_0 = \varphi(t_0)$, are inside the unit circle in the complex plane (i.e., have magnitudes less unity).

The local stability of the periodic solution of a periodically varying state-space model can thus be checked by evaluating (the eigenvalues of) the Jacobian matrix of its Poincaré map.

Next we consider an autonomous system described by the state-space model

$$\dot{x} = f(x) \tag{4.83}$$

where x and f are n-dimensional vectors, and $f(x)$ has a continuous first derivative with respect to x. We again assume that the system (4.83) has a periodic solution $\varphi(t)$ whose stability is our concern here. Let $\phi(t, x_0)$ be the general solution of (4.83) with the initial condition $x(0) = x_0$; we have chosen the initial time to be $t_0 = 0$ without loss of general-ity in this autonomous case, so t_0 no longer appears as a parameter in our notation for the general solution. As with the nonautonomous system (4.77), we define the map $P : R^n \to R^n$ by

$$P(x_0) = \phi(T, x_0) \tag{4.84}$$

The stability properties of the periodic solution of this autonomous system are captured in the following theorem [1,3].

    Theorem 2: Let $P : R^n \to R^n$ be a map defined by (4.84) and let DP be the Jacobian matrix of P with respect to x_0. The Jacobian matrix $DP(\varphi_0)$, where $\varphi_0 = \varphi(0)$, always has one eigenvalue of unity. The periodic solution $\varphi(t)$ of (4.83) is asymptotically stable if the magnitudes of the $n - 1$ other eigenvalues of the Jacobian matrix $DP(\varphi_0)$ are all less than unity.

    The fact the Jacobian matrix $DP(\varphi_0)$ always has one eigenvalue of unity can be understood through the following intuitive explanation (a more mathematical version of this can be given, but is omitted here). Suppose the original system is released from an initial state that is obtained by perturbing x_0 slightly in the direction of the periodic orbit (i.e., along the direction of $\dot{\varphi}(t)$). This simply has the effect of starting the system further along on its periodic orbit than where it started in the unperturbed case. As a consequence, a time T later the system trajectory has returned to the perturbed position from which it started. In other words, $\dot{\varphi}(t)$ is an eigenvector of the Jacobian matrix $DP(\varphi_0)$, with associated eigenvalue of unity.

    Note that the period T of the periodic solution in an autonomous system of the form (4.83) is usually unknown. To accommodate this, the formal definition of the Poincaré map is actually slightly different from what we have used [1]. In the formal definition, we let $\Sigma \subset R^n$ denote a hyperplane of dimension $n - 1$ that the periodic solution $\varphi(t)$ intersects at a point (let the point be $\varphi(0) \in \Sigma$ here), and let $U \subset \Sigma$ be some neighborhood of $\varphi(0)$. The Poincaré map $P : U \to \Sigma$ is defined by replacing (4.84) by

$$P(x_0) = \phi(T', x_0) \tag{4.85}$$

where $T' = T'(x_0)$ is the time taken for the trajectory starting at x_0 to first return to Σ. The Poincaré map thus defined replaces the nth-order continuous autonomous system by an $(n - 1)$th-order discrete-time system (rather than the nth-order discrete-time system we used earlier), because it is defined on the $(n - 1)$-dimensional hyperplane Σ. Its Jacobian matrix corresponds to the $(n - 1) \times (n - 1)$ matrix obtained by deleting the eigenspace of the unity eigenvalue from the Jacobian matrix $DP(\varphi_0)$ in Theorem 2. For more details and proofs of Theorems 1 and 2, see [1,3].

4.5.3 Computer Methods to Analyze Stability

    This subsection describes numerical methods to check the stability of periodic solutions of nonautonomous and autonomous systems of the form (4.77) and (4.83), based on Theorems 1 and 2 respectively. Recall first that stability of an *equilibrium point* x^* of the system

$$\dot{x} = f(x) \tag{4.86}$$

is investigated by checking the eigenvalues of the coefficient matrix of its linearized model at the point x^*: $\dot{\xi} = Df(x^*)\xi$, where $\xi = x - x^*$. The procedure to check the stability is: (i) obtain the equilibrium point x^*; (ii) compute the Jacobian matrix $Df(x^*)$ of $f(x)$ at the the equilibrium point $x = x^*$; and finally; (iii) evaluate the eigenvalues of $Df(x^*)$. The equilibrium point x^* is obtained by solving the nonlinear equation

$$f(x) = 0 \tag{4.87}$$

The Newton-Raphson method is usually utilized to solve it numerically, by carrying out the iteration

$$x^{i+1} = x^i - [Df(x^i)]^{-1}f(x^i) \tag{4.88}$$

where the superscripts denote the iteration index.

Note that the Jacobian matrix $Df(x^*)$ in (ii) corresponds to that in the iteration (4.88) when it converges ($x^i \to x^*$). This means that the equilibrium point x^* and the Jacobian matrix $Df(x^*)$ in (i) and (ii) are obtained at the same time by using the Newton-Raphson method.

In order to numerically check the stability of *periodic solutions* of nonautonomous and autonomous systems (4.77) and (4.83), Theorems 1 and 2 lead us to the following natural extension of the procedure used to compute and test the stability of equilibrium points:

Step 1: Obtain the periodic solution $\varphi(t)$.

Step 2: Compute the Jacobian matrix $DP(\varphi_0)$ of the Poincaré map at the periodic solution.

Step 3: Evaluate the eigenvalues of $DP(\varphi_0)$.

In Step 1, obtaining the steady-state solution of (4.77) or (4.83) is usually done by simply integrating the system equations (4.77) or (4.83) from some initial states until the transient response appears to be negligible and the solution becomes periodic. However, since many power electronics circuits and systems are very stiff due to high-frequency switching operations, this integration could extend over many periods, which requires extensive computing time. Furthermore, this integration cannot reach an unstable steady-state solution (although reverse-time integration may reach an unstable steady-state solution).

As in the case of computing equilibrium points, the Newton-Raphson method can be utilized to obtain steady-state periodic solutions in Step 1 [2,4]. Note that a periodic solution with period T corresponds to a fixed point of the Poincaré map defined in (4.79) or (4.84):

$$x_0 - P(x_0) = 0 \tag{4.89}$$

The above equation can be considered a nonlinear equation in x_0. The Newton-Raphson iteration to obtain its solution is

$$x_0^{i+1} = x_0^i - [I - DP(x_0^i)]^{-1}[x_0^i - P(x_0^i)] \tag{4.90}$$

In order to perform the above iteration, one has to compute the terms $P(x_0^i)$ and $DP(x_0^i)$ with reasonable accuracy and efficiency. Note that the period T of the periodic solutions in the nonautonomous system (4.77) is usually known *a priori*. For power electronics circuits and systems the periods can be determined by periods of voltage or current sources, or of clock pulses used in the control logic. On the other hand, the period of periodic solutions of the autonomous system (4.83) is usually unknown *a priori*.

The term $P(x_0^i)$ in (4.90) can be obtained by integrating the system equation (4.77) starting from the initial condition $x(0) = x_0^i$ for $0 \le t \le T$. The Jacobian matrix $DP(x_0^i)$ can be computed by applying sensitivity analysis methods, which will be discussed in the next subsection. Note that the Jacobian matrix $DP(\varphi_0)$ in Theorem 1 corresponds to that in the iteration (4.90) when it converges ($x_0^i \to \varphi_0$). This means that, as in the case of equilibrium points, the steady-state periodic solution $\varphi(t)$ in Step 1 and the Jacobian matrix $DP(\boldsymbol{\varphi}_0)$ in Step 2 are obtained at the same time. Therefore, the stability of the periodic solution $\boldsymbol{\varphi}(t)$ can be checked by evaluating the eigenvalues of the Jacobian matrix $DP(\boldsymbol{\varphi}_0)$ in the Newton-Raphson iteration when it converges.

In the autonomous case where T is not known, the above method does not work because the number of unknown variables, $\boldsymbol{\varphi}_0$ and T, total $n + 1$, whereas (4.89) comprises just n equations. To overcome the problem, the method is modified as follows [4]. Note that any point on the periodic orbit $\boldsymbol{\varphi}(t)$ will suffice as the initial point $\boldsymbol{\varphi}_0$ because the system is autonomous. With this in mind, let us assume that one element of $\boldsymbol{\varphi}_0$, say φ_{0k}, is known. (For instance, if the kth state variable is known or believed to cross through 0 at some point of its steady-state operation, then we could pick $\varphi_{0k} = 0$.) Now define a new unknown vector $\boldsymbol{v}_0 \in R^n$:

$$\boldsymbol{v}_0 = [x_{01}, x_{02}, \ldots x_{0,k-1}, T, x_{0,k+1}, \ldots x_{0n}]^T \tag{4.91}$$

Using the new unknown vector, we rewrite (4.89) as

$$\widetilde{\boldsymbol{x}}_0 - \widetilde{P}(\boldsymbol{v}_0) = 0 \tag{4.92}$$

where $\widetilde{\boldsymbol{x}}_0 = [x_{01}, x_{02}, \ldots x_{0,k-1}, \varphi_{0k}, x_{0,k+1}, \ldots, x_{0n}]^T$ and $\widetilde{P}(\boldsymbol{v}_0) = P(\widetilde{\boldsymbol{x}}_0) = \boldsymbol{\phi}(T, \widetilde{\boldsymbol{x}}_0)$. The Newton-Raphson iteration (4.90) now becomes

$$\boldsymbol{v}_0^{i+1} = \boldsymbol{v}_0^i - [\widetilde{I}_k - D\widetilde{P}(\boldsymbol{v}_0^i)]^{-1} [\widetilde{\boldsymbol{x}}_0^i - \widetilde{P}(\boldsymbol{v}_0^i)] \tag{4.93}$$

where $\widetilde{I}_k$ is the $n \times n$ identity matrix with its kth diagonal element being 0: $\widetilde{I}_k = diag\{1, 1, \ldots 1, 0, 1, \ldots 1\}$. The term $\widetilde{P}(\boldsymbol{v}_0^i)$ on the right-hand side of the iteration can be obtained by integrating the system equation (4.83), starting from the initial condition $\boldsymbol{x}(0) = \widetilde{\boldsymbol{x}}_0^i = [x_{01}^i, x_{02}^i, \ldots x_{0,k-1}^i, \varphi_{0k}, x_{0,k+1}^i, \ldots, x_{0n}^i]^T$ for $0 \leq t \leq T^i$. Since the kth component of the vector $\boldsymbol{v}_0^i$ is T^i, the kth column of the Jacobian matrix $D\widetilde{P}(\boldsymbol{v}_0^i)$ in the iteration (4.93) is

$$\frac{\partial \widetilde{P}(\boldsymbol{v}_0^i)}{\partial v_{0k}^i} = \frac{\partial \widetilde{P}(\boldsymbol{v}_0^i)}{\partial T^i} \tag{4.94}$$

By the definition of the Poincaré map (4.84), this column can be obtained as follows.

$$\frac{\partial \widetilde{P}(\boldsymbol{v}_0^i)}{\partial T^i} = \boldsymbol{f}(\boldsymbol{\phi}(T^i, \widetilde{\boldsymbol{x}}_0^i)) \tag{4.95}$$

The other columns can also be computed by applying sensitivity analysis methods, which are discussed in the next subsection. The Jacobian matrix $DP(\boldsymbol{\varphi}_0)$ in Theorem 2 can be evaluated by the obtained periodic solution $\boldsymbol{\varphi}(t)$ and its period T if the iteration (4.93) converges.

4.5.4 Computation of the Jacobian Matrix

The crucial step of computing the Jacobian matrix of the Poincaré map can be done by applying sensitivity analysis methods, using either the sensitivity equations or adjoint equations developed below. The Jacobian matrix $DP(\boldsymbol{x}_0^i)$ in (4.90) is expressed as

$$P(x_0) = \left[\frac{\partial\boldsymbol{\phi}(t_0 + T, t_0, x_0)}{\partial x_{01}}, \frac{\partial\boldsymbol{\phi}(t_0 + T, t_0, x_0)}{\partial x_{02}}, \ldots \right.$$
$$\left. \frac{\partial\boldsymbol{\phi}(t_0 + T, , t_0, x_0)}{\partial x_{0k}}, \ldots, \frac{\partial\boldsymbol{\phi}(t_0 + T, t_0, x_0)}{\partial x_{0n}} \right] \tag{4.96}$$

where $x_0 = [x_{01}, x_{02}, \ldots x_{0k}, \ldots x_{0n}]^T$. The iteration number i is omitted in (4.96) for notational simplicity. The kth column of the Jacobian matrix is computed as follows. Differentiating both sides of (4.77) with respect to x_{0k}, we have

$$\frac{\partial \dot{x}}{\partial x_{0k}} = Df \frac{\partial x}{\partial x_{0k}}, \qquad \frac{\partial x(t_0)}{\partial x_{0k}} = e^k \tag{4.97}$$

where $e^k \in R^n$ is the kth unit vector, with 1 in its kth position and 0's elsewhere. For simplicity, introduce the notation $z^k = \partial x / \partial x_{0k}$. Then (4.97) becomes

$$\dot{z}^k = [Df]z^k, \qquad z^k(t_0) = e^k. \tag{4.98}$$

This is the sensitivity equation for obtaining kth column of the Jacobian matrix. Integrating the sensitivity equation with the given initial condition for $t_0 \leq t \leq t_0 + T$ yields $\partial\boldsymbol{\phi}(t_0 + T, t_0, x_0)/\partial x_{0k}$. Therefore $n + 1$ integrations over one period T (one for the system equation (4.77) and n for the sensitivity equations in (4.98) ($k = 1, 2, \ldots, n$)) produce the whole Jacobian matrix.

This Jacobian matrix of the Poincaré map can also be computed by the adjoint of the sensitivity equations, as described next. One reason for discussing the adjoint approach is that it forms the basis for a very convenient sensitivity analysis procedure for power electronic systems, one that uses a circuit simulator; this procedure is described in the next subsection. Introduce a new variable $\hat{z}$, called the adjoint of the sensitivity variable z^k. Taking the inner product between the vector $\hat{z}$ at any time and each term on the two sides of the sensitivity equation (4.98) at that time, we obtain

$$\left\langle \frac{d}{dt}z^k, \hat{z} \right\rangle = \left\langle [Df]z^k, \hat{z} \right\rangle, \quad (k = 1, 2, \ldots n) \tag{4.99}$$

where $\langle \cdot, \cdot \rangle$ denotes the inner product of vectors. Substituting the equality

$$\left\langle \frac{d}{dt}z^k, \hat{z} \right\rangle = \frac{d}{dt}\left\langle z^k, \hat{z} \right\rangle - \left\langle z^k, \frac{d}{dt}\hat{z} \right\rangle$$

into (4.99) and rearranging the result yields

$$\frac{d}{dt}\left\langle z^k, \hat{z} \right\rangle = \left\langle z^k, \frac{d}{dt}\hat{z} \right\rangle + \left\langle [Df]z^k, \hat{z} \right\rangle$$
$$= \left\langle z^k, \frac{d}{dt}\hat{z} \right\rangle + \left\langle z^k, [Df]^T\hat{z} \right\rangle \tag{4.100}$$
$$= \left\langle z^k, \frac{d}{dt}\hat{z} + [Df]^T\hat{z} \right\rangle$$

where $[Df]^T$ is the transpose of the Jacobian matrix Df. Noting the right-hand side of the above equation, we choose to impose the condition

$$\frac{d}{dt}\hat{z} = -[Df]^T\hat{z} \tag{4.101}$$

which (apart from initial conditions) serves to define $\hat{z}$. This system of differential equations for the adjoint vector $\hat{z}$ is called the *adjoint system*. With the above choice, (4.100) becomes

$$\frac{d}{dt}\langle z^k, \hat{z}\rangle = 0, \quad (k = 1, 2, \ldots n) \tag{4.102}$$

Integrating both sides of the above equation for $t_0 \le t \le t_0 + T$ yields

$$\langle z^k(t_0 + T), \hat{z}(t_0 + T)\rangle - \left\langle z^k(t_0), \hat{z}(t_0)\right\rangle = 0, \quad (k = 1, 2, \ldots n) \tag{4.103}$$

We are now ready to show how to compute the Jacobian matrix of the Poincaré map via the adjoint equation (4.101). Recalling $z^k(t_0) = e^k$ and specifying $\hat{z}(t_0 + T) = e^j$, (4.103) becomes

$$z^k_j(t_0 + T) = \hat{z}_k(t_0), \quad (k = 1, 2, \ldots n) \tag{4.104}$$

This implies that the reverse-time integration of the adjoint equation (4.101) from $t = t_0 + T$ to $t = t_0$ yields the jth row of the Jacobian matrix $DP(x_0^i)$ in (4.90). Defining a new time variable by $\tau = t_0 + T - t$ and rewriting the adjoint equation (4.101), we have

$$\frac{d}{d\tau}\hat{z} = [Df]^T\hat{z}, \quad \hat{z}(t_0) = e^j \tag{4.105}$$

Once again, $n + 1$ integrations over one period T (one forward integration of the system equation (4.77) and n reverse-time integrations of the adjoint equation (4.105) $(k = 1, 2, \ldots n)$) produce the whole Jacobian matrix. The Jacobian matrix of the Poincaré map of an autonomous system can be computed similarly.

4.5.5 Analysis Method Based on Transient Simulator

We now present a method to implement the stability analysis method outlined above, by using a transient simulator for a power electronic circuit. In many power electronic circuits and systems it is not easy to write down the system equations and to derive their sensitivity or adjoint equations. The approach described here makes it possible to analyze the stability automatically without the labor of deriving these various equations.

Consider a general power electronic circuit, denoted by N, that is made up of resistors (R), capacitors (C), uncoupled inductors (L), coupled inductors (M), ideal transformers (T), power semiconductor switches, independent voltage sources (E) and current sources (J), and dependent sources, namely voltage-controlled voltage sources $(VCVS)$, current-controlled current sources $(CCCS)$, voltage-controlled current sources $(VCCS)$, and current-controlled voltage sources $(CCVS)$. The discussion here can be easily extended to more general power electronic systems which include non-linear elements, for instance, electromechanical elements such as ac and dc motors. The power semiconductor devices are treated as ideal switching elements, and three types of such devices are considered: power diodes (D), thyristors (Th), and forced switches (Sw). The forced switches Sw are idealized models of power semiconductor switches whose turn-on and turn-off operations both can be externally controlled; examples of forced switches are GTOs and power transistors. We assume that the power electronic circuit has a periodic solution of *known* period T, where T is the period of the clock input. We are interested in computing and determining the stability of this periodic solution.

We first determine the periodic steady-state solution of the circuit. Let us choose the state vector

$$x = [i_L, v_C, i_M]^T \tag{4.106}$$

which corresponds to the variables associated with energy storage elements. Let the dimension of x be n: $x \in R^n$. In steady state, the following condition holds:

$$x(0) = x(T) \tag{4.107}$$

Note that $x(T)$ is a function of $x(0)$, denoted by

$$x(T) = G(x(0)) \tag{4.108}$$

which corresponds to the Poincaré map. Equation (4.107) can be rewritten as the following nonlinear equation with respect to $x_0 = x(0)$:

$$x_0 - G(x_0) = 0 \tag{4.109}$$

Applying the Newton-Raphson method, we obtain:

$$x_0^{i+1} = x_0^i - [I - DG(x_0^i)]^{-1}[x_0^i - G(x_0^i)] \tag{4.110}$$

When the iteration converges, we obtain the initial condition x_0^* that generates the steady-state periodic solution, the stability of which can be checked by evaluating the eigenvalues of the associated Jacobian matrix $DG(x_0^*)$.

In the iteration (4.110), $G(x_0^i)$ can be computed using a transient simulator for power electronic circuits. We also need to compute the Jacobian matrix $DG(x_0^i)$ to obtain x_0^{i+1}. This Jacobian matrix can be expressed as

$$DG(x_0^i) = \begin{bmatrix} \frac{\partial i_L(T)}{\partial i_L(0)}(x_0^i) & \frac{\partial i_L(T)}{\partial v_C(0)}(x_0^i) & \frac{\partial i_L(T)}{\partial i_M(0)}(x_0^i) \\ \frac{\partial v_C(T)}{\partial i_L(0)}(x_0^i) & \frac{\partial v_C(T)}{\partial v_L(0)}(x_0^i) & \frac{\partial v_C(T)}{\partial i_M(0)}(x_0^i) \\ \frac{\partial i_M(T)}{\partial i_L(0)}(x_0^i) & \frac{\partial i_M(T)}{\partial v_C(0)}(x_0^i) & \frac{\partial i_M(T)}{\partial i_M(0)}(x_0^i) \end{bmatrix} \tag{4.111}$$

To compute the Jacobian matrix, we can apply the adjoint network approach [5], which is a circuit representation of the adjoint equation approach derived in the previous subsection. A direct circuit-based derivation and implementation of the adjoint network approach can be obtained by application of Tellegen's theorem, as sketched out next. We begin with a definition of the adjoint circuit $\hat{N}$ of a given power electronic circuit N [6,7].

Definition: The circuit $\hat{N}$ is said to be the adjoint circuit of the original power electronic circuit N if

1. The circuits N and $\hat{N}$ have the same topology.
2. Each element in $\hat{N}$ is related to its matching element in N as follows:
   - Independent voltage and current sources E and J in N are replaced in $\hat{N}$ by a short-circuited branch and an open-circuited branch, respectively.

- Elements R, C, L, T, and M in N are retained intact in $\hat{N}$, except that the inductance matrix M gets transposed to M^T.
- Dependent sources $VCCS$, $CCVS$, $CCCS$, and $VCVS$ in N are replaced in $\hat{N}$ by $VCCS$, $CCVS$, $VCVS$, and $CCCS$ respectively.
- D, Th, and Sw in N are replaced in $\hat{N}$ by forced switches Sw whose on and off transitions are synchronized to those of their matching switches in N.

Figure 4.44 shows the branch characteristics of the adjoint circuit $\hat{N}$ corresponding to the original circuit N.

Tellegen's theorem can now be applied to compute the sensitivity $DG(x_0^i)$. Suppose $\Delta x(T)$ is the deviation when $x(0)$ is perturbed by $\Delta x(0)$. By Tellegen's theorem, and following the derivation in [6], we can obtain the relation

$$\sum_L \left[\hat{i}_L(0)L\Delta i_L(T) - \hat{i}_L(T)L\Delta i_L(0) \right]$$

$$- \sum_C [\hat{v}_C(0)C\Delta v_C(T) - \hat{v}_C(T)C\Delta v_C(0)] \tag{4.112}$$

$$+ \sum_M \left[\tilde{i}_M^T(0)M\Delta i_M(T) - \tilde{i}_M^T(T)M\Delta i_M(0) \right] = 0$$

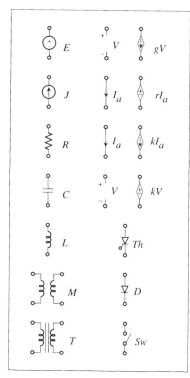

Original Network N

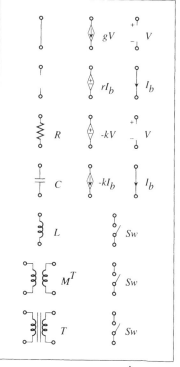

Adjoint Network $\hat{N}$

Figure 4.44 Power electronic circuit and its adjoint circuit.

where the time axis τ of the adjoint circuit is defined by $\tau = T - t$. Using (4.112), the Jacobian matrix (4.111) can be obtained row by row. For example, the kth row of the Jacobian matrix (4.111) corresponding to the kth inductor L_k is computed as follows. The initial condition of the kth inductor in the adjoint network is set to $\hat{i}_{L_k}(0) = -1/L_k$ and the rest of the initial conditions are all set to zero. Substituting these initial states into (4.112), we obtain

$$
\Delta i_{L_k}(T) = - \sum_L \hat{i}_L(T) L \Delta i_L(0) + \sum_C \hat{v}_C(T) C \Delta v_C(0)
$$
$$
- \sum_M \tilde{i}_M^T(T) M \Delta i_M(0) \tag{4.113}
$$

Then the kth row of the Jacobian matrix (4.111) is obtained as:

$$
\Big[-\hat{i}_{L_1}(T) L_1, \; -\hat{i}_{L_2}(T) L_2, \ldots \hat{v}_{C_1}(T) C_1, \hat{v}_{C_2}(T) C_2, \ldots
$$
$$
\ldots - \tilde{i}_{M_1}^T(T) M_1, -\tilde{i}_{M_2}^T(T) M_2, \ldots \Big] \tag{4.114}
$$

Similarly, the row of the Jacobian matrix (4.111) corresponding to the kth capacitor C_k is obtained by specifying fixing the initial state of kth capacitor at the value $\hat{v}_{C_k}(0) = 1/C_k$ and setting the rest of the initial values to zero. Now (4.112) simplifies to

$$
\Delta v_{C_k}(T) = - \sum_L \hat{i}_L(T) L \Delta i_L(0) + \sum_C \hat{v}_C(T) C \Delta v_C(0)
$$
$$
- \sum_M \tilde{i}_M^T(T) M \Delta i_M(0) \tag{4.115}
$$

Thus one single-period simulation of the original power electronic circuit and n single-period simulations of its adjoint circuit give us the whole Jacobian matrix; the entire computation is carried out using a transient simulator, without writing down any circuit descriptions analytically [6,7]. It should be noted that, for power electronic circuits with closed-control loop, additional sensitivity calculations (sensitivities with respect to on or off times of switching devices) are required to obtain the Jacobian matrix. For details on these refinements, see [8].

4.5.6 Computer Method to Analyze Bifurcation Phenomena

In this subsection we discuss a computer-aided method for analyzing bifurcation phenomena in nonlinear dynamical systems. In particular, we seek to determine the values of system parameters where bifurcations occur in a system.

Classification of Bifurcations

Here we give a brief introduction to those aspects of bifurcation theory that are required in this subsection (see also Sections 3.1 and 3.3). Consider a nonlinear dynamical system described by a discrete-time model depending on a parameter vector μ:

$$
x(t+1) = f_\mu(x(t)) \tag{4.116}
$$

with a fixed point x^*, so

$$x^* = f_\mu(x^*) \tag{4.117}$$

or a nonlinear dynamical system described by a continuous-time model:

$$\frac{d}{dt}x(t) = f_\mu(x(t)) \tag{4.118}$$

with an equilibrium point x^*, so

$$f_\mu(x^*) = 0 \tag{4.119}$$

The discrete-time model may be the result of constructing the Poincaré map for a continuous-time dynamical system. Note also that the fixed points x^* of (4.116) (or the equilibrium points of (4.118)) depend on the parameter μ. In what follows we consider bifurcations of x^* with respect to μ. There are several kinds of bifurcations in nonlinear dynamical systems. They are generally classified into three types, depending on the conditions that hold at the point of bifurcation.

Let $Df_\mu(x)$ be the Jacobian matrix of f_μ with respect to x. A bifurcation is said to occur in the system described by (4.116) (or (4.118)) at $\mu = \mu^*$ if one of the eigenvalues λ of $Df_\mu(x^*)$ satisfies one of the following conditions as μ is varied [1,9]:

Type 1: The eigenvalue λ is real and passes through the point $(+1, 0)$ (or $(0, 0)$ for (4.118)) in the complex plane at $\mu = \mu^*$.

Type 2: The eigenvalue λ is real and passes through the point $(-1, 0)$ in the complex plane at $\mu = \mu^*$.

Type 3: The eigenvalue λ is complex and, together with its conjugate $\bar{\lambda}$, passes through the unit circle at points other than $(+1, 0)$ and $(-1, 0)$ (or through the imaginary axis at points other than $(0, 0)$ for (4.118)) at $\mu = \mu^*$.

A Type 1 bifurcation corresponds to a *saddle-node* bifurcation, a *transcritical* bifurcation or a *pitchfork* bifurcation, these being differentiated from each other by additional conditions [1]. A Type 2 bifurcation is called a *period-doubling* or *subharmonic* bifurcation, in which the stability of x^* changes at $\mu = \mu^*$ and a new orbit which is not a fixed point of f_μ but has period-2 $(x^* \neq f_\mu(x^*)$ but $x^* = f_\mu\{f_\mu(x^*)\})$ appears. The bifurcation of Type 3 is called the *Neimark* bifurcation (for discrete-time model) or the *Hopf* bifurcation (for continuous-time model), where the stability of x^* changes and a limit cycle surrounding the equilibrium point x^* emerges. Note that the bifurcations of Type 1 and Type 3 occur in both the continuous- and discrete-time models (4.116) and (4.118), but the bifurcation of Type 2 does not occur in the continuous-time model. In order to analyze Type 2 bifurcations in continuous-time systems, therefore, it is necessary to derive some discretized model of them.

Method to Determine Bifurcation Values

We now discuss a computer method to determine values of system parameters at which bifurcations occur for the periodic solution of the system (4.77). Recall that the periodic solution of the system (4.77) satisfies (4.89), rewritten below in a form that explicitly shows the dependence on the system parameters μ:

$$x_0 - P_\mu(x_0) = 0 \tag{4.120}$$

From the above conditions, bifurcations of Types 1, 2, and 3 occur at $x_0 = x_0^*$ and $\mu = \mu^*$ if a pair (x_0, μ) satisfies the following equations:

$$\det\{I - DP_\mu(x_0)\} = 0 \quad \text{for Type 1} \tag{4.121}$$

$$\det\{-I - DP_\mu(x_0)\} = 0 \quad \text{for Type 2} \tag{4.122}$$

$$\det\{e^{j\beta} I - DP_\mu(x_0)\} = 0 \quad \text{for Type 3} \tag{4.123}$$

where β is the angle around the unit circle at which the eigenvalue λ crosses the unit circle. The problem of determining bifurcation values of the system parameters is now reduced to determining pairs (x_0, μ) that satisfy both the fixed point condition (4.120) and the appropriate bifurcation condition: (4.121) for Type 1; (4.122) for Type 2; or (4.123) for Type 3. For instance, the bifurcation values of Type 2 (period-doubling bifurcation) are determined by solving the following nonlinear equation with respect to x_0 and μ:

$$F(\mu, x_0) = 0 \tag{4.124}$$

where

$$F(\mu, x_0) = \begin{bmatrix} P_\mu(x_0) - x_0 \\ \det\{-I - DP_\mu(x_0)\} \end{bmatrix} \tag{4.125}$$

The Newton-Raphson algorithm can be applied to solve the above equation [10]. The Jacobian matrix in the Newton-Raphson algorithm can be calculated by solving the sensitivity equations or adjoint equations of the original system equations with respect to both initial conditions and parameters. The bifurcation values of Type 3 (Hopf bifurcation) can be determined similarly, with the problem being reduced to determining x_0, μ and β that satisfy (4.120) and (4.123). Note that, in obtaining the bifurcation values for Type 1, this method cannot be applied directly because $I - DP_\mu(x_0)$ becomes singular; additional bifurcation conditions are needed to determine the bifurcation values [10] in this case.

REFERENCES

[1] J. Guckenheimer and P. Holms, *Nonlinear Oscillations, Dynamical Systems, and Bifurcations of Vector Field.* New York: Springer-Verlag, 1983.

[2] T. J. Aprille and T. Trick, Steady-state analysis of nonlinear circuits with periodic inputs, *Proc. IEEE*, vol. 60, no. 1, pp. 108–114, 1972.

[3] E. A. Coddington and N. Levinson, *Theory of Ordinary Differential Equations.* New York: McGraw-Hill, 1955.

[4] T. J. Aprille and T. Trick, A computer algorithm to determine the steady-state response of nonlinear oscillators, *IEEE Trans. on Circuit Theory*, vol. CT-19, no. 4, pp. 354–360, 1972.

[5] L. O. Chua and P. Lin, *Computer-Aided Analysis of Electronic Circuits: Algorithms and Computational Techniques.* Englewood Cliffs, NJ: Prentice-Hall, 1975.

[6] Y. Kuroe, H. Haneda, and T. Maruhashi, General steady-state analysis program ANASP for thyristor circuits based on adjoint-network approach, *Proc. IEEE Power Electronics Specialists' Conference*, pp. 180–189, 1980.

[7] H. Haneda, Y. Kuroe, and T. Maruhashi, Computer-aided analysis of power-electronic dc-motor drives: Transient and steady-state analysis, *Proc. IEEE Power Electronics Specialists' Conference*, pp. 128–139, 1982.

[8] Y. Kuroe, T. Maruhashi, and T. Kanayama, Computation of sensitivities with respect to conduction time of power semiconductors and quick determination of steady state for closed loop power electronic systems, *Proc. IEEE Power Electronics Specialists' Conference*, pp. 756–764, June 1988.

[9] K. Hirai, Bifurcation phenomena and chaos in nonlinear systems, *Systems and Control*, vol. 28, no. 8, pp. 502–512, 1984 (in Japanese).

[10] H. Kawakami, Bifurcation of periodic response in forced dynamic nonlinear circuit: Computation of bifurcation values of the system parameters, *IEEE Trans. on Circuits Syst.*, vol. CAS-31, no. 3, pp. 248–260, 1984.

[11] Y. Kuroe and T. Maruhashi, Stability analysis of power electronic induction motor drive system, *Proc. IEEE 1987 International Symposium on Circuits and Systems*, pp. 1009–1013, 1987.

4.6 COMPUTATION OF OPERATING-MODE BOUNDARIES

Yasuaki Kuroe
Toshiji Kato
George C. Verghese

4.6.1 Introduction

A power electronic circuit or system in conventional periodic steady-state operation goes through a cyclic sequence of topological states or network configurations, corresponding to the various configurations of the switches. The particular sequence of topological states—or the *operating mode*—depends on circuit parameters such as duty ratio and load. If these parameters change, the circuit may move from one operating mode to another. This section discusses two computer-aided approaches to finding how operating modes change with circuit parameters, determining the boundaries (in parameter space) among the various operating modes. Such an analysis provides a more comprehensive picture of converter operation, and allows one to select circuit parameters that keep the converter in desirable operating modes.

We restrict ourselves throughout to the case of two parameters, so the parameter space is two-dimensional (and can therefore be easily represented graphically); extensions to higher dimensions can in principle be carried out along similar lines. We also assume that there is a unique periodic steady-state operating mode associated with each combination of parameters; in other words, we do not treat cases where there is more than one stable periodic solution for a given choice of parameters.

The conventional method for finding the operating-mode boundaries in power electronic circuits is to derive analytical expressions for conditions on circuit variables and parameters that hold at the boundaries, and to extract from these—by analytical or numerical methods—the parameter relations that comprise the operating mode boundaries. However, tractable analytical expressions are in general hard to obtain for all but the simplest low-order circuits. Even for a second-order circuit, it is almost impossible to derive the requisite analytical expressions if the circuit contains nonlinear or parasitic elements. Therefore, numerical methods are required that can avoid the need for analytical expressions. The first part of this section describes an approach [1] to such computer-aided tracing of the boundary, which is inspired by homotopy continuation methods [2,3]. This method needs only to know what conditions on the system variables define the boundaries, but does not require actual analytical expressions for these conditions.

The second part of this section is devoted to a computer-based method [4] that defines operating-mode boundaries without knowing what conditions on the system variables define the boundaries, but rather by efficiently finding and classifying steady-state operating waveforms numerically at a set of points in parameter space, chosen with a resolution that can be improved selectively and recursively.

4.6.2 What Is Operating-Mode Analysis?

Consider as an introductory example the basic buck-boost converter shown in Figure 4.45(a). The circuit model in (a) can be simplified by assuming that the output voltage of the converter is constant at a value V_o, as shown in (b). It is known that this converter has two operating modes, the continuous-conduction mode (or simply continuous mode) and the discontinuous-conduction mode (or discontinuous mode), depending on whether the inductor current i_L becomes zero at some part of the cycle. Figure 4.46 shows the waveforms of i_L for one periodic cycle in the two respective modes. The continuous mode involves two topological states of the circuit over the course of the cycle: Sw on and D off; then Sw off and D on. In the discontinuous mode these two topological states are followed by a third, namely Sw and D both off.

In this simple buck-boost converter, it is possible to find the boundary between continuous and discontinuous conduction analytically, by analyzing the case in which the inductor current drops down to 0 precisely at the end of a cycle, so $i_L(T) = 0$. If we neglect the (usually small) resistance R in series with the inductor, and denote the duty ratio of the switch by d, then the condition that defines the boundary is easily seen to be

$$Ed - V_o(1 - d) = 0 \qquad (4.126)$$

(A somewhat more involved expression results if the original circuit model is analyzed, but the idea is the same.) If the parameters of interest are d and V_o, then a plot of d versus V_o displays the operating mode boundary very clearly; see Figure 4.49 (where E is set at a normalized value of unity).

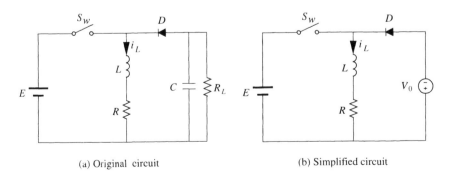

(a) Original circuit   (b) Simplified circuit

Figure 4.45 A buck-boost converter.

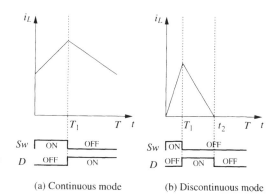

Figure 4.46 Operating waveforms of the inductor current i_L of the buck-boost converter.

(a) Continuous mode   (b) Discontinuous mode

In carrying out such an operating-mode analysis for more complicated converters, the tasks are still the same as in the above example:

1. Find all possible operating modes.
2. Determine the boundaries in parameter space among these operating modes.

In this section we take up Problem 2 for a general class of power electronic circuits.

4.6.3 Computation of Operating-Mode Boundary by Curve Tracing

In this subsection we describe our first method, based on computer-aided tracing of the boundary defined by conditions on system variables and avoiding the need for an analytical expression to define this boundary. In general, the analytical expression ends up being a nonlinear algebraic or even transcendental equation in the circuit parameters. For the two-parameter case (which is all that we treat here), the solution of such an equation can be represented as a curve in the parameter space (in this case, the parameter *plane*). We describe a method to trace such a solution curve using the concept of a homotopy continuation method.

Conditions that Define Operating-Mode Boundaries

The transition from one operating mode to another occurs when a particular topological state appears or disappears. Thus the condition that defines the boundary between one operating mode and another is just a limiting instance of the condition that determines the transition from one topological state to another. These topological transitions are in turn determined by the switching events that involve the various power semiconductor devices: power diodes D, thyristors Th, and forced switches Sw. The forced switches Sw are idealized models of switching devices such as GTOs and power transistors, whose turn-on and turn-off operations both can be externally controlled.

Switching events are typically determined by threshold conditions on currents or voltages that govern the power semiconductor devices. Specifically, Ds and Ths turn off when their currents become zero ($i_D = 0$ or $i_{Th} = 0$), and Ds turn on when their voltages become zero ($v_D = 0$). Similarly, the turn-on operation of Ths as well as the turn-on and turn-off operations of Sws are controlled by their triggering signals, which are generally themselves the result of certain currents or voltages reaching desired threshold values ($i_o = I^d$ or $v_o = V^d$). Writing down the appropriate threshold event for the appropriate limiting case yields the condition that defines the boundary between two operating modes. Rather than describing the procedure in general notation, we return to the specific example of the buck-boost converter in Figure 4.45 and indicate how a limiting case of a threshold condition leads to specification of the desired boundary condition.

    The inductor current i_L is a function of time t and also depends on the parameters V_o and d, so we can denote it by $i_L(t; V_o, d)$. In continuous conduction mode, the transition from the topological state in which the diode conducts to the one in which it no longer conducts occurs at $t = T$, when $i_L = i_L(T; V_o, d) \geq 0$. The limiting instance of this condition in continuous conduction mode is reached when i_L actually *becomes zero* at $t = T$:

$$i_L(T; V_o, d) = 0 \qquad (4.127)$$

This condition identifies the operating mode boundary with discontinuous conduction from the viewpoint of the continuous conduction mode.

    To obtain a detailed analytical expression that defines the boundary, one would have to now derive a detailed analytical expression for i_L and impose the condition (4.127), thereby producing an expression such as (4.126). However, for our numerical approach, this is not done. We instead consider how the same operating mode boundary looks when expressed from the viewpoint of the discontinuous conduction mode. The reason for requiring the complementary viewpoint is that our numerical solution will inevitably lie on one side or the other of this boundary, and we will need the appropriate form of the boundary condition in order to move toward the boundary.

    In discontinuous conduction mode, the transition from the topological state in which the diode conducts to the one in which it no longer conducts occurs at a time $t_1 \leq T$ at which the inductor current first hits zero. This time t_1 is implicitly defined via a relation of the form

$$i_L(t_1; V_o, d) = 0 \qquad (4.128)$$

so we can denote it by $t_1(V_o, d)$. From the point of view of the discontinuous conduction mode, the boundary with continuous conduction is reached when the time t_1 becomes equal to the period T:

$$t_1(V_o, d) - T = 0 \qquad (4.129)$$

    Equations (4.127) and (4.129) yield the same relation between d and V_o, of course, and define an operating mode boundary curve in the $V_o - d$ parameter plane that demarcates the boundary between the continuous and discontinuous modes; see Figure 4.49. Either equation would suffice for an analytical solution. For our numerical solution, we use both together.

    We now describe how to numerically solve and trace the solution curves defined by the boundary conditions.

Numerical Tracing of Boundary Curves

As discussed above, boundary conditions among operating modes in power electronics circuits are given by nonlinear equations in the parameters of interest. For two parameters, a general boundary condition takes the form

$$f(p_1, p_2) = 0 \qquad (4.130)$$

where p_1 and p_2 are the parameters of interest and f is a continuously differentiable function (mapping R^2 to R). Suppose that the solution of (4.130) is a continuous curve in the parameter space (p_1, p_2). Suppose also that a point on the solution curve is given *a priori* as a starting point for the tracing of the boundary; this may be a point that is easily computed analytically (e.g., the case $d = 0$ for the buck-boost converter). We now discuss a method to trace the solution curve continuously, starting from the given point, by applying the homotopy continuation method [2,3].

The approach we use consists of two phases at each iteration of the tracing process, a predictor phase and corrector phase, as shown in Figure 4.47. In the predictor phase, starting from a known solution point denoted by $A : (p_{1,j}, p_{2,j})$, we move a small distance along the tangent line at the point A, arriving at the point $B : (p_{1,j+1}^{(0)}, p_{2,j+1}^{(0)})$. Denoting by s the distance (or arc length) measured along the solution curve, the point $B : (p_{1,j+1}^{(0)}, p_{2,j+1}^{(0)})$ is given by the following equation:

$$
\begin{aligned}
p_{1,j+1}^{(0)} &= p_{1,j} + \frac{dp_{1,j}}{ds} \Delta s \\
p_{1,j+1}^{(0)} &= p_{2,j} + \frac{dp_{2,j}}{ds} \Delta s
\end{aligned}
\qquad (4.131)
$$

where Δs is a specified small distance. The tangent direction $\left(\frac{dp_{1,j}}{ds}, \frac{dp_{2,j}}{ds} \right)$ can be obtained by solving the following equation:

$$
\begin{cases}
\dfrac{\partial f}{\partial p_{1,j}} \cdot \dfrac{dp_{1,j}}{ds} + \dfrac{\partial f}{\partial p_{2,j}} \cdot \dfrac{dp_{2,j}}{ds} = 0 \\
\left(\dfrac{dp_{1,j}}{ds} \right)^2 + \left(\dfrac{dp_{2,j}}{ds} \right)^2 = 1
\end{cases}
\qquad (4.132)
$$

The first of these equations is simply the result of differentiating (4.130), while the second just normalizes the length of the tangent vector.

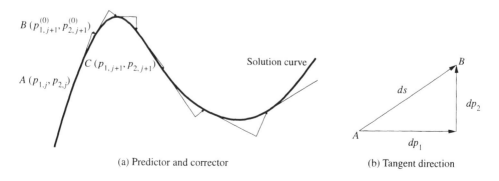

(a) Predictor and corrector   (b) Tangent direction

Figure 4.47 Determination of operating-mode boundaries by curve tracing.

Since the point $B(p_{1,j+1}^{(0)} p_{2,j+1}^{(0)})$ does not necessarily satisfy (4.130), we require the corrector phase. This phase involves solving (4.130) using Newton-Raphson iterations, with the initial guess being the point B. However, an additional constraint is required to solve (4.130) by the Newton-Raphson method. The constraint we impose is that the search direction for the Newton-Raphson step be perpendicular to that of the predictor phase (i.e., perpendicular to the tangent direction of the solution curve at the point A). The result is an efficient search process. The iteration in the corrector phase is as follows:

$$\begin{cases} p_{1,j+1}^{(i+1)} = p_{1,j+1}^{(i)} + \Delta p_{1,j+1}^{(i+1)} \\ p_{2,j+1}^{(i+1)} = p_{2,j+1}^{(i)} + \Delta p_{2,j+1}^{(i+1)} \end{cases} \tag{4.133}$$

where i is the iteration number, and $\Delta p_{1,j+1}^{(i+1)}$ and $\Delta p_{2,j+1}^{(i+1)}$ are obtained by solving

$$\begin{bmatrix} \dfrac{\partial f}{\partial p_{1,j}^{(i)}} & \dfrac{\partial f}{\partial p_{2,j}^{(i)}} \\[2mm] \dfrac{dp_{1,j}}{ds} & \dfrac{dp_{2,j}}{ds} \end{bmatrix} \begin{bmatrix} \Delta p_{1,j+1}^{(i+1)} \\[2mm] \Delta p_{2,j+1}^{(i+1)} \end{bmatrix} = \begin{bmatrix} -f\left(p_{1,j+1}^{(i)}, p_{2,j+1}^{(i)}\right) \\[2mm] 0 \end{bmatrix} \tag{4.134}$$

The first equation in (4.134) is associated with the nonlinear equation (4.130) and the second equation corresponds to the additional constraint on the search direction (perpendicular search). The iteration converges to a point $C:(p_{1,j+1}p_{2,j+1})$ that is on the solution curve and in the neighborhood of point A.

The solution curve of (4.130) can be traced by repeated alternation between the predictor phase (4.131), (4.132) and the corrector phase (4.133), (4.134). Note that in (4.132) and (4.134), we are required to compute $\partial f/\partial p_1$ and $\partial f/\partial p_2$. In the case of the buck-boost converter, where the function f denotes either the inductor current defined in (4.127) or the turn-off instant of the diode implicitly defined in (4.128), these partial derivatives correspond to the parameter sensitivities of the inductor current or the diode turn-off instant, respectively. In the next section an efficient method for *numerically* computing such sensitivities is outlined.

Computation of Steady-State Sensitivities

As noted above, the boundary curve tracing algorithm requires us to numerically compute the sensitivities of currents, voltages, and/or switching times with respect to the parameters: $\frac{\partial i_j}{\partial p}$, $\frac{\partial v_j}{\partial p}$, and/or $\frac{\partial t_s}{\partial p}$. These sensitivities are rather difficult to compute because of the following facts: (1) while the sensitivities of voltage and current values can be computed in well-established ways, the sensitivities of switching times are more involved to calculate; (2) the sensitivities of interest are those in steady state, which requires efficient computation of steady states; and (3) all switching actions, including the switching involved at the boundary of the operating mode, have to be considered. This subsection outlines an effective computation method for the required sensitivities, keeping in mind the above difficulties.

Suppose that, for a given operating mode, there occur n switching actions during one steady-state cycle, with switching instants denoted by $t_1, t_2, \ldots t_n$, all of which can depend on p. Note that these times will also depend in general on the initial conditions. The equations that explicitly or implicitly determine these times can then be expressed as follows:

$$S_i(t_i; \mathbf{x}_0^*, p) = 0 \quad (i = 1, 2, \ldots n) \tag{4.135}$$

where $\mathbf{x}_0$ is the initial value of the state vector $\mathbf{x}$ (which is composed of the variables associated with the energy-storage elements) and $*$ denotes the steady state.

In the periodic steady state, the following condition holds:

$$\mathbf{x}_0^* - \mathbf{x}_T(\mathbf{x}_0^*, p) = 0 \tag{4.136}$$

where $\mathbf{x}_T$ above denotes the state after one period T, when the system starts in the state $\mathbf{x}_0^*$. Differentiating both sides of (4.135) and (4.136), we have

$$\begin{bmatrix} \frac{\partial S_1}{\partial t_1} & \frac{\partial S_1}{\partial t_2} & \cdots & \frac{\partial S_1}{\partial t_n} & \frac{\partial S_1}{\partial \mathbf{x}_0^*} \\ \frac{\partial S_2}{\partial t_1} & \frac{\partial S_2}{\partial t_2} & \cdots & \frac{\partial S_2}{\partial t_n} & \frac{\partial S_2}{\partial \mathbf{x}_0^*} \\ \vdots & \vdots & \ddots & \vdots & \vdots \\ \frac{\partial S_n}{\partial t_1} & \frac{\partial S_n}{\partial t_2} & \cdots & \frac{\partial S_n}{\partial t_n} & \frac{\partial S_n}{\partial \mathbf{x}_0^*} \\ -\frac{\partial \mathbf{x}_T}{\partial t_1} & -\frac{\partial \mathbf{x}_T}{\partial t_2} & \cdots & -\frac{\partial \mathbf{x}_T}{\partial t_n} & I - \frac{\partial \mathbf{x}_T}{\partial \mathbf{x}_0^*} \end{bmatrix} \begin{bmatrix} \frac{\partial t_1^*}{\partial p} \\ \frac{\partial t_2^*}{\partial p} \\ \vdots \\ \frac{\partial t_n^*}{\partial p} \\ \frac{\partial \mathbf{x}_0^*}{\partial p} \end{bmatrix} = \begin{bmatrix} -\frac{\partial S_1}{\partial p} \\ -\frac{\partial S_2}{\partial p} \\ \vdots \\ -\frac{\partial S_n}{\partial p} \\ \frac{\partial \mathbf{x}_T}{\partial p} \end{bmatrix} \tag{4.137}$$

The above equations are simultaneous linear equations, the solution of which gives us the sensitivities $\frac{\partial i_j}{\partial p}$, $\frac{\partial v_j}{\partial p}$ and/or $\frac{\partial t_i}{\partial p}$, which are required in the tracing algorithm. All the elements of the matrix and the vector on the right in (4.137) can be obtained by sensitivity analysis using the adjoint network; see Section 4.5.5 (see also [5]).

Numerical Examples

The above algorithm can be implemented using a general-purpose simulator for power electronic circuits. Figure 4.48 shows the flowchart of an operating-mode analysis program. In order to demonstrate the basic performance of the method, the buck-boost converter shown in Figure 4.45 is taken up again. The parameters are chosen as: $E = 1$ (normalized units), $L = 100\mu H$, $R = 0.1\Omega$, and switching frequency $f = 1/T = 20(\text{kHz})$. The operating-mode boundary of the converter on the $V_o - d$ parameter space is now traced, starting from the point $(V_o, d) = (0, 0)$. Figure 4.49 shows the result obtained with one particular choice of the step size Δs. The solid line tracks the movement of the computed parameter values through the prediction and correction phases, the asterisks $*$ indicate the obtained boundary points after the convergence of each correction phase, and the dashed line shows the theoretically computed boundary. We chose Δs to be rather large in this example so that the tracing process can be seen clearly.

The second example is a clamped-mode series resonant dc/dc converter, shown in Figure 4.50, whose operating-mode boundaries have been studied in [6]. The circuit consists of a full-bridge inverter feeding a series resonant circuit and a bridge rectifier circuit with a resistor load. We assume that the capacitance C_F is large enough so that the output voltage V_o can be considered constant. This converter circuit has seven topological states, labeled $N0$ to $N6$ in Figure 4.51, and has six operating modes, labeled Mode I to Mode VI. Figure 4.52 shows, as examples, the waveforms of the inductor current i_L^* corresponding to Modes I, II, and III, respectively, with $v_s^* = v_1^* -$

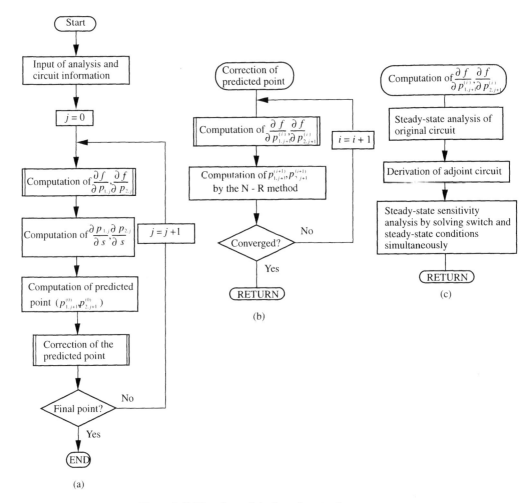

Figure 4.48 Flowchart of the boundary-tracing program.

v_2^* denoting the output voltage of the inverter. Suppose the parameters of interest are the output voltage V_o and the pulse width β of the output voltage of the inverter portion. The forms taken by the boundary conditions among the operating modes can be seen in the representative examples below:

Boundary Condition Between I and II:

$$v_C(t_z; V_o, \beta) - (E + V_o) = 0$$
$$\text{with } i_L(t_z; V_o, \beta) = 0 \tag{4.138}$$

Boundary Condition Between II and III (approached from II):

$$t_z(V_o, \beta) - \frac{T}{2} = 0$$
$$\text{with } i_L(t_z; V_o, \beta) = 0 \tag{4.139}$$

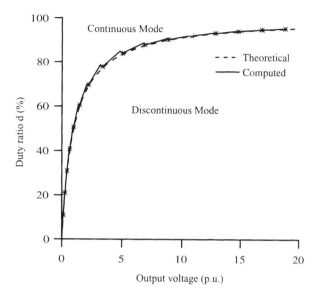

Figure 4.49 Computed and theoretical operating-mode boundaries for the buck-boost converter.

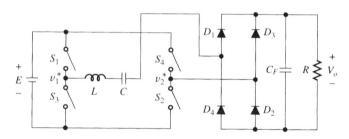

Figure 4.50 A clamped-mode series resonant dc/dc converter.

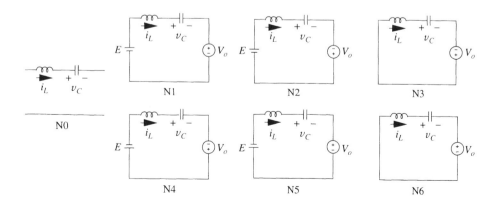

Figure 4.51 Topological states of the clamped-mode resonant converter.

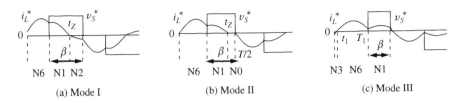

Figure 4.52 Operating waveforms of the inductor current i_L of the clamped-mode
resonant converter.

Boundary Condition Between II and III (approached from III):

$$i_L\left(\frac{T}{2}; V_o, \beta\right) = 0 \tag{4.140}$$

The boundary conditions among the other operating modes can be (derived and) written similarly. Choosing $E = 1$ (normalized units), $L = 100\mu\text{H}$, $C = 0.15\mu\text{F}$, and switching frequency $f = 1/T = 25k\text{Hz}$, we obtain the operating-mode boundaries shown in Figure 4.53. These results agree with those in [6].

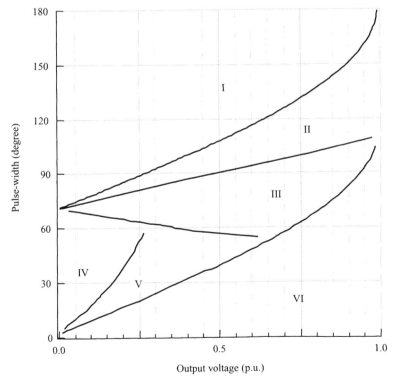

Figure 4.53 Computed operating-mode boundaries for the clamped-mode resonant
converter.

4.6.4 Computation of Operating-Mode Boundaries by a Binary-Box Method

In this subsection we explain the second method, which is based on finding and classifying steady-state operating waveforms numerically at a set of points in parameter space. The method is distinguished by its parsimonious choice of these points. The algorithm is based on a binary recursive scheme, which permits one to selectively increase the resolution with which boundaries are computed.

Basic Approach

In principle, the periodic steady-state behavior and hence the operating mode corresponding to a specific set of parameters can be determined numerically. The sequence of topological states associated with this steady-state solution is assessed to determine the operating mode, either recognizing the sequence as one that has already been encountered and labeled, or noting that it is the first representative of a new operating mode, which then gets a new label. This procedure can then be repeated at other points in the parameter space, and—if done at a fine enough resolution—will eventually lead to labeling of all the operating modes and delineation of the boundaries between them.

The result of applying this method to the buck-boost converter example in Figure 4.45 is shown in Figure 4.54. The lower figure corresponds to simulation at all grid points (17 × 17 points) in parameter space. The upper figure corresponds to simulation at far fewer grid points, selected according to the binary-box method described below. The mode classifications are shown for each grid point. The results are consistent with the analytical results of Figure 4.49.

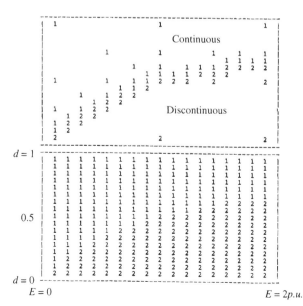

Figure 4.54 Mode boundary for the buck-boost case (1 = continuous, 2 = discontinuous).

Binary-Box Method

The direct approach above is quite general, as long as the circuit analysis program can handle the converter. However, for computational efficiency it is critical to avoid computations at unnecessary grid points. The heart of our binary-box method is a strategy for picking only grid points near boundaries. The locations of the boundaries are not known initially, of course, but they can be learned with increasing resolution as the analysis advances. The results obtained for the buck-boost example by this efficient method are shown in the upper part of Figure 4.54. It is clear that the number of grid points needed to define the boundary at a given level of resolution is considerably reduced.

The basic idea is illustrated in Figure 4.55. The grid search proceeds via binary refinement of the resolution in those regions where refinement is required. The search alternates between two patterns of refinement as it goes from one resolution to the next, alternately picking the odd-step and even-step patterns shown in Figure 4.56. At each new level of refinement, one selects only those points whose classifications are ambiguous; points that lie entirely within a single operating mode are not selected. Also, among the ambiguous points, those that are farthest from old grid points are tested first.

Application to More Complicated Converters

We now apply the binary-box method to the clamped-mode series resonant dc/dc converter treated earlier via the curve-tracing approach. The result is shown in Figure

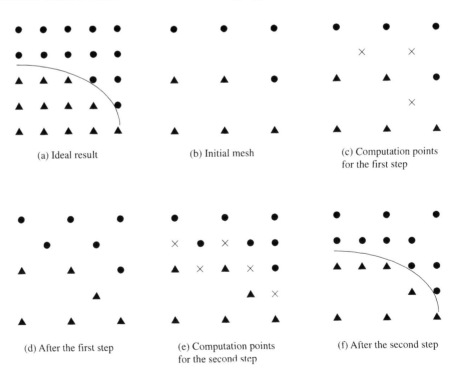

Figure 4.55 A simple example of the binary-box method.

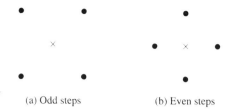

Figure 4.56 Two patterns of refinement in the binary-box method.

(a) Odd steps   (b) Even steps

4.57 and is again consistent with the analytical results in [6]. Note that there are isolated points near the boundaries, labeled 6, 7, and 9-22, where the mode classification routine declared the presence of distinct operating modes. This is the result of there being spurious topological states caused by numerical errors.

For a final example, we consider a series resonant circuit with clamped tank capacitor voltage [7], shown in Figure 4.58(a). The circuit is a modification of the previous circuit in that clamping diodes have been placed across the capacitor in a bridge arrangement, and the inductor has been split into two. These modifications improve circuit behavior in several respects. The output is controlled by the phase difference ϕ between the two inverter arms, v_1 and v_2. We treat the simplified circuit shown in Figure 4.58(b), where the source voltage is a square wave representing the output of the inverter, and the output capacitor and load are replaced with a voltage source V_l, as in the previous examples. Even this simplified circuit has 108 possible topological states! The computed operating-mode map is shown in Figure 4.59. The parameter space is divided into 32 areas (numbered from 1 to 32) whose 13 boundaries are rather intricate curves (numbered from 1 to 13 and circled). Such a calculation is practically impossible by conventional analytical methods.

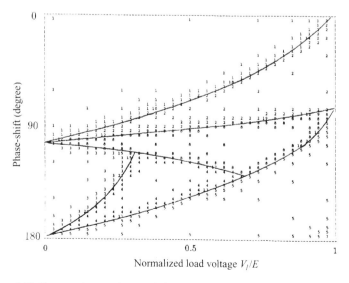

Figure 4.57 Computed operating-mode boundaries for a clamped-mode series resonant dc-dc converter.

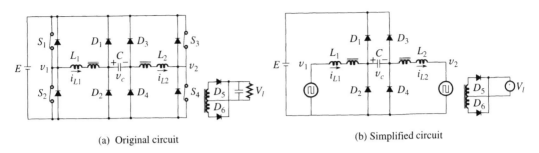

(a) Original circuit   (b) Simplified circuit

Figure 4.58 Series resonant dc/dc converter with clamped tank capacitor voltage.

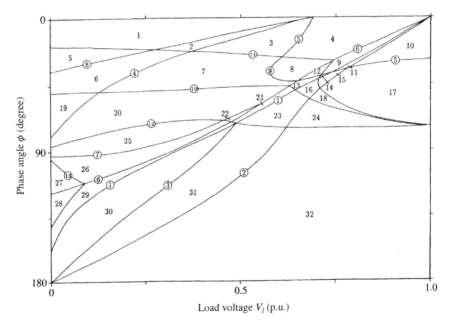

Figure 4.59 Operating modes and their boundaries.

REFERENCES

[1] Y. Kuroe and T. Kato, A computer aided method for determining operating-mode boundaries in power electronic circuits, *PESC'97 Record*, pp. 1345–1351, 1997.

[2] C. B. Garcia and W. I. Zangwill, *Pathways to Solutions, Fixed Points and Equilibria.* Englewood Cliffs, NJ: Prentice-Hall, 1981.

[3] Y. Kuroe, Efficient method to determine steady-state solution of power electronic systems by continuation method, *Proc. Int. Power Electronics Conf.*, pp. 235–241, 1990.

[4] T. Kato and G. C. Verghese, Efficient numerical determination of boundaries between operating modes of a power converter, *3rd IEEE Workshop on Computers in Power Electronics*, pp. 205–216, 1992.

[5] Y. Kuroe, T. Maruhashi, and N. Kanayama, Computation of sensitivities with respect to conduction time of power semiconductors and quick determination of steady state, *PESC'88 Record*, pp. 756–764, 1988.

[6] F. Tsai, P. Materu, and F. C. Lee, Constant-frequency clamped-mode resonant converters, *IEEE Trans. on Power Electronics*, vol. 3, no. 4, pp. 460–473, 1988.

[7] B. S. Jacobson and R. A. DiPerma, Series resonant converter with clamped tank capacitor voltage, *IEEE APEC'90 Record*, pp. 137–146, 1990.

Chapter 5

NONLINEAR PHENOMENA IN DC/DC CONVERTERS

5.1 BORDER COLLISION BIFURCATIONS IN THE CURRENT-MODE-CONTROLLED BOOST CONVERTER

Soumitro Banerjee
Priya Ranjan

In this section we consider a peak-current-controlled boost converter (Figure 5.1) operating in continuous conduction mode (CCM). The nonlinear phenomena in this converter have been investigated extensively in [1,2,3,4]. In this section we explain the observed bifurcation phenomena in light of the theory [5,6] presented in Section 3.3.

In the current-mode control logic, the switch is turned on by clock pulses that are spaced T seconds apart. When the switch is closed, the inductor current increases until it reaches the specified reference value I_{ref}. The switch opens when $i = I_{\text{ref}}$. Any clock pulse arriving during the *on* period is ignored. Once the switch has opened, the next clock pulse causes it to close.

5.1.1 Modeling and Analysis

The evolution of the state variables i and v_c during the *on* and *off* intervals is described by linear differential equations:

$$\text{``On'' interval:} \quad \frac{di}{dt} = \frac{V_{\text{in}}}{L}, \qquad \frac{dv_c}{dt} = -\frac{v_c}{CR} \qquad (5.1)$$

$$\text{``Off'' interval:} \quad \frac{di}{dt} = \frac{V_{\text{in}}}{L} - \frac{v_c}{L}, \qquad \frac{dv_c}{dt} = \frac{i}{C} - \frac{v_c}{CR} \qquad (5.2)$$

The variables can be normalized using the following definitions (assuming I_{ref}, R, L and C are not zero):

$$I = \frac{i}{I_{\text{ref}}}$$

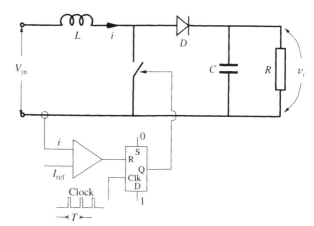

Figure 5.1 Schematic circuit diagram of the peak-current-controlled boost converter.

$$V = \frac{v_c}{I_{\text{ref}} R}$$

$$\zeta = \frac{1}{2R} \sqrt{\frac{L}{C}}$$

$$T_0 = \frac{T}{\sqrt{LC}}$$

$$\rho = \frac{V_{\text{in}}}{I_{\text{ref}} R}$$

$$\omega = \sqrt{1 - \zeta^2}$$

$$\gamma = \frac{T}{RC}$$

We adopt sampled data modeling in the form of *stroboscopic map*, where the state variables are observed in synchronism with the clock. Let the normalized state variables at a clock instant be I_n, V_n, and those at the next clock instant be I_{n+1}, V_{n+1}. There are two ways in which the state can move from one clock instant to the next (see Figure 5.2). A clock pulse may arrive before the current reaches I_{ref} (Figure 5.2(a)). In that case the map can be obtained by solving the equations for the *on*-state with V_n and I_n as the initial conditions. This yields

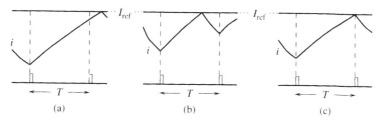

Figure 5.2 (a) and (b) The two possible types of evolution between two consecutive clock instants, and (c) the borderline case.

$$I_{n+1} = \frac{\rho T_0}{2\zeta} + I_n \tag{5.3}$$

$$V_{n+1} = V_n e^{-\gamma} \tag{5.4}$$

If the inductor current reaches I_{ref} before the arrival of the next clock pulse (Figure 5.2(b)), the map would include an *on* and an *off* interval. In this case the *on*-period t_{on} is obtained by using the final value $I = 1$ and the states at the next clock instant are found by solving the *off*-time equations for the interval $t_{off} = (T - t_{on})$. This gives the map

$$I_{n+1} = e^{-\zeta\tau_n'}\left(C_1 \cos \omega\tau_n' + C_2 \sin \omega\tau_n'\right) + \rho \tag{5.5}$$

$$V_{n+1} = \rho - e^{-\zeta\tau_n'}\left(K_1 \cos \omega\tau_n' + K_2 \sin \omega\tau_n'\right) \tag{5.6}$$

where

$$\tau_n = \frac{2\zeta}{\rho}(1 - I_n)$$

$$\tau_n' = T_0 - \tau_n$$

$$V_f = V_n e^{-2\zeta\tau_n}$$

$$C_1 = 1 - \rho$$

$$C_2 = \frac{\frac{\rho - V_f}{2\zeta} + \zeta C_1}{\omega}$$

$$K_1 = \rho - V_f$$

$$K_2 = \frac{2\zeta}{\omega}\left(\frac{V_f - \rho}{2} - C_1\right)$$

The borderline (Figure 5.2(c)) between the two cases is given by the value of I_n for which the inductor current reaches I_{ref} exactly at the arrival of the next clock pulse,

$$I_{border} = 1 - \frac{\rho T_0}{2\zeta} \tag{5.7}$$

Thus (5.3) and (5.4) apply if $I_n \leq I_{border}$, while (5.5) and (5.6) apply for $I_n \geq I_{border}$.

In the above expressions we have assumed an oscillatory solution ($\zeta < 1$) of the second-order differential equation in the *off* period. The normal design procedures, based on obtaining continuous conduction mode and low output voltage ripple, usually give parameter values which satisfy this condition. However, in order to investigate nonstandard regions of the parameters' space, one would need to obtain the map from the nonoscillatory solution also. Such a map, including the parasitic elements such as resistances of the inductor and capacitor, has been derived in [2].

Inspection of the above expressions of the stroboscopic maps reveals that the state space is divided into two distinct regions, with two different expressions for the map. A *borderline* in the state space then divides these two regions. The map is continuous throughout the state space. But the derivatives are continuous only in the two regions and are discontinuous at the borderline. The stroboscopic map of the boost converter is therefore *piecewise smooth*.

5.1.2 Analysis of Bifurcations

In this circuit it has been observed [2,4] that the asymptotic orbit of the system changes with the change in a bifurcation parameter. In the normalized map there are three primary bifurcation parameters ζ, ρ, and T_0. Moreover, the overall bifurcation structure changes with a *secondary bifurcation parameter*. We show the bifurcation diagrams in Figure 5.3, where ρ has been taken as the primary bifurcation parameter and γ is taken as the secondary bifurcation parameter.

To obtain the bifurcation diagram, we vary the bifurcation parameter ρ in steps and plot 100 values of I_n after elimination of the transients for each ρ. We also plot the value of I_{border} on the bifurcation diagram to identify border collisions. Every time I_n crosses I_{border}, there is a border collision bifurcation which is clearly visible in the bifurcation diagram.

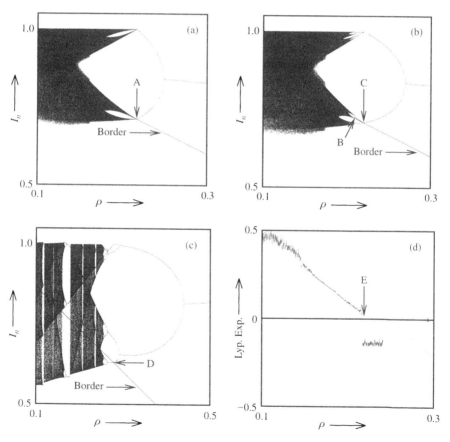

Figure 5.3 Bifurcation diagrams of the boost converter for different values of γ. (a) $\gamma = 0.125$; (b) $\gamma = 0.25$; (c) $\gamma = 1.0$ and (d) The spectrum of the maximal Lyapunov exponent corresponding to (a).

In all the bifurcation diagrams, a period-1 orbit undergoes period doubling. But the subsequent bifurcation structures are different. In Figure 5.3(d), we present the Lyapunov spectrum corresponding to Figure 5.3(a), calculated with the program Dynamics [7]. It shows that the maximal Lyapunov exponent touches zero at the period-doubling bifurcation but remains negative. Subsequently, the period-2 orbit directly bifurcates into a chaotic orbit (point A in Figure 5.3(a)). In Figure 5.3(d) we see that the Lyapunov exponent jumps from a negative value to a positive value discontinuously at the border collision (point E) rather than crossing it with a zero value continuously.

To analyze the bifurcations, we adopt the method presented in Section 3.3. Wherever a border collision bifurcation is detected, we calculate the eigenvalues of the Jacobian matrix before and after the border collision, compute the traces (τ_L and τ_R) and determinants (δ_L and δ_R) at the two sides of the border, and analyze the bifurcations as per the classification developed in Section 3.3.

To compute the eigenvalues, we note that for $I_n > I_{\text{border}}$, the Jacobian is given by

$$a_{11} = e^{-\zeta\tau_n'}\left(\frac{K_1}{\gamma}\cos\omega\tau_n' - \frac{\zeta(2-\gamma+V_f)}{\gamma\omega}\sin\omega\tau_n'\right)$$

$$a_{12} = -e^{-\zeta(\tau_n'+2\tau_n)}\frac{\sin\omega\tau_n'}{2\zeta\omega}$$

$$a_{21} = \frac{2\zeta e^{-\zeta\tau_n'}}{\gamma}\left(\frac{(\gamma - V_f - 2\zeta^2)}{\omega}\sin\omega\tau_n' + 2\zeta\cos\omega\tau_n'\right)$$

$$a_{22} = e^{-\zeta(\tau_n'+2\tau_n)}\left(\cos\omega\tau_n' - \frac{\zeta}{\omega}\sin\omega\tau_n'\right)$$

otherwise

$$a_{11} = 1$$

$$a_{12} = 0$$

$$a_{21} = 0$$

$$a_{22} = e^{-\gamma}$$

TABLE 5.1   The parameter values corresponding to Figure 5.3.

Parameter	$\gamma = 0.125$	$\gamma = 0.25$	$\gamma = 1$
R(Ω)	40	20	10
L(mH)	1.5	1.5	1.5
C(μF)	10	20	20
T(μSec.)	50	100	200
ω_0	0.4082	0.5773	1.1547
ζ	0.1531	0.2165	0.43301

The Jacobian of a period-n fixed point is calculated by multiplying the Jacobians of each fixed point depending on their locations.

The bifurcation diagram in Figure 5.3(a) shows that one of the points of the period-2 orbit collided with the border at $\rho = 0.2165$ (point A). Computation of the properties of the period-2 fixed point before and after the bifurcation gives $\tau_L = 1.3192, \delta_L = 0.5756, \tau_R = -0.6079$, and $\delta_R = -0.6697$. This means that it is a border collision bifurcation from a spiral attractor to a flip saddle and, as shown in Section 3.3, we expect that in the second iterate a fixed point would bifurcate into a chaotic attractor.

For $\gamma = 0.25$ (Figure 5.3(b)), after the period-doubling bifurcation at $\rho = 0.2691$, eigenvalues of the linearly attracting period-2 orbit approach each other and at $\rho = 0.2527$ they become complex. This spirally attracting orbit collides with the border at $\rho = 0.2165$ (point C). The computation of eigenvalues before and after border collision gives $\tau_L = 0.6701, \delta_L = 0.3126, \tau_R = -1.0655$, and $\delta_R = -0.4354$. Thus at this bifurcation a spiral attractor turns into a flip saddle and the parameters of the normal form satisfy the conditions of the border collision period-doubling bifurcation. Note that at point C, the bifurcated orbits do not emerge orthogonally to the ρ axis, confirming that this is period doubling of the border collision type.

The resulting period-4 orbit undergoes normal period-doubling and the period-8 orbit again collides with border at $\rho = 0.2085$ (point B). In this case we compute the eigenvalues of the eighth iterate before and after border collision and obtain $\tau_L = 1.0188, \delta_L = 0.0274, \tau_R = -1.9752$, and $\delta_R = -0.0358$. This is a bifurcation from regular attractor to flip saddle, where we expect a periodic orbit to bifurcate into a robust chaotic orbit.

Now we come to Figure 5.3(c) where $\gamma = 1$. Here we see two successive period-doubling bifurcations of the normal kind, at $\rho = 0.4398$ and $\rho = 0.2921$. After that the period-4 orbit collides with the border at $\rho = 0.2772$ (point D), but no bifurcation occurs. Only the path of the orbit changes a bit. Computation of the eigenvalues of the period-4 fixed point gives $\tau_L = 0.4093, \delta_L = 6.0176 \times 10^{-6}, \tau_R = 0.4287$, and $\delta_R = 4.4103 \times 10^{-5}$. Thus in the fourth iterate, it is a bifurcation from a regular attractor to a regular attractor and the behavior is as expected from the theory.

Note that the parameter γ appears in the exponent in (5.4) and in the Jacobian for $I < I_{border}$. Therefore it changes the character of the fixed point after border collision, which, in turn, affects the subsequent bifurcation phenomena. This explains why γ acts as a secondary bifurcation parameter changing the bifurcation structures.

Experimental work reported by Tse et al. [8] also revealed similar bifurcation structures (Figure 5.4). Since I_{ref} was used as the bifurcation parameter, the maximum value of i_n shows a linearly increasing trend with increase of the bifurcation parameter. Otherwise the qualitative bifurcation structures are the same in Figure 5.3 and Figure 5.4. This implies that border collision events organize the bifurcation structures in this converter and the results presented in this section are not specific to the particular parameter values chosen in the simulation.

We also conclude that *robust chaos* (absence of coexisting attractors and periodic windows) can occur in this converter topology when $T/CR \ll 1$. This result will be useful in designing converters where reliable operation under chaos is desirable.

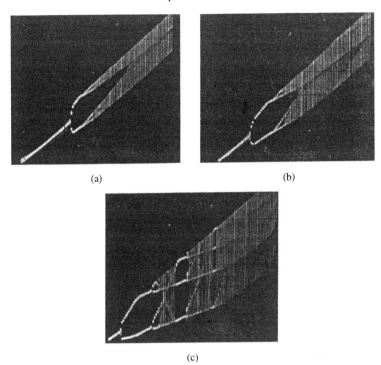

(a)   (b)

(c)

Figure 5.4 Experimental bifurcation diagrams of the current-mode-controlled boost
converter [8]. Here I_{ref} is used as the bifurcation parameter (varied from
0A to 1A). The other parameters are: frequency = 5kHz, $L = 1.5$mH,
$R = 40\Omega$, $V_{in} = 5$V. The secondary bifurcation parameter $\gamma = T/CR$ is
set by varying the value of the capacitor: (a) $\gamma = 0.025$; (b) $\gamma = 0.256$; and
(c) $\gamma = 2.564$.

REFERENCES

[1] J. H. B. Deane, Chaos in a current-mode controlled boost dc-dc converter, *IEEE Trans. on Circuits and Systems—I*, vol. 39, pp. 680–683, August 1992.

[2] S. Banerjee and K. Chakrabarty, Nonlinear modeling and bifurcations in the boost converter, *IEEE Trans. on Power Electronics*, vol. 13, no. 2, pp. 252–260, 1998.

[3] J. L. R. Marrero, J. M. Font, and G. C. Verghese, Analysis of the chaotic regime for dc-dc converters under current mode control, in *IEEE Power Electronics Specialists' Conference*, pp. 1477–1483, 1996.

[4] W. C. Y. Chan and C. K. Tse, Study of bifurcations in current programmed dc/dc boost converters: from quasiperiodicity to period doubling, *IEEE Trans. on Circuits and Systems—I*, vol. 44, no. 12, pp. 1129–1142, 1997.

[5] S. Banerjee, M. S. Karthik, G. H. Yuan, and J. A. Yorke, Bifurcations in one-dimensional piecewise smooth maps: Theory and applications in switching circuits, *IEEE Trans. on Circuits and Systems—I*, vol. 47, no. 3, 2000.

[6] S. Banerjee, P. Ranjan, and C. Grebogi, Bifurcations in two-dimensional piecewise smooth maps: Theory and applications in switching circuits, *IEEE Trans. on Circuits and Systems—I*, vol. 47, no. 5, 2000.

[7] H. E. Nusse and J. A. Yorke, *Dynamics: Numerical Explorations.* New York: Springer-Verlag, 1997.

[8] C. K. Tse and W. C. Y. Chan, Experimental verification of bifurcations in current-programmed dc/dc boost converters: From quasi-periodicity to period-doubling, in *European Conference on Circuit Theory and Design* (Budapest), pp. 1274–1279, September 1997.

5.2 BIFURCATION AND CHAOS IN THE VOLTAGE-CONTROLLED BUCK CONVERTER WITH LATCH

Soumitro Banerjee
Debaprasad Kastha
Santanu Das

5.2.1 Overview of Circuit Operation

In this section we will explore the dynamics of the buck converter with duty cycle controlled by output voltage feedback. A schematic diagram of the circuit is shown in Figure 5.5. The controlled switch S (generally realized by a MOSFET) opens and closes in succession, thus chopping the dc input into a square wave that alternates between the input voltage V_{in} and zero. The pulsed waveform is then low-pass filtered by a simple LC network, removing most of the switching ripple and delivering a relatively smooth dc output voltage v to the load resistance R. The diode D provides a path for the continuation of the inductor current during the *off* period. The dc output voltage can be easily varied by changing the duty ratio (i.e., the fraction of time that the switch is closed in each cycle).

In practice it is often necessary to regulate v against changes in the input voltage and the load current. This can be achieved by controlling the switch S by voltage feedback. In general the feedback loop may have dynamic elements to shape the stability properties, but in this section we will consider the simplest case: a proportional controller as shown in Figure 5.5. In this controller, a constant reference voltage V_{ref} is subtracted from the output voltage and the error is amplified with gain A to form a control signal $v_{con} = A(v - V_{ref})$. The switching signal is generated by comparing the control signal with a periodic sawtooth (ramp) waveform (v_{ramp}). S turns on whenever

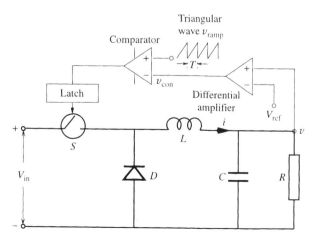

Figure 5.5 The schematic circuit diagram of the voltage-controlled buck converter.

v_{con} goes below v_{ramp} and a latch keeps it on until the end of the ramp cycle, at which point a clock pulse turns off the switch. (In this scheme, therefore, the switch is closed during the *last* part of each clock cycle rather than the first part, if it closes at all in the cycle.) Without the latch, there may be multiple on-off switchings within a single ramp cycle—a situation which will be taken up in detail in the Section 5.3.

Though this circuit or its variants are used in a large number of practical applications requiring a regulated dc power supply, it has been demonstrated [1,2,3] that the system can exhibit bifurcations and chaos for a large portion of the parameter space. To investigate the dynamics analytically, we obtain a two-dimensional Poincaré map by sampling the inductor current and capacitor voltage at the end of each ramp cycle. This is the *stroboscopic map*, introduced in Section 2.2.

Because of the transcendental form of the equations, the map cannot be determined in closed form. In simulation, the map has to be obtained numerically. It is, however, possible to infer the form of the map. Figure 5.6 shows that there are three compartments in the state space, separated by two borderlines. Since the map is continuous in the compartments as well as across the borderlines (see Figure 3.23 in Chapter 3 for illustration), the map is piecewise smooth.

5.2.2 Experimental Results

Section 3.3 has shown that some atypical bifurcations are predicted to occur when a fixed point crosses the border. These include:

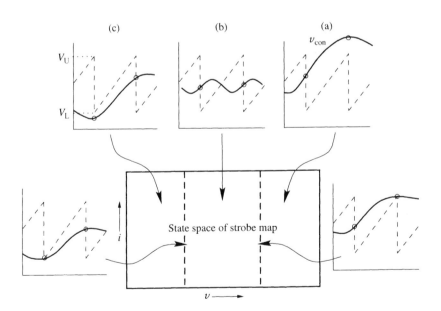

Figure 5.6 There are three possible courses of evolution from the beginning of one ramp cycle to the next: (a) the trajectory is above the ramp and the switch remains off; (b) the trajectory crosses the ramp waveform and there is an off phase and an on phase; (c) the trajectory remains below the ramp and the switch remains on through the ramp cycle. In the stroboscopic map, these three possibilities would be represented by three regions in the state space, separated by borderlines representing grazing conditions.

- A periodic orbit directly bifurcating into a chaotic orbit
- Period doubling caused by border collision, where the two bifurcated orbits do not emerge orthogonally from the earlier orbit

Here we show experimental evidence of the above bifurcation phenomena.

An experimental bifurcation diagram is shown in Figure 5.7(a). The y-axis represents one of the variables—in this case the error signal—sampled at the clock instants. The observed variable was fed to the y-channel while a periodic impulse signal synchronized with the falling edge of the triangular wave was fed to the z-modulation port of a digital storage oscilloscope which is set to the x-y mode. The input voltage V_{in}, obtained from a regulated dc power supply, was swept manually over the range 35V to 75V, and was fed to the x-channel. As a result, for each value of the input voltage, the sampled values of the error signal appear on a vertical line, and the sweep of the x-channel signal allows the resulting image to capture the change in system behavior as the input voltage is changed.

Figure 5.7(a) shows that for low values of the input voltage, the system orbit is period-1. This is the situation when the ripple in the inductor current and output voltage oscillate with the same period as the ramp waveform. As the input voltage is increased, the period-1 fixed point becomes unstable. Consequently two period-2 fixed points become stable and their paths diverge at an angle of $90°$ from the path of the period-1 fixed point. This is the typical period-doubling bifurcation. After the period-

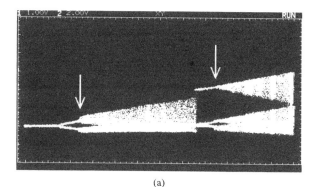

(a)

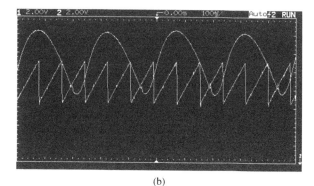

Figure 5.7 Experimental bifurcation diagram of the buck converter for: $R = 23.5\Omega$, $C = 5\mu F$, $L = 2.96mH$; triangular wave: $V_l = 8.43V$, $V_L = 3.62V$, frequency 12kHz. Bifurcation parameter V_{in} varied from 35 to 75V. (b) Time plots of v_{con} and v_{ramp} corresponding to the second arrow in (a).

(b)

doubling bifurcation, the periodic-2 orbit directly bifurcates into a chaotic orbit (shown with an arrow).

At a higher value of the parameter, a period-3 orbit comes into existence. Such emergence of a periodic window is known to occur due to saddle-node bifurcation. As the parameter is increased further, this periodic attractor again bifurcates directly into a chaotic attractor (also marked with an arrow). Such bifurcations from a periodic orbit directly into chaos have been observed in other systems also [4,5]

To establish that these are indeed border collision events, we present in Figure 5.7(b) the continuous time plots of v_{con} and the triangular wave voltage at the bifurcation point shown by the second arrow, where a period-3 orbit bifurcates into a three-piece chaotic orbit. It is seen that the v_{con} waveform grazes the top of the triangular wave, which means that a border collision bifurcation has occurred. Note that there are three cycles of v_{ramp} in a cycle of v_{con} and for that reason the stroboscopic map shows it as a period-3 fixed point. If we consider the third iterate of the map, namely $x_{n+3} = f \circ f \circ f(x_n)$, this is a bifurcation from a period-1 orbit to a one-piece chaotic orbit. We have seen in Figures 3.32 and 3.33 and that such a bifurcation can occur if an attracting fixed point moves across a border and turns into a flip saddle at the other side of the border.

Another experimental bifurcation diagram for this system is shown in Figure 5.8(a). The arrow shows a period-doubling bifurcation, but the two bifurcated orbits do not diverge at an angle of $90°$ from the path of the fixed point before the critical parameter value. This is therefore not a standard (smooth) period-doubling bifurcation. Figure 5.8(b) gives the continuous time plots of v_{con} and the triangular wave voltage just after the bifurcation, and shows that the period-doubling occurred at a border collision. In the second iterate $f \circ f(x_n)$ this is a bifurcation from a period-1 attractor to a period-2 attractor. We have seen in Figures 3.32 and 3.33 that this kind of period doubling can occur if a flip attractor or a spiral attractor moves across a border and becomes a flip saddle.

5.2.3 Coexisting Attractors and Crises

It has been found that in this system multiple attractors may exist for the same parameter values. When such coexisting attractors occur, the state often flips between the two attractors due to ambient noise. The occurrence of coexisting attractors, therefore, is a matter of engineering importance. Figure 5.9 is an oscilloscope photograph showing coexisting attractors. During the 1/30 second exposure, the orbit flipped between the two attractors and spent more time in the smaller attractor—making it brighter.

It is known that in smooth systems such coexisting orbits are born through saddle-node bifurcation. In piecewise-smooth systems, new orbits can also be born on the borderline through border collision pair bifurcation. There are four possibilities resulting from a border collision pair bifurcation: the creation of a period-1 orbit, a period-2 orbit, a chaotic orbit, or an unstable chaotic orbit. Therefore, in a piecewise-smooth system there are many possible mechanisms for creation of a coexisting orbit.

An important consequence of the occurrence of coexisting orbits is the phenomenon called *crisis* (see Section 3.6). Figure 5.10 shows a typical bifurcation diagram of the converter as the input voltage is varied from 10V to 35V. This bifurcation structure of the buck converter was first reported, through experimental and numerical investiga-

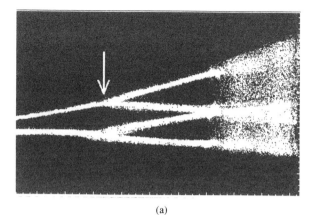

(a)

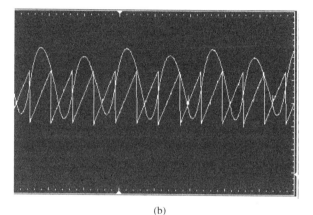

Figure 5.8 (a) Experimental bifurcation diagram of the buck converter. The parameter values are: $R = 28.9\Omega$, $C = 5\mu\text{F}$, $L = 2.96\text{mH}$; triangular wave: $V_U = 8.43\text{V}$, $V_L = 3.62\text{V}$; frequency 8kHz. Bifurcation parameter V_{in} varied from 50 to 70V. (b) Time plots of v_{con} and v_{ramp} corresponding to the arrow in (a).

(b)

tion, by Deane and Hamill [3]. Further investigation proved this structure to be prevalent over a large range of parameter values.

We find that for low values of the input voltage, the system orbit is period-1, which subsequently undergoes a typical (smooth) period-doubling bifurcation. As we have seen in Chapter 3, such a bifurcation occurs when one of the eigenvalues of the period-1 fixed point leaves the unit circle along the negative real line.

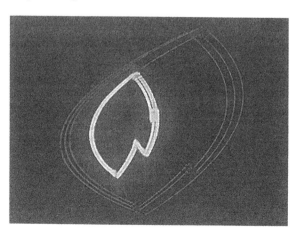

Figure 5.9 Coexisting attractors in the voltage-controlled buck converter.

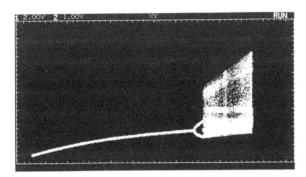

Figure 5.10 An experimental bifurcation diagram of the buck converter. Parameter values are: $R = 86\Omega$, $C = 5\mu\text{F}$, $L = 2.96\text{mH}$, $V_U = 8.5\text{V}$, $V_L = 3.6\text{V}$, clock speed 11.14kHz.

As the input voltage is further increased, the orbit suddenly becomes a much larger chaotic orbit. This chaotic orbit did not form through a period-doubling cascade; instead the sudden increase of the size of the attractor is due to a crisis.

The question is, where does the chaotic orbit come from? Such sudden expansion into a chaotic orbit has been studied in detail by dynamical system theorists and the general scenario is illustrated in Figure 5.11. In piecewise-smooth systems two more possibilities exist, which are shown in Figure 5.12.

First let us explain the general scenario (Figure 5.11) that can happen in a smooth map [6,7]. While the main attractor shows a periodic orbit, at a certain parameter value another coexisting attractor is created through a saddle-node bifurcation. This is a stable orbit with its own basin of attraction. But it will not be visible in experiment unless the state somehow moves to a point in its basin of attraction due to noise in the system. Unless that happens, its existence will remain hidden. The existence of such coexisting attractors can, however, be detected in numerical simulations if a large number of initial conditions, spread over the state space, are used.

As the parameter is changed, this new attractor undergoes the usual period-doubling route to chaos. At the same time, the basin boundaries also change with the change of parameter. At some critical parameter value, the boundary of the basins

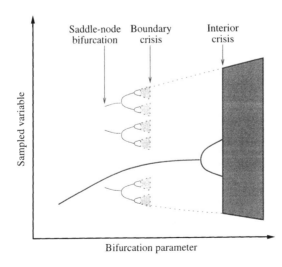

Figure 5.11 Schematic diagram explaining the sudden expansion of the attractor into a chaotic orbit.

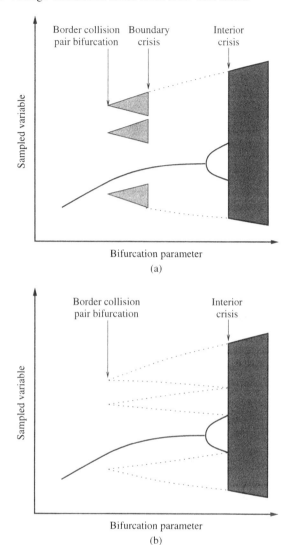

Figure 5.12 In piecewise-smooth systems two other possibilities exist: (a) birth of a chaotic orbit through border collision pair bifurcation–boundary crisis–interior crisis, and (b) direct birth of an unstable chaotic orbit through border collision pair bifurcation–interior crisis. Both these types may occur in the buck converter.

of the main attractor and the coexisting attractor touches the coexisting attractor. This is a *boundary crisis*. After the boundary crisis, the secondary attractor becomes unstable. It is then called an *unstable chaotic orbit* or *chaotic saddle*. Any initial condition falling on this orbit moves chaotically for some time and then eventually settles in the main periodic attractor. Therefore this unstable chaotic orbit is not visible in the bifurcation diagram—neither in experiment nor in simulation. In an experimental bifurcation diagram, the trajectory is generally locked to the main attractor and the coexisting orbit remains hidden. In simulation, if an initial condition falls in the region of the unstable periodic orbit, it moves chaotically for a large number of iterates before settling on the main attractor. If a sufficient number of preiterates is not eliminated, the bifurcation diagram may show lines in the region of the unstable chaotic orbit.

As the parameter is changed further, at another critical parameter value the main attractor touches the unstable chaotic orbit. At this crisis point (known as interior crisis), the preexisting unstable chaotic orbit becomes stable and merges with the main attractor. At this point we see a sudden expansion of the attractor on the bifurcation diagram. This is the standard scenario leading to the sudden enlargement of a system's trajectory.

Though the above bifurcation sequence has been studied in the context of smooth maps, similar bifurcation sequences can also occur in piecewise-smooth maps if the bifurcations occur in the smooth regions. In this converter configuration, this phenomenology has been reported in [8].

In dynamical systems represented by piecewise-smooth maps [9] there are two more possibilities, shown schematically in Figure 5.12(a),(b). We have seen in Section 3.3 that new orbits can be born on the borderline through border collision pair bifurcation. We have also seen that a border collision pair bifurcation can lead to the creation of a period-1 orbit, a period-2 orbit, a chaotic orbit, or an unstable chaotic orbit. In case a period-1 orbit is created on the border, the subsequent bifurcation structures would look like Figure 5.11. If a chaotic orbit is directly created through a border collision pair bifurcation, then the bifurcation structure would be like Figure 5.12(a). At a critical parameter value the chaotic orbit will disappear through a boundary crisis, and at another parameter value this unstable chaotic orbit will merge with the main attractor causing a sudden enlargement of the attractor.

A more exotic situation may occur if an *unstable chaotic orbit* is born on the border through a border collision pair bifurcation—a situation illustrated in Figure 3.15 as region 3. At another parameter value this unstable orbit may merge with the main attractor and cause the sudden enlargement of the attractor, as shown in Figure 5.12(b). In this case experimental or numerical investigations will reveal no coexisting attractor in the bifurcation diagram. However, long chaotic transients may be observed in numerical experiments if the initial condition falls in the unstable chaotic orbit.

It may be noted that in the buck converter, the bifurcation structure as depicted in Figure 5.10 occurs over a large range of parameter values, and a large number of coexisting attractors come into existence and go out of existence as a parameter is varied. All the situations depicted in Figure 5.11 and Figure 5.12 may occur in specific parameter ranges. The next section will show that if the latch is not included in the control logic, allowing multiple on-off sequences within one clock period, then other mechanisms can also lead to the sudden expansion of the chaotic attractor.

We thus see that in the latched voltage-controlled buck converter, many types of bifurcation phenomena can occur. Some can be understood and explained in terms of the well-known theory of smooth maps. There are also bifurcation phenomena observed in this converter configuration that can be explained only in terms of the theory of piecewise-smooth maps as presented in Chapter 3 of this book. In Figure 5.13 we present a collection of experimentally obtained bifurcation diagrams showing a wealth of bifurcations and crises, and the reader may wish to attempt explanation of these in terms of the theory presented in Chapter 3.

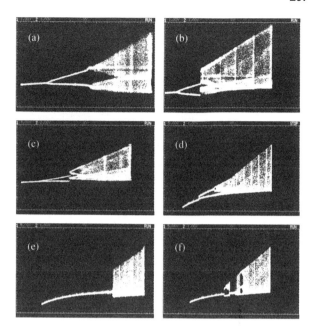

Figure 5.13 A collection of experimentally obtained bifurcation diagrams of the buck converter for various parameter ranges, showing the incredibly rich variety of nonlinear phenomena exhibited by this converter topology.

REFERENCES

[1] J. H. B. Deane and D. C. Hamill, Instability, subharmonics, and chaos in power electronic circuits, *IEEE Trans. on Power Electronics*, vol. 5, no. 3, pp. 260–268, 1990.

[2] D. C. Hamill, Power electronics: A field rich in nonlinear dynamics, in *3rd Int. Specialists' Workshop on Nonlinear Dynamics of Electronic Systems* (University College, Dublin), pp. 165–178, 1995.

[3] D. C. Hamill and J. H. B. Deane, Modeling of chaotic dc-dc converters by iterated nonlinear mappings, *IEEE Trans. on Power Electronics*, vol. 7, no. 1, pp. 25–36, 1992.

[4] C. K. Tse and W. C. Y. Chan, Instability and chaos in a current-mode controlled Ćuk converter, in *IEEE Power Electronics Specialists' Conference*, 1995.

[5] M. Ohnishi and N. Inaba, A singular bifurcation into instant chaos in a piecewise linear circuit," *IEEE Trans. on Circuits and Systems—I*, vol. 41, no. 6, pp. 433–442, 1994.

[6] C. Grebogi, E. Ott, and J. A. Yorke, Crises, sudden changes in chaotic attractors and transient chaos, *Physica D*, vol. 7, p. 181, 1993.

[7] Y. C. Lai, C. Grebogi, and J. A. Yorke, Sudden change in the size of chaotic attractors: How does it occur?, in *Applications of Chaos*, pp. 441–456. New York: Wiley Interscience, 1992.

[8] S. Banerjee, Coexisting attractors, chaotic saddles and fractal basins in a power electronic circuit, *IEEE Trans. on Circuits and Systems—I*, vol. 44, no. 9, pp. 847–849, 1997.

[9] S. Banerjee and C. Grebogi, Border collision bifurcations in two-dimensional piecewise smooth maps, *Physical Review E*, vol. 59, no. 4, pp. 4052–4061, 1999.

5.3 ROUTES TO CHAOS IN THE VOLTAGE-CONTROLLED BUCK CONVERTER WITHOUT LATCH

Mario di Bernardo
Gerard Olivar
Francesco Vasca

A dc/dc buck converter controlled by constant-frequency pulse-width modulation in continuous conduction mode gives rise to a great variety of behavior, depending on the values of the input and the circuit parameters. Subharmonics, bifurcations, and the presence of strange attractors are commonly observed as a bifurcation parameter is varied. The main antecedent in the study of chaos in power electronics circuits can be found in the work reported almost simultaneously by Deane and Hamill, on one hand [1,8], and Wood on the other [13]. The three papers illustrate how chaos can occur in dc-to-dc switching converters of the buck type operating in the continuous conduction mode and being controlled by a pulse-width modulator (PWM). These dc-to-dc switching converters of the buck type are widely used in a large area of applications, ranging from switching converters for telecommunications to airborne radar power supplies, due to their high efficiency and simplicity of control using PWM techniques. Unlike the classical examples that have been proved to show chaotic dynamics, such as Lorenz's equations or Chua's circuit, different behavior can be expected in the buck converter due to its inherent discontinuities.

5.3.1 Buck Converter Modeling Under Voltage-Mode Control

Figure 5.14 shows the block diagram of a buck regulator that uses a PWM voltage loop. We assume throughout that the components in the circuit are ideal. The com-

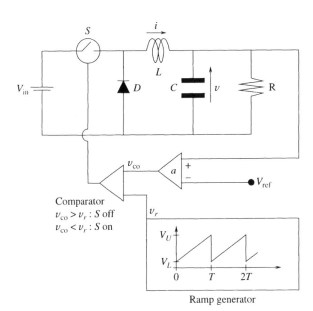

Comparator
$v_{co} > v_r$: S off
$v_{co} < v_r$: S on

Ramp generator

Figure 5.14 Block diagram of the buck converter.

parator has infinite gain, the switch S and diode D have zero on and infinite off resistance, and can switch instantly. During the interval when switch S is on, D is off and the input provides energy to the load as well as to the inductor. During the interval when switch S is off, the inductor current, which flows through the diode D, transfers some of its stored energy to the load. One of the methods for controlling the output voltage employs switching at a constant frequency (hence, a constant switching time period $T = t_{on} + t_{off}$), and adjusting the on-duration of the switch to control the average output voltage. In this method, called *pulse-width modulation* (PWM) switching, the switch duty ratio, which is defined as the ratio of the on-duration to the switching time period, is varied. Considering that the linear amplifier has gain a, we can write

$$v_{con}(t) = a(v(t) - V_{ref}) \tag{5.8}$$

Then, both v_{con} and the sawtooth voltage signal v_r are applied to the comparator, and every time the output difference changes its sign, the position of the switch S is commuted in such a way that S is open (and consequently D will close) when the control voltage exceeds the ramp voltage; otherwise S is closed (and D will open).

Differential Equations

Due to the fact that the discontinuous-conduction mode does not take place, the converter can be represented by a piecewise-linear vector field, described by two systems of differential equations as follows:

$$\frac{d}{dt}\begin{pmatrix} v(t) \\ i(t) \end{pmatrix} = \begin{pmatrix} -1/(RC) & 1/C \\ -1/L & 0 \end{pmatrix}\begin{pmatrix} v(t) \\ i(t) \end{pmatrix} + \begin{pmatrix} 0 \\ V_{in}/L \end{pmatrix}q(t) \tag{5.9}$$

where $q(t) = 0$ (and this defines system 1) when $v_{con}(t) > v_r(t)$ (i.e., during the OFF phase) while $q(t) = 1$ (and this defines system 2) when $v_{con}(t) < v_r(t)$ (i.e., during the ON phase).

Due to the linearity of the equations, analytical solutions can be easily obtained for each converter configuration, yielding

$$\begin{pmatrix} v(t) \\ i(t) \end{pmatrix} = e^{-k(t-t_0)}[I\cos(\omega(t-t_0)) + A\sin(\omega(t-t_0))]$$

$$\left[\begin{pmatrix} v_0 \\ i_0 \end{pmatrix} - V_{in}\begin{pmatrix} 1 \\ 1/R \end{pmatrix}q(t)\right] + V_{in}\begin{pmatrix} 1 \\ 1/R \end{pmatrix}q(t) \tag{5.10}$$

where t_0 is the considered initial time instant, $v_0 = v(t_0)$, $i_0 = i(t_0)$ and

$$k = \frac{1}{2RC}, \quad \omega = \sqrt{\frac{1}{LC} - k^2}, \quad I = \begin{pmatrix} 1 & 0 \\ 0 & 1 \end{pmatrix}, \quad A = \begin{pmatrix} -k/\omega & 1/(C\omega) \\ -1/(L\omega) & k/\omega \end{pmatrix}$$

Let us define the normalized time variable $\alpha = t/T$, and introduce the matrix operator

$$N(\alpha) = e^{-kT\alpha}[I\cos(\omega T\alpha) + A\sin(\omega T\alpha)] \tag{5.11}$$

which has the following properties [7]:

a. $N(0) = I$.

b. $N(\alpha + \beta) = N(\alpha)N(\beta)$.

c. $det(N(\alpha)) = e^{-2kT\alpha}$.

Then, the solutions, over one period of the modulating ramp signal, can be rewritten as

$$\mathbf{x}(\alpha) = N(\alpha - \alpha_0)\mathbf{x}_0 + V_{\text{in}}[I - N(\alpha - \alpha_0)]\mathbf{b}q(\alpha), \tag{5.12}$$

where

$$\mathbf{x}(\alpha) = \begin{pmatrix} v(\alpha) \\ i(\alpha) \end{pmatrix}, \quad \mathbf{x}_0 = \begin{pmatrix} v(\alpha_0) \\ i(\alpha_0) \end{pmatrix}, \quad \mathbf{b} = \begin{pmatrix} 1 \\ 1/R \end{pmatrix} \tag{5.13}$$

It follows that, between two consecutive commutation time instants, we know exactly the state variables of the system. Essentially, they are a combination of exponential and sinusoidal functions. Thus, the evolution of each system, considered separately, is trivial, and corresponds to damped oscillations around the equilibrium point of each of the linear systems. But when both systems are taken together as a global one, as in the buck converter, the behavior is radically different. When a trajectory is near the equilibrium point of system 1, one has $v_{\text{con}} < v_r$, the circuit switches its topology, and the orbit is attracted to the equilibrium point of system 2, because this is the system that is working. When the trajectory, attracted to the equilibrium point of system 2, moves near it, system 1 begins to work, attracting the trajectory to the equilibrium point of system 1. This wandering between the two equilibrium points of the separate systems produces a highly nontrivial evolution, without any equilibrium point in the global system. Notice that multiple pulsing can occur, where the switches change the topology of the circuit many times per ramp cycle. This is very undesirable in practice as it greatly increases the switching losses. One way to avoid it is to use a latch. With a latched PWM, multiple pulsing is eliminated but subharmonics and chaos are still possible [2,10] as has been shown in Section 5.2.

5.3.2 Discrete-Time Map and Periodic Orbits

Usually, the first step which must be taken in order to study a dynamical system is to find the equilibrium points and the periodic orbits. Since our system has no equilibrium points, we begin with the periodic orbits. Thus, we start by constructing a Poincaré map, which in this case is chosen to be a stroboscopic map due to the T-periodicity imposed by the ramp signal. The stroboscopic map P, closely studied in Chapter 2, is obtained by considering the current and voltage at every T-switching. We recall its definition

$$(v_n, i_n) \longmapsto (v_{n+1}, i_{n+1}) \tag{5.14}$$

where $v_n = v(nT)$ and $i_n = i(nT)$.

It is relevant to point out that the structure of the stroboscopic map changes according to the number of switchings in a given cycle of the ramp. Thus, the analytical form of the mapping can be derived once the system evolution between the two T-switchings has been specified. We write $v_l = V_L + V_{\text{ref}}/a$. Let $\{\alpha_j\}_{j=1,\dots,m}$ be the normalized switching instants and $\mathbf{x}_0 = (v_0, i_0)$ the initial conditions. The matrix expressions (see Chapter 2) for the six possibilities, depending on the parity of the number of crossings in the ramp and the initial value for the voltage, are the following (see also Figure 5.15):

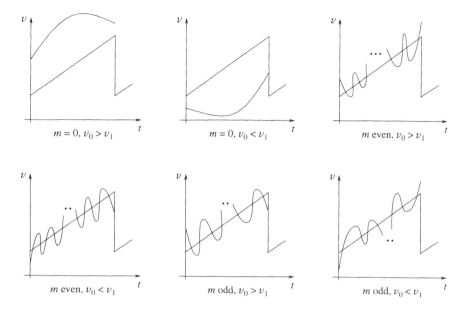

Figure 5.15 The six different possibilities for the stroboscopic map, depending on the parity (m even or m odd) of the number of crossings in the ramp and the initial voltage ($v_0 < v_l$ or $v_0 > v_l$).

$$m = 0, v_0 > v_l : \quad P(\mathbf{x}_0) = N(1)\mathbf{x}_0$$

$$m = 0, v_0 < v_l : \quad P(\mathbf{x}_0) = N(1)\mathbf{x}_0 + V_{\text{in}}[I - N(1)]\mathbf{b}$$

$$m \text{ even}, \ v_0 > v_l : \quad P(\mathbf{x}_0) = N(1)\mathbf{x}_0 + V_{\text{in}}\left[\sum_{j=1}^{m}(-1)^j N(1 - \alpha_j)\right]\mathbf{b}$$

$$m \text{ even}, \ v_0 < v_l : \quad P(\mathbf{x}_0) = N(1)\mathbf{x}_0 + V_{\text{in}}\left[I - \sum_{j=1}^{m}(-1)^j N(1 - \alpha_j) - N(1)\right]\mathbf{b}$$

$$m \text{ odd}, \ v_0 > v_l : \quad P(\mathbf{x}_0) = N(1)\mathbf{x}_0 + V_{\text{in}}\left[I + \sum_{j=1}^{m}(-1)^j N(1 - \alpha_j)\right]\mathbf{b}$$

$$m \text{ odd}, \ v_0 < v_l : \quad P(\mathbf{x}_0) = N(1)\mathbf{x}_0 + V_{\text{in}}\left[-\sum_{j=1}^{m}(-1)^j N(1 - \alpha_j) - N(1)\right]\mathbf{b}$$

Then, construction of the map requires the solution of the following conditions with respect to the normalized switching instants $\{\alpha_j\}_{j=1,\dots,m}$:

$$v(\alpha_j) = v_l + \eta T \alpha_j \quad j = 1, \dots m \qquad (5.15)$$

where $v_l + \eta T \alpha$ stands for the equivalent ramp voltage ($\eta = \frac{V_l - V_L}{aT}$).

Therefore, it is possible to derive a closed form for the stroboscopic map to obtain analytically the Jacobian, and perform the investigation of the stability of the system orbits via its eigenvalues. Alternative mappings which have been explained in Chapter 2 can also be used. In this case the different possible elementary map configurations to be considered reduce to just three possibilities (see [5,6] for further details).

Since our dynamical system contains several parameters which can be varied, the problem of finding out all the periodic orbits with all the possible periods would require an enormous work and computer processing. Thus, we restrict our investigation to some types of periodic orbits which appear as the input voltage is varied and the rest of parameters are fixed. As in [1], the numerical simulations will be performed with the following parameter values (see Figure 5.14): $L = 20$mH, $C = 47\mu$F, $R = 22\Omega$, $a = 8.4$, $V_{ref} = 11.3$V, $V_L = 3.8$V, $V_U = 8.2$V and $T = 400\mu$s.

Different Types of Periodic Orbits

Here we have restricted our study to rT-periodic orbits with all the parameters fixed with the exception of the input voltage. It should be noted that there can be many types of T- and $2T$-periodic orbits. This is due to the number of switchings that can occur in a ramp cycle, from zero to theoretically infinity, and to the state of the system at the beginning of each ramp cycle ($v_{con} < V_L$ or $v_{con} > V_L$). Thus, there are an infinite number of combinations that produce different type of orbits. For each special configuration of periodic orbit, one can construct its Poincaré map, which allows easy computation when there are a small number of switchings. When this number rises up, as it will be shown, the Poincaré map gets more and more complicated, and the numerical computation gets more and more extensive.

Analytical Study of Periodic Orbits: Existence and Stability

To show how the Poincaré map is specialized to a single type of periodic orbit, we will analyze the T- and $2T$-periodic orbits which cross the voltage ramp once per cycle, and later we will study their stability by computing the characteristic multipliers associated with each one.

Having fixed the values of the parameters, we will study for which values of the input voltage V_{in} we may have periodic orbits. We will find an rT-periodic orbit if the following conditions are satisfied:

$$\begin{cases} v(rT) &= v(0) \\ i(rT) &= i(0) \end{cases} \qquad (5.16)$$

that is, an orbit of the phase space (v, i) which repeats after r cycles of the ramp. These conditions are equivalent to

$$\begin{cases} v_{con}(rT) &= v_{con}(0) \\ i(rT) &= i(0) \end{cases} \qquad (5.17)$$

A T-periodic orbit with one switching per ramp cycle must start with $v_{con}(0) \in [V_L, V_U]$. Thus, the initial part of the orbit is system 1 (the one with switch OFF) that initially draws the orbit; at some time $t_1 < T$, v_{con} crosses the ramp and system 2 (the one with switch ON) enters into action until $t = T$, when one must have

$$\begin{cases} v_{con}(T) &= v_{con}(0) \\ i(T) &= i(0) \end{cases} \qquad (5.18)$$

The time $t_1 \in [0, T]$ must satisfy

$$\begin{pmatrix} v(t_1) \\ i(t_1) \end{pmatrix} = e^{-kt_1}[I \cos \omega t_1 + A \sin \omega t_1] \begin{pmatrix} v(0) \\ i(0) \end{pmatrix} \tag{5.19}$$

and

$$a(v(t_1) - V_{\text{ref}}) = V_L + \frac{V_U - V_L}{T} t_1 \tag{5.20}$$

From t_1 to T, system 2 gives the dynamics and so we have

$$\begin{pmatrix} v(0) \\ i(0) \end{pmatrix} = \begin{pmatrix} v(T) \\ i(T) \end{pmatrix} = V_{\text{in}} \mathbf{b} + e^{-k(T-t_1)}[I \cos \omega(T - t_1)$$

$$+ A \sin \omega(T - t_1)] \begin{pmatrix} v(t_1) - V_{\text{in}} \\ i(t_1) - V_{\text{in}}/R \end{pmatrix} \tag{5.21}$$

In terms of N, if we write $t_1 = \alpha_1 T$, $\alpha_1 \in (0, 1)$ (and thus α_1 is the duty cycle), the conditions we have imposed are equivalent to

$$f_1(V_{\text{in}}, \alpha_1) = 0 \tag{5.22}$$

where

$$f_1(V_{\text{in}}, \alpha_1) \equiv V_{\text{ref}} + \frac{V_L}{a} + \frac{V_U - V_L}{a} \alpha_1 - (1, 0) \cdot N(\alpha_1) \begin{pmatrix} v_0(V_{\text{in}}, \alpha_1) \\ i_0(V_{\text{in}}, \alpha_1) \end{pmatrix} \tag{5.23}$$

with

$$\begin{pmatrix} v_0(V_{\text{in}}, \alpha_1) \\ i_0(V_{\text{in}}, \alpha_1) \end{pmatrix} = V_{\text{in}}[N(0) - N(1)]^{-1}[N(0) - N(1 - \alpha_1)]\mathbf{b} \tag{5.24}$$

After substitution of $v_0(V_{\text{in}}, \alpha_1)$ and $i_0(V_{\text{in}}, \alpha_1)$ in $f_1(V_{\text{in}}, \alpha_1)$, we have to solve (5.22) numerically for V_{in} and α_1. Figure 5.16(a) shows the duty cycle α_1 obtained when V_{in} is varied from 12V to 50V.

Notice that the conditions we imposed to obtain periodic orbits are necessary, but they are not sufficient: it could be that t_1 is not the first time the control voltage crosses the ramp or, being the first, it could be that it is not the last. Thus, after numerically computing the values of (v_0, i_0), we must check that the obtained trajectory is really of the desired kind.

Something similar can be done for the $2T$-periodic orbits that cross the ramp once per cycle. If we denote by $t_1 = \alpha_1 T$ the first time that the control voltage crosses the ramp, we will have, with the notation previously introduced

$$\begin{pmatrix} v(t_1) \\ i(t_1) \end{pmatrix} = N(\alpha_1) \begin{pmatrix} v_0 \\ i_0 \end{pmatrix} \tag{5.25}$$

and

$$a(v(t_1) - V_{\text{ref}}) = V_L + \frac{V_U - V_L}{T} t_1 \tag{5.26}$$

After t_1, system 2 acts until the end of the period, and we obtain

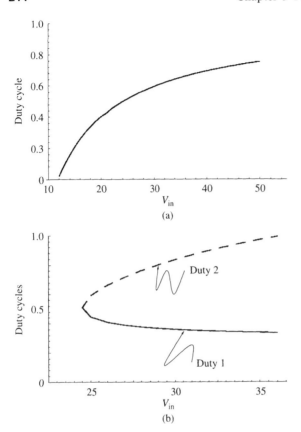

Figure 5.16 Evolution of the duty cycle for the T-periodic and $2T$-periodic orbits.

$$\begin{pmatrix} v(T) \\ i(T) \end{pmatrix} = N(1 - \alpha_1)\begin{pmatrix} v(t_1) - V_{in} \\ i(t_1) - V_{in}/R \end{pmatrix} + V_{in}\mathbf{b} \tag{5.27}$$

During the second period, the ramp is crossed at $t = t_2 = \alpha_2 T$, and we will have

$$\begin{pmatrix} v(t_2) \\ i(t_2) \end{pmatrix} = N(\alpha_2)\begin{pmatrix} v(T) \\ i(T) \end{pmatrix} \tag{5.28}$$

and

$$a(v(t_2) - V_{ref}) = V_L + \frac{V_U - V_L}{T}t_2 \tag{5.29}$$

During the second part of the second period the dynamics is given again by system 2 and

$$\begin{pmatrix} v(2T) \\ i(2T) \end{pmatrix} = N(1 - \alpha_2)\begin{pmatrix} v(t_2) - V_{in} \\ i(t_2) - V_{in}/R \end{pmatrix} + V_{in}\mathbf{b} \tag{5.30}$$

Finally, to get a $2T$-periodic orbit, we must impose

$$\begin{pmatrix} v(2T) \\ i(2T) \end{pmatrix} = \begin{pmatrix} v_0 \\ i_0 \end{pmatrix} \tag{5.31}$$

In a similar way, as in the case of the T-periodic orbits we obtain

$$\begin{pmatrix} v_0 \\ i_0 \end{pmatrix} = V_{in}[N(0) - N(2)]^{-1}[N(0) - N(1 - \alpha_2) + N(1) - N(2 - \alpha_1)]\mathbf{b} \qquad (5.32)$$

with additional conditions (5.26) and (5.29).

If we substitute the values of $v(t_1)$ and $v(t_2)$ in terms of v_0 and i_0, conditions (5.26) and (5.29) are equivalent to

$$\begin{cases} f_1(V_{in}, \alpha_1, \alpha_2) = 0 \\ f_2(V_{in}, \alpha_1, \alpha_2) = 0 \end{cases} \qquad (5.33)$$

where

$$\begin{aligned} f_1(V_{in}, \alpha_1, \alpha_2) = V_{ref} &+ \frac{V_L}{a} + \frac{V_U - V_L}{a}\alpha_1 \\ &- (1, 0) \cdot N(\alpha_1) \begin{pmatrix} v_0(V_{in}, \alpha_1, \alpha_2) \\ i_0(V_{in}, \alpha_1, \alpha_2) \end{pmatrix} \end{aligned} \qquad (5.34)$$

and

$$\begin{aligned} f_2(V_{in}, \alpha_1, \alpha_2) = V_{ref} &+ \frac{V_L}{a} + \frac{V_U - V_L}{a}\alpha_2 - (1, 0) \\ &\cdot \left[N(1 + \alpha_2)\begin{pmatrix} v_0(V_{in}, \alpha_1, \alpha_2) \\ i_0(V_{in}, \alpha_1, \alpha_2) \end{pmatrix} + V_{in}(N(\alpha_2) - N(1 - \alpha_1 + \alpha_2))\mathbf{b} \right] \end{aligned}$$

$$(5.35)$$

We now propose to study the stability of the T- and $2T$-periodic orbits that we have found. The characteristic multipliers test is used to decide whether a periodic orbit is stable. It has been shown in Chapter 3 that if the norms of all the characteristic multipliers (i.e., the eigenvalues of the discrete map) of a periodic orbit are less than 1, then the orbit is stable, while one characteristic multiplier with norm greater than 1 suffices to render the periodic orbit unstable.

To compute the characteristic multipliers we first need to obtain the Jacobian of the map of the orbit under investigation. Let $\mathbf{x}_0 = (v_0, i_0)$ be the initial conditions for a fixed point of the stroboscopic map P and let us indicate with DP the corresponding Jacobian. The eigenvalues of $DP(\mathbf{x}_0)$, or characteristic multipliers, m_1, m_2 must be the roots of the equation

$$z^2 - \text{tr}(DP(\mathbf{x}_0))z + \det(DP(\mathbf{x}_0)) = 0 \qquad (5.36)$$

where tr and det stands for the trace and the determinant of the matrix, respectively.

The equations for an orbit with a switching at a normalized time α_1 in the first cycle are

$$\begin{aligned} \mathbf{x}_1 &= N(\alpha_1)\mathbf{x}_0 \\ \mathbf{x}_T &= V_{in}[I - N(1 - \alpha_1)]\mathbf{b} + N(1)\mathbf{x}_0 \end{aligned} \qquad (5.37)$$

with $\mathbf{x}_1 = (v_1, i_1)$ the point in the phase space which corresponds to the switching. To get a T-periodic $\mathbf{x}_T = \mathbf{x}_0$ must be imposed, and so the following expression for $\mathbf{x}_0$ is obtained

$$\mathbf{x}_0 = V_{in}[I - N(1)]^{-1}[I - N(1 - \alpha_1)]\mathbf{b} \qquad (5.38)$$

together with the switching condition

$$a(v_1 - V_{\text{ref}}) = V_L + \alpha_1(V_U - V_L).$$ (5.39)

Now, differentiation with respect to $\mathbf{x}_0$ of equation

$$P(\mathbf{x}_0) = N(1)\mathbf{x}_0 + V_{\text{in}}[I - N(1 - \alpha_1)]\mathbf{b}$$ (5.40)

which gives the image of $\mathbf{x}_0$ by the stroboscopic map yields

$$DP(\mathbf{x}_0) = N(1) + V_{\text{in}}[N'(1 - \alpha_1)]\mathbf{b}\frac{d\alpha_1}{d\mathbf{x}_0}$$ (5.41)

Also, differentiation of equation (5.37) respect to $\mathbf{x}_0$ yields

$$\frac{d\mathbf{x}_1}{d\mathbf{x}_0} = N(\alpha_1) + N'(\alpha_1)\mathbf{x}_0\frac{d\alpha_1}{d\mathbf{x}_0}$$ (5.42)

Finally, differentation of equation (5.39) yields

$$a\frac{dv_1}{d\mathbf{x}_0} = (V_U - V_L)\frac{d\alpha_1}{d\mathbf{x}_0} \Rightarrow \frac{d\alpha_1}{d\mathbf{x}_0} = \frac{a}{V_U - V_L}\frac{dv_1}{d\mathbf{x}_0}$$ (5.43)

Thus,

$$\frac{d\mathbf{x}_1}{d\mathbf{x}_0} = N(\alpha_1) + N'(\alpha_1)\mathbf{x}_0\frac{a}{V_U - V_L}\frac{dv_1}{d\mathbf{x}_0},$$ (5.44)

and this equation gives

$$\frac{dv_1}{dv_0} \quad \text{and} \quad \frac{dv_1}{di_0}$$

as functions of α_1, v_0, i_0. Using also (5.43), this can be put into equation (5.41), and an expression for $DP(\mathbf{x}_0)$ is obtained, which depends on $V_{\text{in}}, \alpha_1, v_0, i_0$.

In Figure 5.17(a) we plot the evolution of the characteristic multipliers of the T-periodic orbits in the complex plane when V_{in} sweeps the range from 12V to 25V. First,

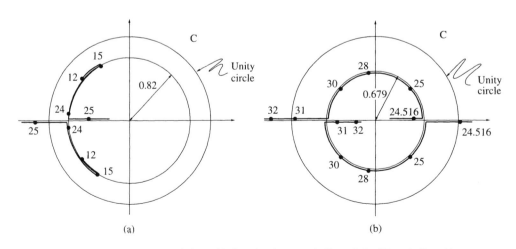

(a)   (b)

Figure 5.17 Characteristic multipliers for the T-periodic and the $2T$-periodic orbits as a function of V_{in} (the numbers are the values of V_{in}).

we observe that the characteristic multipliers are complex conjugates which move on a circle of radius $r \approx 0.82$, and thus the orbit is asymptotically stable. Near $V_{in} = 24\text{V}$, both characteristic multipliers become real, and when V_{in} reaches a certain value between $V_{in} = 24\text{V}$ and $V_{in} = 25\text{V}$ one of the characteristic multipliers has norm greater than 1, and the periodic orbit becomes unstable. After $V_{in} = 25\text{V}$, it remains unstable.

Figure 5.17(b) shows the same kind of diagram in the case of $2T$-periodic orbits. Near $V_{in} = 25\text{V}$ both multipliers enter the circle of radius $r = 1$, yielding stable orbits. Next, they move on the circle of radius $r = 0.679$ until $V_{in} = 31\text{V}$ is reached. After $V_{in} = 31\text{V}$, one of the multipliers goes out of the unit circle and the stability is lost and not recovered.

5.3.3 One-Dimensional Bifurcation Diagrams

One of the fundamental means by which nonlinear phenomena in a dynamic system can be studied is the bifurcation diagram. With this aim, taking V_{in} as the bifurcation parameter, a bifurcation diagram for P is plotted in the region of interest. Having fixed the other parameters, and for a set of initial conditions (v_0, i_0), successive iterates of P are taken. To avoid the transient dynamics we eliminate some preiterations and we only plot the last ones. This process is repeated for every discrete value of the bifurcation parameter in the range of interest.

The Main Bifurcation Diagram

The voltage-controlled dc/dc buck converter exhibits several bifurcations when the input voltage is varied. Bifurcations usually apply to fixed or mT-periodic points $\mathbf{x}_0$ (which is then a fixed point of the mth iteration of the map), but they also attain more complex sets like chaotic attractors and chaotic saddles. The type of fixed point $\mathbf{x}_0$ is determined through calculation of the eigenvalues of the Jacobian $DP(\mathbf{x}_0)$. It can be either stable (sink, S) or unstable (saddle) depending on whether both eigenvalues, or only one of them, stay inside the unit circle in the complex plane. By varying the bifurcation parameter V_{in}, regular or flip unstable periodic solutions are established through saddle-node (SN) or period-doubling (PD) smooth bifurcations, respectively. Also, as shown in a previous section, nonsmooth bifurcations of border-collision and grazing type can also occur, which correspond to a jump in the eigenvalues or to a singularity of the Jacobian matrix, respectively. Note that as outlined in [4] a grazing phenomenon (i.e., when the control voltage touches the ramp tangentially) is actually associated with infinite local stretching on the phase plane (i.e., the Jacobian of the map becomes infinite). Thus, this phenomenon offers one of the fundamental mechanisms which are likely to yield the onset of chaotic evolutions. The notations S_j^m, R_j^m, and F_j^m, will be used to denote mT-periodic sinks, regular saddles and flip saddles, respectively, the subscript $j = 1, 2, \ldots m$ referring to different image points of the periodic solution. Chaotic attractors made of a finite number m of disconnected pieces are obtained in several situations. They are denoted as C_j^m attractors, the subscript $j = 1, 2, \ldots m$ referring to the different pieces contained in subdomains of the whole basin.

Figure 5.18 represents the bifurcation diagram for the parameter values which were taken in the previous sections. The main T-periodic branch bifurcates into a $2T$-periodic at $V_{in} = 24.516\text{V}$ as experimentally and numerically observed in [1]. Successive period doublings can be found at $V_{in} = 31.121\text{V}$, $V_{in} = 32.095\text{V}$,

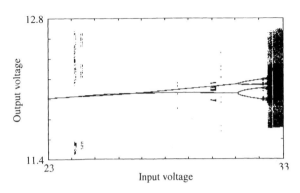

Figure 5.18 (After [4]) Buck bifurcation diagram. The input voltage is the bifurcation parameter in the range (23,33)V. Several secondary bifurcations and attractors are outlined.

$V_{in} = 32.239$V, $V_{in} = 32.270$V, $V_{in} = 32.277$V and $V_{in} = 32.278$V. Once chaos is established in the system, merging bands begin to occur at $V_{in} = 32.279$V approximately. Then, while four-band chaos is present, there is a sudden enlargement of the attractor at $V_{in} = 32.336$V. This latter phenomenon was recently shown to be due to a nonsmooth bifurcation [3].

It is also worth noticing what happens at

$$V_{in} = V_{ref} + \frac{V_L}{a} = 11.752\text{V}$$

which corresponds to the lower voltage of the ramp. As the bifurcation parameter passes through this value, a stable equilibrium point in the phase space (which exists due to the fact that only one of the linear topologies is in action) turns into a stable T-periodic orbit. If the trajectories are inspected in the three-dimensional cylindrical space $R^2 \times S^1$, the T-periodic orbit exists even before the bifurcation value, but its projection in the phase space is seen as an equilibrium point because the trajectory in $R^2 \times S^1$ is simply a circumference contained in a horizontal surface below the ramp surface. At the bifurcation, the T-periodic orbit begins to torsion due to the collision with the ramp surface, and then its projection in the phase space is seen as a T-periodic orbit instead of an equilibrium point (see Figure 5.19).

Then a period-doubling cascade follows, ending in a narrow four-piece chaotic attractor. Such evolution occurs via successive interior crises of merging type. At each of them, due to the collision with a flip saddle F_j^n, the $2n$ pieces of a chaotic attractor merge two by two, giving rise to a n-piece chaotic attractor.

After the period-doubling cascade, near $V_{in} = 32.279$V, merging bands and periodic windows follow, the latter corresponding to saddle-node bifurcations followed by interior crises, when the merging attractor touches the saddle born at the saddle-node bifurcation. Between $V_{in} = 32.280$V and $V_{in} = 32.281$V a $5T$-periodic window is also detected [4].

Apart from the main period-doubling route to chaos, other secondary phenomena are also present in the bifurcation diagram. Intervals of three-piece chaos in the range [24.16, 25.01], period-doubling route to chaos starting on $6T$-periodic and $12T$-periodic orbits and a $5T$-periodic window are clearly seen in Figure 5.18. Moreover, other secondary phenomena not appearing in the bifurcation diagram were detected, such as a nonsmooth route to chaos starting on a $3T$-periodic orbit.

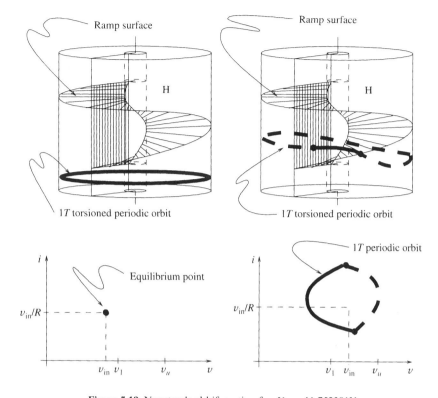

Figure 5.19 Nonstandard bifurcation for $V_{in} = 11.752381$V.

Secondary Bifurcations

The bifurcation diagram has shown the existence of many other dynamical behaviors coexisting with those belonging to the main period-doubling route to chaos previously described. The richness of this dynamical scenario motivates the search for possible links between the new dynamical evolutions outlined above, and some well-known dynamics of the converter. For instance, the analysis of $5T$-periodic orbits can provide an explanation for the five-zones chaotic attractor exhibited by the system after the jump to *larger* chaos [4,7]. It is important to point out that due to the competition between different dynamical evolutions, the numerical computation of the bifurcation diagram, Figure 5.18, has been carried out by considering a cluster of initial conditions for each value of the input voltage. Namely, the cluster was chosen as the main diagonal of the phase plane $v \in (11.75, 12.70)$, $i \in (0.43, 0.75)$. Assuming a time interval of $5000T$, for each pair of initial conditions the system evolution is simulated and the last 100 stroboscopic points are stored so that transient has settled down.

5.3.4 Chaotic Attractors in the Buck Converter

3T-Periodic Orbits and the Three-Piece Chaotic Attractor

As mentioned before, a stable chaotic regime exists for $V_{in} \in (24.160, 25.010)$V, and also for $V_{in} \in (13.542, 13.880)$V. The existence of these three-piece chaotic attractors (one of them is depicted in Figure 5.20 and the bifurcation diagram is shown in Figure 5.21) suggests that they might originate from a branch of unstable $3T$-periodic orbits [4]. Although they are unstable, they can provide a deeper insight into the dynamics of a chaotic system, as outlined in [11].

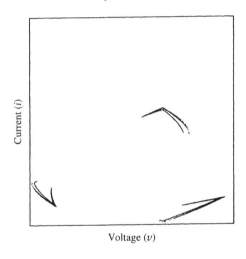

Voltage (v)

Figure 5.20 (After [4]) Three-piece chaotic attractor in the (v, i) phase space for $V_{in} = 26.14$V; v range is (11.41, 12.70) and i range is (0.42, 0.75).

Invariant Manifolds and Basins of Attraction

One of the points at which invariant manifolds are worth computing is $V_{in} = 11.752$V. A stable equilibrium point which exists before this value turns into a stable T-periodic orbit plus two saddle $3T$-periodic orbits. Invariant manifolds can be computed for these two saddles. Figure 5.22(a),(b) shows the stable and unstable manifolds for the saddles when $V_{in} = 11.8$V.

Apart from the many-fingered shape of the manifolds, transversal homoclinic orbits also exist. Thus an invariant set with horseshoe dynamics can be found at an early value in this small region of the phase space. At this point instant transversal homoclinic orbits are involved: a stable equilibrium point before $V_{in} = 11.752$V turns into a stable T-periodic orbit and a chaotic saddle, which probably includes the two unstable $3T$-periodic orbits after the bifurcation.

As the parameter is increased, this many-fingered tangle, which is small near $V_{in} = 11.8$V, widens and changes its shape for $V_{in} = 13.5$V approximately, when a first nonsmooth bifurcation takes place. At this point, trajectories are able to follow the ramp signal upward since the fixed point of one of the two topologies $(V_{in}, V_{in}/R)$ is high enough up the upper voltage of the ramp, and so different behavior can be expected. Some snapshots of the invariant manifolds are taken for the $3T$-periodic saddle in the first $3T$-branch. While the main stable T-periodic orbit exists, the interior

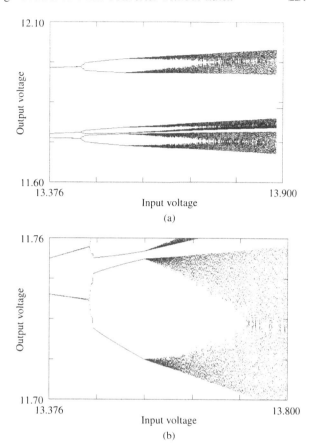

Figure 5.21 (a) Bifurcation of the $3T$-attractor starting near $V_{in} = 13.376$V. After the nonsmooth bifurcation creating the $3T$-periodic orbit, a period-doubling bifurcation occurs. Next, border collision bifurcation into a six-piece chaotic attractor is observed. An interior crisis of first kind producing merging bands turns the six-piece chaotic attractor into a three-piece chaotic attractor, and finally a boundary crisis destroys it. (b) Detail of the border collision bifurcation producing instant chaos.

unstable half-manifold spirals toward it, and each of the exterior half-manifolds transversely intersects the stable manifold (see Figure 5.22 and Figure 5.23). On the other hand, something different occurs with the invariant manifolds of the second $3T$-branch. None of the unstable half-manifolds spiral to the main stable T-periodic orbit. They intersect the corresponding stable manifolds instead, providing a homoclinic tangle, and since this is nonattracting, they also provide a chaotic saddle for the system.

If V_{in} is increased before the first period-doubling of the main T-periodic attractor, the manifolds of the $3T$-saddles also change their shape (see Figure 5.22 and Figure 5.23). While the interior unstable half-manifold is spiraling to the main stable T-periodic orbit, the exterior unstable half-manifold also spirals and despirals around three zones in the phase space, which coincide with the three zones where the $3T$-saddle orbits accumulate. Thus, the manifolds approximate to the infinite-stretching point. As the manifolds pass near this zone of high number of crossings, they are highly twisted and folded, and as the manifolds are invariant sets, this twist and fold propagates all along the manifolds. That is why their shape is like *dense spiraling islands*; they seem to open and widen their fingers, also spiraling to three zones in the phase space.

Around $V_{in} = 13.376$V, a three-piece chaotic attractor was found following a path of $3T$-periodic stable orbits, which is later destroyed at a boundary crisis. Consequently, a coexisting attractor with the main T-periodic branch exists.

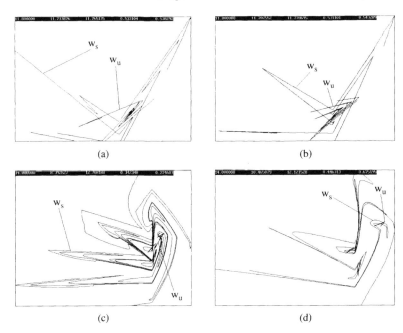

Figure 5.22 Invariant manifolds for the $3T$-saddles in the $3T$-branches:
(a) $V_{in} = 11.8$V, first $3T$-branch; (b) $V_{in} = 11.8$V, second $3T$-branch;
(c) $V_{in} = 14.0$V, first $3T$-branch; (d) $V_{in} = 14.0$V, second $3T$-branch.

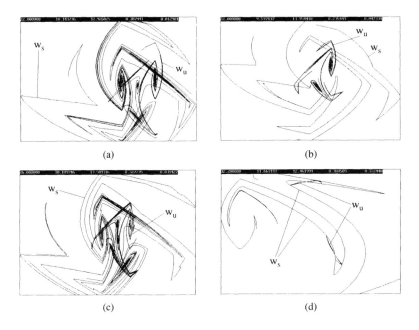

Figure 5.23 Invariant manifolds for the $3T$-saddles of the $3T$-branches and for the
$2T$-periodic flip saddle created at the second period doubling of the main
branch attractor: (a) $V_{in} = 22.0$V, first $3T$-branch; (b) $V_{in} = 22.0$V, second $3T$-branch; (c) $V_{in} = 26.0$V, first $3T$-branch; (d) $V_{in} = 32.2$V.

Furthermore, a three-piece chaotic attractor is also present near $V_{in} = 24.16$V, and until $V_{in} = 25.01$V. Their basins of attraction can be computed following a cell-to-cell mapping algorithm [9]. The stable manifold of the $3T$-saddle corresponds to the basin boundary of the attractors, while the unstable manifold is the closure of the chaotic attractor. This can be seen in Figure 5.24.

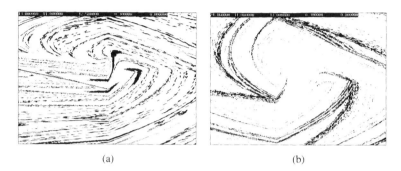

(a)   (b)

Figure 5.24 (a) Basins of attraction for $V_{in} = 13.8$V. Black corresponds to the $3T$-attractor and white to the main T-attractor; (b) Basin of attraction for $V_{in} = 24.16$V. Black stands for the T-periodic basin, white for the three-piece chaotic attractor basin.

As the shapes of the manifolds change with increasing V_{in}, the future large chaotic attractor at V_{in} near 32.34V is also apparent. Thus it is conjectured that the five-zones chaotic attractor, which exists for large values of V_{in}, is in fact an evolution from the three-zones chaotic saddle from the intersection of the manifolds and the period-doubling evolution of the main branch attractor itself. For example, when the first period doubling has taken place at $V_{in} = 26.0$V, the central part of the invariant manifold has widened and begins to connect the other three branches (see Figure 5.23).

Invariant manifolds for the T-periodic flip saddle can also be computed. It can be shown that additional transverse homoclinic orbits exist for $V_{in} = 30.0$V after the invariant manifolds have touched themselves. For $V_{in} = 29.0$V, the stable manifold is the boundary of the two subdomains of the stable $2T$-periodic orbit. The corresponding unstable manifold leads to this orbit, but in a way that anticipates the next $6T$ saddle-node bifurcation to appear. Homoclinic tangles also take place for the invariant manifolds of the $2T$-periodic flip saddle created at the second period doubling of the main attractor. The unstable manifold begins to fold around $V_{in} = 32.0$V. This is more clearly seen at $V_{in} = 32.2$V (Figure 5.23), and the homoclinic tangency occurs at a value between 32.2V and 32.4V, probably at the same time as the small attractor becomes larger for V_{in} near 32.336V. Thus this homoclinic tangle could be responsible for the sudden expansion of the attractor.

6T-Periodic Orbits

In addition to the main attractor branch and the $3T$-periodic orbits discussed, there also exists a secondary attractor in a neighborhood of $V_{in} = 30.0$V, generated after a saddle-node bifurcation at $V_{in} = 29.906$V. This is located by computing the characteristic multipliers of the $6T$-periodic, one switching per cycle, stable orbit, and its corresponding regular saddle. Taking initial conditions on the regular saddle, and letting the system evolve, one recovers the stable $6T$-sink, and thus the $6T$-regular saddle belongs to the boundary of the basin of attraction of the $6T$-sink all over the range of existence of the sink. One of the multipliers takes unitary modulus at $V_{in} = 29.906$V. This is also confirmed by the solution of the conditions of existence of the $6T$-periodic orbit.

The orbit follows a bifurcation path into chaos through a standard period-doubling cascade, which ends in a six-piece chaotic attractor coexisting with the main $2T$-periodic stable orbit. Further details can be found in [4].

12T-Periodic Orbits

Coexisting with the $8T$-periodic stable orbit generated at the third period doubling of the main T-periodic attractor, in the neighborhood of $V_{in} = 32.136$V, a $12T$-periodic stable orbit also appears. This is independent from the $12T$-periodic orbits bifurcating from the $6T$-periodic ones and is born after a saddle-node bifurcation, occuring when the input voltage gets near 32.1365V. It is worth pointing out that this $12T$-periodic orbit is organized in a different way from the $12T$-periodic orbit previously generated at the period doubling of the $6T$. The $12T$ period-doubled orbit is organized in six pairs around the previous existing $6T$-periodic sink, while the $12T$ orbit born at the saddle-node bifurcation is arranged in four trios around the main $8T$-periodic stable orbit (see Figure 5.25).

This orbit continues and it experiences a period-doubling cascade, which leads to a twelve-piece chaotic attractor coexisting with the stable $8T$-periodic orbit. Later, successive period doublings are found, leading to chaos. Successive merging bands from 48 to 24 and 24 to 12 pieces can be observed, leading to a twelve-piece chaotic attractor. Crisis in each of the twelve bands can also be distinguished for $V_{in} = 32.163$V. Again, the chaotic attractor disappears through a boundary crisis near this value, where the regular saddle originated in the saddle-node bifurcation collides with the attractor. In

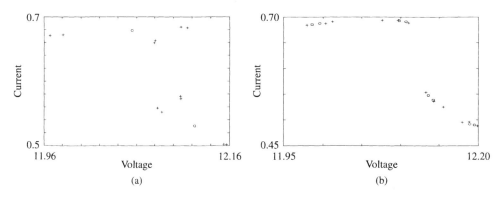

Figure 5.25 (a) Organization of the period-doubled $12T$-periodic orbits. Crosses stand for the $12T$, and circles for the coexisting $2T$-stable periodic orbit. (b) Organization of the $12T$-periodic orbits born at the saddle-node bifurcation. Crosses stand for the $12T$-, and circles for the coexisting $8T$-stable periodic orbit.

each of the twelve subbranches, full chaos, crisis, and $3T$-periodic and $5T$-periodic windows (and thus, $36T$-periodic and $60T$-periodic windows in the full domain) exist.

5T-Periodic Orbits and the Jump to Larger Chaos

As pointed out above and shown in Figure 5.18, the main branch of the bifurcation diagram of the buck converter undergoes a period-doubling cascade which is suddenly interrupted when bands of chaos are merging around $V_{in} = 32.336$V. At this value of the input voltage, the amplitude of the voltage oscillations is abruptly enlarged. This phenomenon, which was observed experimentally for the buck in [1], has been detected in other systems [12] and is typically the result of nonsmooth bifurcations such as border collisions and grazings. A detailed simulation of the bifurcation diagram [4] indicates that when the jump occurs, the system is already evolving along a small scale two-zones chaotic attractor originated from the period-doubling cascade of the main branch (see Figure 5.26). Then, as the voltage is increased, the system starts to evolve along the five-zones attractor depicted in Figure 5.26.

Analytical investigation of this phenomenon is particularly difficult since the system is already chaotic when the sudden enlargement occurs. Nevertheless, it will be analyzed in the following using the analytical and simulation tools presented before.

The fact that the large-scale chaos is organized around five zones suggests a connection between the attractor itself and some $5T$-periodic solutions, as in the case of the three-piece chaos analyzed before. This type of orbit is indeed detected as a window of periodicity embedded in the large-scale chaotic attractor around $V_{in} = 32.55$V, as shown in Figure 5.18. The orbits were continued backward by imposing the necessary conditions for their existence. In so doing, an entire branch of different types of unstable $5T$-periodic orbits was detected in the intervals $(32.53, 32.59)$ and $(34.76, 34.78)$ approximately. These orbits are characterized by one switching in the first cycle, one or two switchings in the second cycle, a skipping in the third, one switching in the fourth, and three switchings in the last cycle (Figure 5.27). It is important to point out that each stroboscopic point corresponds to the *barycenter* of each zone of the attractor, confirming the hypothesis that chaotic evolutions of the buck are organized around this

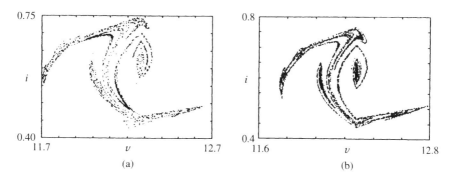

Figure 5.26 (After [4]) (a) Phase plane portrait just after the enlargement has occurred. The small-scale two-zones chaotic attractor can be still clearly observed. (b) Large scale five-zones chaotic attractor.

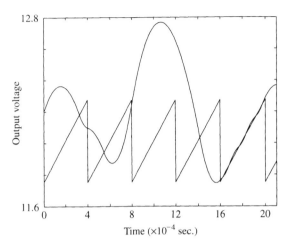

Figure 5.27 (After [4]) 5T-periodic organizing orbit.

type of 5T-periodic orbits when V_{in} grows beyond 32.336V (see Figure 5.26). This is also confirmed by the changing in the average number of switchings per cycle as the input voltage is varied, which is reported in Figure 5.28.

Up to the range of input voltages where the attractor enlargement takes place, all system dynamics are characterized by one switching per cycle. Then, when the jump occurs, the average number of switchings per cycle suddenly becomes different from one and starts to grow. This is predictable since the 5T-periodic solutions are characterized by multiple switchings in the last cycle. Moreover, the stroboscopic point preceding the multiple switching cycle is near to the condition of infinite local stretching presented in [5]. Hence, the number of switchings in the last cycle is highly influenced by the initial condition at this point (theoretically infinite when the infinite local stretching condition is perfectly satisfied).

Therefore, as V_{in} increases, the system states at the fourth stroboscopic point match the infinite stretching condition better and better, and more and more switchings occur in the last cycle of the 5T-periodic orbit. This yields a corresponding peak in the

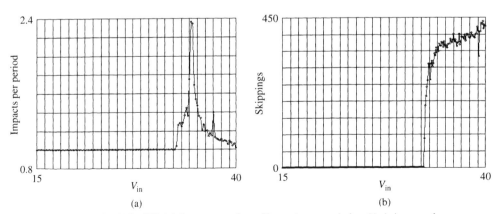

(a)   (b)

Figure 5.28 (After [4]) (a) Average number of impacts per period as V_{in} is increased.
(b) Total number of skippings over 2000 periods for different V_{in} values.

diagram reported in Figure 5.28. In fact, a $5T$-unstable periodic orbit fulfilling the infinite stretching condition can be found near

$$V_{in} = 34.34V$$

giving infinite switchings in the last cycle.

Finally, the jump from the small-scale chaotic evolution, characterized by one switching per cycle, to the five-zones attractor is expected to take place when the first skipping occurs. This is confirmed by Figure 5.28, where the number of skippings over 2000 cycles is plotted against the input voltage. We notice that the number of skippings becomes different from zero exactly when the large-scale chaos appears, quickly settling down around 400 when the input voltage is increased. It is relevant to point out that the $5T$-periodic orbit, described above, is characterized by one skipping every five cycles [7] (i.e., 400 skippings over 2000 periods). This confirms again that the large-scale attractor is organized around this type of unstable periodic solution. Recently, the nature of the attractor resulting from the sudden enlargement described above has been related to the occurrence in the voltage-controlled buck converter of *sliding* solutions. These solutions can be seen heuristically as characterized by an infinite number of switchings in the last ramp cycle and have been shown to organize peculiar bifurcation diagrams having the shape of an intertwined double spiral (see Figure 5.29). These diagrams describe the

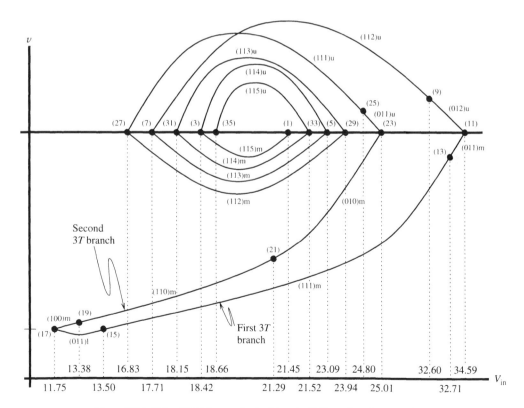

Figure 5.29 (After [4]) Scheme of a double-spiral bifurcation diagram of $3T$-periodic unstable orbits. Dots stand for nonsmooth bifurcations: the numbers are given accordingly to different types of orbits.

accumulation of periodic solutions characterized by an increasing number of switchings onto a sliding solution and have been shown to be particularly relevant to organizing the dynamics of the buck converter. A complete analytical study of their occurrence and shape can be found in [3], where the topology of the resulting chaotic attractors is also described in further detail.

REFERENCES

[1] J. H. B. Deane and D. C. Hamill, Analysis, simulation and experimental study of chaos in the buck converter, *IEEE Power Electronics Specialists Conf.*, pp. 491–498, June 1990.

[2] J. H. B. Deane and D. C. Hamill, Chaotic behavior in current-mode controlled dc-dc convertor, *Electron. Lett.*, vol. 27, no. 13, pp. 1172–1173, June 1991.

[3] M. Di Bernardo, C. Budd, and A. Champneys, Grazing, skipping and sliding: analysis of the non-smooth dynamics of the DC/DC buck converter, *Nonlinearity*, vol. 11, no. 4, pp. 858–890, 1998.

[4] M. Di Bernardo, E. Fossas, G. Olivar, and F. Vasca, Secondary bifurcations and high periodic orbits in voltage controlled buck converter, *Int. J. Bif. Chaos*, vol. 7, no. 12, pp. 2755–2771, 1997.

[5] M. Di Bernardo, F. Garofalo, L. Glielmo, and F. Vasca, Switchings, bifurcations and chaos in dc/dc converters, *IEEE Trans. on Circuits and Systems—I*, vol. 45, no. 2, pp. 133–141, 1998.

[6] M. Di Bernardo and F. Vasca, On discrete time maps for the analysis of bifurcations and chaos in dc/dc converters, *IEEE Trans. on Circuits and Systems—I*, vol. 47, no. 2, pp. 130–143, 2000.

[7] E. Fossas and G. Olivar, Study of chaos in the buck converter, *IEEE Trans. on Circuits and Systems—I*, vol. 43, no. 1, pp. 13–25, 1996.

[8] D. C. Hamill, J. H. B. Deane, and D. J. Jefferies, Modeling of chaotic dc-dc converters by iterated nonlinear mappings, *IEEE Trans. on Power Electron.*, vol. 7, pp. 25–36, January 1992.

[9] C. Hsu, Global analysis by cell mapping, *Int. J. Bif. Chaos*, vol. 2, pp. 727–771, 1992.

[10] S. Miles and M. Di Bernardo, Preventing multiple switchings in power electronic circuits: Effects of the latch on the nonlinear dynamics of the dc-dc buck converter, *Int. J. Bifurcations and Chaos*, vol. 10, no. 2, pp. 431–442, 2000.

[11] M. Ogorzalek and L. O. Chua, Exploring chaos in Chua's circuit via unstable periodic orbits, *Proc. ISCAS'93*, pp. 2608–2611, 1993.

[12] E. Ott, *Chaos in Dynamical Systems*. Cambridge: Cambridge University Press, 1993.

[13] J. R. Wood, Chaos: A real phenomenon in power electronics, *1989 IEEE Applied Power Electron. Conf. Rec.*, pp. 115–124, 1989.

5.4 SADDLE-NODE AND NEIMARK BIFURCATIONS IN PWM DC/DC CONVERTERS

Chung-Chieh Fang
Eyad H. Abed

5.4.1 Introduction

In PWM dc/dc converters, the nominal operating condition is a periodic steady state (i.e., in the language of nonlinear dynamics, a limit cycle). Three typical local bifurcations of the limit cycle are period-doubling bifurcation, saddle-node bifurcation, and Neimark bifurcation. These bifurcations are best explained by the sampled-data model instead of the averaged model.

The averaging method has been a popular approach for stability analysis of PWM dc/dc converters. Here, the nominal periodic steady state of a PWM converter is averaged to an equilibrium. The periodic steady state in high switching operation has small amplitude (ripple), and averaging is therefore a reasonable approach. However, close to the onset of instability, the *periodic* nature of the steady-state operating condition needs to be considered in order to obtain accurate results. Indeed, it has been reported [1,2] that averaging leads to erroneous conclusions regarding the onset of instability. In this section we demonstrate the occurrence of saddle-node bifurcation and Neimark bifurcation in dc/dc converters through sampled-data models and show that input filter instability is closely related to the Neimark bifurcation.

5.4.2 General Sampled-Data Model for Closed-Loop PWM Converters

The methodology of sampled-data modeling has been presented in Chapter 2. Here we'll first recount the essence of the method to get the reader primed for the analysis that is to follow [3,4]. Without loss of generality, only continuous-conduction mode (CCM) is considered. This model is applicable both to voltage-mode control and current-mode control.

A block diagram model for a PWM converter in continuous-conduction mode is shown in Figure 5.30. In the diagram, $\mathbf{A}_1, \mathbf{A}_2 \in I\!R^{N \times N}$, $\mathbf{B}_1, \mathbf{B}_2 \in I\!R^{N \times 2}$, $\mathbf{C}, \mathbf{E}_1, \mathbf{E}_2 \in I\!R^{1 \times N}$, and $\mathbf{D} \in I\!R^{1 \times 2}$ are constant matrices, $\mathbf{x} \in I\!R^N$, $y \in I\!R$ are the state and the feedback signal, respectively, and N is the state dimension, typically given by the number of energy storage elements in the converter. For example, $N = 3$ for a typical buck converter with a first-order error amplifier. The input voltage is V_{in}, and the output voltage is v_o. The notation V_{ref} denotes the reference signal, which could be a voltage or current reference. Let $\mathbf{u} = (v_{\text{in}}, V_{\text{ref}})^T \in I\!R^{2 \times 1}$. The signal $h(t)$ is a T-periodic ramp with the lowest value $h(0) = V_L$ and the highest value $h(T^-) = V_U$. In current-mode control, it is used to model a slope-compensating ramp. The clock has the same frequency $f_s = 1/T$ as the ramp. Within a clock period, the dynamics is switched between the two stages S_1 and S_2. The system is in S_1 immediately following a clock pulse, and switches to S_2 at instants when $y(t) = h(t)$.

Typical waveforms in current- and voltage-mode control are shown in Figure 5.31. In Figure 5.31(a), the ramp has positive slope, instead of negative slope as commonly seen in most literature, in order to be consistent with the case of voltage-mode control.

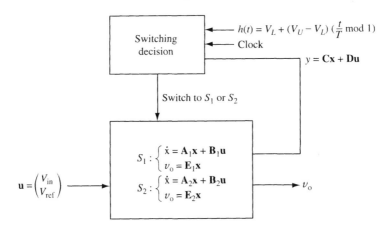

Figure 5.30 Block diagram model for PWM converter operation in continuous-conduction mode.

Consider the cycle from $t = nT$ to $t = (n + 1)T$. Let $\mathbf{x}_n = \mathbf{x}(nT)$ and $v_{on} = v_o(nT)$. In the sampled-data dynamics, the state $\mathbf{x}_n$, undergoing the two stages S_1 and S_2, is mapped to $\mathbf{x}_{n+1}$. Switching from S_1 to S_2 is implicitly determined by the instant when $y(t)$ and $h(t)$ intersect. Denote this switching instant within the cycle by $nT + d_n$, i.e., $y(nT + d_n) = h(nT + d_n)$. Then, the system in Figure 5.30 has the following (large-signal) sampled-data dynamics:

$$
\begin{aligned}
\mathbf{x}_{n+1} &= f(\mathbf{x}_n, d_n) \\
&= e^{\mathbf{A}_2(T-d_n)}\left(e^{\mathbf{A}_1 d_n}\mathbf{x}_n + \int_0^{d_n} e^{\mathbf{A}_1(d_n-\sigma)}d\sigma\mathbf{B}_1\mathbf{u}\right) \\
&\quad + \int_{d_n}^T e^{\mathbf{A}_2(T-\sigma)}d\sigma\mathbf{B}_2\mathbf{u}
\end{aligned}
\tag{5.45}
$$

$$
\begin{aligned}
g(\mathbf{x}_n, d_n) &= y(nT + d_n) - h(nT + d_n) \\
&= \mathbf{C}\left(e^{\mathbf{A}_1 d_n}\mathbf{x}_n + \int_0^{d_n} e^{\mathbf{A}_1(d_n-\sigma)}d\sigma\mathbf{B}_1\mathbf{u}\right) + D\mathbf{u} - h(d_n) \\
&= 0
\end{aligned}
\tag{5.46}
$$

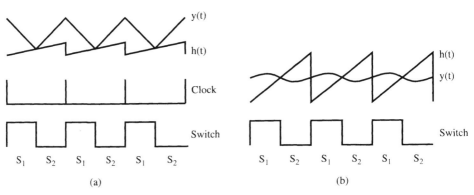

Figure 5.31 Waveforms of a PWM converter in CCM: (a) under current mode control; (b) under voltage mode control.

It is a constrained nonlinear discrete-time dynamics. The constraint equation $g(\mathbf{x}_n, d_n) = 0$ determines the switching instant d_n. An illustration of mapping from $\mathbf{x}_n$ to $\mathbf{x}_{n+1}$ is shown in Figure 5.32.

A periodic solution $\mathbf{x}^0(t)$ in Figure 5.30 corresponds to a fixed point $\mathbf{x}^0(0)$ in the sampled-data dynamics (5.45) and (5.46). The fixed point $(\mathbf{x}_n, d_n) = (\mathbf{x}^0(0), d)$ can be obtained by Newton's method. The periodic solution $\mathbf{x}^0(t)$ is then obtained:

$$\mathbf{x}^0(t) = \begin{cases} e^{\mathbf{A}_1 t}\mathbf{x}^0(0) + \int_0^t e^{\mathbf{A}_1(t-\sigma)}d\sigma\mathbf{B}_1\mathbf{u} & \text{for } t \in [0, d) \\ e^{\mathbf{A}_2(t-d)}\mathbf{x}^0(d) + \int_d^t e^{\mathbf{A}_2(t-\sigma)}d\sigma\mathbf{B}_2\mathbf{u} & \text{for } t \in [d, T) \\ \mathbf{x}^0(t \bmod T) & \text{for } t \geq T \end{cases} \tag{5.47}$$

A typical periodic solution $x^0(t)$ is shown in Figure 5.33, where $\dot{\mathbf{x}}^0(d^-) = \mathbf{A}_1\mathbf{x}^0(d) + \mathbf{B}_1\mathbf{u}$ and $\dot{\mathbf{x}}^0(d^+) = \mathbf{A}_2\mathbf{x}^0(d) + \mathbf{B}_2\mathbf{u}$ denote the time derivative of $\mathbf{x}^0(t)$ at $t = d^-$ and d^+, respectively.

Using a hat ($\hat{\,}$) to denote small perturbations (e.g., $\hat{\mathbf{x}}_n = \mathbf{x}_n - \mathbf{x}^0(0)$), the system (5.45), (5.46) has the linearized (small-signal) dynamics [3,4]:

$$\hat{\mathbf{x}}_{n+1} = \boldsymbol{\Phi}\hat{\mathbf{x}}_n \tag{5.48}$$

where

$$\begin{aligned} \boldsymbol{\Phi} &= e^{\mathbf{A}_2(T-d)}\left(\mathbf{I} - \frac{((\mathbf{A}_1 - \mathbf{A}_2)\mathbf{x}^0(d) + (\mathbf{B}_1 - \mathbf{B}_2)\mathbf{u})\mathbf{C}}{\mathbf{C}(\mathbf{A}_1\mathbf{x}^0(d) + \mathbf{B}_1\mathbf{u}) - \dot{h}(d)}\right)e^{\mathbf{A}_1 d} \\ &= e^{\mathbf{A}_2(T-d)}\left(\mathbf{I} - \frac{(\dot{\mathbf{x}}^0(d^-) - \dot{\mathbf{x}}^0(d^+))\mathbf{C}}{\mathbf{C}\dot{\mathbf{x}}^0(d^-) - \dot{h}(d)}\right)e^{\mathbf{A}_1 d} \end{aligned} \tag{5.49}$$

Local *orbital* stability of the periodic solution $\mathbf{x}^0(t)$ is determined by the eigenvalues of $\boldsymbol{\Phi}$. The periodic solution $\mathbf{x}^0(t)$ in the original continuous-time system of Figure 5.30 is asymptotically orbitally stable if all of the eigenvalues of $\boldsymbol{\Phi}$ are inside the unit circle of the complex plane. Moreover, if $\mathbf{x}^0(t)$ is asymptotically orbitally stable, then no eigenvalues of $\boldsymbol{\Phi}$ lie outside the unit circle.

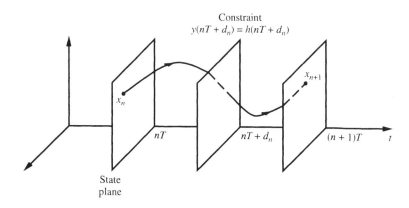

Figure 5.32 Illustration of sampled-data dynamics of PWM converter.

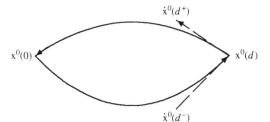

Figure 5.33 A typical periodic solution $\mathbf{x}_0(t)$ of a PWM dc/dc converter in state space.

5.4.3 Periodic Solution Before and After Local Bifurcation

In the PWM dc/dc converter, instabilities involve bifurcations of the *periodic* solution. In the saddle-node bifurcation, a stable T-periodic solution collides with an unstable one at the bifurcation point, and no periodic solution exists after the bifurcation. This may explain some jump phenomena, and sudden appearance or disappearance of the nominal periodic solution in dc/dc converters. An illustration of such a bifurcation is shown in Figure 5.34.

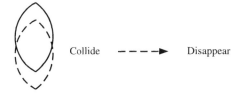

Collide   - - - - ►   Disappear

Figure 5.34 Periodic solution before and after saddle-node bifurcation (solid line for stable solution and dashed line for unstable solution).

An illustration of a (supercritical) Neimark bifurcation is given in Figure 5.35. After the bifurcation, the steady-state trajectory is on a torus (with the time axis circled as another dimension). The two angular frequency vectors (ω_s and ω_f) of the trajectory in the figure are perpendicular to each other. One of them is the same as the angular switching frequency $\omega_s = 2\pi f_s$. Another one (ω_f) can be determined from the bifurcation point where the eigenvalue trajectory of $\boldsymbol{\Phi}$ crosses the unit circle of the complex plane. Its value is $f_s \cdot \angle\sigma(\boldsymbol{\Phi})$, f_s times the argument (i.e., phase) of the complex conjugate pair of eigenvalues of $\boldsymbol{\Phi}$ crossing the unit circle. The state trajectory will be periodic (phase-locking) if these two frequencies are commensurate; otherwise it will be quasi-periodic.

5.4.4 Saddle-Node Bifurcation in Buck Converter Under Discrete-Time Control

Consider a buck converter with a discrete-time controller. The resulting system diagram is shown in Figure 5.36. The system parameters are $T = 400\mu s$, $L = 20mH$, $C = 47\mu F$, $R = 22\Omega$. Let V_{in} be the bifurcation parameter and it is varied from 18.5V to 20.5V. For duty cycle 0.7, the nominal inductor current is about $I_p = 0.6785$ and the nominal output voltage is about $V_p = 14.0263$. The switching decision in the cycle, $t \in [nT, (n+1)T)$, is designed as follows (similar to a leading-edge modulation where the switch is off first and then on in a cycle): the switch is turned off at $t = nT$ and turned on at $t = nT + d_n$. The switching instant d_n is updated by $d_n = \ell(0.3T - k_i$

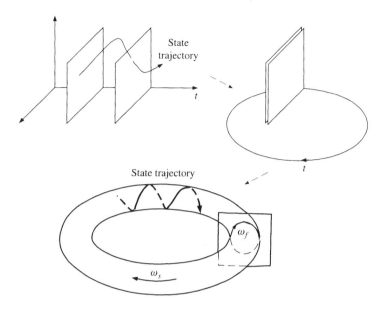

Figure 5.35 State trajectory after Neimark bifurcation.

$(i_n - I_p) - k_v(v_n - V_p))$, where $k_i = -8.574 \times 10^{-4}$ and $k_v = 5.53 \times 10^{-5}$ are feedback gains, and ℓ is a limiter (to limit the duty cycle within 1):

$$\ell(t) = \begin{cases} 0 & \text{for } t \le 0 \\ t & \text{for } t \in (0, T] \\ T & \text{for } t > T \end{cases} \tag{5.50}$$

This discrete-time control law produces different static and periodic solutions for different V_{in}. First, the switch being always on is a possible operation under some circumstances. When the switch is always on, $d_n = 0$ for any n. The steady-state solutions are constant instead of being periodic: $v_o(t) = V_{\text{in}}$ and $i(t) = V_{\text{in}}/R$. From Eq. (5.50), the following inequality needs to hold in order to make $d_n = 0$:

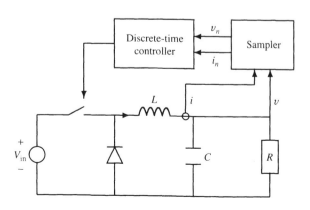

Figure 5.36 System diagram for the circuit in Section 5.4.4.

$$0.3T - k_i(i_n - I_p) - k_v(v_n - V_p) = 0.3T - k_i\left(\frac{V_{in}}{R} - I_p\right) - k_v(V_{in} - V_p)$$

$$= 0.3T + k_i I_p + k_v V_p - \left(\frac{k_i}{R} + k_v\right)V_{in}$$

$$\leq 0$$

Therefore, for $V_{in} > (\frac{k_i}{R} + k_v)/(0.3T + k_i I_p + k_v V_p) = 19.213$, the switch can be always on.

However, the switch being always on is not the only possible operation for $V_{in} > 19.213$. For $V_{in} \in (19.213, 20)$, there are two other periodic solutions: one is stable, the other is unstable.

Take $V_{in} = 19.9$, for example. Performing steady-state analysis [3,4], one stable periodic solution with duty cycle 0.6267 and one unstable periodic solution with duty cycle 0.7878 can be obtained. They are shown as the solid line and dashed line respectively in Figure 5.37. The stable one has output voltage around 12.5V; the unstable one has output voltage around 15.7V. As V_{in} is further increased, these two periodic solutions become closer and collide when $V_{in} = 20$. For $V_{in} = 20$, one eigenvalue of Φ is 1 and a saddle-node bifurcation occurs. If V_{in} is increased a little bit above 20, the operation suddenly jumps to the situation where the switch is always on and the output voltage jumps from 14V to 20V.

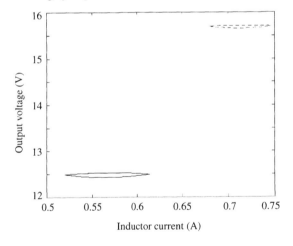

Figure 5.37 Stable periodic solution (solid line) and unstable periodic solution (dashed line) for $V_{in} = 19.9$V.

The circuit is simulated for $V_{in} \in [18.5, 20.5]$ and the resulting bifurcation diagram is shown in Figure 5.38. In the figure, the upper solid line is for the stable *static* solution when the switch is always on (hence duty cycle is 1 and $d_n = 0$); the dashed line and the lower solid line are for unstable and stable *periodic* solutions respectively with duty cycle less than 1. For V_{in} below 19.213, there is only one stable periodic solution and the output voltage is regulated below 11V.

5.4.5 Neimark Bifurcation in Buck Converter Under Voltage-Mode Control

Consider the example [5, p. 228] of a buck converter under voltage-mode control shown in Figure 5.39. The system parameters are $f_s = 15$kHz, $L = 0.9$mH, $C = 22\mu$F, $R = 20\Omega$, $V_{ref} = 5$V, $R_1 = R_2 = 7.5$kΩ, $R_3 = 60$kΩ, $C_2 = 0.4\mu$F, $V_{in} = 30$V,

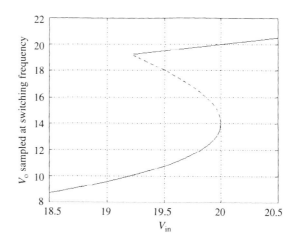

Figure 5.38 Bifurcation diagram of the buck converter in Section 5.4.4; solid lines for stable solutions and the dashed line for an unstable solution; bifurcation parameter is input voltage V_{in}.

$V_L = 2.8$V, $V_U = 8.2$V (then $h(t) = 2.8 + 5.4[\frac{t}{T} \bmod 1]$). All parasitic resistances are ignored. Here, under voltage-mode control, the output voltage is sensed, scaled by 0.5, compared with $V_{ref} = 5$V, and input into the error amplifier. Therefore, an output voltage around 10V is expected.

Let the state $\mathbf{x} = (i, v, v_{c2})^T$. In terms of the representation in Figure 5.30, one has

$$\mathbf{A}_1 = \mathbf{A}_2 = \begin{bmatrix} 0 & \frac{-1}{L} & 0 \\ \frac{1}{C} & \frac{-1}{RC} & 0 \\ 0 & \frac{1}{R_1 C_2} & \frac{-1}{R_3 C_2} \end{bmatrix}$$

$$\mathbf{B}_1 = \begin{bmatrix} \frac{1}{L} & 0 \\ 0 & 0 \\ 0 & \frac{-1}{C_2}\left(\frac{1}{R_1} + \frac{1}{R_2}\right) \end{bmatrix} \qquad \mathbf{B}_2 = \begin{bmatrix} 0 & 0 \\ 0 & 0 \\ 0 & \frac{-1}{C_2}\left(\frac{1}{R_1} + \frac{1}{R_2}\right) \end{bmatrix}$$

$$\mathbf{C} = \begin{bmatrix} 0 & 0 & -1 \end{bmatrix} \qquad\qquad \mathbf{D} = \begin{bmatrix} 0 & 1 \end{bmatrix}$$

$$\mathbf{E}_1 = \mathbf{E}_2 = \begin{bmatrix} 0 & 1 & 0 \end{bmatrix}$$

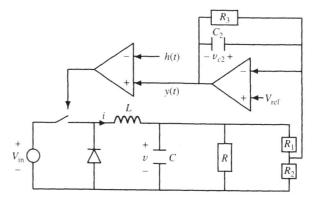

Figure 5.39 System diagram for the circuit in Section 5.4.5.

The fixed point $\mathbf{x}^0(0)$ is $(0.2539, 10.0053, 0.3918)^T$. The eigenvalues of $\boldsymbol{\Phi}$ are 0.8799 and $0.8797 \pm 0.4474i$, which are inside the unit circle. Thus the periodic solution is asymptotically orbitally stable.

As V_{in} is increased from 30V, the magnitude of the complex pair of eigenvalues begins to grow. For $V = 36.9$V, the eigenvalues $(0.8897 \pm 0.4567i)$ exit the unit circle. Thus a Neimark bifurcation occurs. A low oscillating frequency $\frac{f_s}{2\pi} \angle (0.8897 + 0.4567i)$ $= 1132$Hz modulating the original one f_s is expected ($\angle$ denotes the angle in radians). Since these two frequencies are not commensurate, the steady state is quasiperiodic.

For $V_{\text{in}} = 30$V (before the bifurcation), the stable periodic solution $\mathbf{x}^0(t)$ is shown as a solid line in Figure 5.40(a), which becomes unstable following the Neimark bifurcation at $V_{\text{in}} > 36.9$V. The unstable periodic solution for $V_{\text{in}} = 50$V is shown as a dashed line. Following the Neimark bifurcation, a quasiperiodic state trajectory comes into existence, coexisting with the unstable periodic solution. For $V_{\text{in}} = 50$V, the quasiperiodic state trajectory is shown in Figure 5.40(b).

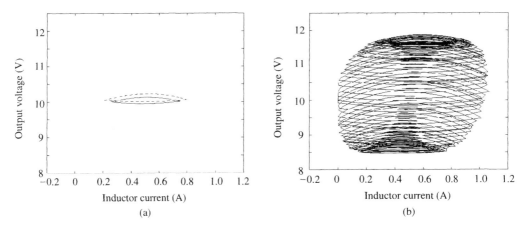

Figure 5.40 (a) Stable periodic solution (solid line) for $V_{\text{in}} = 30$V becomes unstable periodic solution (dashed line) for $V_{\text{in}} = 50$V. (b) Quasiperiodic state trajectory in state space for $V_{\text{in}} = 50$V.

Output voltage waveforms of the quasiperiodic steady state and the unstable periodic solution are shown as solid line (with larger amplitude) and dashed line respectively in Figure 5.41 for $V_{\text{in}} = 50$V. It is seen that the quasiperiodic steady state has two oscillating frequencies as expected: f_s modulated by a lower frequency around 1132Hz.

5.4.6 Neimark Bifurcation in Buck Converter with Input Filter Under Voltage-Mode Control

The system diagram is shown in Figure 5.42. The system parameters are $T = 400\mu$s, $L = 20$mH, $C = 47\mu$F, $R = 22\Omega$, $V_{\text{in}} = 15.8$V, $V_{\text{ref}} = 11.3$V, $g_1 = 8.4$, $V_L = 3.8$V, $V_U = 8.2$V, (then $h(t) = 3.8 + 4.4[\frac{t}{T} \bmod 1]$), $L_f = 2.5$mH, $C_f = 160\mu$F, and R_ρ is varied from 1Ω to 100Ω to adjust the damping. The angular resonance frequency of input filter is $\omega_r = 1/\sqrt{L_f C_f} = 1581.1$. Here under voltage-mode control, the output voltage is regulated around $V_{\text{ref}} = 11.3$V.

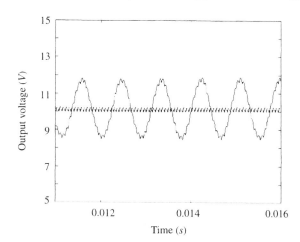

Figure 5.41 Waveforms of quasiperiodic output voltage (solid line with larger amplitude) and unstable (normally unobserved) periodic output voltage (dashed line), both for $V_{\text{in}} = 50\text{V}$.

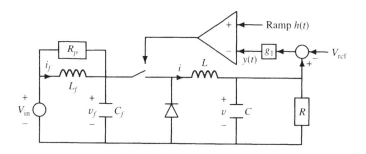

Figure 5.42 System diagram for the circuit in Section 5.4.6.

Let the state be $\mathbf{x} = (i, v, i_f, v_f)^T$, where i_f and v_f are the inductor current and capacitor voltage in the input filter, respectively. In terms of the block diagram model in Figure 5.30, one has (S_1 is the off stage and S_2 is the on stage)

$$
\mathbf{A}_1 = \begin{bmatrix} 0 & \frac{-1}{L} & 0 & 0 \\ \frac{1}{C} & \frac{-1}{RC} & 0 & 0 \\ 0 & 0 & 0 & \frac{-1}{L_f} \\ 0 & 0 & \frac{1}{C_f} & \frac{-1}{R_p C_f} \end{bmatrix} \quad
\mathbf{A}_2 = \begin{bmatrix} 0 & \frac{-1}{L} & 0 & \frac{1}{L} \\ \frac{1}{C} & \frac{-1}{RC} & 0 & 0 \\ 0 & 0 & 0 & \frac{-1}{L_f} \\ \frac{-1}{C_f} & 0 & \frac{1}{C_f} & \frac{-1}{R_p C_f} \end{bmatrix}
$$

$$
\mathbf{B}_1 = \mathbf{B}_2 = \begin{bmatrix} 0 \\ 0 \\ \frac{1}{L_f} \\ \frac{1}{R_p C_f} \end{bmatrix}
$$

$$\mathbf{C} = \begin{bmatrix} 0 & g_1 & 0 & 0 \end{bmatrix} \qquad\qquad \mathbf{D} = \begin{bmatrix} 0 & -g_1 \end{bmatrix}$$

$$\mathbf{E}_1 = \mathbf{E}_2 = \begin{bmatrix} 0 & 1 & 0 & 0 \end{bmatrix}$$

The eigenvalue trajectory of $\boldsymbol{\Phi}$ as R_p varies is shown in Figure 5.43. One pair of eigenvalues is almost fixed at $-0.5963 \pm 0.5301i$, while the other pair moves as R_p varies. A Neimark bifurcation occurs when $R_p = 38.85$, where a pair of eigenvalues $0.8087 \pm 0.5883i$ crosses the unit circle. After the bifurcation, another oscillating angular frequency $f_s[\angle(0.81 + 0.59i)] = 1581.1$ is expected. This angular frequency has the same value as ω_r. After the Neimark bifurcation the original oscillating frequency (i.e., the angular switching frequency $\omega_s = 2\pi f_s$) is modulated by the resonance frequency of input filter (ω_r). Since these two frequencies are not commensurate, the state trajectory is quasiperiodic.

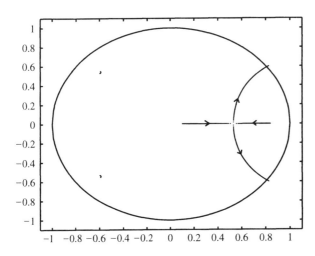

Figure 5.43 Eigenvalue trajectory of Φ as R_p varies from 1 to 100.

5.4.7 Neimark Bifurcation in Buck Converter with Input Filter Under Current-Mode Control

The system diagram [6, p. 96] is shown in Figure 5.44, where $f_s = 30\text{kHz}$, $V_{\text{in}} = 15\text{V}$, $R = 10.4\Omega$, $L = 0.48\text{mH}$, $C = 30\mu\text{F}$, $R_L = 0.6\Omega$ with input filter parameters $R_{L1} = 0.25\Omega$, $L_f = 0.43\text{mH}$ and $C_f = 10.4\mu\text{F}$. The duty cycle D_c, adjusted by

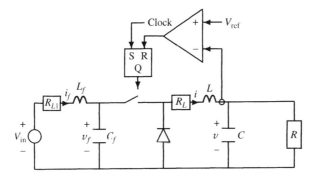

Figure 5.44 System diagram for Section 5.4.7.

V_{ref} (current reference), is used as the bifurcation parameter and varied from 0 to 0.5.

Let the state be $\mathbf{x} = (i, v, i_f, v_f)^T$, where i_f and v_f are the inductor current and capacitor voltage in the input filter, respectively. In terms of the block diagram model in Figure 5.30, one has

$$\mathbf{A}_1 = \begin{bmatrix} \dfrac{-R_L}{L} & \dfrac{-1}{L} & 0 & \dfrac{1}{L} \\[2mm] \dfrac{1}{C} & \dfrac{-1}{RC} & 0 & 0 \\[2mm] 0 & 0 & \dfrac{-R_{L1}}{L_f} & \dfrac{-1}{L_f} \\[2mm] \dfrac{-1}{C_f} & 0 & \dfrac{1}{C_f} & 0 \end{bmatrix} \qquad \mathbf{A}_2 = \begin{bmatrix} \dfrac{-R_L}{L} & \dfrac{-1}{L} & 0 & 0 \\[2mm] \dfrac{1}{C} & \dfrac{-1}{RC} & 0 & 0 \\[2mm] 0 & 0 & \dfrac{-R_{L1}}{L_f} & \dfrac{-1}{L_f} \\[2mm] 0 & 0 & \dfrac{1}{C_f} & 0 \end{bmatrix}$$

$$\mathbf{B}_1 = \mathbf{B}_2 = \begin{bmatrix} 0 \\[2mm] 0 \\[2mm] \dfrac{1}{L_f} \\[2mm] 0 \end{bmatrix}$$

$$\mathbf{C} = \begin{bmatrix} 1 & 0 & 0 & 0 \end{bmatrix} \qquad\qquad \mathbf{D} = \begin{bmatrix} 0 & -1 \end{bmatrix}$$

$$\mathbf{E}_1 = \mathbf{E}_2 = \begin{bmatrix} 0 & 1 & 0 & 0 \end{bmatrix} \qquad h(t) = 0$$

The eigenvalue trajectory of $\boldsymbol{\Phi}$ as D_c varies is shown in Figure 5.45. An eigenvalue pair departs the unit circle for the parameter value $D_c = 0.2443$. Again as in the example in Section 5.4.6, the bifurcated solution is on a torus, with angular frequencies, ω_s and $1/\sqrt{L_f C_f}$.

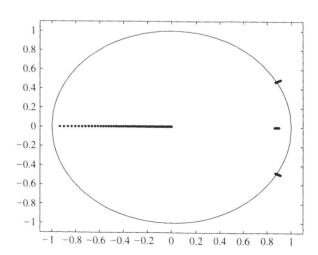

Figure 5.45 Eigenvalues trajectory of Φ as duty cycle D_c varies from 0 to 0.5, where the eigenvalue trajectories go outward.

REFERENCES

[1] D. C. Hamill, Power electronics: A field rich in nonlinear dynamics, in *Nonlinear Dynamics of Electronic Systems*, Dublin, 1995.

[2] B. Lehman and R. M. Bass, Switching frequency dependent averaged models for PWM dc/dc converters, *IEEE Trans. on Power Electronics*, vol. 11, no. 1, pp. 89–98, 1996.

[3] C.-C. Fang and E. H. Abed, Sampled-data modeling and analysis of closed-loop PWM dc/dc converters, in *IEEE Int. Symp. on Circuits and Systems*, 1999, vol. 5, pp. 110-115.

[4] C.-C. Fang and E. H. Abed, Sampled-data modeling and analysis of PWM dc/dc converters I. Closed-loop circuits, Tech. Rep. 98-54, Institute for Systems Research, University of Maryland, College Park, 1998, available at http://www.isr.umd.edu/TechReports/ISR/1998/.

[5] K. K. Tse and H. Chung, Decoupled technique for the simulation of PWM switching regulators using second order output extrapolations, *IEEE Trans. on Power Electronics*, vol. 13, no. 2, pp. 222–234, 1998.

[6] K. M. Smedley, Control art of switching converters. Ph.D. diss., California Institute of Technology, 1990.

5.5 NONLINEAR ANALYSIS OF OPERATION IN DISCONTINUOUS-CONDUCTION MODE

Chi K. Tse

5.5.1 Review of Operating Modes

In terms of circuit operation, a dc/dc converter can be described as a switched electrical circuit whose topology toggles, in some predefined pattern, between a number of linear circuits. In simple dc/dc converters such as the buck, boost, and buck-boost converters, the involved linear circuits are second-order circuits containing an inductor, a capacitor, a switch (externally controlled switch) and a diode (internally controlled switch).

Assuming that the switch and the diode are turned on and off in a complementary fashion, a simple dc/dc converter toggles between two linear circuits, one with the switch closed (diode opened) and the other with the switch opened (diode closed). This happens when the converter satisfies certain conditions which we will discuss. Taking a closer look at the circuit operation we readily observe that the inductor current ramps up and down, respectively, when the switch is on and off. During the off state of the switch, the diode conducts the same current as the inductor, and the inductor current never goes down to zero. As is customary in the literature, such an operating mode is referred to as *continuous-conduction mode*. Figure 5.46(a) shows the inductor current waveform when the converter operates in this mode.

Clearly, continuous-conduction mode can be maintained only if the inductor current does not go down to zero during the off state of the switch. Intuitively this requires either a sufficiently large inductance, a relatively large average inductor current or a relatively short switching period. Hence: if (1) the inductance is too small; or (2) the output load current is too low; or (3) the period is too long, then the inductor current can fall to zero during the off state of the switch. Once it touches zero, it will stay at zero because the diode does not allow current reversal. When this happens, the con-

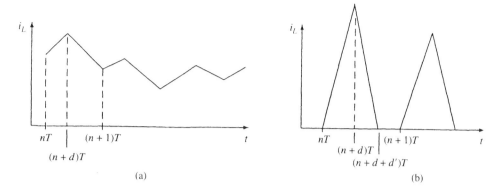

Figure 5.46 Inductor current waveform in (a) continuous-conduction mode, and (b) discontinuous-conduction mode.

verter is said to be operating in discontinuous conduction mode. Figure 5.46(b) shows the inductor current waveform for this operating mode [1,2]. As a matter of practical interest, the discontinuous-conduction mode enjoys smaller inductance and faster dynamical response. However, since the peak current is comparatively high (for the same power level that a continuous-conduction-mode converter would handle), only low-power applications are feasible for discontinuous-conduction mode.

In summary, depending on the relative magnitudes of the storage inductance, output load current, and the switching period, a simple dc/dc converter can operate in either continuous- or discontinuous-conduction mode.

5.5.2 Derivation of Discrete-Time Maps

The simple buck, boost, and buck-boost converters, having two independent storage elements, are second-order systems. However, a close inspection of the inductor current waveform reveals that the inductor current is identically zero at the start of each switching period when operating in discontinuous conduction mode, i.e.,

$$i_L(nT) = 0 \quad \text{for all integers } n \tag{5.51}$$

where T is the switching period. Thus, the inductor current is no longer a dynamic variable. As a result, the converter reduces to first order, with the capacitor voltage serving as the only state variable. We therefore expect that the discrete-time model for a dc/dc converter operating in discontinuous-conduction mode takes the form of a first-order map:

$$v_{n+1} = f(v_n, d_n) \tag{5.52}$$

where subscript n denotes the value at $t = nT$, and other symbols have their usual meanings.

The derivation of the above discrete-time equation for a given dc/dc converter involves a rather straightforward procedure [3]. Let **x** be the state vector, which is defined as $[v, i_L]^T$. In a switching period, the switch is turned on for a fraction d of the period, and is off for the rest of the period. (Thus, as usual, d is the duty cycle.) As mentioned above, during the off state, the diode initially conducts and then opens itself when i_L reaches zero. We let $d'T$ be the duration of the interval during the off period in which the diode is conducting. We may write the state equations as follows:

$$\frac{d\mathbf{x}}{dt} = \begin{cases} A_1\mathbf{x} + B_1 V_{\text{in}} & \text{for } nT \leq t < (n+d)T \\ A_2\mathbf{x} + B_2 V_{\text{in}} & \text{for } (n+d)T \leq t < (n+d+d')T \\ A_3\mathbf{x} + B_3 V_{\text{in}} & \text{for } (n+d+d')T \leq t < (n+1)T \end{cases} \quad (5.53)$$

The solution to each of these state equations can be expressed in terms of the respective transition matrix, and the consecutive solutions are then *stacked* over a switching cycle, resulting in a discrete-time difference equation:

$$\mathbf{x}((n+1)T) = \Phi_3(d''T)\Phi_2(d'T)\Phi_1(dT)\left(\mathbf{x}(nT) + \int_{nT}^{(n+d)T} \Phi_1(nT - \tau)B_1 V_{\text{in}}.d\tau\right)$$

$$+ \Phi_3(d''T)\Phi_2(d'T)\int_{(n+d)T}^{(n+d+d')T} \Phi_2((n+d)T - \tau)B_2 V_{\text{in}}.d\tau \quad (5.54)$$

$$+ \Phi_3(d''T)\int_{(n+d+d')T}^{(n+1)T} \Phi_3((n+d+d')T - \tau)B_3 V_{\text{in}}.d\tau$$

where $d'' = 1 - d - d'$ and the transition matrix $\Phi_i(.)$ is given by

$$\Phi_i(\xi) = 1 + \sum_{n=1}^{\infty} \frac{1}{n!}A_i^n\xi^n \quad \text{for } i = 1, 2, \text{ and } 3 \quad (5.55)$$

Moreover, when operating in discontinuous-conduction mode, the third equation in (5.53) describes only a simple first-order discharge of the output capacitor. Thus, writing the third equation in (5.53) as $\dot{v} = -v/CR$, we have $\Phi_3(\xi) = \exp(-\xi/CR)$. Also, since i_L is identically zero at each switching instant, we may reduce (5.54) to the following first-order equation:

$$v_{n+1} = e^{-d''T/CR}\mathbf{P}\Phi_2(d'T)\Phi_1(dT)\left(\mathbf{P}^T v_n + \int_{nT}^{(n+d)T} \Phi_1(nT - \tau)B_1 V_{\text{in}}.d\tau\right)$$

$$+ e^{-d''T/CR}\mathbf{P}\Phi_2(d'T)\int_{(n+d)T}^{(n+d+d')T} \Phi_2((n+d)T - \tau)B_2 V_{\text{in}}.d\tau \quad (5.56)$$

$$+ e^{-d''T/CR}\int_{(n+d+d')T}^{(n+1)T} e^{-((n+d+d')T-\tau)/CR}B_3 V_{\text{in}}.d\tau$$

where $\mathbf{P} = [1\ 0]$. (A detailed discussion on using $\mathbf{P}$ in formulating reduced-dimension equations can be found in Chapter 6.) When all transition matrices have been computed and substituted in (5.56), the required discrete-time equation is obtained. Moreover, volt-time balance of the inductor gives an equation relating d' and d. For example, for the buck converter, we have

$$\frac{d'}{d} = \frac{V_{\text{in}} - v}{v} \quad (5.57)$$

and we can find a similar relation for any other converter. Using this d-d' relation and a truncated series for $\Phi_i(.)$, we can eliminate d' and derive an approximate discrete-time map for a dc/dc converter operating in discontinuous-conduction mode. We will skip the algebra, which is straightforward, and state the results for the boost and buck converters as follows:

$$v_{n+1} = \alpha v_n + \frac{\beta d_n^2 V_{\text{in}}(V_{\text{in}} - v_n)}{v_n} \quad \text{for buck converter} \tag{5.58}$$

$$v_{n+1} = \alpha v_n + \frac{\beta d_n^2 V_{\text{in}}^2}{v_n - V_{\text{in}}} \quad \text{for boost converter} \tag{5.59}$$

where

$$\alpha = 1 - \frac{T}{CR} + \frac{T^2}{2C^2R^2} \tag{5.60}$$

$$\beta = \frac{T^2}{2LC} \tag{5.61}$$

In practice, dc/dc converters are controlled via a feedback mechanism. The usual control objective is to keep the output voltage fixed. For simplicity, we consider a proportional feedback which effectively samples the output voltage and generates an error signal from which the value of the duty cycle is derived, i.e.,

$$d_n = H(D + \kappa(v_n - V_{\text{ref}})) \tag{5.62}$$

where D is the steady-state duty cycle, κ is the small-signal feedback gain, and $H(.)$ accounts for the limited range of the duty cycle between 0 and 1:

$$H(x) = \begin{cases} 0 & \text{for } x < 0 \\ 1 & \text{for } x > 1 \\ x & \text{otherwise} \end{cases} \tag{5.63}$$

Combining this control equation with the discrete-time map of the system, we yield a discrete-time map for the closed-loop system. For the buck converter, we get

$$v_{n+1} = \alpha v_n + \frac{\beta(H(D + \kappa(v_n - V_{\text{ref}})))^2 V_{\text{in}}(V_{\text{in}} - v_n)}{v_n} \tag{5.64}$$

Also, D can be found by putting $v_{n+1} = v_n$ in (5.64), i.e.,

$$D = \sqrt{\frac{(1 - \alpha)V_C^2}{\beta V_{\text{in}}(V_{\text{in}} - V_C)}} \tag{5.65}$$

where uppercase letters denote steady-state values as usual.

In the following we will use the discrete-time map (5.64) to study the period-doubling phenomenon in the buck converter operating in discontinuous conduction mode. Readers may refer to Tse [4] for a similar treatment of the boost converter.

5.5.3 Period-Doubling Bifurcation

In Chapter 3 we discuss the use of the Schwarzian derivative in studying period-doubling bifurcations in first-order maps. It has been shown that the small-signal feedback gain plays a crucial role in determining if period doubling may occur. In the following we re-examine the system in terms of stability of the fundamental operation, and attempt to derive a condition for the first period doubling.

We consider small disturbance Δv around the steady-state value V_C. The usual Taylor series expansion can be written as

$$\Delta v_{n+1} = \sum_{k=1}^{\infty} \frac{1}{k!} \frac{\partial^k f(v)}{\partial v^k} \bigg|_{v=V_C} (\Delta v_n)^k \qquad (5.66)$$

If the disturbance is small, the magnitude of $\partial f(v)/\partial v$ at $v = V_C$ determines the stability. This partial derivative is sometimes referred to as the *characteristic multiplier* or *eigenvalue*. For the present 1-D map, it simply corresponds to the slope of $f(x)$ at the fixed point.

We assume that in the neighborhood of the steady-state point the duty cycle does not saturate. Hence, we may consider the discrete-time map (5.64) without the need for applying $H(.)$. Thus, the characteristic multiplier, λ, can be obtained by direct differentiation:

$$\lambda = \frac{\partial f(v)}{\partial v} \bigg|_{v=V_C} = \alpha - \frac{\beta V_{in} D[2\kappa V_C(V_{in} - V_C) + D V_{in}]}{V_C^2} \qquad (5.67)$$

The system remains fundamentally stable if the magnitude of the characteristic multiplier is less than unity, i.e.,

$$|\lambda| = \left| \alpha - \frac{\beta V_{in} D[2\kappa V_C(V_{in} - V_C) + D V_{in}]}{V_C^2} \right| < 1 \qquad (5.68)$$

At the boundary where the characteristic multiplier is -1, a period-doubling bifurcation takes place and v repeats itself every second period. As the characteristic multiplier decreases below -1, v may diverge in an oscillatory fashion or maintain a stable subharmonic oscillation, depending on the higher-order terms in (5.66). The critical value of the small-signal feedback gain can be found by setting the characteristic multiplier to -1:

$$\kappa_c = \frac{(1 + \alpha)V_C^2 - \beta V_{in}^2 D^2}{2\beta V_{in} D V_C(V_{in} - V_C)} \qquad (5.69)$$

Now using (5.64), we can easily arrive at some useful conclusion concerning the behavior of the system near $\kappa = \kappa_c$. As we will see, (5.64) represents a typical unimodal map [5]. A common plan of attack for such maps is as follows. Initially we set κ at a value smaller than κ_c and confirm that the system has a stable fixed point. Then, we increase κ and observe the way the system loses stability and bifurcates into subharmonic orbits of period-2. We further increase κ to observe a typical subharmonic cascade and eventually chaotic motion.

An example will help visualize the situation. Suppose $T/CR = 0.12$, $RT/L = 20$, $V_{in} = 33$V, and $V_C = 25$V. This gives $D = 0.4717$. Also it is readily verified that the value of RT/L is large enough to ensure a discontinuous-conduction-mode operation. Direct substitution gives

$$v_{n+1} = 0.8872 v_n + \frac{1.2 \times 33 \times (33 - v_n) \times H(d_n)^2}{v_n} \qquad (5.70)$$

where $d_n = 0.4717 - \kappa(v_n - 25)$. The characteristic multiplier, as given in (5.68), is

$$\lambda = 0.4220 - 11.9548\kappa \qquad (5.71)$$

Thus the critical value of κ is 0.1189. Figure 5.47 shows the iterative maps corresponding to a subcritical case ($\kappa < \kappa_c$), and two supercritical cases ($\kappa > \kappa_c$). As shown clearly, the system has a stable fixed point in the subcritical case, and a stable subharmonic

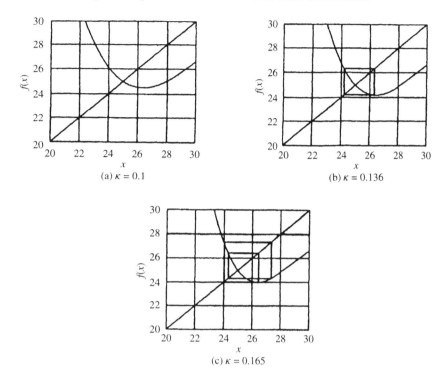

(a) $\kappa = 0.1$

(b) $\kappa = 0.136$

(c) $\kappa = 0.165$

Figure 5.47 Iterative maps showing (a) stable fixed point; (b) period-2 subharmonic solution; and (c) period-4 subharmonic solution.

orbit in supercritical cases. Furthermore, using (5.70), a bifurcation diagram can be generated as shown in Figure 5.48. Reference to this diagram shows that the system becomes chaotic when κ is larger than about 0.17.

5.5.4 Computer Simulations and Experiments

In this subsection, we provide verification of the period-doubling bifurcation using "exact" time-domain simulation of the system. Our simulation is based on a piecewise-

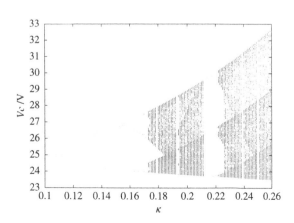

Figure 5.48 Bifurcation diagram using the approximate discrete-time map.

switched model which represents a very accurate description of the system. Essentially the model involves toggling between three linear circuits according to the duty cycle control and circuit condition. The simulation uses the following parameters: $T = 33.33\mu s$, $V_{in} = 33V$, $V_C = 25V$, $L = 208\mu H$, $C = 222\mu F$, $R = 12.5\Omega$.

We have simulated the steady-state waveforms for various values of κ. Figures 5.49(a), 5.50(a), 5.51(a), and 5.52(a) show the steady-state waveforms of the closed-loop system with $\kappa = 0.1$, 0.126, 0.184, and 0.216 respectively. The phase portraits corresponding to these four cases are shown in Figures 5.49(b), 5.50(b), 5.51(b) and 5.52(b), which demonstrate clearly the fundamental, period-2 subharmonic, period-4 subharmonic, and chaotic orbits. We have also summarized in Figure 5.53 the steady-state information in the form of a bifurcation diagram which clearly demonstrates the sequence of period-doubling subharmonics as well as the presence of a period-3 window around $\kappa = 0.245$.

The general appearance of this simulation-based bifurcation diagram resembles that of Figure 5.48. However, some noticeable differences are still observed between them, which can be attributed to the fact that Figure 5.48 is generated from an approximate iterative map whose validity relies very much on the accuracy of the truncated Taylor series. On the other hand, Figure 5.53 represents exact simulated system behavior.

Further evidence of period doubling in discontinuous-mode dc/dc converters can be provided by laboratory tests. Refer to Tse [4,6] for experimental confirmation of the occurrence of period-doubling cascades for boost and buck converters operating in discontinuous-conduction mode.

5.5.5 Remarks and Summary

The same procedure can be used to analyze the period-doubling bifurcation in the boost converter operating in discontinuous-conduction mode [4]. We have in particular identified how the small-signal feedback gain affects the qualitative behavior of such circuits. The result from this analysis is consistent with the theoretical prediction based on Schwarzian derivative described earlier in Section 3.5.

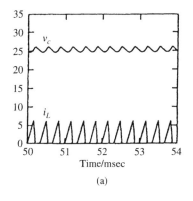
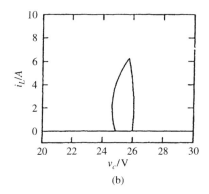

Figure 5.49 (a) Fundamental waveforms from simulation of the exact state equation with $\kappa = 0.1$; (b) phase portrait.

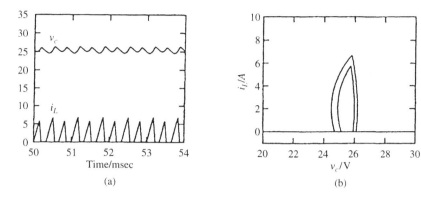

Figure 5.50 (a) Period-2 subharmonic waveforms from simulation of the exact state equation with $\kappa = 0.136$; (b) phase portrait.

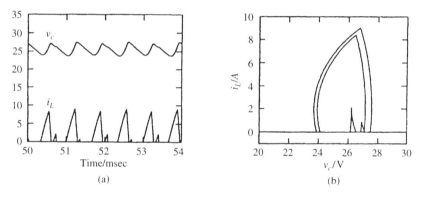

Figure 5.51 (a) Period-4 subharmonic waveforms from simulation of the exact state equation with $\kappa = 0.184$; (b) phase portrait.

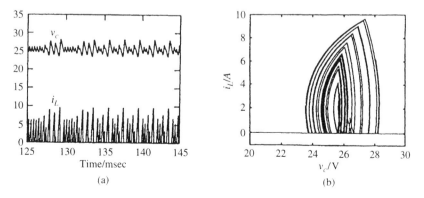

Figure 5.52 (a) Chaotic waveforms from simulation of the exact state equation with $\kappa = 0.216$; (b) phase portrait.

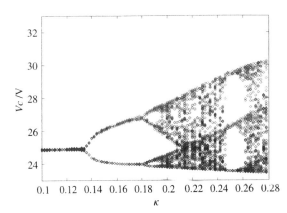

Figure 5.53 Bifurcation diagram from simulations; 500 consecutive points of v after transient are plotted for each κ.

REFERENCES

[1] R. P. Severns and E. J. Bloom, *Modern DC-to-DC Switchmode Power Converter Circuits*. New York: Van Nostrand, 1985.

[2] G. C. Chryssis, *High Frequency Switching Power Supplies*. New York: McGraw-Hill, 1989.

[3] C. K. Tse and K. M. Adams, Qualitative analysis and control of a dc-to-dc boost converter operating in discontinuous mode, *IEEE Trans. on Power Electron.*, vol. 5, no. 3, pp. 323–329, July 1990.

[4] C. K. Tse, Flip bifurcation and chaos in three-state boost switching regulators, *IEEE Trans. on Circ. Syst.—I*, vol. 41, no. 1, pp. 7–13, Jan. 1994.

[5] J. M. T. Thompson and H. B. Stewart, *Nonlinear Dynamics and Chaos*. Chichester: Wiley, 1988.

[6] C. K. Tse, Chaos from a buck switching regulator operating in discontinuous mode, *Int. J. Circ. Theory & Appl.*, vol. 22, no. 4, pp. 263–278, July–Aug. 1994.

5.6 NONLINEAR PHENOMENA IN THE ĆUK CONVERTER

Chi K. Tse

5.6.1 Review of the Ćuk Converter and its Operation

With only one smoothing inductor, simple dc/dc converters cannot provide non-pulsating current for both input and output. The Ćuk converter [1] was proposed originally to overcome this problem by using two inductors. Figure 5.54(a) shows the basic Ćuk converter. For simplicity we will focus on operation in continuous conduction mode, for which only two complementary switch states are involved (i.e., the switch is closed while the diode is open, and vice versa, as shown in Figures 5.54(b) and (c)). Provided the sum of the inductor currents remain positive, the diode conducts current for the whole subinterval during which the switch is off, and the Ćuk converter maintains itself in continuous-conduction mode. Thus, the situations illustrated in Figures 5.54(d) and (e) both belong to continuous-conduction mode. However, we should stress that, unlike other simple dc/dc converters, the Ćuk converter can operate in a number of discontinuous-conduction modes [2].

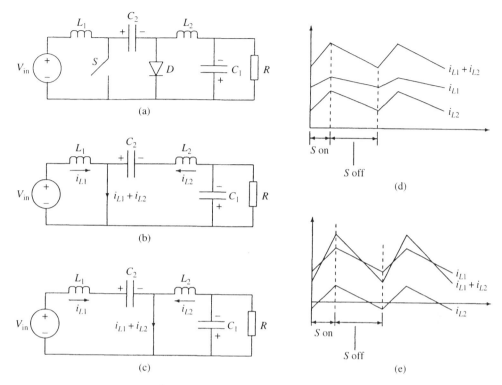

Figure 5.54 (a) The Ćuk converter; (b) equivalent circuit when switch is on and diode is off; (c) equivalent circuit when switch is off and diode is on; (d) inductor current waveforms in continuous mode; (e) inductor current waveforms also in continuous mode.

The Ćuk converter, being a fourth-order system, does not lend itself to any simple analysis. The complexity lies mainly in the modeling of the circuit. If the same iterative map approach can be taken, as for the case of a simple dc/dc converter, we will end up with a more complex fourth-order iterative map. Analysis and simulations become rather complicated. At the time of writing, however, the dynamics of the Ćuk converter remain relatively unexplored, even for the simplest mode of operation in continuous conduction.

5.6.2 Discrete-Time Modeling for Fixed Frequency Operation

As mentioned above, the iterative map approach is applicable to the analysis of the Ćuk converter if one can bear the algebraic tedium. We start with defining a suitable state vector **x**, e.g.,

$$\mathbf{x} = [v_{C1}\ v_{C2}\ i_{L1}\ i_{L2}]^T \tag{5.72}$$

For each switch state, we can write down a state equation in the following form:

$$\frac{d\mathbf{x}}{dt} = A_i\mathbf{x} + B_i V_{\text{in}} \quad \text{for } i = 1, 2 \tag{5.73}$$

where $i = 1$ corresponds to switch being closed and diode being opened, and $i = 2$ corresponds to the complementary state. The usual stacking procedure can be followed to yield a discrete-time map which takes the form:

$$
\mathbf{x}_{n+1} = \Phi_2(T - d_n T)\Phi_1(d_n T)\left(\mathbf{x}_n + \int_{nT}^{nT+d_n T} \Phi_1(nT - \tau)B_1 V_{\text{in}}.d\tau \right)
$$

$$
+ \Phi_2(T - d_n T) \int_{nT+d_n T}^{(n+1)T} \Phi_2(nT + d_n T - \tau)B_2 V_{\text{in}}.d\tau
$$

(5.74)

where all symbols have their usual meanings, and the transition matrix $\Phi(.)$ is

$$
\Phi_i(\xi) = \mathbf{1} + \sum_{k=1}^{\infty} \frac{1}{k!} A_i^k \xi^k, \quad \text{for } k = 1, 2
$$

(5.75)

If a truncated series is used for Φ_i, an approximate iterative map may be obtained which takes the form:

$$
\begin{bmatrix} v_{C1,n+1} \\ v_{C2,n+1} \\ i_{L1,n+1} \\ i_{L2,n+1} \end{bmatrix} = \begin{bmatrix} f_{11}(d_n) & f_{12}(d_n) & f_{13}(d_n) & f_{14}(d_n) \\ f_{21}(d_n) & f_{22}(d_n) & f_{23}(d_n) & f_{24}(d_n) \\ f_{31}(d_n) & f_{32}(d_n) & f_{33}(d_n) & f_{34}(d_n) \\ f_{41}(d_n) & f_{42}(d_n) & f_{43}(d_n) & f_{44}(d_n) \end{bmatrix} \begin{bmatrix} v_{C1,n} \\ v_{C2,n} \\ i_{L1,n} \\ i_{L2,n} \end{bmatrix} + \begin{bmatrix} g_1(d_n) \\ g_2(d_n) \\ g_3(d_n) \\ g_4(d_n) \end{bmatrix} V_{\text{in}}
$$

(5.76)

The Ćuk converter can be controlled in various ways, similar to the other dc/dc converters, giving different dynamical properties. In Tse-Chan [3], for instance, a particular current-mode control scheme has been considered. In this scheme, the sum of the inductor currents is the controlling variable, and the conventional peak-limiting current-mode control is applied. Essentially, the switch is turned on periodically, and is turned off when the sum of the inductor currents reaches a reference value I_{ref}. We can derive the following control equation almost by inspection:

$$
I_{\text{ref}} - (i_{L1} + i_{L2})_n = \left[\frac{V_{\text{in}}}{L_1} + \frac{v_{C2,n} - v_{C1,n}}{L_2} \right] d_n T
$$

(5.77)

where subscript n denotes values at $t = nT$. Hence, we can write

$$
d_n = \frac{I_{\text{ref}} - (i_{L1} + i_{L2})_n}{\left(\dfrac{V_{\text{in}}}{L_1} + \dfrac{v_{C2,n} - v_{C1,n}}{L_2} \right) T}
$$

(5.78)

which combines with (5.76) to give the discrete-time map required.

In fact, we can make use of (5.76) and (5.78) to study the bifurcation phenomena of the Ćuk converter under the above specific current-mode control. Moreover, if a different form of control is used, we need to derive another control equation, in lieu of (5.78), for analyzing the the system. Nonetheless, (5.76) remains applicable. Readers are referred to Tse-Chan [3] for approximate expressions of $f_{ij}(.)$'s and $g_i(.)$'s, and details of analytical and simulated results regarding bifurcation of the current-mode-controlled Ćuk converter operating at a fixed frequency. It has been found that the bifurcation pattern is similar qualitatively to those from other current-mode-controlled dc/dc converters [3].

5.6.3 Free-Running Current-Mode-Controlled Ćuk Converter

Self-oscillating or free-running current-controlled switching converters are often used in low-cost switching power supplies, since they require no external clocks and their constructions are relatively simple. In contrast to their nonautonomous counterparts for which chaos is observed even for the simplest first-order discontinuous-mode converters, free-running converters of order below three cannot exhibit chaos. The essential feature of an autonomous switching converter is the absence of any external driving signal, which is mandatory in the nonautonomous case for periodic switching of the power switch. In this subsection we study the Ćuk converter operating in free-running (autonomous) mode. In particular, we will present the following aspects of study: (1) derivation of describing state equation; (2) stability of the equilibrium state and identification of Hopf bifurcation based on the describing state equation; (3) computer simulations of the circuits revealing the bifurcation from fixed point, through limit cycles and quasiperiodic orbits, and eventually to chaos. (For experimental verification, refer to Chapter 4.)

Autonomous System Modeling

In the free-running Ćuk converter being considered, the switch is turned on and off, in a hysteretic fashion, when the sum of the inductor currents falls below or rises above a certain preset hysteretic or tolerance band [4]. The average value and width of this preset band are adjusted by a feedback Schmitt trigger circuit. Also, the output voltage is fed back to set the average value of the hysteretic band, forcing the control variable to be related by the following control equation:

$$i_{L1} + i_{L2} = g(v_{C1}) \tag{5.79}$$

where $g(.)$ is the control function. For example, a simple proportional control takes the form of

$$\Delta(i_{L1} + i_{L2}) = -\mu \Delta v_{C1} \tag{5.80}$$

where μ is the gain factor. This equation has the following equivalent form, assuming regulated output:

$$i_{L1} + i_{L2} = K - \mu v_{C1} \tag{5.81}$$

where K and μ are the control parameters. Figure 5.55 shows a simplified schematic of the system.

The system can be represented by the following state-space equations where $s = 1$ when the switch is turned on, and $s = 0$ when the switch is off.

$$\begin{cases} \dfrac{di_{L1}}{dt} &= -\dfrac{(1-s)v_{C2}}{L} + \dfrac{V_{in}}{L} \\[2mm] \dfrac{di_{L2}}{dt} &= \dfrac{v_{C2}s}{L} - \dfrac{v_{C1}}{L} \\[2mm] \dfrac{dv_{C1}}{dt} &= \dfrac{i_{L2}}{C} - \dfrac{v_{C1}}{CR} \\[2mm] \dfrac{dv_{C2}}{dt} &= \dfrac{(1-s)i_{L1}}{C} - \dfrac{i_{L2}s}{C} \end{cases} \tag{5.82}$$

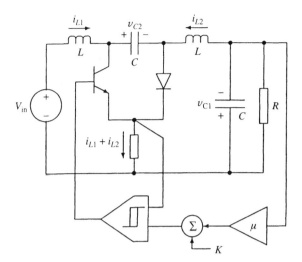

Figure 5.55 Ćuk converter under free-running current-mode control.

The state-space averaged model has the same form as above, with s replaced by the duty cycle d which is the fraction of the switching period for which the switch is turned on.

Since $i_{L1} + i_{L2}$ is related to v_{C1} by a linear algebraic equation, the system reduces its order by one. Specifically, when the control equation (5.81) is taken into account, the system can be reduced to the following third-order system:

$$
\begin{cases}
\dfrac{di_{L2}}{dt} = \dfrac{v_{C2}d}{L} - \dfrac{v_{C1}}{L} \\[2mm]
\dfrac{dv_{C1}}{dt} = \dfrac{i_{L2}}{C} - \dfrac{v_{C1}}{CR} \\[2mm]
\dfrac{dv_{C2}}{dt} = \dfrac{(1-d)(K - \mu v_{C1})}{C} - \dfrac{i_{L2}}{C}
\end{cases}
\tag{5.83}
$$

where d is the duty cycle. Also, from (5.81), $d(i_{L1} + i_{L2})/dt = -\mu dv_{C1}/dt$. Substitution of the involving derivatives gives

$$
d = \frac{1}{2} - \frac{\dfrac{\mu L}{C}i_{L2} - \left(1 + \dfrac{\mu L}{CR}\right)v_{C1} + V_{in}}{2v_{C2}}
\tag{5.84}
$$

which must satisfy $0 < d < 1$. Finally, putting (5.84) into (5.83) results in the following state equations that describe the dynamics of the autonomous system:

$$
\begin{cases}
\dfrac{di_{L2}}{dt} = -\dfrac{\mu i_{L2}}{2C} - \left(1 - \dfrac{\mu L}{CR}\right)\dfrac{v_{C1}}{2L} + \dfrac{v_{C2}}{2L} - \dfrac{V_{in}}{2L} \\[3mm]
\dfrac{dv_{C1}}{dt} = \dfrac{i_{L2}}{C} - \dfrac{v_{C1}}{CR} \\[3mm]
\dfrac{dv_{C2}}{dt} = -\dfrac{i_{L2}}{C} + \left(\dfrac{K - \mu v_{C1}}{2C}\right)\left(1 + \dfrac{\frac{\mu L}{C}i_{L2} - \left(1 + \frac{\mu L}{CR}\right)v_{C1} + V_{in}}{v_{C2}}\right)
\end{cases}
\tag{5.85}
$$

Note that this representation is valid only if $0 < d < 1$. Thus, when analyzing the system numerically we must implement a saturating function such that the value of d is clipped at 0 or 1, as appropriate.

Dimensionless Equations

The earlier-derived state equations can be put in a dimensionless form. Define the dimensionless state variables as follows:

$$x_1 = \frac{Ri_{L2}}{V_{in}}, \quad x_2 = \frac{v_{C1}}{V_{in}}, \quad x_3 = \frac{v_{C2}}{V_{in}} \tag{5.86}$$

Also define the dimensionless time and parameters as follows:

$$\tau = \frac{Rt}{2L}, \quad \xi = \frac{L/R}{CR}, \quad \kappa_1 = \mu R, \quad \kappa_o = \frac{KR}{V_{in}} \tag{5.87}$$

Direct substitution of these new dimensionless variables, time and parameters in the autonomous equations (5.85) yields the following dimensionless autonomous equations:

$$\begin{cases} \dfrac{dx_1}{d\tau} = -\xi\kappa_1 x_1 - (1 - \kappa_1\xi)x_2 + x_3 - 1 \\[2mm] \dfrac{dx_2}{d\tau} = 2\xi(x_1 - x_2) \\[2mm] \dfrac{dx_3}{d\tau} = -2\xi x_1 + \xi(\kappa_o - \kappa_1 x_2)\left(1 + \dfrac{\kappa_1\xi x_1 - (1 + \kappa_1\xi)x_2 + 1}{x_3}\right) \end{cases} \tag{5.88}$$

To complete the model, saturation must be included. Now, d may be written as

$$d = 0.5 - \frac{\kappa_1\xi x_1 - (1 + \kappa_1\xi)x_2 + 1}{2x_3} \tag{5.89}$$

The condition for saturation is

$$d > 1 \Leftrightarrow \kappa_1\xi x_1 - (1 + \kappa_1\xi)x_2 + x_3 + 1 < 0 \tag{5.90}$$

$$\text{or} \quad d < 0 \Leftrightarrow \kappa_1\xi x_1 - (1 + \kappa_1\xi)x_2 - x_3 + 1 > 0 \tag{5.91}$$

By putting $d = 1$ or 0 in (5.83) and performing dimensionless substitution, the state equations for saturation are

$$\begin{cases} \dfrac{dx_1}{d\tau} = 2(x_3 - x_2) \\[2mm] \dfrac{dx_2}{d\tau} = 2\xi(x_1 - x_2) \qquad\qquad \text{for } d > 1 \\[2mm] \dfrac{dx_3}{d\tau} = -2\xi x_1 \end{cases} \tag{5.92}$$

and

$$\begin{cases} \dfrac{dx_1}{d\tau} = -2x_2 \\[2mm] \dfrac{dx_2}{d\tau} = 2\xi(x_1 - x_2) & \text{for } d < 0 \\[2mm] \dfrac{dx_3}{d\tau} = -2\xi x_1 + 2\xi(\kappa_0 - \kappa_1 x_2) \end{cases} \qquad (5.93)$$

The equilibrium point can be calculated by putting $dx_1/d\tau = dx_2/d\tau = dx_3/d\tau = 0$ in (5.88) and considering the restricted sign of X_2. This gives

$$X = \begin{bmatrix} X_1 \\ X_2 \\ X_3 \end{bmatrix} = \begin{bmatrix} X_s \\ X_s \\ X_s + 1 \end{bmatrix} \qquad (5.94)$$

where

$$X_s = \frac{-(1 + \kappa_1) + \sqrt{(1 + \kappa_1)^2 + 4\kappa_o}}{2} \qquad (5.95)$$

Stability of Equilibrium Point and Hopf Bifurcation

The Jacobian matrix, $J(X)$, for the dimensionless system evaluated at the equilibrium point is given by

$$J(X) = \begin{bmatrix} -\kappa_1\xi & -(1 - \kappa_1\xi) & 1 \\ 2\xi & -2\xi & 0 \\ J_{31} & J_{32} & J_{33} \end{bmatrix}$$

where

$$J_{31} = -2\xi + \frac{\kappa_1\xi^2(\kappa_o - \kappa_1 X_s)}{1 + X_s} \qquad (5.96)$$

$$J_{32} = \frac{-2\kappa_1\xi - \xi(1 + \kappa_1\xi)(\kappa_o - \kappa_1 X_s)}{1 + X_s} \qquad (5.97)$$

$$J_{33} = \frac{\xi(\kappa_o - \kappa_1 X_s)(X_s - 1)}{(1 + X_s)^2} \qquad (5.98)$$

From (5.95), $X_s(X_s + 1) = \kappa_o - \kappa_1 X_s$. The Jacobian matrix can hence be simplified to

$$J(X) = \begin{bmatrix} -\kappa_1\xi & -(1 - \kappa_1\xi) & 1 \\ 2\xi & -2\xi & 0 \\ -2\xi + \kappa_1\xi^2 X_s & \dfrac{-2\kappa_1\xi}{1 + X_s} - \xi(1 + \kappa_1\xi)X_s & \dfrac{-\xi X_s(1 - X_s)}{1 + X_s} \end{bmatrix} \qquad (5.99)$$

We shall attempt to study the stability of the equilibrium point and the trajectory around the equilibrium point by deriving the eigenvalues of the system at the equilibrium point. The usual procedure is to solve the following equation for λ:

$$\det[\lambda \mathbf{I} - J(X)] = 0 \qquad (5.100)$$

Upon expanding, we get

$$\lambda^3 + \frac{\xi[(\kappa_1 + 2) + (\kappa_1 + 3)X_s - X_s^2]}{1 + X_s}\lambda^2$$

$$+ \frac{2\xi[2 + (\xi + 2)X_s - \xi(\kappa_1 + 1)X_s^2]}{1 + X_s}\lambda + \frac{4\xi[\kappa_1 + \xi(1 + 2X_s)]}{1 + X_s} = 0 \qquad (5.101)$$

Using this equation, the following conditions are easily verified:

$$\lim_{\lambda \to -\infty} \det[\lambda \mathbf{I} - J(X)] \to -\infty \qquad (5.102)$$

$$\text{and } \det[-J(X)] > 0 \qquad (5.103)$$

Hence, there exists at least one $\lambda \in (-\infty, 0)$ such that $\det[\lambda \mathbf{I} - J(X)] = 0$ (i.e., the system has at least one negative real eigenvalue). Also, numerical calculations of eigenvalues for the practical range of parameters $0 < \kappa_0 < 100$, $0 < \kappa_1 < 10$ and $0.01 < \xi < 10$ reveal that the other two eigenvalues are a complex conjugate pair which have either a positive or negative real part depending upon values of κ_0 and κ_1. In particular the following observations are made:

1. For small values of κ_0, the pair of complex eigenvalues have a negative real part.
2. As κ_0 increases, the complex eigenvalues approach the imaginary axis, and at a critical value of κ_0, the real part changes from negative to positive. Table 5.2 shows a typical scenario of the variation of the eigenvalues.
3. The critical value of κ_0 depends on the values of κ_1 and ξ. Figure 5.56 shows the boundary curves where the sign of the real part of the complex eigenvalues changes. On these curves, the system loses stability via a *Hopf bifurcation* [5,6].

Remark: To establish a Hopf bifurcation formally, one needs to show that, for given ξ and κ_1, there exists κ_0 for which the following conditions are satisfied by the complex eigenvalue pair [7]:

TABLE 5.2 Eigenvalues at $\xi = 0.0136$ showing dependence on κ_0.

	$\kappa = 1$	Remarks
$\kappa_0 = 1$	$-0.0078725 \pm j0.232363$, -0.0274423	Stable equilibrium point
$\kappa_0 = 3$	$-0.00666899 \pm j0.232102$, -0.0275472	Stable equilibrium point
$\kappa_0 = 5$	$-0.00482465 \pm j0.231866$, -0.0275798	Stable equilibrium point
$\kappa_0 = 7$	$-0.0029592 \pm j0.231535$, -0.0276402	Stable equilibrium point
$\kappa_0 = 9$	$-0.0011668 \pm j0.231428$, -0.0276955	Stable equilibrium point
$\kappa_0 = 11$	$0.000538546 \pm j0.231217$, -0.0277466	Unstable equilibrium point

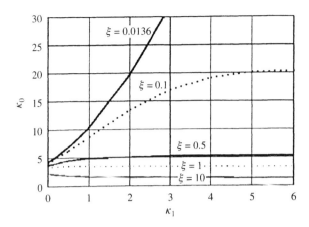

Figure 5.56 Boundary of stability. Area below the curve corresponds to stable equilibrium points, and that above to unstable equilibrium points.

$$\left.\text{Re}(\lambda)\right|_{\kappa_0=\kappa_{0c}} = 0 \tag{5.104}$$

$$\left.\text{Im}(\lambda)\right|_{\kappa_0=\kappa_{0c}} \neq 0 \tag{5.105}$$

$$\left.\frac{d}{d\kappa_0}\text{Re}(\lambda)\right|_{\kappa_0=\kappa_{0c}} \neq 0 \tag{5.106}$$

where κ_{0c} is the critical value of κ_0 at which a Hopf bifurcation occurs. Note that the last condition is necessary to ensure that the complex eigenvalue pair moves from the left side to the right side of the complex plane (preventing it from *locusing* along the imaginary axis). In fact, all the above conditions can be numerically established using (5.101).

Local Trajectories from Describing Equation

We now re-examine the stability in terms of the local trajectories near the equilibrium point. Our aim is to observe, by plotting the local trajectories, the behavior of the system as it goes from a stable region to an unstable region. The trajectory of the system near the equilibrium point can be easily derived from the corresponding eigenvalues and eigenvectors. Suppose the eigenvalues and corresponding eigenvectors are

$$\lambda_r, \sigma \pm j\omega \quad \text{and} \quad \bar{v}_r, \bar{v}_1 \pm j\bar{v}_2 \tag{5.107}$$

The solution in general is given by

$$\mathbf{x}(t) = c_r e^{\lambda_r t}\bar{v}_r + 2c_c e^{\sigma t}[\cos(\omega t + \phi_c)\bar{v}_1 - \sin(\omega t + \phi_c)\bar{v}_2] \tag{5.108}$$

where c_r, c_c and ϕ_c are determined by initial conditions. The geometry of the trajectory is best described in terms of the eigenline L_r which is parallel $\bar{v}_r$, and the eigenplane E_c which is spanned by $\bar{v}_1$ and $\bar{v}_2$, the intersection of L_r and E_c being the equilibrium point. Essentially, since the real eigenvalue is negative, the system moves initially in the direction of L_r going toward E_c. At the same time it moves in a helical motion converging toward or diverging away from L_r, depending on the sign of the real part of the complex eigenvalues. As it lands on E_c, it keeps spiraling along E_c toward or away from the equilibrium point. The following examples illustrate two typical local trajectories, corresponding to a stable and an unstable equilibrium point.

We first examine the stable system with $\xi = \kappa_0 = \kappa_1 = 1$. The Jacobian matrix evaluated at the equilibrium point is

$$J(X) = \begin{pmatrix} -1 & 0 & 1 \\ 2 & -2 & 0 \\ -1.58579 & -2.24264 & -0.171573 \end{pmatrix} \tag{5.109}$$

The eigenvalues, λ, and their corresponding eigenvectors, $\bar{v}$, are found as

$$\lambda = -2.74051, -0.215533 \pm j1.69491$$

$$\bar{v} = \begin{pmatrix} -0.297167 \\ 0.802604 \\ 0.517222 \end{pmatrix}, \begin{pmatrix} 0.185114 \mp j0.399955 \\ -0.114761 \mp j0.339261 \\ 0.823104 \end{pmatrix}$$

Using the INSITE program [8], we can view the trajectory from different perspectives, two of which are shown in Figure 5.57.

We next examine the unstable system with $\xi = \kappa_1 = 1$ and $\kappa_0 = 4$. As shown in Figure 5.56, the system just loses stability. The Jacobian matrix evaluated at the equilibrium point is

$$J(X) = \begin{pmatrix} -1 & 0 & 1 \\ 2 & -2 & 0 \\ -0.763932 & -3.36656 & 0.130495 \end{pmatrix} \tag{5.110}$$

The eigenvalues, λ, and their corresponding eigenvectors, $\bar{v}$, are found as

$$\lambda = -2.9757, 0.0530965 \pm j1.63879$$

$$\bar{v} = \begin{pmatrix} -0.331404 \\ 0.679316 \\ 0.654753 \end{pmatrix}, \begin{pmatrix} 0.233197 \mp j0.362892 \\ -0.033598 \mp j0.326689 \\ 0.840282 \end{pmatrix}$$

Using the INSITE program again, we can view the local trajectory from different perspectives, two of which are shown in Figure 5.58.

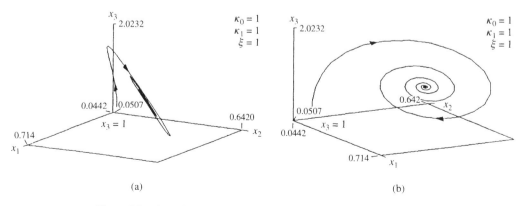

Figure 5.57 Two views of the *stable* local trajectory for $\xi = \kappa_0 = \kappa_1 = 1$ (based on averaged model).

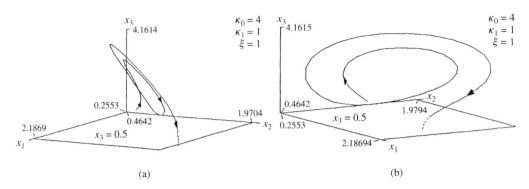

Figure 5.58 Two views of the *unstable* local trajectory for $\xi = \kappa_1 = 1$, $\kappa_0 = 4$ (based on averaged model).

From the above examples, we clearly observe that the system loses stability via Hopf bifurcation as a stable spiral develops into an unstable spiral in the locality of the equilibrium point. In the next section we re-examine the system using exact computer simulations of the actual switching circuit. As we will see, the system develops into a limit cycle as it loses stability, and further develops into quasiperiodic and chaotic orbits.

Computer Simulation Study

Since the foregoing analysis is based on a nonlinear state equation which is derived from an approximate (average) continuous model, it falls short of predicting the details of the bifurcation sequence. Instead of refining the model, we will examine the system using computer simulation which employs an exact piecewise-switched model. Essentially the computer simulation program generates the cycle-by-cycle waveforms of all capacitor voltages and inductor currents by toggling between a set of linear differential equations that describe the constituent linear circuits for all possible switch states. The program also incorporates the free-running current-control algorithm for determining the switch state during simulation.

In our simulation study of the free-running Ćuk converter, we set the input voltage at 15V and the values of the components as follows:

$$L_1 = L_2 = 0.01\text{H}, \quad C_1 = C_2 = 47\mu\text{F}, \quad R = 40\Omega$$

Note that since we are simulating the actual circuit, the parameters used will be μ and K instead of the dimensionless ones used for analysis. In particular, we will focus on the qualitative change of dynamics as the parameters are varied, as hinted from the result of Section 5.6.3.

To see the trend, it suffices to keep μ constant and vary K. A summary of the observed behavior is as follows. A complete view of the effect of ξ, μ, and K on stability of the fundamental equilibrium state will be provided later in this subsection.

1. When K is small, the trajectory spirals into a fixed period-1 orbit, corresponding to a fixed point in the averaged system. Figure 5.59 shows the simulated trajectory.

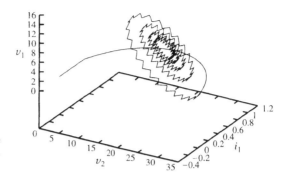

Figure 5.59 Trajectory spiraling into stable period-1 orbit ($K = 0.4$) from exact cycle-by-cycle simulation.

2. For a larger K, the period-1 orbit becomes unstable, and the trajectory spirals outward as shown in Figure 5.60(a), and settles into a limit cycle as shown in Figure 5.60(b).

3. For yet a larger K, a quasiperiodic orbit can be observed, as shown in Figure 5.61(a). A Poincaré section is shown in Figure 5.61(b), which is essentially the points of intersection of the trajectory and the vertical plane $i_1 = 8.2$.

4. Finally, chaos occurs when K is further increased. Figures 5.62(a) and (b) show the measured trajectory and a Poincaré section.

Furthermore, based on a number of simulation runs, we can obtain the boundary of stability similar to Figure 5.56, for different values of ξ. More precisely, the boundary curves define the values of parameters for which a trajectory changes its qualitative behavior from one that spirals into a fixed period-1 orbit (i.e., fixed point corresponding to the case of averaged model) to one that spirals away from it. As shown in Figure 5.63, the stability boundary curves obtained from exact circuit simulations agree with those of Figure 5.56 obtained from the averaged model.

Before we close this section, we refer the reader to Section 4.1 for experimental evidence of the aforementioned and simulated phenomena.

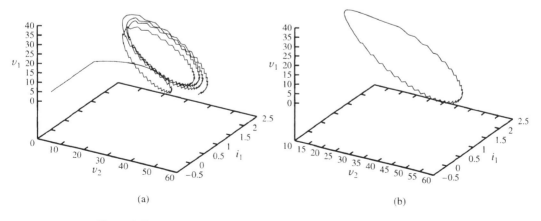

(a)   (b)

Figure 5.60 (a) Trajectory spiraling away from the unstable period-1 orbit; (b) limit cycle ($K = 1.5$), both from exact cycle-by-cycle simulation.

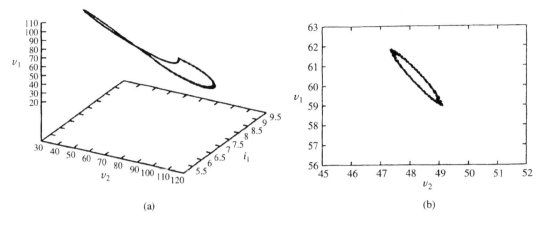

(a)   (b)

Figure 5.61 (a) Quasiperiodic orbit from exact cycle-by-cycle simulation; (b) blow-up
of a Poincaré section taken at $i_1 = 8.2$ ($K = 10.5$).

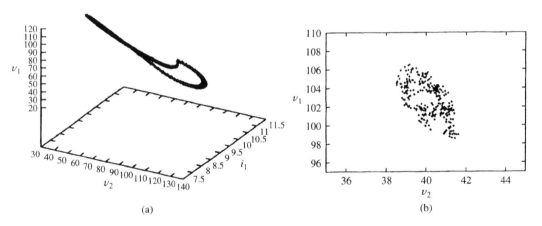

(a)   (b)

Figure 5.62 (a) Chaotic orbit from exact cycle-by-cycle simulation; (b) blow-up of a Poincaré
section taken at $i_1 = 9.5$ ($K = 13$).

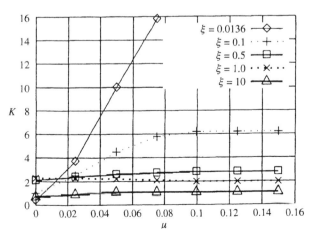

Figure 5.63 Boundary of stability from exact
simulation. Area below the curve corresponds
to stable fundamental operation, and the
above to operations other than stable fundamental operation. These curves agree with
analytical curves shown earlier.

REFERENCES

[1] S. Ćuk and R. D. Middlebrook, A new optimum topology switching dc-to-dc converter, *IEEE Power Electron. Spec. Conf. Rec.*, Palo Alto, pp. 160–179, June 1977.

[2] N. Femia, G. Spagnuolo, and V. Tucci, State-space models and order reduction for dc/dc switching converters in discontinuous mode, *IEEE Trans. on Power Electron.*, vol. 10, no. 6, pp. 640–650, Nov. 1995.

[3] C. K. Tse and W. C. Y. Chan, Chaos from a current-programmed Ćuk converter, *Int. J. Circ. Theory & Appl.*, vol. 23, no 3, pp. 217–225, May–June 1995.

[4] S. C. Wong and Y. S. Lee, SPICE modelling and simulation of hysteretic current-controlled Ćuk converter, *IEEE Trans. on Power Electron.*, vol. 8, no. 4, pp. 580–587, Oct. 1993.

[5] J. Hale and H. Kocak, *Dynamics and Bifurcations.* New York: Springer-Verlag, 1991.

[6] S. H. Strogatz, *Nonlinear Dynamics and Chaos.* Reading, MA: Addison-Wesley, 1994.

[7] K. T. Alligood, T. D. Sauer, and J. A. Yorke, *Chaos: An Introduction to Dynamical Systems.* New York: Springer-Verlag, 1996.

[8] T. S. Parker and L. O. Chua, *Practical Numerical Algorithms for Chaotic Systems.* New York: Springer-Verlag, 1989.

Chapter 6

NONLINEAR DYNAMICS IN
THYRISTOR AND DIODE CIRCUITS

Ian Dobson

6.1 INTRODUCTION

This chapter explains aspects of stability, bifurcation, and nonlinear dynamics in ideal thyristor and diode circuits. Thyristor circuits exhibit many interesting and unusual dynamical features and are of technological importance at high power levels up to MW utility power levels. Highlights of the chapter include useful simplifications in computing stability, the damping inherent in thyristor turn-off, jumps, or bifurcations in switching times and repeated violation of the behavior normally expected in smooth nonlinear dynamical systems. Thyristors and diodes constrain their currents to zero when they are off. This important constraint is accounted for by changing the state space dimension as thyristors or diodes switch on and off. The inhibition of thyristor turn-on until a firing pulse is present has a significant effect on the system dynamics.

The ideas are mainly developed for the regularly fired thyristor circuit for static VAR control introduced in Chapter 2 but almost all results can be generalized. The chapter also outlines the modifications required for diode circuits and for thyristor circuits with feedback control of the firing times. The chapter is mainly based on work by Dobson, Jalali, Rajaraman and Lasseter [4,5,14,13,15,24,25], which, in turn, builds on work by von Lutz, Grötzbach, Chua, Hasler, and Verghese [17,6,1,30] and the pioneering work of Louis [11]. Most of the chapter material first appeared in conferences [13,12,3,23]. Wolf et al. [31] present an alternative account of switching time bifurcations and their effect on the Poincaré map.

We briefly indicate some applications of the ideas and calculations in this chapter. The Poincaré map Jacobian can be used with Newton-Raphson methods to compute periodic orbits of power electronic circuits [6]. Resonances are predicted in an SVC circuit in [4] and more generally in [24]. The dynamics of the thyristor-controlled series capacitor are studied in [13,22,20]. The effects of the thyristor-controlled series capacitor on subsynchronous resonance are studied in [27,21,20]. (The exact dynamics of the thyristor-controlled series capacitor computed in [27] have also been reproduced in [18] using the phasor averaging techniques of Chapter 2.) HVDC dynamics are analyzed in [22].

6.2 IDEAL DIODE AND THYRISTOR SWITCHING RULES

It is useful to idealize the behavior of diodes and thyristors, particularly for high-power utility power electronics and for many systems studies at a range of power levels. This section explains the idealized diode and thyristor models used in this chapter.

Ideal diodes and thyristors are either on and modeled as a short circuit or off and modeled as an open circuit. An off diode has a negative voltage; it switches on when the voltage becomes zero. An on diode has a positive current; it switches off when the current becomes zero. Diode switchings are uncontrolled in that they are completely determined by the circuit waveforms.

A thyristor at this level of idealization is a diode that is inhibited from turn-on until a firing pulse is present. That is, an on thyristor switches off like a diode and an off thyristor switches on when the voltage is nonnegative and the thyristor firing pulse is on. In particular, if a thyristor receives a firing pulse when its voltage is negative, it does not turn on and the thyristor is said to have *misfired*.

Firing pulses are often short (less than $50\mu s$ [10]) and this chapter approximates a short firing pulse by assuming that the firing pulse is on only at one instant of time. Thyristor switch-on is controlled by the firing pulse, but thyristor switch-off is determined by the first zero of the thyristor current waveform encountered after switch-on.

The dependence of thyristor and diode switching times on the circuit currents and voltages causes significant nonlinearity in thyristor and diode circuits. Moreover, in thyristor circuits the system dynamics are augmented by the rule that the thyristor switches on only when a firing pulse is present. This rule has a major effect on the system dynamics and causes behavior quite different from conventional nonlinear dynamical systems. The novel behavior occurs even in the simplified case in which the firing pulses are periodic and unaffected by the dynamics of the rest of the circuit.

6.3 STATIC VAR SYSTEM EXAMPLE

Figure 6.1 shows a single-phase static VAR system consisting of a thyristor-controlled reactor and a parallel capacitor C. This system is connected to an infinite bus behind a power system impedance of inductance L_s and resistance R_s. The thyristor-controlled reactor is modeled as an inductance L_r and resistance R_r in series with back-to-back thyristors. The source voltage $u(t) = \sin(\omega t - 2\pi/3)$ has frequency $\omega = 2\pi60$ rad/s and period $T = 2\pi/\omega$. The per-unit component values are $L_s = 0.195\text{mH}$, $R_s = 0.9\text{m}\Omega$, $L_r = 1.66\text{mH}$, $R_r = 31.3\text{m}\Omega$ and $C = 1.5\text{mF}$. The SVC modeling is further explained and referenced in [14]. For general background on static VAR compensators see [16,8,19].

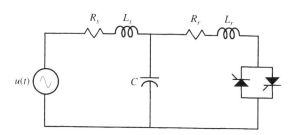

Figure 6.1 Single-phase static VAR system.

The switching element of the thyristor-controlled reactor consists of two back-to-back thyristors which conduct on alternate half-cycles of the supply frequency. Except in Section 6.9, the thyristor firing pulses are assumed to be supplied periodically and the system is controlled by varying the timing delay of the firing pulses. For simplicity, the system is studied in most of the chapter with firing pulse delay as an open loop control parameter. In practice, a closed loop control would modify the firing pulse delay.

The idealized operation of a thyristor-controlled reactor is explained in Figure 6.2. If the thyristors are fired at the point where the voltage $V_c(t)$ is at a peak, full conduction results. The circuit then operates as if the thyristors were shorted out, resulting in a thyristor current which lags the voltage by nearly 90 degrees. If the firing is delayed past the peak voltage, the thyristor conduction time and the fundamental component of reactive current are reduced.

When a thyristor is on, the system state vector $\mathbf{x}(t)$ specifies the thyristor-controlled reactor current, capacitor voltage, and the source current:

$$\mathbf{x}(t) = \begin{pmatrix} I_r(t) \\ V_c(t) \\ I_s(t) \end{pmatrix} \tag{6.1}$$

and the system dynamics are described by the ON linear system:

$$\dot{\mathbf{x}} = A_{on}\mathbf{x} + B_{on}u \tag{6.2}$$

$$\text{where } A_{on} = \begin{pmatrix} -R_r/L_r & 1/L_r & 0 \\ -1/C & 0 & 1/C \\ 0 & -1/L_s & -R_s/L_s \end{pmatrix} \text{ and } B_{on} = \begin{pmatrix} 0 \\ 0 \\ 1/L_s \end{pmatrix} \tag{6.3}$$

During the off time of each thyristor, the circuit state is constrained to lie in the plane of zero thyristor current specified by $I_r = 0$. In this mode, the system state vector $\mathbf{y}(t)$ specifies the capacitor voltage and the source current:

$$\mathbf{y}(t) = \begin{pmatrix} V_c(t) \\ I_s(t) \end{pmatrix} \tag{6.4}$$

and the system dynamics are given by the OFF linear system

$$\dot{\mathbf{y}} = A_{off}\mathbf{y} + B_{off}u \tag{6.5}$$

$$\text{where } A_{off} = \begin{pmatrix} 0 & 1/C \\ -1/L_s & -R_s/L_s \end{pmatrix} \text{ and } B_{off} = \begin{pmatrix} 0 \\ 1/L_s \end{pmatrix} \tag{6.6}$$

Figure 6.3 outlines the system switchings as the system state evolves over a period T. A thyristor starts conducting at time ϕ_0. This mode is described by (6.2) and ends when the thyristor current goes through zero at time τ_0. The ensuing non-conducting

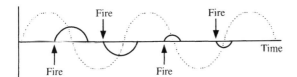

Figure 6.2 Idealized operation of a thyristor-controlled reactor (gray line = capacitor voltage V_c, solid line = thyristor current I_r). V_c is also the voltage across the thyristor-controlled reactor.

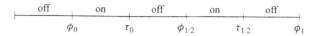

Figure 6.3 Static VAR switchings.

mode is described by (6.5) and continues until the next firing pulse in applied at time ϕ_2. This starts a similar on-off cycle which lasts until the next period starts at time $\phi_1 = \phi_0 + T$.

Define R to be the projection matrix

$$R = \begin{pmatrix} 0 & 1 & 0 \\ 0 & 0 & 1 \end{pmatrix} \quad \text{and} \quad Q = R^{\mathrm{T}} = \begin{pmatrix} 0 & 0 \\ 1 & 0 \\ 0 & 1 \end{pmatrix} \tag{6.7}$$

(The matrix transpose of R is notated as R^{T}.) The state at the switch-on time ϕ_0 is denoted either by the vector $\mathbf{y}(\phi_0)$ or by the vector $\mathbf{x}(\phi_0)$. These representations of the state at the switch-on time are related by

$$\mathbf{x}(\phi_0) = Q\mathbf{y}(\phi_0) \quad \text{or} \quad \begin{pmatrix} I_r(\phi_0) \\ V_c(\phi_0) \\ I_s(\phi_0) \end{pmatrix} = \begin{pmatrix} 0 & 0 \\ 1 & 0 \\ 0 & 1 \end{pmatrix} \begin{pmatrix} V_c(\phi_0) \\ I_s(\phi_0) \end{pmatrix} \tag{6.8}$$

Equation (6.8) expresses the fact that the $\mathbf{x}$ representation of the state at a switch-on is obtained from the $\mathbf{y}$ representation by adding a new first component which has value zero.

The state at the switch-off time τ_0 is similarly denoted either by $\mathbf{x}(\tau_0)$ or $\mathbf{y}(\tau_0)$ and these are related by

$$\mathbf{y}(\tau_0) = R\mathbf{x}(\tau_0) \quad \text{or} \quad \begin{pmatrix} V_c(\tau_0) \\ I_s(\tau_0) \end{pmatrix} = \begin{pmatrix} 0 & 1 & 0 \\ 0 & 0 & 1 \end{pmatrix} \begin{pmatrix} I_r(\tau_0) \\ V_c(\tau_0) \\ I_s(\tau_0) \end{pmatrix} \tag{6.9}$$

The matrix R in (6.9) may be thought of as projecting the vector $\mathbf{x}$ onto the plane in state space of zero thyristor current.

6.4 POINCARÉ MAP

The Poincaré map is a standard tool from dynamical systems theory to study the dynamics of periodic systems [7,29]. The main idea of this approach is to observe the system states once per cycle and define the Poincaré map as the map which advances the system states by one cycle. If the system state at time t_0 is denoted by $\mathbf{x}(t_0)$, then the Poincaré map P maps the state at time t_0 to the state at time $t_0 + T$:

$$P(\mathbf{x}(t_0)) = \mathbf{x}(t_0 + T) \tag{6.10}$$

(This form of the Poincaré map is called a *stroboscopic map* in Chapter 2; see Section 2.2 for other forms of Poincaré map.)

If $\mathbf{x}(t_0)$ is the steady-state value of the state $\mathbf{x}$ at time t_0, then $P(\mathbf{x}(t_0)) = \mathbf{x}(t_0 + T) = \mathbf{x}(t_0)$ and the map P has a fixed point at $\mathbf{x}(t_0)$. Fixed points of the Poincaré map P correspond to periodic orbits of the system. If the state $\mathbf{x}(t_0)$ is perturbed from its steady-state value, there will be a transient. Samples of this transient once per cycle can be obtained by applying the Poincaré map successively to the perturbed state $\mathbf{x}(t_0)$.

That is, the samples are $\mathbf{x}(t_0)$, $P(\mathbf{x}(t_0))$, $P(P(\mathbf{x}(t_0)))$, $P(P(P(\mathbf{x}(t_0))))$, ... The Poincaré map analysis of stability is equivalent to the sampled-data approach [30] and Floquet stability theory [9].

One way to visualize the Poincaré map is to suppose that the switching circuit is running in a dark room and a strobe light flashes once every cycle. Suppose that one can observe by measurements the currents and voltages of the system state. Then the Poincaré map is the operation which takes the system state at one flash and constructs the system state at the next flash. If the switching circuit is in a steady state which is a periodic orbit of period one cycle, one will see a fixed point of the Poincaré map. A stable transient will appear as a system state which approaches the steady state at each flash.

Consider the simple example of the linear system

$$\dot{\mathbf{x}} = A\mathbf{x} + Bu \tag{6.11}$$

with state vector $\mathbf{x}$, state matrix A, and input u of period T. The Poincaré map is computed by integrating (6.11) forward for one period T:

$$P(\mathbf{x}(t_0)) = e^{AT}\mathbf{x}(t_0) + \int_{t_0}^{t_0+T} e^{A(t_0+T-s)}Bu(s)ds \tag{6.12}$$

Now we show how to compute the Poincaré map for the SVC circuit example. The Poincaré map P advances the state by one period T. In particular we choose P to advance the state $\mathbf{y}(t_0)$ at time t_0 to $P(\mathbf{y}(t_0)) = \mathbf{y}(t_0 + T)$ as shown in Figure 6.4. P can be computed by integrating the system equations (6.2) and (6.5) and taking into account the coordinate changes (6.8) and (6.9) when the switchings occur [5]. Given a time interval $[s_1, s_2]$, it is convenient to write F_{s_2,s_1} for the map which advances the state at s_1 to the state at s_2. For example, if a thyristor is on at time s_1 and off at time s_2, then we write

$$\mathbf{y}(s_2) = F_{s_2,s_1}(\mathbf{x}(s_1)) \tag{6.13}$$

If the thyristor is on during all of the time interval $[s_1, s_2]$, we write F_{s_2,s_1} as F_{s_2,s_1}^{on}. Similarly, if the thyristor is off during all of $[s_1, s_2]$, we write F_{s_2,s_1} as F_{s_2,s_1}^{off}. The formulas for F_{s_2,s_1}^{on} and F_{s_2,s_1}^{off} are found by integrating the on and off dynamics (6.2) and (6.5) respectively:

$$F_{s_2,s_1}^{\text{on}}(\mathbf{x}(s_1)) = e^{A_{\text{on}}(s_2-s_1)}\mathbf{x}(s_1) + \int_{s_1}^{s_2} e^{A_{\text{on}}(s_2-s)}B_{\text{on}}u(s)ds \tag{6.14}$$

$$F_{s_2,s_1}^{\text{off}}(\mathbf{y}(s_1)) = e^{A_{\text{off}}(s_2-s_1)}\mathbf{y}(s_1) + \int_{s_1}^{s_2} e^{A_{\text{off}}(s_2-s)}B_{\text{off}}u(s)ds \tag{6.15}$$

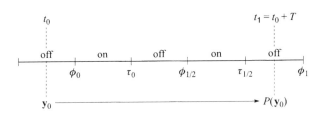

Figure 6.4 Poincaré map P.

By referring to Figure 6.4, the Poincaré map P may be written in terms of F^{on} and F^{off} and the coordinate changes (6.8) or (6.9) described by Q or R at the switching times:

$$P(\mathbf{y}_0) = F_{t_1,t_0}(\mathbf{y}_0)$$
$$= F_{t_1,\tau_{1/2}}^{off}\left(R\, F_{\tau_{1/2},\phi_{1/2}}^{on}\left(Q\, F_{\phi_{1/2},\tau_0}^{off}\left(R\, F_{\tau_0,\phi_0}^{on}\left(Q\, F_{\phi_0,t_0}^{off}(\mathbf{y}_0)\right)\right)\right)\right) \tag{6.16}$$

Different Poincaré maps can be obtained by varying the sample time t_0. One convenient choice is to let t_0 tend to ϕ_0 from below. That is, $t_0 = \phi_0-$. Then the Poincaré map becomes

$$P(\mathbf{y}_0) = F_{\phi_1,\phi_0}(\mathbf{y}_0)$$
$$= F_{\phi_1,\tau_{1/2}}^{off}\left(R\, F_{\tau_{1/2},\phi_{1/2}}^{on}\left(Q\, F_{\phi_{1/2},\tau_0}^{off}\left(R\, F_{\tau_0,\phi_0}^{on}\left(Q\, \mathbf{y}_0\right)\right)\right)\right) \tag{6.17}$$

If, instead, t_0 is chosen in the interval $[\phi_0, \tau_0]$ when a thyristor is conducting, then the Poincaré map becomes

$$P(\mathbf{x}_0) = F_{t_1,t_0}(\mathbf{x}_0)$$
$$= F_{t_1,\phi_1}^{on}\left(Q\, F_{\phi_1,\tau_{1/2}}^{off}\left(R\, F_{\tau_{1/2},\phi_{1/2}}^{on}\left(Q\, F_{\phi_{1/2},\tau_0}^{off}\left(R\, F_{\tau_0,t_0}^{on}(\mathbf{x}_0)\right)\right)\right)\right) \tag{6.18}$$

Although the Poincaré map P varies with the choice of sample time t_0 (even the dimension of the vectors it acts on varies!), the information that we seek to extract from P such as periodic orbit stability does not depend on the sample time.

6.5 JACOBIAN OF POINCARÉ MAP

To study stability, a steady-state operating point must be assumed. In a switching circuit this steady state is often periodic. Here we assume that the steady state is periodic with period T. (More precisely, the steady-state waveforms, sources, and switching times are assumed to be periodic with period T.) Stability analysis of this periodic orbit studies the behavior of the transients which occur when the system is slightly perturbed. Since thyristor switching circuits are nonlinear, the stability depends on the steady state chosen.

We first illustrate the stability computation for a periodic orbit of the simple linear system (6.11). Differentiating (6.12) yields the Jacobian of the Poincaré map $DP = e^{AT}$. This means that a linearized perturbation in state $\delta\mathbf{x}(t_0)$ at time t_0 evolves to a linearized perturbation $\delta\mathbf{x}(t_0 + T) = e^{AT}\delta\mathbf{x}(t_0)$ at the end of the period $t_0 + T$. In this case, the stability of the periodic orbit is usually determined by the Jacobian DP and particularly by the eigenvalues of DP: if all the eigenvalues of DP lie inside the unit circle of the complex plane, then the periodic orbit is asymptotically stable. This stability result applies generally to conventional smooth nonlinear systems.

Now consider a regularly fired thyristor circuit which is linear except for the thyristors. Then each thyristor pattern of on or off yields a linear system and the system switches between these linear systems as the thyristors switch. Although the switch-on occurs at a fixed time in the cycle, the switch-off time varies during a transient because it depends on the time at which the thyristor current becomes zero. This dependence of the switch-off time on the system state makes the thyristor circuit nonlinear. The analysis in the following subsections derives simple and exact formulas for the Jacobian of the Poincaré map. These formulas are assembled simply by multiplying the matrix exponentials which correspond to each linear system and taking account of the varying state space dimension at each switching. The formulas are advantageous for both computations and insight into the stability of thyristor switching circuits.

6.5.1 Thyristor Current Function and Transversality

This subsection defines a thyristor current function f which describes the actual thyristor current while the thyristor is on and the thyristor current that *would have* flowed after the thyristor turn-off time *if the thyristor had not turned off*. The thyristor current function is useful for explaining thyristor stability and bifurcation results.

To introduce the thyristor current function f, it is convenient to consider the first half-cycle of operation of the SVC example circuit and to assume that the thyristor turn-on which begins the half-cycle occurs at time zero. At time zero, the thyristor current $I_r(0) = 0$ and the initial state is given by $\mathbf{x}(0) = (0, V_c(0), I_s(0))^\mathrm{T}$. The thyristor current function $f(t, p)$ is defined to be the thyristor current assuming the thyristor is on for all time:

$$f(t, p) = \begin{pmatrix} 1 & 0 & 0 \end{pmatrix} \left[e^{A_{on}t} \begin{pmatrix} 0 \\ V_c(0) \\ I_s(0) \end{pmatrix} + \int_0^t e^{A_{on}(t-s)} B_{on} u(s) ds \right] \qquad (6.19)$$

Note that $f(0, p) = 0$. p denotes parameters of the SVC circuit; the dependence of f on p is used to describe switching time bifurcations in the following sections.

The thyristor current function $f(t, p)$ can be used to describe the thyristor switch-off rule precisely. If the thyristor switches on at $t = 0$, then the thyristor will switch off at the *first positive* root τ of f:

$$\tau = \min\{t \,|\, f(t, p) = 0, t > 0\} \qquad (6.20)$$

In particular, the switch-off time τ satisfies

$$0 = f(\tau, p) \qquad (6.21)$$

If the slope of the thyristor current is negative at the turn-off time so that

$$\left. \frac{\partial f}{\partial t} \right|_{\tau, p} < 0 \qquad (6.22)$$

then the turn-off is called *transversal*. (A transversal turn-off is a simple root of f.) The transversality condition (6.22) can also be written in terms of the actual thyristor current I_r as

$$\left. \frac{dI_r}{dt} \right|_{\tau-} < 0 \qquad (6.23)$$

where $\tau-$ denotes a limit taken from the left-hand side of τ.

If a switch-off is transversal at time τ, then sufficiently small variations in initial conditions or circuit parameters cause a switch-off time near to τ, and this switch-off time is a smooth function of small variations in initial conditions or circuit parameters. (This is proved in [5]: it is apparent that the transversal root of f near τ is preserved under sufficiently small variations, but it must also be checked that sufficiently small variations prevent any *new* roots of f occurring before the switch-off near τ; according to (6.20), any such root would become the first root of f for $t > 0$ and the thyristor switch-off time.)

Transversality at all the switch-offs in a periodic steady state is essential for computing the Jacobian of the Poincaré map. On the other hand, the transversality condition (6.22) fails when $\frac{\partial f}{\partial t}|_{\tau,p} = 0$ at a switch-off and in this case it is typical for switching times to jump or bifurcate and for steady-state stability to be lost as explained in Section 6.7.1.

6.5.2 Relations Between On and Off Systems

The matrices R and Q were introduced in Section 6.3 to change the dimension of the state of the SVC example circuit at switchings and are

$$R = \begin{pmatrix} 0 & 1 & 0 \\ 0 & 0 & 1 \end{pmatrix} \quad \text{and} \quad Q = R^T = \begin{pmatrix} 0 & 0 \\ 1 & 0 \\ 0 & 1 \end{pmatrix} \tag{6.24}$$

This subsection shows how R and Q relate the on and off system dynamics and also their relation to the thyristor switch-off condition. These relationships are shown here only for the SVC example circuit, but this example accurately reflects structural properties of general thyristor and diode circuits [5].

The on and off system dynamics are specified in Section 6.3 using

$$A_{\text{on}} = \begin{pmatrix} -R_r/L_r & 1/L_r & 0 \\ -1/C & 0 & 1/C \\ 0 & -1/L_s & -R_s/L_s \end{pmatrix} \quad \text{and} \quad B_{\text{on}} = \begin{pmatrix} 0 \\ 0 \\ 1/L_s \end{pmatrix} \tag{6.25}$$

$$A_{\text{off}} = \begin{pmatrix} 0 & 1/C \\ -1/L_s & -R_s/L_s \end{pmatrix} \quad \text{and} \quad B_{\text{off}} = \begin{pmatrix} 0 \\ 1/L_s \end{pmatrix} \tag{6.26}$$

It is easy to check that

$$A_{\text{off}} = R A_{\text{on}} Q \quad \text{and} \quad B_{\text{off}} = R B_{\text{on}} \tag{6.27}$$

The thyristor switch-off condition is zero thyristor current:

$$0 = I_r(\tau) \tag{6.28}$$

and this switch-off condition may be written in terms of the state $\mathbf{x}$ using the row vector $c = (1\ 0\ 0)$:

$$0 = (1 \quad 0 \quad 0) \begin{pmatrix} I_r(t) \\ V_c(t) \\ I_s(t) \end{pmatrix} = c\mathbf{x}(\tau) \tag{6.29}$$

The connection between c and R and Q may be shown by computing

$$
I - QR = \begin{pmatrix} 1 & 0 & 0 \\ 0 & 1 & 0 \\ 0 & 0 & 1 \end{pmatrix} - \begin{pmatrix} 0 & 0 \\ 1 & 0 \\ 0 & 1 \end{pmatrix} \begin{pmatrix} 0 & 1 & 0 \\ 0 & 0 & 1 \end{pmatrix}
$$

$$
= \begin{pmatrix} 1 & 0 & 0 \\ 0 & 0 & 0 \\ 0 & 0 & 0 \end{pmatrix} = \begin{pmatrix} 1 \\ 0 \\ 0 \end{pmatrix} \begin{pmatrix} 1 & 0 & 0 \end{pmatrix} = c^{\mathrm{T}} c
$$

(6.30)

Relationships (6.27) and (6.30) underpin useful simplifications of the Jacobian formulas in the next section.

6.5.3 Derivation of Jacobian Formula

We now derive a simple formula for the Poincaré map Jacobian of a periodic orbit. The approach is to divide one period of operation into subintervals, each of which contains one thyristor switching and to compute the Jacobian of the map which advances the state from the beginning to the end of each subinterval. Then the chain rule is used to compute the Jacobian of the Poincaré map as the product of the Jacobians for the subintervals. It is assumed that the thyristor turns on when the firing pulse is applied; that is, there are no thyristor misfires.

Interval Containing a Switch-On

Let $[s_2, s_3]$ be a time interval including a thyristor switch-on at time ϕ and no other switchings. We write F_{s_3, s_2} for the flow which maps the initial state $\mathbf{y}(s_2)$ at s_2 to the final state $\mathbf{x}(s_3)$ at s_3 so that

$$
\mathbf{x}(s_3) = F_{s_3, s_2}(\mathbf{y}(s_2))
$$

(6.31)

In general the switch-on time ϕ is a function of the initial state $\mathbf{y}(s_2)$ and we write this (with some abuse of notation) as $\phi = \phi(\mathbf{y}(s_2))$. We now compute F_{s_3, s_2} and its Jacobian DF_{s_3, s_2} with respect to $\mathbf{y}(s_2)$.

The thyristor is off in $[s_2, \phi]$ so that integrating the off system equations (6.5) yields

$$
\mathbf{y}(\phi) = F_{\phi, s_2}^{\mathrm{off}}(\mathbf{y}(s_2)) = e^{A_{\mathrm{off}}(\phi - s_2)} \left(\mathbf{y}(s_2) + \int_{s_2}^{\phi} e^{A_{\mathrm{off}}(s_2 - s)} B_{\mathrm{off}} u(s) ds \right)
$$

(6.32)

The equation transforming to the $\mathbf{x}$ coordinates at ϕ is

$$
\mathbf{x}(\phi) = Q\mathbf{y}(\phi)
$$

(6.33)

The thyristor is on in $[\phi, s_3]$ so that integrating the on system equations (6.2) and using (6.33) yields

$$
F_{s_3, s_2}(\mathbf{y}(s_2)) = F_{s_3, \phi}^{\mathrm{on}}(\mathbf{x}(\phi)) = e^{A_{\mathrm{on}}(s_3 - \phi)} Q\mathbf{y}(\phi) + \int_{\phi}^{s_3} e^{A_{\mathrm{on}}(t_3 - s)} B_{\mathrm{on}} u(s) ds
$$

(6.34)

Substitute from (6.32) to obtain

$$
F_{s_3, s_2}(\mathbf{y}(s_2)) = G_{s_3, s_2}(\mathbf{y}(s_2), \phi(\mathbf{y}(s_2)))
$$

(6.35)

where

$$G_{s_3,s_2}(\mathbf{y}(s_2), \phi) = e^{A_{on}(s_3-\phi)}Qe^{A_{off}(\phi-s_2)}\left(\mathbf{y}(s_2) + \int_{s_2}^{\phi} e^{A_{off}(s_2-s)}B_{off}u(s)ds\right)$$
$$+ \int_{\phi}^{s_3} e^{A_{on}(t_3-s)}B_{on}u(s)ds$$

(6.36)

Differentiating (6.35) and writing D for the derivative with respect to $\mathbf{y}(s_2)$ gives

$$DF_{s_3,s_2} = DG_{s_3,s_2} + \frac{\partial G_{s_3,s_2}}{\partial \phi} D\phi$$

(6.37)

$D\phi$ is the derivative of the switch-off time with respect to $\mathbf{y}(s_2)$. In this section we make the simplifying assumption that the thyristor is fired regularly; that is, ϕ is constant, and hence $D\phi = 0$ and

$$DF_{s_3,s_2} = DG_{s_3,s_2} = e^{A_{on}(s_3-\phi)}Qe^{A_{off}(\phi-s_2)}$$

(6.38)

(The case of thyristor firing control or synchronization is briefly treated in Section 6.9.)

Interval Containing a Switch-Off

Let $[s_1, s_2]$ be a time interval including a single transversal thyristor switch-off at time τ and no other switchings. Write F_{s_2,s_1} for the flow which maps the state at s_1 to the state at s_2 so that

$$\mathbf{y}(s_2) = F_{s_2,s_1}(\mathbf{x}(s_1))$$

(6.39)

The switch-off time τ is a function of the initial state $\mathbf{x}(s_1)$ and we write this as $\tau = \tau(\mathbf{x}(s_1))$. We now compute F_{s_2,s_1} and its Jacobian DF_{s_2,s_1} with respect to $\mathbf{x}(s_1)$.

The thyristor is on in $[s_1, \tau]$ so that integrating the on system equations (6.2) yields

$$\mathbf{x}(\tau) = F_{\tau,s_1}^{on}(x(s_1)) = e^{A_{on}(\tau-s_1)}\left(\mathbf{x}(s_1) + \int_{s_1}^{\tau} e^{A_{on}(s_1-s)}B_{on}u(s)ds\right)$$

(6.40)

The transformation to $\mathbf{y}$ coordinates at the switch-off time τ is

$$\mathbf{y}(\tau) = R\mathbf{x}(\tau)$$

(6.41)

The thyristor is off in $[\tau, s_2]$ so that integrating the off system equations (6.5) with initial condition $\mathbf{y}(\tau)$ gives

$$F_{s_2,s_1}(\mathbf{x}(s_1)) = F_{\tau,s_1}^{off}(\mathbf{y}(\tau)) = e^{A_{off}(s_2-\tau)}R\mathbf{x}(\tau) + \int_{\tau}^{s_2} e^{A_{off}(s_2-s)}B_{off}u(s)ds$$

(6.42)

Substituting for $\mathbf{x}(\tau)$ from (6.40) yields

$$F_{s_2,s_1}(\mathbf{x}(s_1)) = G_{s_2,s_1}(\mathbf{x}(s_1), \tau(\mathbf{x}(s_1)))$$

(6.43)

where

$$G_{s_2,s_1}(\mathbf{x}(s_1), \tau) = e^{A_{off}(s_2-\tau)}Re^{A_{on}(\tau-s_1)}\left(\mathbf{x}(s_1) + \int_{s_1}^{\tau} e^{A_{on}(s_1-s)}B_{on}u(s)ds\right)$$
$$+ \int_{\tau}^{s_2} e^{A_{off}(s_2-s)}B_{off}u(s)ds$$

(6.44)

The transversality of the switch-off was assumed above and it implies that τ is a smooth function of $\mathbf{x}(s_1)$ [5] and hence that F_{s_2,s_1} is a smooth function of $\mathbf{x}(s_1)$. Differentiating (6.43) and writing D for the derivative with respect to $\mathbf{x}(s_1)$ gives

$$DF_{s_2,s_1} = DG_{s_2,s_1} + \frac{\partial G_{s_2,s_1}}{\partial \tau} D\tau \tag{6.45}$$

Differentiating (6.44) yields

$$
\begin{aligned}
\frac{\partial G_{s_2,s_1}}{\partial \tau} &= e^{A_{\mathrm{off}}(s_2-\tau)}(RA_{\mathrm{on}} - A_{\mathrm{off}}R)e^{A_{\mathrm{on}}(\tau-s_1)}\left(\mathbf{x}(s_1) + \int_{s_1}^{\tau} e^{A_{\mathrm{on}}(s_1-\tau)}B_{\mathrm{on}}u(\tau)d\tau \right) \\
&\quad + e^{A_{\mathrm{off}}(s_2-\tau)}(RB_{\mathrm{on}} - B_{\mathrm{off}})u(\tau) \\
&= e^{A_{\mathrm{off}}(s_2-\tau)}(RA_{\mathrm{on}} - A_{\mathrm{off}}R)\mathbf{x}(\tau) + e^{A_{\mathrm{off}}(s_2-\tau)}(RB_{\mathrm{on}} - B_{\mathrm{off}})u(\tau)
\end{aligned}
\tag{6.46}
$$

But the structural relations (6.27) in Section 6.5.2 state that $B_{\mathrm{off}} = RB_{\mathrm{on}}$ and $A_{\mathrm{off}} = RA_{\mathrm{on}}Q$ so that

$$\frac{\partial G_{s_2,s_1}}{\partial \tau} = e^{A_{\mathrm{off}}(s_2-\tau)}RA_{\mathrm{on}}(I - QR)\mathbf{x}(\tau) \tag{6.47}$$

Using the relation (6.30) linking $I - QR$ and $c = (1\ 0\ 0)$ from Section 6.5.2,

$$
\begin{aligned}
\frac{\partial G_{s_2,s_1}}{\partial \tau} &= e^{A_{\mathrm{off}}(s_2-\tau)}RA_{\mathrm{on}}c^{\mathrm{T}}c\mathbf{x}(\tau) \\
&= 0
\end{aligned}
\tag{6.48}
$$

since the thyristor switch-off condition is $0 = I_r(\tau) = 1\ 0\ 0\,\mathbf{x}(\tau) = c\,\mathbf{x}(\tau)$. Equation (6.48) states that the final state $y(s_2)$ is independent of the switch-off time τ to first order!

Hence we obtain the surprising and simple result

$$DF_{s_2,s_1} = DG_{s_2,s_1} = e^{A_{\mathrm{off}}(s_2-\tau)}Re^{A_{\mathrm{on}}(\tau-s_1)} \tag{6.49}$$

Assembling the Jacobian

The Poincaré map formula (6.16) is rewritten omitting the thicket of brackets:

$$
\begin{aligned}
P(\mathbf{y}_0) &= F^{\mathrm{off}}_{t_1,\tau_{1/2}}\left(R\,F^{\mathrm{on}}_{\tau_{1/2},\phi_{1,2}}\left(Q\,F^{\mathrm{off}}_{\phi_{1/2},\tau_0}\left(R\,F^{\mathrm{on}}_{\tau_0,\phi_0}\left(Q\,F^{\mathrm{off}}_{\phi_0,t_0}(\mathbf{y}_0) \right) \right) \right) \right) \\
&= F^{\mathrm{off}}_{t_1,\tau_{1/2}}R\,F^{\mathrm{on}}_{\tau_{1/2},\phi_{1/2}}Q\,F^{\mathrm{off}}_{\phi_{1/2},\tau_0}R\,F^{\mathrm{on}}_{\tau_0,\phi_0}Q\,F^{\mathrm{off}}_{\phi_0,t_0}\mathbf{y}_0
\end{aligned}
\tag{6.50}
$$

Choose times s_2 in the interval $(\phi_{1/2}, \tau_{1/2})$, s_1 in $(\tau_0, \phi_{1/2})$, and s_0 in (ϕ_0, τ_0). Then

$$P(\mathbf{y}_0) = F_{t_1,s_2}F_{s_2,s_1}F_{s_1,s_0}F_{s_0,t_0}\mathbf{y}_0 \tag{6.51}$$

Each of the time intervals corresponding to the decomposition of the Poincaré map in (6.51) contains exactly one switching. Differentiating (6.51) with the chain rule and using the results of Section 6.5.3 gives

$$DP = DF_{t_1,s_2} DF_{s_2,s_1} DF_{s_1,s_0} DF_{s_0,t_0}$$

$$= \left(e^{A_{\text{off}}(t_1-\tau_{1/2})} Re^{A_{\text{on}}(\tau_{1/2}-s_2)}\right)\left(e^{A_{\text{on}}(s_2-\phi_{1/2})} Qe^{A_{\text{off}}(\phi_{1/2}-s_1)}\right)$$

$$\left(e^{A_{\text{off}}(s_1-\tau_0)} Re^{A_{\text{on}}(\tau_0-s_0)}\right)\left(e^{A_{\text{on}}(s_0-\phi_0)} Qe^{A_{\text{off}}(\phi_0-t_0)}\right) \qquad (6.52)$$

$$= e^{A_{\text{off}}(t_1-\tau_{1/2})} Re^{A_{\text{on}}(\tau_{1/2}-\phi_{1/2})} Qe^{A_{\text{off}}(\phi_{1/2}-\tau_0)} Re^{A_{\text{on}}(\tau_0-\phi_0)} Qe^{A_{\text{off}}(\phi_0-t_0)}$$

If, for convenience, the sample time of the Poincaré map is changed so that $t_0 = \phi_0-$, then the Jacobian becomes

$$DP = e^{A_{\text{off}}(\phi_1-\tau_{1/2})} Re^{A_{\text{on}}(\tau_{1/2}-\phi_{1/2})} Qe^{A_{\text{off}}(\phi_{1/2}-\tau_0)} Re^{A_{\text{on}}(\tau_0-\phi_0)} Q \qquad (6.53)$$

If the sample time t_0 is chosen when a thyristor is on so that, for example, $t_0 = \phi_0+$, then the Jacobian becomes

$$DP = Qe^{A_{\text{off}}(\phi_1-\tau_{1/2})} Re^{A_{\text{on}}(\tau_{1/2}-\phi_{1/2})} Qe^{A_{\text{off}}(\phi_{1/2}-\tau_0)} Re^{A_{\text{on}}(\tau_0-\phi_0)} \qquad (6.54)$$

The Jacobian in (6.54) has one more row and column than the Jacobian in (6.53). However, it can be shown that (6.54) and (6.53) have the same eigenvalues except that (6.54) has an additional zero eigenvalue. Thus (6.54) and (6.53) describe exactly the same periodic orbit stability information in different forms.

6.5.4 Discussion of Jacobian Formula

We summarize the outcome of the preceding subsections in deriving the Poincaré map Jacobian for the SVC circuit example. The thyristor firing pulses and hence, assuming no misfire, the thyristor switch-on times are assumed to occur at fixed times ϕ_0 and $\phi_{1/2}$ in the cycle. It is convenient to choose the Poincaré map sample time at the turn-on ϕ_0. Then the Poincaré map P advances the state $\mathbf{y}(\phi_0)$ at turn-on to $P(\mathbf{y}) = \mathbf{y}(\phi_0 + T)$, where T is the period. A thyristor switches off at τ_0 and $\tau_{1/2}$ and these switchings are assumed to be transversal in order to guarantee that the Poincaré map is differentiable. The Jacobian of the Poincaré map is

$$DP = e^{A_{\text{off}}(\phi_1-\tau_{1/2})} Re^{A_{\text{on}}(\tau_{1/2}-\phi_{1/2})} Qe^{A_{\text{off}}(\phi_{1/2}-\tau_0)} Re^{A_{\text{on}}(\tau_0-\phi_0)} Q \qquad (6.55)$$

Suppose that the circuit has a periodic orbit passing through $\mathbf{y}_0$ at time ϕ_0 so that $\mathbf{y}_0$ is a fixed point of P and $P(\mathbf{y}_0) = \mathbf{y}_0$. The stability of the periodic orbit is the same as the stability of $\mathbf{y}_0$ and is given (except in marginal cases) by the eigenvalues of DP, the Jacobian of the Poincaré map evaluated at $\mathbf{y}_0$. (Here we continue to assume that there are no misfires and that all switch-offs are transversal.)

One of the interesting and useful consequences of formula (6.55) is that DP and the stability of the periodic orbit only depend on the state and the input via the thyristor nonconduction times $\phi_1 - \tau_{1/2}$ and $\phi_{1/2} - \tau_0$ and the thyristor conduction times $\tau_{1/2} - \phi_{1/2}$ and $\tau_0 - \phi_0$. It is remarkable that (6.55) is also the formula that would be obtained for fixed switch-off times τ_0 and $\tau_{1/2}$; the varying switch-off times introduce no additional complexity in the formula, but the nonlinearity of the circuit is clear since τ_0 and $\tau_{1/2}$ vary as a function of the periodic orbit.

If the periodic orbit is assumed to be half wave symmetric, then $\phi_{1/2} = \phi_0 + T/2$ and $\tau_{1/2} = \tau_0 + T/2$ and (6.55) simplifies to

$$DP = \left(e^{A_{\text{off}}(\phi_{1/2}-\tau_0)} Re^{A_{\text{on}}(\tau_0-\phi_0)} Q\right)^2 \qquad (6.56)$$

which can also be expressed in terms of the thyristor conduction time $\sigma = \tau_0 - \phi_0$ as

$$DP = \left(e^{A_{\text{off}}(T/2-\sigma)} R e^{A_{\text{on}}\sigma} Q \right)^2 \tag{6.57}$$

The action of (6.57) on a linearized perturbation $\delta\mathbf{y}$ for the first half-period may be informally described as follows: change to $\mathbf{x}$ coordinates with the matrix Q, let the on system act for time σ, project to the off coordinates with the matrix R and let the off system act for time $T/2 - \sigma$. The action of (6.57) on $\delta\mathbf{y}$ for the whole period is equivalent to the action of two successive half-periods.

6.6 SWITCHING DAMPING

This section analyzes thyristor switch-off as a source of damping. This damping is a dynamic effect which damps transients; the effect has nothing to do with static or steady-state performance. The switching damping occurs in addition to other sources of circuit damping such as resistance or control loops. In particular, switching damping occurs in regularly fired thyristor circuits with no resistance or control loop.

6.6.1 Simple Example

The damping caused by thyristor switch-off can be most easily demonstrated in an example from [15]: consider the circuit of Figure 6.5 which consists of a sinusoidal voltage source, a thyristor fired regularly once a period, and an inductor, all in series. This example is simple, but does contain the essence of the switching-damping phenomenon in general thyristor circuits.

The periodic steady state $I(t)$ of the circuit current is shown by the gray line in Figure 6.6. If there is a disturbance ϵ which perturbs the current I at time zero, then a transient $I(t) + \Delta I(t)$ with $\Delta I(0) = \epsilon$ ensues as shown by the solid line in the upper portion of Figure 6.6. The transient persists until shortly after the thyristor switches off at time τ_0. It is clear that by the next period, the transient has been damped to nothing. That is, the Poincaré map P zeros the disturbance:

$$P(I(0) + \epsilon) = I(0) \tag{6.58}$$

(In discrete time control, this would be called deadbeat damping.)

Computing the Poincaré map Jacobian confirms the analysis: The on equation has one state I in the one-dimensional state space $\mathbb{R}$ and is $\dot{I} = -\sin t$. The off equation has the degenerate, zero-dimensional state space $\mathbb{R}^0 = \{0\}$, which consists only of the origin.

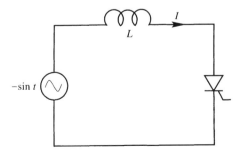

Figure 6.5 Simple thyristor circuit showing switching damping.

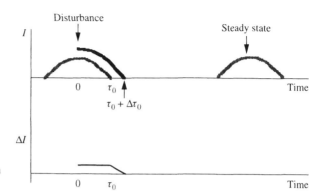

Figure 6.6 Damping of a disturbance $\Delta I(t)$ in simple thyristor circuit.

Thus, as expected, $I = 0$ when the thyristor is off. Since $R = 0$, the Jacobian calculation yields $DP = e^{A_{on}\tau_0} R e^{A_{off}(T-\tau_0)} = 0$.

The size of the disturbance $\Delta I(t)$ is given by the difference between the disturbed trajectory and the steady state. It is this difference which is damped to zero. To see how the thyristor switch-off accomplishes the damping, observe that the disturbance remains constant until the thyristor of the steady-state trajectory switches off; the disturbance is damped to zero during the time in which the thyristor of the disturbed trajectory is on and the thyristor of the steady-state trajectory is off.

One might be tempted to neglect the change in state-space dimension in the problem formulation; that is, to assume that the off equation was $\dot{I} = 0$ with I in the one-dimensional state space $\mathbb{R}$ and neglect the projection R. After all, the periodic solution can be correctly calculated with this formulation since I is zero at the beginning of the switch-off mode. However, in this formulation, linearizing either the on equation or the off equation gives $\dot{\delta I} = 0$ and the solution is $\delta I(t) = \delta I(0)$. This implies that, to first order, the disturbance is preserved and not damped at the end of the cycle (that is, $DP = 1$). This is plainly wrong. Essentially the same mistake of neglecting the change in state-space dimension can be made in general thyristor and diode circuits. For another example, see Section 6.8.2.

6.6.2 Switching Damping in the SVC Example

We compute damping in the SVC example circuit with the thyristor conduction time σ treated as a parameter. A half-wave symmetric periodic steady state is assumed. Then, according to (6.57), the Poincaré map Jacobian is given by

$$DP = (DH)^2 \tag{6.59}$$

where

$$DH = e^{A_{off}(T/2-\sigma)} R e^{A_{on}\sigma} Q \tag{6.60}$$

H may be regarded as the map $F_{\phi+T/2,\phi}$ advancing the state by half a period:

$$\mathbf{y}(\phi + T/2) = H(\mathbf{y}(\phi)) = F_{\phi+T/2,\phi}(\mathbf{y}(\phi)) \tag{6.61}$$

Since the eigenvalues of DP are the eigenvalues of DH squared, the eigenvalues of DH determine the stability of periodic orbits of the circuit.

It is straightforward to use (6.60) to compute the locus of eigenvalues of DH as σ varies over its range of 0 to 180 degrees and the results are shown in Figure 6.7(a). (The gap in results for $60° < \sigma < 90°$ is due to the switching time bifurcations explained in Section 6.7.1; the half-wave symmetric periodic orbit disappears in this range.)

When all the eigenvalues are inside the unit circle, the circuit periodic orbit is asymptotically stable and the system damps out any small perturbations. This damping cannot be entirely attributed to resistance in the circuit. Indeed, if the circuit resistances are set to zero then the eigenvalue locus of Figure 6.7(b) is obtained and the switching damping for most values of σ is evident. (The exceptional points on the unit circle of zero damping are due to a resonance effect explained in [4].)

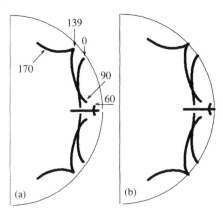

Figure 6.7 Eigenvalues of DH: (a) circ with resistance; (b) circuit without resistan The numbers show the thyristor conducti time σ in degrees.

6.6.3 Variational Equation

We consider the first-order variation [2] about periodic orbits to get another view of how the Poincaré map Jacobian works. The first-order variation is a linear differential equation which propagates forward in time linearized deviations from the periodic orbit. Propagation with the variational equation of an initial linearized deviation for one period T is equivalent to the action of the Poincaré map Jacobian on the initial linearized deviation.

First we examine the first-order variation for the simple example circuit shown in Figure 6.5. Write τ_0 for the switch-off time of the steady-state trajectory $I(t)$ and $\tau_0 + \Delta\tau_0$ for the switch-off time of the disturbed trajectory $I(t) + \Delta I(t)$ as shown in Figure 6.6. Here Δ denotes deviation; shortly we will use δ to denote linearized deviation. By inspection,

$$\Delta I(t) = \begin{cases} \epsilon & ; \quad 0 \le t \le \tau_0 \\ 0 & ; \quad \tau_0 + \Delta\tau_0 \le t \end{cases} \qquad (6.62)$$

Since $\Delta\tau_0 \to 0$ as $\epsilon \to 0$,

$$\left.\frac{\partial \Delta I(t)}{\partial \epsilon}\right|_{\epsilon=0} = \begin{cases} 1 & ; \quad 0 \le t \le \tau_0 \\ 0 & ; \quad \tau_0 < t \end{cases} \qquad (6.63)$$

and (6.63) is the solution to the variational equation when the initial disturbance is 1. The general solution $\delta I(t)$ to the variational equation when the initial disturbance is $\delta I(0)$ is

$$\delta I(t) = \begin{cases} \delta I(0) & ; \quad 0 \le t \le \tau_0 \\ 0 & ; \qquad \tau_0 < t \end{cases} \tag{6.64}$$

The zeroing of the linearized disturbance $\delta I(0)$ at switch-off τ_0 in (6.64) corresponds exactly to the action of the projection R in the Poincaré map formula. A circuit model of the variational equation is the inductor of the simple circuit in series with an ideal switch which opens and zeros the current at time τ_0.

Now we consider the variational equation of the SVC circuit about a half-wave symmetric periodic orbit. The linearized deviations propagated by the variational equations are $\delta\mathbf{x}(t) = (\delta I_r(t), \delta V_c(t), \delta I_s(t))^{\mathrm{T}}$ and $\delta\mathbf{y}(t) = (\delta V_c(t), \delta I_s(t))^{\mathrm{T}}$. The variational equations for the first half-period are

$$\delta\mathbf{x}(\phi_0) = Q\delta\mathbf{y}(\phi_0) \quad \text{(switch opening at } \phi_0) \tag{6.65}$$

$$\dot{\delta\mathbf{x}} = \mathbf{A}_{\mathrm{on}}\delta\mathbf{x} \quad \text{(switch closed)} \tag{6.66}$$

$$\delta\mathbf{y}(\tau_0) = R\delta\mathbf{x}(\tau_0) \quad \text{(switch closing at } \tau_0) \tag{6.67}$$

$$\dot{\delta\mathbf{y}} = A_{\mathrm{off}}\delta\mathbf{y} \quad \text{(switch open)} \tag{6.68}$$

and these equations are repeated for the second half-cycle. Integrating the variational equations over one half-cycle produces the Jacobian of the half-wave map H:

$$DH = e^{A_{\mathrm{off}}(T/2-\sigma)} R e^{A_{\mathrm{on}}\sigma} Q \tag{6.69}$$

and integrating the variational equations over one complete cycle produces DP, the Jacobian of the Poincaré map:

$$DP = \left(e^{A_{\mathrm{off}}(T/2-\sigma)} R e^{A_{\mathrm{on}}\sigma} Q \right)^2 \tag{6.70}$$

A circuit model of the variational equations is shown in Figure 6.8; it is obtained from the SVC example circuit by shorting the source and replacing the thyristor by an ideal switch which opens at times τ_0 and $\tau_{1/2}$. In general the switch current is nonzero when the switch opens at τ_0 and the switch opening is assumed to immediately zero the inductor current. This somewhat nonphysical event is described in (6.67) by the projection of the current $\delta\mathbf{x}(\tau_0)$ onto the plane $\delta I_r = 0$ by R. This projection or ideal switch opening is the source of switching damping.

We consider the case of zero resistances so that equations (6.66) and (6.68) are simply lossless oscillators. At the beginning of the cycle the switch turns on and the on oscillation proceeds for time σ. Then the state is projected onto the plane of zero thyristor current and the off oscillation proceeds for time $T/2 - \sigma$. Since we have

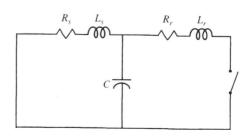

Figure 6.8 Circuit model for the variational equations.

assumed zero resistance, the oscillators are lossless and the damping in DH is entirely accounted for by the projection onto the plane of zero thyristor current.

It is straightforward to use an energy or Lyapunov method to show that DH is never unstable and that its eigenvalues always lie inside or on the unit circle. Consider the incremental energy [5,26,28]

$$\delta E(t) = \frac{1}{2} L_r (\delta I_r)^2 + \frac{1}{2} L_s (\delta I_s)^2 + \frac{1}{2} C (\delta V_c)^2 \tag{6.71}$$

δE measures the square of the size of the perturbation $(\delta I_r, \delta I_s, \delta V_c)^T$. δE is preserved at the switch opening (6.65) because the reactor current δI_r is zero when the switch opens (the first component of $Q\delta y(\phi_0)$ is always zero). At the switch closing (6.68), the incremental energy δE decreases by the nonnegative quantity $\frac{1}{2} L_r (\delta I_r(\tau_0))^2$ because the effect of the projection R is to zero the incremental thyristor current δI_r.

In the case of zero circuit resistances, equations (6.66) and (6.68) are simply lossless oscillators. Then δE is preserved at switch-on, is constant while the lossless oscillators act, and decreases or is preserved at switch-off. Since δE is a Lyapunov function for the discrete time system $\delta y^{k+1} = DH\delta y^k$, $k = 0, 1, 2, 3\ldots$, DH must be stable. If the circuit resistances are included, then δE is strictly decreasing when the oscillators act and δE is a strict Lyapunov function and DH is asymptotically stable. The stability or asymptotic stability of DP follows from the stability or asymptotic stability of DH.

In summary, for the case of no resistance we have shown stability of DP by Lyapunov methods and for the case of circuit resistance we have shown asymptotic stability of DP both by direct calculation of the eigenvalues and by Lyapunov methods. It would now seem routine for the case of circuit resistance to conclude from the Jacobian asymptotic stability that the periodic orbit is always asymptotically stable. However, this conclusion is false: the periodic orbit can sometimes lose stability! The following section explains how this happens.

6.7 SWITCHING TIME BIFURCATIONS

6.7.1 Switching Time Bifurcations and Instability

This subsection explains how switching time bifurcations cause a loss of steady-state stability.

Thyristor switching circuits initially operating at a periodic orbit can lose stability when switching times jump or bifurcate as a parameter is slowly varied. One example of such a parameter is the conduction time σ of the thyristor. (σ for a periodic orbit can be varied by varying the thyristor firing time.)

The switching time bifurcations can be explained using the thyristor current function. It is convenient to assume that a thyristor turns on at time zero. Recall from Section 6.5.1 that the thyristor current function $f(t, \sigma)$ is defined to be the thyristor current assuming the thyristor is on for all time. In the case of the SVC circuit example,

$$f(t, \sigma) = \begin{pmatrix} 1 & 0 & 0 \end{pmatrix} \left[e^{A_{on} t} \begin{pmatrix} 0 \\ V_c(0) \\ I_s(0) \end{pmatrix} + \int_0^t e^{A_{on}(t-\tau)} B_{on} u(\tau) d\tau \right] \tag{6.72}$$

It is important to remember that the thyristor current function is identical to the actual thyristor current before the switch-off τ and a useful mathematical fiction after the switch-off τ.

If the thyristor turns on at time zero, the next thyristor switch-off time is at the *first* positive root τ of f. Switching time bifurcations are bifurcations of the roots of f which alter which root is the first positive root. The switching time bifurcations occur in practice when the harmonic components of the thyristor current are large and the thyristor current function becomes distorted [14].

Figure 6.9 explains a switching time bifurcation in which the periodic orbit loses stability as a thyristor current zero disappears. Figure 6.9(a) shows the thyristor current function for a stable periodic solution; there is a transversal switch-off at time τ. As the thyristor conduction time σ is slowly varied, the dip in the thyristor current function after τ rises until, passing through the critical parameter value σ^*, the current zero disappears and a new, later zero of the thyristor current applies (see Figure 6.9(b)). The switch-off time of the thyristor has suddenly increased in a switching time bifurcation and stability has suddenly been lost. As soon as the switching time bifurcates and system stability is lost, a transient starts.

Another manifestation of a switching time bifurcation is shown in Figure 6.10. Suppose that harmonic distortion produces a dip in the thyristor current as shown in Figure 6.10(a). The periodic steady state is stable in Figure 6.10(a). As the thyristor conduction time σ is varied, the dip lowers until, at a critical parameter value σ^*, a new, earlier zero of the thyristor current is produced (Figure 6.10(b)). The switch-off time of the thyristor suddenly decreases and the stable operation of the system at the previous periodic steady state is lost. As soon as the switching time bifurcates, a transient starts.

In both Figures 6.9 and 6.10 the disappearance or appearance of the switching time occurs by a fold (or saddle-node) bifurcation of the zeros of f in which zeros coalesce. At the bifurcating zero, f has zero gradient and the transversality condition (6.22) is violated. Figures 6.9 and 6.10 are qualitative representations of switching time bifurcations; detailed simulation and experimental results on a single-phase SVC circuit are presented in [14].

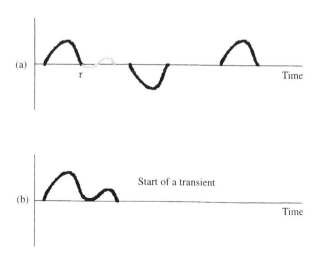

Figure 6.9 Two thyristor current zeros disappear (a) $\sigma < \sigma^*$; (b) $\sigma \approx \sigma^*$, $\sigma > \sigma^*$.

(a)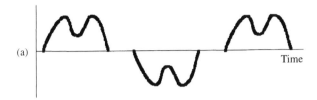

Time

(b)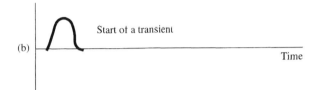

Start of a transient

Time

Figure 6.10 A new earlier thyristor current zero appears: (a) $\sigma > \sigma^*$; (b) $\sigma = \sigma^*$.

Switching time bifurcations are essentially bifurcations of the constraint condition determining the switching time and they differ in some respects from conventional bifurcations. For example, the eigenvalues of the Poincaré map Jacobian evaluated at the periodic orbit give no indication of either of the switching time bifurcations in Figures 6.9 and 6.10. In the case of the switching time bifurcations shown in Figure 6.10, the condition for the switching time bifurcation is the zero gradient of current at the new, earlier current zero and this condition has no relation to the Poincaré map Jacobian. (There is more subtlety in the case of the switching time bifurcations shown in Figure 6.9. At the switching time bifurcation $D\tau$ becomes infinite and it might be expected from formula (6.45) that this would imply that DF_{s_2,s_1} and hence the Poincaré map Jacobian DP would have large eigenvalues outside the unit circle near the bifurcation. But this is false: the simplification (6.48) and formula (6.49) show that $D\tau$ has no effect on DP.)

Since the switching time is discontinuous at a switching time bifurcation and the Poincaré map depends on the switching time, the Poincaré map is also discontinuous at a switching time bifurcation [23,25,31] (see Section 6.7.4). The switching time bifurcation can also be understood as the fixed point of the Poincaré map encountering the Poincaré map discontinuity. This aspect is emphasized by Wolf et al. [31].

6.7.2 Switching Time Bifurcations for Transients

Switching time bifurcations also occur during transients and have interesting effects on the system dynamics as explained in the following subsections. This subsection and the remainder of Section 6.7 explain these phenomena by summarizing mostly numerical results from [25] on the SVC example circuit specified in Section 6.3. More detailed results and discussion may be found in [25]. All results assume that the thyristor firing pulses have a constant phase delay of $120°$ relative to the sinusoidal voltage–source voltage crossings.

It is convenient to assume that a thyristor turns on at time zero. The initial state at time zero is given by $p - \mathbf{y}(0) = (V_c(0), I_s(0))^{\mathrm{T}}$ and is considered to be a parameter in order to study bifurcations of transients. The thyristor current function (6.19) or (6.72) is written $f(t, p)$.

The thyristor switching off time τ in the first half-cycle is plotted as a function of the initial state $p = (V_c(0), I_s(0))^T$ in Figure 6.11. Discontinuities of the switching time are apparent as sharp changes in the plot. These discontinuities can be understood by examining the roots of $f(t, p)$. Figure 6.12 shows a "slice" of Figure 6.11 obtained by plotting several roots of $f(t, p)$ versus the initial capacitor voltage $V_c(0)$ for a fixed initial source current $I_s(0) = 9$. The switching time τ is indicated by circles in the plot. As can be seen from Figure 6.12, a discontinuity in the switching time occurs near $V_c(0) = 5.1$ where the first and second roots of $f(t, p)$ coalesce and disappear in a fold bifurcation so that what was previously the third positive root becomes the first positive root and the switching time τ.

Graphs of f corresponding to a fold bifurcation as $V_c(0)$ varies and $I_s(0) = 4$ are shown in Figure 6.13. When $p = p_1 = (3.2, 4)^T$, the transversality condition is satisfied at the thyristor switch-off at $\tau(p_1)$ as shown in Figure 6.13(a). There is a second root of f near $\tau(p_1)$ and a third root of f at a later time. When $p = p_* = (4.2, 4)^T$ as in Figure 6.13(b), f has zero gradient at the double root at $\tau(p_*)$ and the transversality condition is not satisfied. When p changes to a new value $p_2 = (5.2, 4)^T$ near p_* as shown in Figure 6.13(c), the previous first and second root have disappeared and the previous third root has suddenly become the first root.

6.7.3 Misfire Onset as a Transcritical Bifurcation

A thyristor misfires at a switch-on time if the thyristor voltage is negative when the gate turn-on pulse arrives. Consider the thyristor firing at $t = 0$ (the analysis is similar for misfire at $t = T/2$). Just before the gate pulse arrives, at time $t = 0-$, the thyristor voltage is the capacitor voltage $V_c(0)$ (see Figure 6.1). From the system equations (6.2), $V_c(0)$ is the gradient of the thyristor current at $t = 0+$ or, equivalently, $\frac{\partial f}{\partial t}(0, p)$, the gradient of f at zero. Misfiring is described in the sequence of diagrams in Figure 6.14 which are plots of $f(t, p)$ versus t as $V_c(0)$ varies and $I_s(0) = 4$. In Figure 6.14(a), $V_c(0) = 1$, the circuit is operating normally, and the gradient of f at zero is positive. In Figure 6.14(b), $V_c(0) = 0$ and the gradient of f at zero has decreased to zero and this is the onset of misfire. In Figure 6.14(c), $V_c(0) = -1$ and the thyristor will misfire since the gradient of f at zero is negative. If we define root -1 to be the first negative root of $f(t, p) = 0$, then root -1 increases through the root at zero and becomes relabeled as the

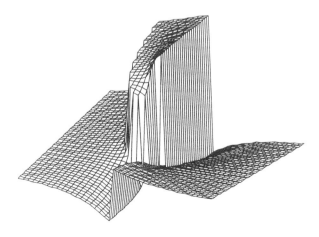

Figure 6.11 3-D plot of switch-off time τ versus $p = (V_c(0), I_s(0))$.

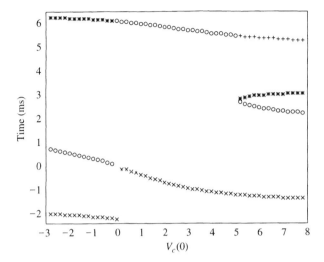

Figure 6.12 Roots of $f(t,p)$: $\circ$ = switch time τ, $*$ = root 2, $+$ = root 3, $\times$ = root -1.

first root when it becomes positive. The onset of the misfire occurs when root -1 coalesces with the root at zero. Since the root at zero is fixed, this is a transcritical bifurcation of roots of f. Also note that when $I_s(0) = 9$, a transcritical bifurcation diagram is evident at the origin of Figure 6.12. (A transcritical bifurcation generically occurs in a conventional dynamical system when two equilibria coalesce under the condition that one of the equilibria is fixed in position by the structure of the system. The bifurcation diagram is similar to two intersecting lines and the equilibria exchange stability when they coalesce. See [7,29] for a detailed description of transcritical bifurcation.)

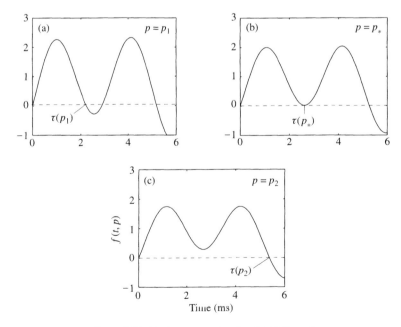

Figure 6.13 $f(t,p)$ versus t showing fold bifurcation.

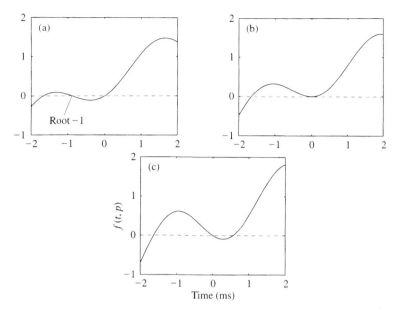

Figure 6.14 $f(t, p)$ versus t showing transcritical bifurcation.

6.7.4 Noninvertibility and Discontinuity of the Poincaré Map

This section discusses how irreversibility of trajectories and switching time bifurcations can make the Poincaré map P respectively noninvertible and discontinuous.

Trajectories in thyristor circuits are sometimes reversible. Suppose that the thyristor is off and we wish to integrate the trajectory backward in time. Regard the previous switch-off time τ as a variable to be solved for. A constraint on τ can be found by requiring that integrating the on system backward in time from time τ yields a current zero at the time of a firing pulse at which the thyristor turned on. If we can solve this constraint for τ (and also confirm that there was no misfire at the turn-on), then a solution integrating the trajectory backward in time until the previous switch-on has been found. The problem is that the solution for τ need not be unique; there can be two distinct trajectories leading to the same state. A consequence of this irreversibility is that the Poincaré map is sometimes not invertible. This can be seen in the SVC example circuit by computing the Poincaré map for a segment of a semicircular disk of initial conditions and observing the overlapping portion of Figure 6.15.

The switching time bifurcations of Section 6.7.2 cause the switching times in some transients to vary discontinuously as the initial condition is varied. That is, a switching time bifurcation can cause two initially nearby trajectories to separate greatly because a portion of one trajectory occurs in a circuit with a thyristor on while the same portion of the second trajectory occurs in a circuit with a thyristor off. Thus the switching time bifurcations cause discontinuities of the Poincaré map. If the initial condition $\mathbf{y}(0)$ is such that a switching time bifurcation occurs at one of the switching times in the period, then P is discontinuous at $\mathbf{y}(0)$ [23,25,31]. We write Θ for the set of discontinuities of P. As shown in Section 6.5 and proved in [5], P is smooth away from Θ. Wolf et al. [31] show interesting graphs of the Poincaré map discontinuities.

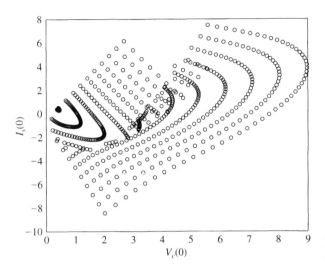

Figure 6.15 Poincaré map of semicircular disk.

6.7.5 Multiple Attractors and Their Basin Boundaries

Numerical experiments for the SVC example circuit [23,25] show that there are two asymptotically stable fixed points y_1 and y_2 of the Poincaré map; these correspond to two asymptotically stable periodic orbits of the circuit. This occurrence of multiple attractors can be surprising (in contrast, the corresponding diode circuits with suitable resistive damping containing diodes, voltage sources of period T, resistors, capacitors, and inductors have unique attractors that are globally asymptotically stable).

A steady state (periodic orbit or fixed point) may be asymptotically stable, but a large disturbance may cause stability to be lost. The set of initial states which return to the steady state is called the *basin of attraction* of the steady state. (Think of a marble rolling in a bowl: the basin of attraction of the stable point at the bottom of the bowl is the inside of the bowl. Disturbances which perturb the state inside the bowl will decay as the marble eventually returns to bottom of the bowl. Larger disturbances will cause the marble to leave the bowl and roll away elsewhere. The boundary of the basin of attraction is the edge of the bowl.) The basin of attraction of a steady state is used to quantify the robustness to perturbations of that steady state. One way to describe the basins of attraction is to instead describe the boundaries of these basins.

In smooth dynamical systems, the essential mechanism which separates nearby trajectories so that they can tend to attractors in different basins is the saddle-type behavior of unstable fixed points or periodic orbits in the basin boundary. The SVC circuit example has a basin boundary ∂B separating the basins of attraction of the two fixed points y_1 and y_2. However, there are no unstable fixed points in ∂B. ∂B interacts with the set of Poincaré map discontinuities Θ and the essential mechanism separating nearby trajectories to tend to attractors in different basins is the switching time bifurcations associated with Θ.

This is supported by numerical results for a portion of ∂B [26]: Figure 6.16(a) shows the fine structure of the Poincaré map discontinuities Θ. Θ is composed of three curves C_1, C_2, C_3. Initial conditions on C_1 yield switching time bifurcations in

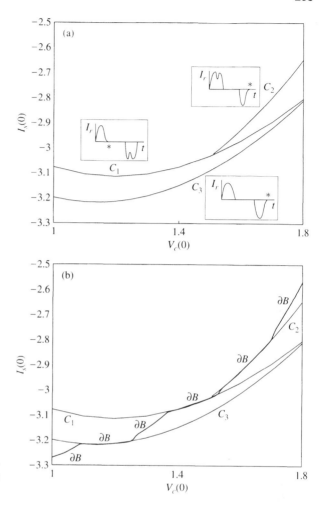

Figure 6.16 (a) Detail of Poincaré map discontinuities Θ. (b) Detail of basin boundary ∂B and Θ.

the first half-cycle as shown in the corresponding inset of Figure 6.16(a) (the inset shows the qualitative form of thyristor current $I_r(t)$ for initial conditions on C_1 and ∗ indicates a switching time bifurcation). C_2 and C_3 correspond to switching time bifurcations in the second half-cycle as shown in the corresponding insets of Figure 6.16(a).

Figure 6.16(b) shows how the basin boundary ∂B intersects with Θ. The points in $\partial B \cap \Theta$ are initial conditions on the basin boundary which encounter a switching time bifurcation during the next cycle. The discontinuity in P caused by switching time bifurcation is the mechanism by which nearby initial conditions on either side of ∂B tend to different fixed points under iterations of P. Numerical results show that the points in ∂B not in Θ eventually map to $\partial B \cap \Theta$. That is, the basin boundary ∂B consists of initial conditions which are either on Θ or eventually map to Θ. However, there are also many initial conditions in Θ but not in ∂B whose forward trajectories encounter switching time bifurcations.

The intricate and nonconventional structure of the basin boundary ∂B shows the great influence of switching time bifurcations and the thyristor switching rules on the system dynamics.

6.8 DIODE CIRCUITS

This section considers circuits with ideal diodes. The analysis is useful in applications such as diode rectifier circuits and can be adapted to analyze circuits containing diodes and other switching elements, such as dc/dc converters in a discontinuous mode of operation. The dynamics of diode circuits are generally simpler than the dynamics of thyristor circuits. Instead of describing the dynamics of diode circuits from scratch, it is convenient to describe the dynamics of diode circuits by comparing them to the dynamics of thyristor circuits.

6.8.1 Transversality and Poincaré Map Jacobian Formula

Diode switch-off is the same as thyristor switch-off and can be analyzed as described earlier in the chapter. Diode switch-on occurs when diode voltage becomes positive and is analogous or dual to diode switch-off. More precisely, diode switch-on occurs at the first zero of diode voltage which is after the diode switch-off. Similarly to Section 6.5.1, the switch-on is transversal if the gradient of the diode voltage is positive (taking the limit from below) at the switching-on time.

Similarly to the case of diode or thyristor switch-off in Section 6.5.3, it is shown in [5] that at a transversal diode switch-on, the map advancing the state over a time interval containing only this switch-on is smooth and that its derivative is computed using formula (6.38). It follows by assembling the Jacobian as in Section 6.5.3 that, if all the diode switchings in a cycle are transversal, the Poincaré map is smooth and its Jacobian is calculated by a formula such as (6.53). Formula (6.53) applies to the case of four diode switchings per cycle and the number of terms should be changed for other cases. For example [5], a symmetrical three-phase diode bridge circuit in steady-state Mode 1 operation with ac line impedance represented has Poincaré map Jacobian

$$DP = \left(e^{A_2(T/6-\mu)} Re^{A_3\mu} Q \right)^6 \tag{6.73}$$

where A_2 corresponds to the circuit with two diodes conducting, A_3 corresponds to the commutating circuit with three diodes conducting, and μ is the commutation time. In this case, the symmetry of the circuit and the six time intervals containing noncommutating and commutating circuits gives an expression to the sixth power in (6.73).

6.8.2 Poincaré Map Jacobian for the DC/DC Buck-Boost Converter in Discontinuous Mode

To illustrate the effect of the diode switching in the dc/dc buck-boost converter we compute the Poincaré map Jacobian. Our notation follows that of Chapter 2. For simplicity, the switch is assumed to be switched once on and once off at fixed times in the cycle. Suppose the switch turns on at time 0 and that the Poincaré map sample time is chosen to be 0+. Then, as explained in detail in Chapter 2, the cycle consists of an interval of duration $\delta_1 T$ with the switch on with state $\mathbf{x} = \begin{pmatrix} i \\ v \end{pmatrix}$ and matrix

$$A_1 = \begin{pmatrix} -R_l/L & 0 \\ 0 & -1/RC \end{pmatrix} \tag{6.74}$$

followed by an interval of duration $\delta_2 T$ with the switch off with state $\mathbf{x} = \begin{pmatrix} i \\ v \end{pmatrix}$ and matrix

$$A_2 = \begin{pmatrix} -R_l/L & 1/L \\ -1/C & -1/RC \end{pmatrix} \qquad (6.75)$$

followed by an interval of duration $\delta_3 T$ with the diode off with state $\mathbf{y} = v$ and matrix

$$A_3 = -1/RC \qquad (6.76)$$

It is necessary to correctly model the change in state-space dimension when the diode switches off in order to obtain correct stability results. At the diode switch-off, the matrix projecting the two-dimensional state to the one-dimensional state is

$$R = \begin{pmatrix} 0 & 1 \end{pmatrix} \qquad (6.77)$$

At the switch-on, the matrix augmenting the state-space dimension from one to two is

$$Q = \begin{pmatrix} 0 \\ 1 \end{pmatrix} \qquad (6.78)$$

The state is unchanged in dimension and continuous at the switch turn-off.

Computing the Poincaré map Jacobian formula gives

$$DP = Qe^{A_3\delta_3 T} Re^{A_2\delta_2 T} e^{A_1\delta_1 T} \qquad (6.79)$$

$$= \begin{pmatrix} 0 \\ 1 \end{pmatrix} e^{-\delta_3 T/RC} \begin{pmatrix} 0 & 1 \end{pmatrix} e^{A_2\delta_2 T} e^{A_1\delta_1 T} \qquad (6.80)$$

$$= \begin{pmatrix} 0 & 0 \\ 0 & e^{-\delta_3 T/RC} \end{pmatrix} e^{A_2\delta_2 T} e^{A_1\delta_1 T} \qquad (6.81)$$

It is clear that DP has a zero eigenvalue and that $DP\begin{pmatrix} \delta i \\ \delta v \end{pmatrix}$ has first component zero for any linearized perturbation $\begin{pmatrix} \delta i \\ \delta v \end{pmatrix}$. This means that the circuit damps any initial perturbation δi in inductor current to zero at the end of the cycle; this is clearly correct since the inductor current is constrained to be zero at the end of the cycle because of the diode being off.

Neglecting the change in state-space dimension by assuming that, when the diode is not conducting, $\mathbf{x} = \begin{pmatrix} i \\ v \end{pmatrix}$ and

$$A_3 = \begin{pmatrix} 0 & 0 \\ 0 & -1/RC \end{pmatrix} \qquad (6.82)$$

would yield

$$DP = e^{A_3\delta_3 T} e^{A_2\delta_2 T} e^{A_1\delta_1 T} \qquad (6.83)$$

$$= \begin{pmatrix} 1 & 0 \\ 0 & e^{-\delta_3 T/RC} \end{pmatrix} e^{A_2\delta_2 T} e^{A_1\delta_1 T} \qquad (6.84)$$

which is wrong. In particular, (6.84) would generally imply a nonzero current perturbation at the end of a cycle.

Instead of choosing the Poincaré map sample time at 0+, which is done above so that it can be more directly compared with the wrong calculation (6.84), it is usually more convenient to choose the Poincaré map sample time at 0−. Then the Poincaré map Jacobian becomes the scalar

$$DP = e^{A_3 \delta_3 T} Re^{A_2 \delta_2 T} e^{A_1 \delta_1 T} Q \qquad (6.85)$$

For example, the Poincaré map sample time at 0− is used in Section 5.5.2.

6.8.3 Poincaré Map Continuity and Switching Time Bifurcations

Consider a circuit of ideal diodes, time-dependent sources, and resistors, inductors, and capacitors. It can be shown [5] using incremental energy methods that the Poincaré map is continuous. The diode switch-offs decrease incremental energy and provide switching damping to the circuit as in the case of thyristor switch-offs. (Diode switch-ons also decrease incremental energy, but this is a second-order effect.)

The Poincaré map is not differentiable when one of the switchings is not transversal (the Jacobian has a discontinuity) and a switching time bifurcation can then occur. However, the typical consequences of the switching time bifurcation differ from those in thyristor circuits. In diode circuits a new diode switch-off generally appears or disappears together with a closely following diode switch-on [5]. The consequence of the switching time bifurcation is then a mode change in the circuit in which a short time interval with a particular pattern of diode conduction appears or disappears.

This behavior occurs in the simple diode circuit shown in Figure 6.17 when the circuit resistance is positive. When the constant bias p of the voltage source lies in the interval $(-1, 0)$, there is a unique and asymptotically stable periodic orbit in which the diode switches twice per cycle. In this mode, the Poincaré map at the periodic orbit with initial time when the diode is off simply maps zero incremental current to zero incremental current and the Jacobian of the Poincaré map is zero. That is, a small perturbation in one cycle vanishes before the next cycle.

If p increases through 0, the periodic orbit persists and remains asymptotically stable, but the diode never turns off. The stability of the periodic orbit is now governed by the resistor so that the Jacobian of the Poincaré map changes discontinuously when p increases through zero. If p decreases through −1, the periodic orbit becomes a constant zero current and the diode never turns on. In both these switching time bifurcations, the two switching times coalesce and disappear. The effect of the switching time bifurcations is a mode change in the circuit and asymptotic stability is not lost. Note, however, that in the extreme case of zero circuit resistance, the periodic orbit, although stable for $p \leq 0$, disappears for $p > 0$ and the circuit trajectory becomes unbounded.

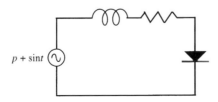

Figure 6.17 Simple diode circuit showing mode changes.

6.9 FIRING ANGLE CONTROL

Practical high-power thyristor circuits control the firing angle both to regulate performance with closed loop controls and to synchronize the firing with the ac waveform. Since these important effects are omitted from the previous analysis in the chapter, this section briefly introduces an example of the appropriate modifications to the Jacobian formula for the SVC example circuit. Control and synchronization both have important effects on circuit stability [13].

We proceed by modifying the analysis from Section 6.5.3 to account for changes in the thyristor firing angle. Section 6.5.3 considered a time interval $[s_2, s_3]$ including a thyristor switch-on at time ϕ and no other switchings. F_{s_3, s_2} is the flow which maps the initial state $\mathbf{y}(s_2)$ at s_2 to the final state $\mathbf{x}(s_3)$ at s_3 and Section 6.5.3 computed that

$$DF_{s_3, s_2} = DG_{s_3, s_2} + \frac{\partial G_{s_3, s_2}}{\partial \phi} D\phi \tag{6.86}$$

where G_{s_3, s_2} is given by (6.36). Differentiating (6.36) yields

$$\frac{\partial G_{s_3, s_2}}{\partial \phi} = e^{A_{on}(s_3 - \phi)}(QR - I)(A_{on}\mathbf{x}(\phi) + B_{on}u(\phi))$$
$$= e^{A_{on}(s_3 - \phi)}c^T c\dot{\mathbf{x}}(\phi+) \tag{6.87}$$

where $c = \begin{pmatrix} 1 & 0 & 0 \end{pmatrix}'$.[1] Now (6.86) becomes

$$DF_{s_3, s_2} = e^{A_{on}(s_3 - \phi)}Qe^{A_{off}(\phi - s_2)} + e^{A_{on}(s_3 - \phi)}c^T c\dot{\mathbf{x}}(\phi+)D\phi \tag{6.89}$$

and the effect of the controlled switch-on is captured in the second term of (6.89).

It remains to compute $D\phi$ according to the control or synchronization represented. Obtaining a formula for $D\phi$ can be difficult, particularly if the control depends on a filtered version of the past system states. However, some cases are tractable [13] and here we illustrate the calculation of $D\phi$ for current synchronization. The row vector $D\phi$ is the gradient of the turn-on time ϕ with respect to $\mathbf{y}(s_2)$.

Synchronizing the firing with respect to the zeros of the line current I_s is modeled by

$$\phi = \zeta + (\pi - \sigma_{req})/2 \tag{6.90}$$

where ζ is the time of the current zero and σ_{req} is the required thyristor conduction time. ζ satisfies the constraint

$$0 = I_s(\zeta) = n y(\zeta) = n G_{\zeta, s_2}(\mathbf{y}(s_2), \zeta(\mathbf{y}(s_2))) \tag{6.91}$$

where $n = (0, 1, 0)$. ζ is assumed to occur when both thyristors are off.

Differentiating the constraint (6.91) yields

$$0 = n DG_{\zeta, s_2} + n \frac{\partial G_{\zeta, s_2}}{\partial \zeta} D\zeta \tag{6.92}$$

---

1 An interesting alternate expression is

$$\frac{\partial G_{s_3, s_2}}{\partial \phi} = e^{A_{on}(s_3 - \phi)}(Q\dot{\mathbf{y}}(\phi-) - \dot{\mathbf{x}}(\phi+)) \tag{6.88}$$

but this is not necessary here.

Now differentiation of (6.90) and solving (6.92) yields

$$D\phi = D\zeta = \frac{-nDG_{\zeta,s_2}}{n\dfrac{\partial G_{\zeta,s_2}}{\partial \zeta}} = \frac{-ne^{A_{\text{off}}(\zeta-s_2)}}{n\dot{\mathbf{y}}(\zeta)} \qquad (6.93)$$

REFERENCES

[1] L. O. Chua, M. Hasler, J. Neirynck, and P. Verburgh, Dynamics of a piecewise-linear resonant circuit, *IEEE Trans. on Circuits and Systems*, vol. 29, no. 8, pp. 535–547, August 1982.

[2] E. A. Coddington and N. Levinson, *Theory of Ordinary Differential Equations*, Chapter 13, Section 2, reprint. Malabar, FL: Krieger, 1984.

[3] I. Dobson and S. G. Jalali, Surprising simplification of the Jacobian of a switching circuit, *IEEE Int. Symposium on Circuits and Systems* (Chicago, IL), pp. 2652–2655, May 1993.

[4] I. Dobson, S. G. Jalali, and R. Rajaraman, Damping and resonance in a high power switching circuit, in *Systems and Control Theory for Power Systems* (J. H. Chow, P. V. Kokotovic, R. J. Thomas, eds.), IMA vol. 64 in mathematics and its applications. New York: Springer Verlag, pp. 137–156, 1995.

[5] I. Dobson, Stability of ideal thyristor and diode switching circuits, *IEEE Trans. on Circuits and Systems, Part 1*, vol. 42, no. 9, September 1995, pp. 517–529.

[6] M. Grötzbach and R. von Lutz, Unified modelling of rectifier-controlled dc-power supplies, *IEEE Trans. on Power Electronics*, vol. 1, no. 2, April 1986, pp. 90–100.

[7] J. Guckenheimer and P. Holmes, *Nonlinear Oscillations, Dynamical Systems and Bifurcations of Vector Fields*. New York: Springer-Verlag, 1986.

[8] L. Gyugyi, Power electronics in electric utilities: Static var compensators, *Proc. IEEE*, vol. 76, no. 4, April 1988, pp. 483–494.

[9] J. K. Hale, *Oscillations in Nonlinear Systems*. New York: McGraw Hill, 1963.

[10] J. G. Kassakian, M. F. Schlecht, and G. C. Verghese, *Principles of Power Electronics*. Reading, MA: Addison-Wesley, 1992.

[11] J. P. Louis, Non-linear and linearized models for control systems including static convertors, *Proc. Third Int. Federation on Automatic Control Symposium on Control in Power Electronics and Electrical Drives* (Lausanne, Switzerland), pp. 9–16. September 1983.

[12] S. G. Jalali, I. Dobson, and R. H. Lasseter, Instabilities due to bifurcation of switching times in a thyristor controlled reactor, *IEEE Power Electronics Specialists' Conference* (Toledo, Spain), pp. 546–552, July 1992.

[13] S. G. Jalali, R. H. Lasseter, and I. Dobson, Dynamic response of a thyristor controlled switched capacitor, *IEEE Trans. on Power Delivery*, vol. 9, no. 3, pp. 1609–1615, July 1994.

[14] S. G. Jalali, I. Dobson, R. H. Lasseter, and G. Venkataramanan, Switching time bifurcations in a thyristor controlled reactor, *IEEE Trans. on Circuits and Systems—Part 1*, vol. 43, no. 3, pp. 209–218, March 1996.

[15] S. G. Jalali, Harmonics and instabilities in thyristor based switching circuits. PhD diss., University of Wisconsin–Madison, 1993.

[16] P. Kundur, *Power System Control and Stability*. New York: McGraw Hill, 1994.

[17] R. von Lutz and M. Grötzbach, Straightforward discrete modelling for power convertor systems, *IEEE Power Electronics Specialists' Conference* (Toulouse, France), pp. 761–770, June 1985.

[18] P. Mattavelli, A. M. Stankovic, and G. C. Verghese, SSR analysis with dynamic phasor model of thyristor-controlled series capacitor, *IEEE Trans. on Power Systems*, vol. 14, no. 1, pp. 200–208, February 1999.

[19] T. J. E. Miller, ed., *Reactive Power Control in Electric Systems*. New York: Wiley, 1982.

[20] H. A. Othman and L. Ängquist, Analytical modeling of thyristor-controlled series capacitors for SSR studies, *IEEE Trans. on Power Systems*, vol. 11, no. 1, pp. 119–127, February 1996.

[21] B. K. Perkins and M.R. Iravani, Dynamic modeling of a TCSC with application to SSR analysis, *IEEE Trans. on Power Systems*, vol. 12, no. 4, pp. 1619–1625, November 1997.

[22] B. K. Perkins and M.R. Iravani, Dynamic modeling of high power static switching circuits in the dq-frame, *IEEE Trans. on Power Systems*, vol. 14, no. 2, pp. 678–684, May 1999.

[23] R. Rajaraman, I. Dobson, and S. G. Jalali, Nonlinear dynamics and switching time bifurcations of a thyristor controlled reactor, *IEEE Int. Symposium on Circuits and Systems* (Chicago), pp. 2180–2183, May 1993.

[24] R. Rajaraman, Damping of subsynchronous resonance and nonlinear dynamics in thyristor switching circuits. PhD diss., Univ. of Wisconsin–Madison, 1996.

[25] R. Rajaraman, I. Dobson, and S. Jalali, Nonlinear dynamics and switching time bifurcations of a thyristor controlled reactor circuit, *IEEE Trans. on Circuits and Systems—Part 1*, vol. 43, no. 12, pp. 1001–1006, December 1996.

[26] R. Rajaraman and I. Dobson, Damping and incremental energy in thyristor switching circuits, *IEEE Int. Symposium on Circuits and Systems* (Seattle), pp. 291–294, May 1995.

[27] R. Rajaraman, I. Dobson, R. H. Lasseter, and Y. Shern, Computing the damping of subsynchronous oscillations due to a thyristor controlled series capacitor, *IEEE Trans. on Power Delivery*, vol. 11, no. 2, pp. 1120–1127, April 1996.

[28] S. R. Sanders and G. C. Verghese, Lyapunov-based control for switched power converters, *IEEE Trans. on Power Electronics*, vol. 7, no. 1, pp. 17–24, Jan. 1992.

[29] S. Strogatz, *Nonlinear Dynamics and Chaos: With Applications in Physics, Biology, Chemistry, and Engineering*. Reading, MA: Addison-Wesley, 1994.

[30] G. C. Verghese, M. E. Elbuluk, and J. G. Kassakian, A general approach to sampled-data modeling for power electronic circuits, *IEEE Trans. on Power Electronics*, vol. 1, no. 2, pp. 76–89, April 1986.

[31] D. M. Wolf, M. Varghese, and S. R. Sanders, Bifurcation of power electronic circuits, Section 5, *J. Franklin Institute—Engineering and Applied Mathematics*, vol. 331B, no. 6, pp. 957–999, November 1994.

Chapter 7

NONLINEAR PHENOMENA IN OTHER POWER ELECTRONIC SYSTEMS

7.1 MODELING A NONLINEAR INDUCTOR CIRCUIT

Jonathan H. B. Deane

7.1.1 Introduction

A nonlinear element that appears in many power electronic circuits is the inductor whose core saturates and displays hysteresis. In order to simulate such systems, it is necessary to have a good mathematical model for a nonlinear saturating inductor. In this section, we discuss in detail one particular hysteresis model, due to Jiles and Atherton [1,2,3], in relation to a simple nonlinear circuit that displays all the usual nonlinear behavior, including, apparently, chaos. The circuit we model is sometimes known as the *ferroresonant circuit* [4] (see Figure 7.1). It is not a power electronics circuit *per se*. However, it is a good illustration of the Jiles-Atherton (J–A) model and of the sort of behavior that can occur when core saturation and hysteresis are both present.

Many different models of hysteresis ([2] contains a review) have been proposed. None of these was devised expressly for the purpose of modeling a chaotic circuit, but the J–A model with the correct choice of parameters gives excellent agreement with experimental results [4]. The inclusion of hysteresis is necessary to obtain this agreement. This is of particular interest as the model was developed more for the purposes of describing hysteresis loops (i.e., essentially at zero frequency, and certainly up to no more than 10Hz [5]). Here it is shown to work in a dynamic application, albeit at rather low frequencies (< 2kHz), with small parameter adjustments from the zero-frequency case.

7.1.2 The Circuit

Referring to Figure 7.1, the capacitor C and resistor R are linear, and R includes the inductor winding losses. For the simulations and experiments discussed, $C = 100$nF and $R = 8.85\Omega$. The core is toroidal in shape and is made of 3C85 ferrite. Its average diameter D is 2.4×10^{-2}m and its cross-sectional area A is 4.54×10^{-5}m^2. The winding consists of $n = 230$ turns of copper wire of diameter 1.5×10^{-4}m. Experimentally, neither period multiplication nor chaos are observed when the circuit is driven by a sine wave, but these phenomena readily occur with a square-wave drive.

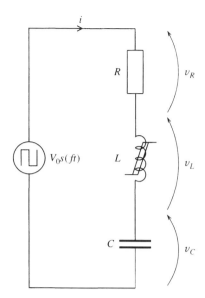

Figure 7.1 A circuit containing a nonlinear inductor.

7.1.3 Saturating and Hysteretic Inductor Modeling

The J–A model of ferromagnetic hysteresis is itself a first-order nonlinear differential equation which can be solved, at least numerically, to give the magnetization, M, as a function of the applied magnetic field, H. The differential equation is

$$\frac{dM(H)}{dH} = (1 - c)\delta_M \frac{M_{an}(H_e) - M(H)}{k(1 - c)\text{sign}(\dot{H}) - \alpha[M_{an}(H_e) - M(H)]} + c \left(\frac{\partial M_{an}(H_e)}{\partial H} \right)_M \quad (7.1)$$

The effective field is given by

$$H_e = H + \alpha M(H) \quad (7.2)$$

and the two step functions sign(.) and δ_M are defined as

$$\text{sign}(x) = \begin{cases} 1 & \text{if } x \geq 0 \\ -1 & \text{otherwise} \end{cases} \quad (7.3)$$

and

$$\delta_M = \begin{cases} 0 & \text{if } \dot{H} < 0 \text{ and } M_{an}(H_e) - M(H) \geq 0 \\ 0 & \text{if } \dot{H} \geq 0 \text{ and } M_{an}(H_e) - M(H) \leq 0 \\ 1 & \text{otherwise} \end{cases} \quad (7.4)$$

This is a slightly modified form of the equation as it appeared in [3]. There are three parameters appearing in (7.1). Fuller definitions are given in [2] and [3], but in brief, c (dimensionless) is the ratio of the initial normal to the initial anhysteretic differential susceptibility; α (dimensionless) is an experimentally obtained mean field parameter, representing the coupling between domains; and k (A/m) is a measure of hysteresis. Since k multiplies sign($\dot{H}$), it gives a measure of the width of the hysteresis loop, and $k = 0$ means that there is no hysteresis.

The overall shape of the hysteresis curve is determined by the anhysteretic magnetization function, $M_{an}(H_e)$. In [1,2,3], the following form, a modified Langevin function, was suggested

$$M_{an}(H_e) = M_s[\coth(H_e/a) - a/H_e] \tag{7.5}$$

where M_s (A/m) is the saturation magnetization and a (A/m) is a scaling factor. The parameters used to model the circuit result in the M–H curves given in Figure 7.2.

7.1.4 Differential Equation for the Circuit

It is now possible to write the differential equation for the circuit. Kirchhoff's voltage law gives

$$V_0 s(ft) = v_R + v_L + v_C \tag{7.6}$$

where $V_0 s(ft)$ denotes a symmetrical square wave of frequency f and single-peak amplitude V_0. The resistor, inductor, and capacitor voltages are, respectively, v_R, v_L, and v_C.

Ampère's circuital law gives $\pi DH = ni$, assuming that the magnetic field is confined to the core, which is thin in comparison to its diameter. This leads to

$$H = \frac{ni}{\pi D} \tag{7.7}$$

Now

$$v_L = \mu_0 nA \frac{d}{dt}(H + M) = \frac{\mu_0 n^2 A}{\pi D}\frac{di}{dt}\left[1 + \frac{dM}{dH}\right] \tag{7.8}$$

which, together with (7.6), gives

$$\frac{di}{dt} = \frac{\pi D}{\mu_0 n^2 A}\frac{V_0 s(ft) - v_C - iR}{1 + \frac{dM}{dH}} \tag{7.9}$$

where we have used Ohm's law, $v_R = iR$. From the definition of capacitance

$$\frac{dv_C}{dt} = \frac{i}{C} \tag{7.10}$$

and finally, using (7.7),

$$\frac{dM}{dt} = \frac{dH}{di}\frac{di}{dt}\frac{dM}{dH} = \frac{n}{\pi D}\frac{di}{dt}\frac{dM}{dH} \tag{7.11}$$

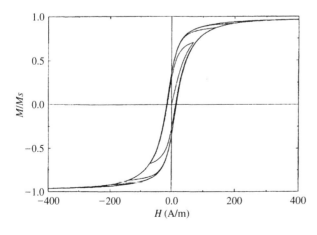

Figure 7.2 Computed $M - H$ curves; H was swept to the following values: 0, 70, −70, 140, −140, 200, −200, 270, −270, 340, −340, 400, −400, 400A/m. $H = 400$A/m corresponds to $i \approx 130$mA. The Jiles-Atherton model parameters were $M_s = 2.75 \times 10^5$A/m, $a = 14.1$A/m, $k = 17.8$A/m, $\alpha = 5 \times 10^{-5}$ and $c = 0.55$. These parameters were used in the simulations for all subsequent figures.

Equations (7.9)–(7.11) with dM/dH given by (7.1) in conjunction with (7.5), describe the circuit. The state vector has three components—i, v_C, and M—so the system is of third order and nonautonomous. The extra dimension arises because when hysteresis is present, the $M - H$ loop encloses nonzero area and so M is not uniquely determined for a given H. Hence, M and H are independent variables. The capacitor voltage, v_c, is the third state variable.

7.1.5 Results

The differential equation for the circuit is an example of a stiff system, since it exhibits behavior at different time scales: typical v_C waveforms have a time constant of order 1ms for the parameter values used here, while the corresponding current waveforms generally consist of spikes of duration around $10\mu s$—see Figure 7.3. This needs to be borne in mind when choosing a numerical method for solving the differential equation.

We now show the measure of agreement between bifurcation diagrams and Poincaré sections of chaotic solutions, produced experimentally and by simulation. A comparison between v_C, i phase-plane diagrams is given in [4].

It would be expected that the major difficulty in obtaining convergence between experiment and simulation would be in finding the "right" set of parameters M_s, a, k, α, and c for the J–A model, and this is indeed the case. Many different sets can be tried before good agreement is obtained. Two sets of parameters are given in the table below; the first column contains the best available measured values at $25°C$, and the second the set that gave the best fit between experiment and simulation.

Measured	Best Fit
$M_s = 3.8–3.98 \times 10^5 A/m$	$M_s = 2.75 \times 10^5 A/m$
$a = 27 A/m$	$a = 14.1 A/m$
$k = 16–30 A/m$	$k = 17.8 A/m$
$\alpha = 5 \times 10^{-5}$	$\alpha = 5 \times 10^{-5}$
$c = 0.55$	$c = 0.55$

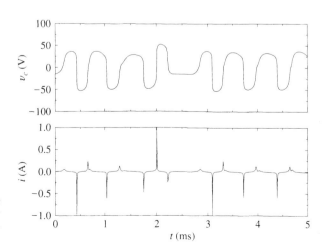

Figure 7.3 Computed capacitor voltage (upper) and current (lower) waveforms, for $V_0 = 15V$ and $f = 1.5kHz$. The resulting solution is chaotic.

Discrepancies exist between the values of M_s and a. These could be due to several factors, including manufacturing tolerances, high frequency (e.g., dynamic broadening of the hysteresis loop) and temperature effects in the ferrite, the skin effect in the windings, and the choice of anhysteretic magnetization function. With regard to the latter, it is explicitly stated in [2] that the modified Langevin function is not the only possible form for $M_{an}(H_e)$, although it has been found to model most anhysteretic curves.

The frequency of the square-wave drive voltage was 1.5kHz for all the measurements and simulations reported below, although both were also carried out at 1 and 2kHz and a similar degree of agreement noted. For the bifurcation diagrams and Poincaré sections, sampled state variables were used; in all cases, samples were taken once per cycle of the square wave, at the point when the drive voltage crosses zero in a negative-going direction.

Bifurcation Diagram Comparison

The experimental bifurcation diagram was obtained by steadily sweeping V_0 from one value to another, while simultaneously sampling v_c once per drive cycle at a fixed phase angle. Sweep times were typically one minute, from which it is clear that the experimental bifurcation diagrams show an accumulation of around 90,000 points—more by a factor of 30 than would be feasible in the simulations because of the prohibitively large amount of computer time needed.

Figure 7.4(a) shows an experimental bifurcation diagram and Figure 7.4(b) shows the corresponding simulation result. In both cases, V_0 is swept from 0 to 27V.

Figures 7.5(a) and (b) show bifurcation diagrams when V_0 is swept from 27V down to 0V. The diagrams for increasing/decreasing V_0 are different because several different solutions to the differential equation can coexist at a given value of V_0; which one is seen at a given V_0 depends on which solution was exhibited at the previous instant.

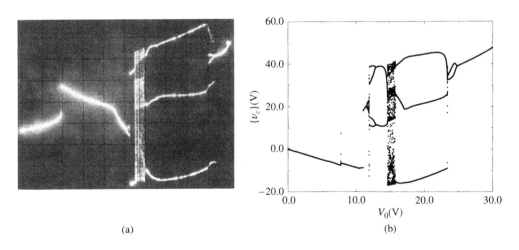

(a)   (b)

Figure 7.4 (a) Experimental bifurcation diagram—V_0 (x-axis) was increased from 0 to 27V. Sampled values of v_C were plotted vertically. The drive frequency was 1.5kHz, for this and all subsequent figures. Horizontal sensitivity: 2.7V/div. Vertical sensitivity: 10V/div. (b) The corresponding computed bifurcation diagram.

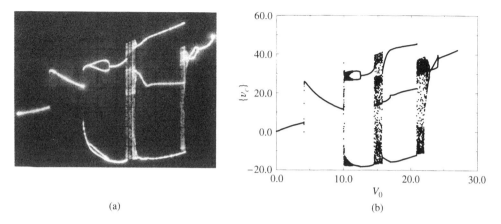

(a)   (b)

Figure 7.5 (a) Experimental bifurcation diagram with V_0 is decreased from 27V to
0V. (b) The corresponding computed bifurcation diagram.

Agreement between the bifurcation diagrams is significant as it shows that the J–A
model can be built into a description of the circuit that is not simply valid for one set of
parameters, but which holds over a whole range.

Poincaré Section Comparison

Comparison between experimental and simulated Poincaré sections is also possible. The state variables chosen were v_C and i. For a nonperiodic solution, the
section has a complicated structure. Figure 7.6(a) shows an experimentally obtained
Poincaré section for $V_0 = 17$V, and Figure 7.6(b) shows the simulation result for
$V_0 = 15$V.

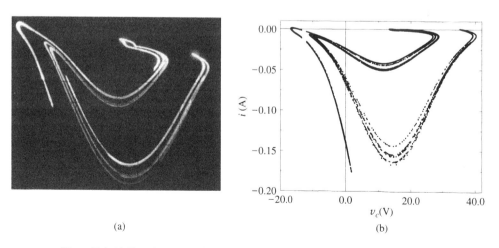

(a)   (b)

Figure 7.6 (a) Experimental Poincaré section. $V_0 = 17$V. The photograph shows the
accumulation of around 3,000,000 points. Plotted horizontally is v_C
(sensitivity 6.25V/div) and i (25mA/div) is plotted vertically; (b) Computed Poincaré section. $V_0 = 15$V. About 15,000 points are shown.

7.1.6 Conclusions

One model that includes the effects of core saturation and hysteresis, and which is easy to implement numerically, is the J–A model. This can be incorporated into the differential equation that describes a simple nonlinear circuit, and has been shown to be capable of displaying all the nonlinear behavior observed in practice, including subharmonic generation and chaotic behavior. The fact that such behavior can occur in even a simple circuit indicates the possibility of chaos in more complex power electronics systems when they contain inductors whose cores can saturate.

REFERENCES

[1] D. C. Jiles and D. L. Atherton, Ferromagnetic hysteresis, *IEEE Trans. on Magnetics*, vol. MAG-19, no. 5, pp. 2183–2185, September 1983.

[2] D. C. Jiles and D. L. Atherton, Theory of ferromagnetic hysteresis, *J. Magnetism and Magnetic Materials*, vol. 61, pp. 48–60, 1986.

[3] D. C. Jiles, J. B. Thoelke, and M. K. Devine, Numerical determination of hysteresis parameters for modeling of magnetic properties using the theory of ferromagnetic hysteresis, *IEEE Trans. on Magnetics*, vol. 28, no. 1, pp. 27–35, January 1992.

[4] J. H. B. Deane, Modeling the dynamics of nonlinear inductor circuits, *IEEE Trans. on Magnetics*, vol. 30, no. 5, pp. 2795–2801, September 1994.

[5] E. C. Snelling, *Soft Ferrites*, 2nd ed. Sevenoaks, Kent, UK: Butterworths, 1988.

7.2 INVERTERS UNDER TOLERANCE BAND CONTROL

Andreas Magauer

7.2.1 Introduction

Switched-mode dc-to-ac inverters are used in ac motor drives and uninterruptable ac power supplies (UPS) where the objective is to produce a sinusoidal ac output whose magnitude and frequency can both be controlled. The dc voltage is obtained by rectifying and filtering the line voltage. Such inverters are referred to as *voltage source inverters* (VSIs). VSIs are generally used with single-phase and three-phase outputs and can be divided in three categories [1]:

1. *Pulse-width modulation (PWM) of inverters.* The input voltage is essentially constant in magnitude; an uncontrolled diode rectifier is used to generate the dc voltage. The inverter must control magnitude and frequency of the ac output voltages. This is achieved by PWM of the inverter switches. There are various schemes to pulse-width-modulate the inverter switches in order to shape the output ac voltages to be as close to a sine wave as possible.

2. *Square-wave inverters.* The dc input voltage is controlled in order to control the magnitude of the ac voltage. The inverter has only to control the frequency of the output voltage. The ac output voltage has a waveform similar to a square wave.

3. *Single-phase inverters with voltage cancellation.* In the case of single-phase inverters it is possible to control the magnitude and frequency of the ac output voltage, even

though the dc input voltage of the inverter is constant and the inverter switches are not pulse-width modulated. This type of inverter combines the characteristics of the previous two inverters. Its behavior shows a typical three-position control, while categories 1 and 2 provide only two positions.

For category 1 various PWM control schemes are available, each suitable for particular applications [1,2]. Sinusoidal PWM, selected harmonic elimination, delta modulation, space vector modulation, tolerance band control, etc., are well-known methods in power electronics. The sinusoidal PWM, also known as the *triangulation, subharmonic,* or *suboscillation* method, is very popular in industrial applications. Here the switching signal is generated by comparing a sinusoidal reference signal and a triangular wave with constant period. A general categorization for PWM control schemes in applications to variable frequency drives can be found in [3].

In this section we investigate the dynamics of the tolerance band control technique. This technique is suitable for PWM and square-wave mode as above related to categories 1 and 2. It is also applied to current-controlled ac inverters for variable-frequency drives and uninterruptable power supply systems. Also modern concepts for voltage and current control in three-phase ac-inverters with decoupling of state-space variables include this method [4].

7.2.2 Functioning Principle

The principle of the tolerance band method is illustrated in Figure 7.7. The variable frequency is produced by variation of the frequency of the reference signal. The magnitude of the ac output voltage is varied by the amplitude of the reference signal. The dc input voltage is produced by the line voltage using an uncontrolled rectifier with (inductance capacitance) LC ripple filter. To eliminate the higher harmonics of the load-current, a further LC-filter is included. In most cases the character of load is resistive-inductive. In Figure 7.7, various possible feedback variables are shown. The feedback can be generated by the load-voltage (1), the load-current (2), or the inverter output current (3) depending on the control strategy. In contrast to sinusoidal PWM, no timebase exists. The switching action is controlled by the upper and lower limit of an on-off controller with hysteresis, as shown in Figure 7.8. One of the system variables is compared with the tolerance band around the reference waveform. If the state variable tends to go above or below the tolerance band, appropriate switching action takes place and the variable is forced to follow the reference waveform within the tolerance band. The dc/dc converters can also use this principle, where the sinusoidal reference is substituted by a constant voltage. In

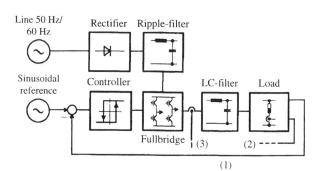

Figure 7.7 Block scheme of the tolerance band method.

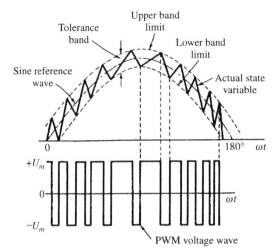

Figure 7.8 The tolerance band PWM principle.

the following subsections the chaotic behavior of the tolerance band method and its application in ac-inverters will be demonstrated and discussed; see also [5].

7.2.3 System Model and Equation

Figure 7.9 shows the control loop of the tolerance band method for one phase with resistive load. In the following discussion this simplification is assumed. From the point of view of control engineering the system in Figure 7.9 shows a nonautonomous relay connected to a linear plant of second order. Because of the switching characteristic of the controller and the linearity of the plant, the nonlinearity can be reduced to piecewise linearity. This allows analytic integration of the differential equation in each piece. The dependence of the signals on time is described by the normalized time variable τ, given by $\tau = \omega_0 t$, where $\omega_0 = \left(\sqrt{LC}\right)^{-1}$ is the resonant frequency of the linear system of second order. Also the frequency of the sinusoidal drive signal ω_w is normalized to Ω when the time t is substituted (i.e., $\omega_w t = \frac{\omega_w}{\omega_0}\tau = \Omega\tau$). Further, the correcting variable of the controller $u(\tau)$, the output voltage $x(\tau)$, the drive signal $w(\tau)$ and the corresponding amplitude a, the error signal $e(\tau)$, and the hysteresis or the width of the tolerance band of the controller h are normalized to the maximum value of the on-off controller output voltage u_m:

$$U(\tau) = \frac{u(\tau)}{2u_m} = \pm\frac{1}{2}, \quad X(\tau) = \frac{x(\tau)}{2u_m}, \quad W(\tau) = \frac{w(\tau)}{2u_m}, \quad E(\tau) = \frac{e(\tau)}{2u_m},$$

$$A = \frac{a}{2u_m}, \quad H = \frac{h}{2}\frac{1}{2u_m}$$

(7.12)

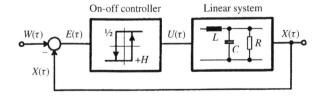

Figure 7.9 The control loop of the tolerance band method.

In normalized notation the expression $2H$ is the width of the tolerance band. H represents the threshold value where the switching action takes place. The piecewise-linear differential equation in normalized notation is as follows:

$$X''(\tau) + 2DX'(\tau) + X(\tau) = U(E(\tau)) \tag{7.13}$$

where

$$E(\tau) = W(\tau) - X(\tau)$$
$$W(\tau) = A\cos[\Omega\tau + \Phi] \tag{7.14}$$

$$U(E(\tau)) = +\frac{1}{2} \quad \text{for } E(\tau) > -H$$

$$U(E(\tau)) = -\frac{1}{2} \quad \text{for } E(\tau) < +H \tag{7.15}$$

The value of the damping factor $D = \frac{1}{2R}\sqrt{\frac{L}{C}}$, which follows from the linear passive components R, L, and C in Figure 7.9, is chosen less than unity in practical circuits [4]. This gives a complex conjugate pole pair in the transfer function of the linear system, implying damped oscillations in the time domain. Moreover, the differential equation is formulated by use of the error signal $E(\tau)$ instead of $X(\tau)$, which effects no qualitative difference; but the equation is more handsome. If $E(\tau)$ is chaotic $X(\tau)$ is also chaotic, because the difference between the signals is just the sinusoidal reference function $W(\tau)$. To simplify the model, we define new variables:

$$E_1 = E(\tau), \quad E_2 = E'(\tau), \quad E_3 = \Omega\tau + \Phi \tag{7.16}$$

E_3 in this nonautonomous system characterizes the dependence on time. Using the appropriate substitution of (7.16) the nonautonomous differential (7.13) can be written as three equations of first order as follows:

$$E_1' = E_2$$
$$E_2' = -U(E_1) - 2DE_2 - E_1 + AG(\Omega)^{-1}\cos(E_3 - \Psi_G(\Omega)) \tag{7.17}$$
$$E_3' = \Omega$$

where $G(\Omega)^{-1}$ is the absolute value of the inverted transfer function and $\Psi_G(\Omega)$ the corresponding phase displacement of the linear system, given by

$$G(\Omega)^{-1} = \sqrt{(1 - \Omega^2)^2 + (2D\Omega)^2}, \tag{7.18}$$

$$\Psi_G(\Omega) = -\arctan\left(\frac{2D\Omega}{1 - \Omega^2}\right) \tag{7.19}$$

Equations (7.17) are linear until $U(E_1)$ changes its value. Because of the relay characteristic (7.15) only two linear regions are obtained. The trajectory in state space changes at the crossing of the threshold from one linear region to the other. The complete solution of the equation must be assembled from two parts, which alter from switching point to switching point. The final condition of one switching phase is the initial condition of the next phase. The solution for the linear regions are as follows:

$$E_1(\tau) = -\frac{1}{2}(-1)^k + A\cos(\Omega\tau + E_{3k})$$

$$+ e^{-D\tau}\cos\left(\tau\sqrt{D^2 - 1}\right)\left(E_{1k} + \frac{1}{2}(-1)^k - A\cos E_{3k}\right)$$

$$+ \frac{e^{-D\tau}\sin\left(\tau\sqrt{D^2 - 1}\right)}{\sqrt{D^2 - 1}}$$

$$\left(E_{2k} + DE_{1k} + D\frac{1}{2}(-1)^k - A(D\cos E_{3k} - \Omega\sin E_{3k})\right)$$

$$= f(E_{1k}, E_{2k}, E_{3k}, \tau) \tag{7.20}$$

E_{1k}, E_{2k}, E_{3k} are the initial values of the kth switching instant and τ is the time after the switching. With the factor $(-1)^k$ the actual state of the controller output is signalized. k, the number of switching instants, can be even or odd—so one of the two linear regions is specified.

Amplitude A and frequency Ω are the parameters of excitation. In the following analysis, while frequency Ω is varied for studying the system, the other parameter values are kept fixed at $A = 0.4$, $D = 0.25$ and $H = 0.005$.

7.2.4 Poincaré Map

A discrete model of the system (7.17) is obtained as a Poincaré map. Though it is possible to establish a rigorous analysis of Poincaré maps for some piecewise-linear autonomous switched circuits with hysteresis [6,7], it has been shown that in many power electronics circuits, the transcendental form of the equations make it impossible to obtain the Poincaré map in closed form [8]. Since the system under consideration falls in the second category, the map has to be evaluated numerically. The numerical algorithm for obtaining the Poincaré map uses the exact analytical solution of the system equation at the linear regions:

$$E_{1(k+1)} = (-1)^k H = f\left(E_{1k}, E_{2k}, E_{3k}, \tau_k\right),$$

$$E_{2(k+1)} = g\left(E_{1k}, E_{2k}, E_{3k}, \tau_k\right), \tag{7.21}$$

$$E_{3(k+1)} = E_{3k} + \Omega\tau_k$$

Equations (7.21) represent the condition for the $(k + 1)$th switching instant, where the threshold H is reached and the trajectory changes from one linear region to the other. The time τ_k between the switching instants must be evaluated numerically to find the solution of the transcendental equation. The computation is done using the Anderson-Bjoerk method for high convergence rate of the numerical process. Thereafter the system variables at the $(k + 1)$th instant can be calculated, where g is the derivation of time of function f in (7.20). To simplify the generation of Poincaré maps using the experimental circuit, the discrete observations were made periodically at the zero crossing of the drive signal $W(\tau)$ with positive slope instead of the switching instant of the controller. To facilitate the comparison of results, the numerical evaluation of the Poincaré map was done in the same way. In this representation the discrete state variables at the nth zero crossing are E_{1n} and E_{2n}. The number n of zero crossings and k of switching instants are independent of each other. Within one period of $W(\tau)$

several switching instants may take place depending on the oscillation mode of the system. Figure 7.10 shows such a time-domain waveform of the signals $E(\tau)$ and $U(\tau)$ of a regular PWM mode. The time τ here is continuously scaled and does not start at each switching instant. If the condition $E_{3(k+1)} > E_{3n} = 2\pi n - \frac{\pi}{2} > E_{3k}$ is satisfied within the actual switching phase, a zero crossing of the drive signal will occur. Therefore, a Poincaré section is obtained using the following algorithm:

$$\Omega \tau_n = 2\pi n - \frac{\pi}{2} - E_{3k}$$

$$E_{1(n+1)} = f\left(E_{1k}, E_{2k}, E_{3k}, \tau_n\right),$$

$$E_{2(n+1)} = g\left(E_{1k}, E_{2k}, E_{3k}, \tau_n\right)$$

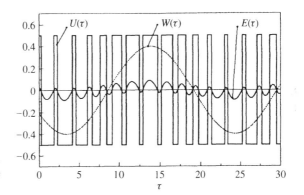

Figure 7.10 Time-domain waveforms of signals $W(\tau)$, $E(\tau)$, and $U(\tau)$ in PWM mode for $\Omega = 0.3$.

7.2.5 Circuit Realization

For obtaining experimental Poincaré maps, an electronic circuit was built. The controller was implemented using a 74HC08 AND-gate as comparator as shown in Figure 7.11. Hysteresis h of the controller follows from the positive feedback of the voltage divider resistors $100k\Omega$ and $1k\Omega$ and the linear system was implemented by an LCR-resonant circuit. ($u_m = 2.5$V, $h = 50$mV, $L = 0.185$H, $C = 1.36$nF, $R = 24k\Omega$, $f_0 = \frac{\omega_0}{2\pi} \approx 10$kHz.) Setting the right value of the damper ($D = 0.25$), the resonant circuit requires a slightly higher value of the resistor R to account for the iron losses in the inductor L. The drive signal with variable frequency was obtained from an audio sine wave generator ($f_w = \frac{\omega_w}{2\pi} \approx [13\text{kHz}, 15\text{kHz}]$, $a = 2$V). The reduction of the frequency to lower values (e.g., 60Hz) as used in some practical applications [4], is possible without loss of validity. Only the values of the components of the LCR-resonant circuit have to be changed.

7.2.6 Mode of Oscillations, Bifurcations, and Crises

At regular operation the signal $U(\tau)$ shows a PWM mode and signal $E(\tau)$ is quasiperiodic, as shown in Figure 7.10. An intersection of the torus in state space is shown in Figure 7.12(a) by the corresponding numerically obtained Poincaré map and in Figure 7.12(b) by the experimentally produced one from the electronic circuit. The scaling factors of Figure 7.12(b) for both channels correspond to the values before normalization $e(t)$ and $e'(t)$. The quasiperiodic behavior occurs due to the interference

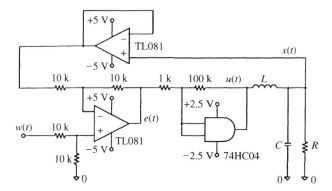

Figure 7.11 Circuit realization of the controller for the tolerance band PWM method.

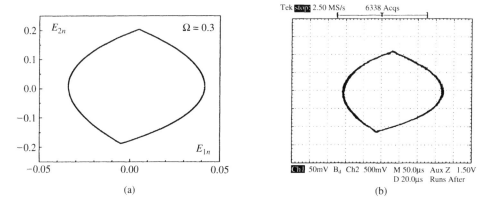

Figure 7.12 Poincaré section $E_{2n}(E_{1n})$ in quasiperiodic PWM mode at $\Omega = 0.3$ obtained (a) numerically and (b) experimentally.

of the autonomous oscillation of the unforced system (where $W(\tau) = 0$) and the reference function with frequency of the drive signal Ω. At high levels of frequency Ω or amplitude A, the signal $U(\tau)$ runs in a square-wave mode with a duty cycle of 50% and fundamental limit cycles of signal $E(\tau)$. The reference signal $W(\tau)$ dominates the behavior of oscillation. Subharmonic oscillations and chaotic behavior can be found in an intermediate region within the parameter range $\Omega \approx [1.30, 1.40]$. The bifurcation phenomena observed in this intermediate region are described in the following subsections.

Chaotic Mode

We observe a folding of the torus as Ω is increased. The first fold appears at $\Omega \approx 1.30$. The end of the intermediate region is observed at $\Omega \approx 1.40$. Within this range of frequency, fundamental and subharmonic oscillations interrupt the chaotic regions. Figure 7.13(a) shows the numerically evaluated and Figure 7.13(b) the experimentally produced Poincaré section for $\Omega = 1.399$ in chaotic mode. The scaling factors

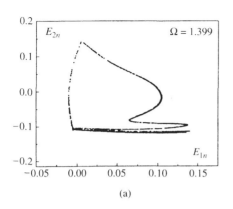 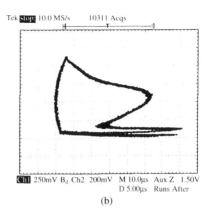

(a)   (b)

Figure 7.13 Poincaré section $E_{2n}(E_{1n})$ in chaotic mode at $\Omega = 1.399$ and
$\lambda_1 = +0.234$, obtained (a) numerically and (b) experimentally.

of Figure 7.13(b) for both channels correspond to the values before normalization $e(t)$
and $e'(t)$. The chaotic behavior is confirmed by evaluating the maximal Lyapunov
exponent, which assumes a value of $\lambda_1 = +0.234$. It has been numerically calculated
from the discrete Poincaré map at the switching point.

The phenomena in the intermediate region are shown in the numerically obtained
bifurcation diagram given in Figure 7.14. In the diagram, E_{1n} represents one component
of the Poincaré map at the zero crossing of the driving signal with positive slope. The
critical values of the parameter Ω where important bifurcation phenomena occur are
marked in the figure. In this system we observe coexisting attractors a_1 and a_2, and
separate bifurcation diagrams are given in Figure 7.14(a) and (b) to illustrate the
evolution of these attractors.

Symmetry-Breaking Bifurcation

In the system under study the Poincaré section is obtained by observing the state
variables E_{1n}, E_{2n} at the zero crossing of the sinusoidal reference signal. Since the states
can be observed either at the zero crossing with positive slope (Ψ_{n0}) or at the zero
crossing with negative slope ($\Psi_{n\pi} = \Psi_{n0} + \pi$), there is a possibility of having two complementary solutions of the system.

In a sinusoidally forced system (7.17), if the Poincaré map at the observation angle
$\Psi_{n0} = E_{3n} = 2n\pi - \frac{\pi}{2}$ is in rotation symmetry with the map at $\Psi_{n\pi}$, we call the attractor
invariant under the symmetry group Γ. In this case, the Poincaré section at $\Psi_{n\pi}$, under a
change of sign (i.e., $-E_{1n\pi}$, $-E_{2n\pi}$), coincides with the section at angle Ψ_{n0}. In other
words, the map at Ψ_{n0} coincides with the map at $\Psi_{n\pi}$ rotated by 180 degrees.

In the actual system there is no preferred direction of deflection of the system
variables. Equation (7.13) represents a relay system with a symmetrical relay characteristic, and the driving term contains no steady component. The driving term, the nonlinear relay characteristic, and the linear part are centered on the origin. Therefore, the
occurrence of attractors invariant under the group Γ can be expected.

It can be observed in the bifurcation diagrams of Figure 7.14(a) and (b) that at
some bifurcation points the attractor is replaced by a pair of coexisting attractors a_1

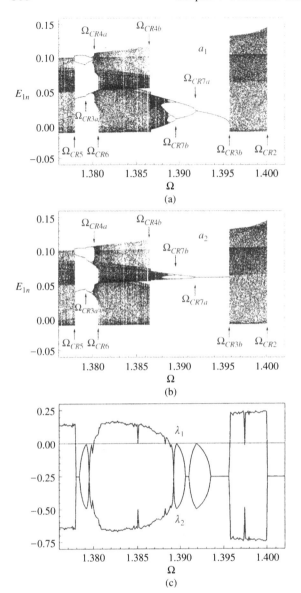

Figure 7.14 The bifurcation diagrams of the attractors (a) a_1 and (b) a_2 for the range $\Omega = [1.376, 1.402]$. (c) The Lyapunov spectra.

and a_2. Both attractors of the bifurcated system have their own basins of attraction. They are of reduced symmetry and only the set $a_1 \cup a_2$ is symmetric under the group Γ. This phenomenon is called *symmetry breaking*. An explanation of the symmetry-breaking phenomenon can be found in [9]. Such paired solutions have also been observed in the Duffing and the Lorenz systems [10].

Figure 7.15 shows the Poincaré section at the parameter value $\Omega = 1.387$ where a pair of coexisting chaotic attractors a_1 and a_2 can be seen. Figure 7.15(a) shows a_1 and a_2 when observed at Ψ_{n0} and Figure 7.15(b) shows the same attractors at $\Psi_{n\pi}$. The attractors a_1 and a_2 are not invariant under the symmetry group Γ—they do not coincide with themselves by rotating one of the maps by 180 degrees. But the attractor

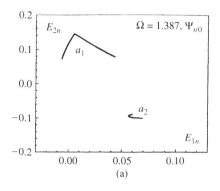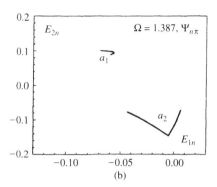

Figure 7.15 The Poincaré section for (a) the attractors a_1 and a_2 at the angle Ψ_{n0} and (b) their complementary solutions at the angle $\Psi_{n\pi}$.

a_2 at $\Psi_{n\pi}$ coincides with the attractor a_1 at Ψ_{n0} by a rotation of 180 degrees. These are the less-symmetric attractors. Only in the set $a_1 \cup a_2$ a rotation of 180 degrees will coincide and the symmetry under the group of Γ is guaranteed.

Therefore, a simple method can be established for evaluating the pair of attractors, if one of the less-symmetric attractors of the set $a_1 \cup a_2$ is known. From the known attractor, the map at the observation angles Ψ_{n0} and $\Psi_{n\pi}$ has to be obtained. By rotating the map at $\Psi_{n\pi}$ by 180 degrees, we obtain the second attractor at Ψ_{n0}. We have used this method for calculating the bifurcation diagram. Only one pass of the algorithm is required to find both coexisting solutions.

As the frequency is increased, the first symmetry breaking occurs at a pitchfork bifurcation at Ω_{CR3a}, and a pair of less-symmetric attractors of the same periodicity is born. These attractors undergo period-doubling bifurcations as Ω is increased and become chaotic.

Another symmetry-breaking phenomenon occurs at a saddle-node bifurcation at frequency Ω_{CR3b}. As the frequency decreases and passes this critical point from above, two stable period-1 attractors are born. The attractors undergo period-doubling cascade until Ω_{CR4b}.

Figure 7.14(c) shows the corresponding Lyapunov spectra, where the positive λ_1 in some parameter ranges confirm the existence of chaos. It may be noted that for this system $\lambda_1 = -2D - \lambda_2$, which follows from the constant contraction rate in the linear regions. Further, we can see that at the symmetry-breaking bifurcation points (Ω_{CR3a} and Ω_{CR3b}) the corresponding Lyapunov exponent λ_1 is zero. This confirms the loss of stability at these points.

Merging Crisis

At Ω_{CR4a} and Ω_{CR4b}, the two less-symmetrical attractors merge and there is a sudden expansion of the chaotic attractor. This is a *merging crisis* [11]. In the range $\Omega > \Omega_{CR4b}$, the two attractors coexist and at $\Omega = \Omega_{CR4b}$, both of them collide with their own basin boundaries. For Ω slightly smaller than Ω_{CR4b}, an orbit typically spends long stretches of time moving chaotically in the region of one of the old attractors, after which it abruptly switches to the region of the other attractor, intermittently switching between the two. Figure 7.16 shows the merged attractor at frequency $\Omega = 1.386$. The long stretches can be recognized by the accumulation of data points at the region of the old attractors a_1 and a_2. The above-mentioned considerations are also valid for Ω_{CR4a} by increasing the frequency.

Interior Crisis

A sudden increase in the size of a chaotic attractor occurs when the periodic orbit with which the chaotic attractor collides is in the interior of its basin and is called an *interior crisis* [11]. In the system under consideration, a stable period-3 orbit is born by a saddle-node bifurcation at Ω_{CR5}. An unstable period-3 orbit is also born at this frequency. The unstable branches collide at frequency Ω_{CR6} with the chaotic attractor, which results in interior crisis.

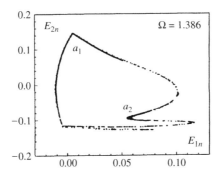

Figure 7.16 The merged attractors at $\Omega = 1.386$.

Saddle-Node Bifurcation and Square-Wave Mode

The *square-wave mode* is an effect caused by a *saddle-node bifurcation*. An exact critical amplitude bound A_{CR} can be evaluated analytically, at which a saddle-node bifurcation takes place. Above this bound, a pair of stable and unstable symmetrical limit cycles with frequency of the drive signal $W(\tau)$ exists, and the external forcing is synchronized to the system. At this synchronized steady state the signal $U(\tau)$ shows a symmetric periodic rectangular function with a duty cycle of 50%. Therefore, the main question to be answered is the following: If frequency Ω of the drive signal is fixed, which minimum value of $A > A_{CR}$ is required to obtain a stable fundamental symmetric limit cycle of the signal $E(\tau)$?

An analytical method due to Tsypkin [12] can be used for the characterization of the relay system's limit cycles [12]. The existence of limit cycles is predicted by the fulfilment of algebraic conditions at the switching point. The positive rising edge of $U(\tau)$ is defined at $\tau_0 = 0$ without loss of generality. Therefore, switching actions take place periodically at time $\tau_k = \frac{k\pi}{\Omega}$. Figure 7.17 shows the timing chart of the relevant signals of the stable limit cycle in square-wave mode. With these assumptions, the switching and switching direction conditions of the controller at the switching instant have to be satisfied:

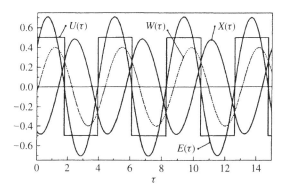

Figure 7.17 Time-domain waveforms of signals $E(\tau)$, $U(\tau)$, $W(\tau)$, and $X(\tau)$ in square-wave mode.

$$\tilde{E}_1(\tau_k) = (-1)^k H$$

$$\tilde{E}_2(\tau_{2k}) = -\tilde{E}_2(\tau_{2k+1}) > 0$$

where $\sim$ characterizes periodicity. The absolute value of the slope $\left|\tilde{E}_2(\tau_k)\right|$ at symmetric limit cycles is constant at each switching instant. The sign alternates from switching point to switching point.

The term $\tilde{E}_1(\tau)$ is derived partly from $\tilde{X}(\tau)$ and partly from external action $W(\tau)$ (7.14). For the switching instant $k = 0$ at $\tau_0 = 0$ and all even values of k follow:

$$\tilde{E}_1(\tau_0) = W(\tau_0) - \tilde{X}(\tau_0) = +H, \quad E_2(\tau_0) = W'(\tau_0) - \tilde{X}'(\tau_0) > 0 \tag{7.22}$$

For the switching instant at τ_0 it is necessary that the value of the slope $\tilde{E}_2(\tau_0)$ is positive to result in a positive slope of the signal $\tilde{U}(\tau)$.

Further, the symmetric rectangular function $\tilde{U}(\tau)$ at the controller output can be expanded in a Fourier series as:

$$\tilde{U}(\tau) = \frac{2}{\pi} \sum_{i=1,3}^{\infty} \frac{\sin[i\Omega\tau]}{i}$$

The linear system filters each harmonic part of the Fourier series by the ith absolute value of the transfer function $G(i\Omega)$ (7.18). Therefore, at the output of the linear system the signal $\tilde{X}(\tau)$ can be assembled by harmonic parts with amplification and phase displacement (7.19).

$$\tilde{X}(\tau) = \frac{2}{\pi} \sum_{i=1,3}^{\infty} G(i\Omega) \frac{\sin[i\Omega\tau + \Psi_G(i\Omega)]}{i} \tag{7.23}$$

Through this formula the result at τ_0 and all even numbers of switching instants for the signal $\tilde{X}(\tau)$ and its slope $\tilde{X}(\tau)$ are obtained as:

$$\tilde{X}(\tau_0) = \tilde{X}_0 = \frac{2}{\pi}\sum_{i=1,3}^{\infty}\frac{v(i\Omega)}{i} = \frac{D\sin\left[\frac{\sqrt{1-D^2}}{\Omega}\pi\right] - \sqrt{1-D^2}\sinh\left[\frac{D}{\Omega}\pi\right]}{2\sqrt{1-D^2}\left(\cosh\left[\frac{D}{\Omega}\pi\right] + \cos\left[\frac{\sqrt{1-D^2}}{\Omega}\pi\right]\right)}$$

$$\tilde{X}'(\tau_0) = \tilde{X}_0' = \frac{2\Omega}{\pi}\sum_{i=1,3}^{\infty}u(i\Omega) = -\frac{\sin\left[\frac{\sqrt{1-D^2}}{\Omega}\pi\right]}{2\sqrt{1-D^2}\left(\cosh\left[\frac{D}{\Omega}\pi\right] + \cos\left[\frac{\sqrt{1-D^2}}{\Omega}\pi\right]\right)}$$

where $u(i\Omega)$ represents the real part and $v(i\Omega)$ the imaginary part of the ith harmonic of the complex transfer function of the linear system. Furthermore, the closed-form presentation has been obtained from its discrete Laplace transform, as shown in [12].

Now the conditions (7.22) have to be satisfied by using $\tilde{X}_0$, $\tilde{X}_0'$, $W(\tau)$ from (7.14) at $\tau = \tau_0$ and its derivative:

$$W'(\tau_0) = -A\Omega\sin[\Omega\tau_0 + \Phi]$$

For a given Ω the amplitude A and the phase angle Φ have to be solved. If equation (7.22) can be satisfied, there are two solutions in respect to the phase angle of the drive signal $E_{3S}(\tau_0) = \pi - E_{3U}(\tau_0) = \Omega\tau_0 + \Phi$.

Both solutions together are the effect of the saddle-node bifurcation, and only one of these solutions is stable. Stability of such predicted limit cycles is clarified by a special linearization at the bias point, as suggested by Tsypkin [12]. An application of the simplified Nyquist criterion to sampling systems results in the final inequalities for the two cases:

$$\tilde{X}_0' < 0 : A > A_{CR} = \left|\tilde{X}_0 + H\right|$$

$$\tilde{X}_0' > 0 : A > A_{CR} = \sqrt{\left(\tilde{X}_0 + H\right)^2 + \left(\frac{\Omega\sin\left[\frac{\sqrt{1-D^2}}{\Omega}\pi\right] - \tilde{X}_0'}{2\Omega\sqrt{1-D^2}\left(\cosh\left[\frac{D}{\Omega}\pi\right] - \cos\left[\frac{\sqrt{1-D^2}}{\Omega}\pi\right]\right)}\right)^2}$$

These are the *Tsypkin bounds* of the given system. The derived relation for A_{CR} gives an explicit bound for the amplitude. Above this bound stable symmetrical limit cycles exist.

The functions $\tilde{X}_0$ and $\tilde{X}_0'$ depend on frequency Ω and damper D. If the slope $\tilde{X}_0'$ is positive, the amplitude A must be of a higher value to finally result in a positive slope $\tilde{E}_2(\tau_0)$; see also (7.22). This is necessary to get the assumed positive rising edge of the signal $\tilde{U}(\tau)$ at τ_0. The function $\tilde{X}_0'$ alters its sign several times depending on the value of Ω.

It should be noted that the attractors a_1 and a_2 in Figure 7.14 with period number one show no square-wave mode of the signal $\tilde{U}(\tau)$. Here one rising edge and more than one falling edge of signal $\tilde{U}(\tau)$ occurs within one period of the driving signal. Otherwise it would be an inconsistency in the evaluation of the Tsypkin's bound.

If the amplitude A_{CR} is known and the frequency Ω is of interest, a numerical solution of the Tsypkin bound has to be performed. For the selected critical amplitude $A_{CR} = 0.4$, a frequency $\Omega_{CR1} = 1.3343745$ is obtained for the saddle-node bifurcation.

The stable attractor caused by this bifurcation is termed a_3. Figure 7.18 shows the evolution of the attractor a_3 and its unstable branch in a bifurcation diagram. The value of E_{1n} of the attractor a_3 at the zero crossing $\Omega \tau_n = 2\pi n - \frac{\pi}{2} - \Phi$ of the drive signal $W(\tau)$ has been analytically evaluated using formula (7.23), where $E_{1n} = -\tilde{X}(\tau_n)$. Note that the saddle-node bifurcation takes place at a frequency inside the intermediate region. Above the critical value Ω_{CR1}, a_3 coexists with other attractors resulting from the folded torus discussed in Subsection 7.2.6 and these can be accessed through variation of initial conditions. The coexisting attractor(s) can be chaotic, subharmonic, or fundamental harmonic, as shown in Figure 7.18. Each attractor has its own basin of attraction. This is also valid for the chaotic attractor in Figure 7.13, which has a frequency above the critical value Ω_{CR1}, so that the attractor a_3 coexists. Beyond the intermediate region at frequency values above Ω_{CR2}, the limit cycle caused by the saddle-node bifurcation becomes the unique solution.

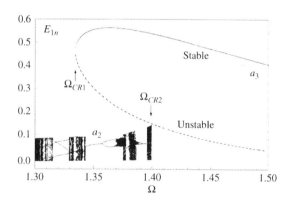

Figure 7.18 Bifurcation diagram for the attractor a_3 and its unstable branch.

Boundary Crisis

As the parameter Ω is raised, the distance between the chaotic attractor and its basin boundary decreases until at a critical $\Omega = \Omega_{CR2}$ value they touch. This is a *boundary crisis* [11]. At this point the attractor also touches an unstable periodic orbit that was on the basin boundary before crisis. For $\Omega > \Omega_{CR2}$ the chaotic attractor no longer exists, and is replaced by a chaotic transient. Above Ω_{CR2} only square-wave mode exists with a fundamental limit cycle. The numerical value of the critical frequency obtained by simulation is $\Omega_{CR2} \approx 1.40$. Because of the very long time constant of the chaotic transient, the value can be only approximately specified.

In Figure 7.19, the situation is shown in the Poincaré surface of section. The chaotic attractor is embedded in its basin of attraction. The saddle fixed point lies on the basin boundary which is built from the stable manifolds. One of the unstable manifolds runs into the stable fixed point, which represents the attractor a_3. The other manifold leads to the chaotic attractor. At the critical value of frequency Ω_{CR1}, the two fixed points are born and diverge with the increase of frequency as shown in Figure 7.18. Thereby the unstable saddle moves in the direction of the chaotic attractor. As the saddle fixed point makes contact with the chaotic attractor at $\Omega = \Omega_{CR2}$, the attractor is destroyed.

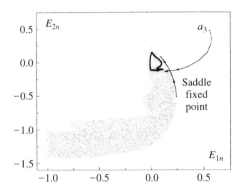

Figure 7.19 The chaotic attractor, its basin of attraction (the gray area), and the stable and unstable manifolds at the saddle fixed point for $\Omega = 1.385$.

Period-Doubling Bifurcations

The standard period-doubling route to chaos can also be found in some parameter ranges. In Figure 7.14(a), a backward *period-doubling bifurcation* occurs at frequency Ω_{CR7a} and Ω_{CR7b}. The period doubles when Ω is decreased.

7.2.7 Conclusions

In this section we have demonstrated the chaotic behavior and the bifurcation scenario of an inverter under tolerance band PWM control. We have shown that the chaotic behavior in this system results from a folded torus in the state space. If the frequency of the external reference signal is varied, the chaotic behavior can be found in an intermediate range between the regular quasiperiodic PWM mode and the square-wave mode of the controller. If chaos is to be avoided for certain industrial applications, this intermediate range of frequency specifies the forbidden zone in the parameter space.

With the analytically evaluated Tsypkin bound the frequency range can be calculated in which a regular PWM operation is reliable. The choice of control strategy (type of feedback), the transfer function G of the linear plant and the hysteresis H of the controller are the important parameters in the design of such a system.

Further, the coexistence of attractors within the intermediate range has been confirmed, which results in a rich variety of bifurcation phenomena including the occurrence of saddle-node bifurcations, boundary crises, merging crises, interior crises, symmetry breaking, and period-doubling bifurcations. We have presented a complete overview of the nonlinear phenomena that can occur in the tolerance band PWM technique—a knowledge necessary for reliable design and operation of such converters and for application of chaos in obtaining flexible switchings.

REFERENCES

[1] N. Mohan, T. M. Undeland, and W. P. Robbins, *Power Electronics*. New York: Wiley, 1995.

[2] B. K. Bose, *Modern Power Electronics*. New York: IEEE Press, 1992.

[3] B. K. Bose, *Power Electronics and Variable Frequency Drives*. New York: IEEE Press, 1997.

[4] O. Wasynczuk, S. D. Sudhoff, T. D. Tran, D. H. Clayton, and H. J. Hegner, A voltage control strategy for current-regulated pwm inverters, *IEEE Trans. on Power Electronics*, vol. 11, no. 1, pp. 7–15, 1996.

[5] A. Magauer and S. Banerjee, Bifurcations and chaos in the tolerance band PWM technique, *IEEE Trans. on Circuits and Systems—I*, Vol. 47, No 2, pp. 254–259, 2000.

[6] T. Saito, An approach toward higher dimensional hysteresis chaos generators, *IEEE Trans. on Circuits and Systems—I*, vol. 37, no. 3, pp. 399–409, 1990.

[7] T. Saito, Reality of chaos in four-dimensional hysteresis circuits, *IEEE Trans. on Circuits and Systems—I*, vol. 38, no. 12, pp. 1517–1524, 1991.

[8] D. C. Hamill, J. H. B. Deane, and D. J. Jeffries, Modelling of chaos dc-dc converters by iterated nonlinear mappings, *IEEE Trans. on Power Electronics*, vol. 7, no. 1, pp. 25–36, 1992.

[9] H. Troger and A. Steindl, *Nonlinear Stability and Bifurcation Theory*. New York: Springer, 1991.

[10] J. Argyris, G. Faust, and M. Haase, *Die Erforschung des Chaos*. Braunschweig, Germany: Vieweg, 1994.

[11] E. Ott, *Chaos in Dynamical Systems*. Cambridge, UK: Cambridge University Press, 1993.

[12] J. S. Tsypkin, *Relay Control Systems*. Cambridge, UK: Cambridge University Press, 1984.

7.3 NONLINEAR NOISE EFFECTS IN POWER CONVERTERS

Philip T. Krein
Pallab Midya

7.3.1 Introduction

Noise has nonlinear effects on PWM controllers used in power electronics. In this section, we examine the basis of nonlinear effects of noise in open-loop and closed-loop power electronic converters. The nature of switching noise and random noise effects are discussed. Much of the open loop treatment is taken from [1], while the basis of closed-loop noise analysis can be found in [2,3].

Nonlinear noise effects in power electronics ultimately derive from two characteristics:

1. Power converter control processes exhibit time asymmetry both in the generation of noise and in the sensitivity to noise effects.

2. In closed-loop systems, noise effects propagate from cycle to cycle, with end results that strongly depend on the feedback configuration.

Time asymmetry is represented in part by switching noise—parasitic ringing and similar effects associated with switch action. By their nature, switching noise effects depend on the time proximity between a switching action and subsequent events. Time asymmetry is also present in the PWM process. In open-loop PWM, a latch is normally used to prevent multiple pulsing. This eliminates chattering and helps to avoid high-frequency chaos. On the other hand, noise signals imposed before the latch is set have a much different impact from those after the latch is set. In closed-loop converters, time asymmetry has complicated effects, as noise alters the output and is fed back to the input.

7.3.2 Discussion of Switching Noise

Noise in a power converter can result either from internal generation or from external interference. For external interference, it is reasonable to model noise as a random external source. Internal interference appears as a result of switch action. Capacitive and inductive parasitics of the switching devices, layouts, and other circuit elements exchange energy during and after a switch acts. In most cases, the rates of change are high enough to produce *radiated* electromagnetic interference (EMI) as well as *conducted* EMI in the circuit. The unusual feature of switching noise is its timing connection with switch action. In a typical situation, high-frequency oscillation is excited by a switch, then damps out quickly.

Figure 7.20 shows a typical switching noise profile—the current measured in a diode and excited by diode reverse recovery. We might attempt to characterize the noise with the oscillation frequency and the damping. Notice that we need to know the time relative to the moment of switch action to determine the noise signal value. The parameters are very sensitive to layout and component parasitics, so good results demand either very detailed simulation or careful experiments. Switching noise is nonergodic, and also differs from other noise sources in terms of autocorrelation. For a switching period T, and a signal $y(t)$ associated with switching, the switching-period autocorrelation $r(t)$ is

$$r(t) = \int_0^t y(s)y(s - T)ds \qquad (7.24)$$

and will have a high value whatever the oscillation frequency and damping. Random noise shows zero correlation for any nonzero delay.

Nonergodocity limits the applicability of conventional statistical approaches to switching noise. Analytical methods are left to future work. However, some important general statements can be made:

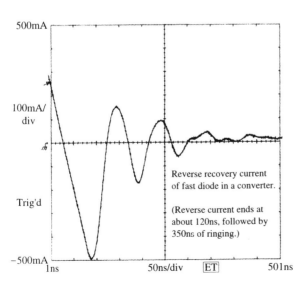

Reverse recovery current of fast diode in a converter.

(Reverse current ends at about 120ns, followed by 350ns of ringing.)

Figure 7.20 Diode reverse recovery waveform, showing switching noise.

- Switching noise effects are large when switch commands follow each other in quick succession. For instance, the effects will be larger at a very low duty ratio than at duty ratios of about 50%, since noise might alter the timing of subsequent switch controls.
- Switching noise injects a ringing frequency each time a switch acts. This is likely to have a substantial influence on chaotic converter controls.
- Switching noise effects in feedback controls can be reduced through careful sample timing. If data samples are taken just prior to switch action, the samples are relatively unlikely to be disturbed.

It might be possible to model switching noise with a duty-ratio-dependent random noise generator. This would capture the first of the above statements—a situation likely to be encountered during dynamic control or when the system is an inverter.

7.3.3 External Noise Effects in Open-Loop Converters

External Noise Action

For PWM control in a power electronic system, an analog signal is compared to a sawtooth to generate switching functions. The signal-to-noise properties of PWM information and its transmission have been examined in detail in the communications literature [4,5]. In a power converter, the switch gate drive is a very-low-impedance source, so the *transmission channel* between the PWM controller and the switch is nearly immune to noise effects. Instead, noise injected at the PWM input is the concern—an issue not addressed in communications. Noise at the PWM input alters the switching times, and therefore converter behavior. A general PWM process is illustrated in Figure 7.21.

The analog control signal is compared to a sawtooth, then a latch prevents multiple transitions per cycle. The usual practice is to set the latch with a clock and reset the latch with the comparator. The set (turn-on) and reset (turn-off) operations can be exchanged with minor modifications to the analysis. When noise is injected, the turn-off time t_{off} becomes a random variable. In the case of a fixed switching period T, and resetting the time origin at the beginning of each cycle, the duty ratio can be defined as

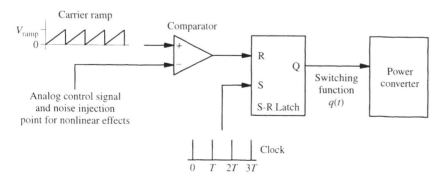

Figure 7.21 Model of the PWM process.

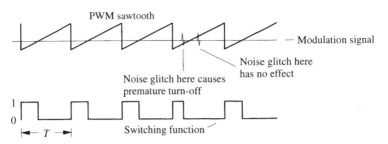

Figure 7.22 PWM waveforms when a noise pulse alters an open-loop modulator.

the ratio t_{off}/T. The duty ratio therefore becomes a random variable proportional to the turn-off time. Figure 7.22 shows the switching function and associated waveforms when a single pulse of injected noise disturbs an open-loop PWM process.

Background of Analysis

Here we introduce a general procedure for the analysis of random noise effects in a PWM converter. Closed-form expressions of external noise effects can be generated if Gaussian statistics are assumed, so we make this assumption here with minimal loss of generality. Let us begin with a band-limited Gaussian random process, of low amplitude, represented as a noise signal $Y(t)$. The bandwidth limitation reflects the fact that low-pass filtering is explicit in most converters, or implicit because of the comparator's bandwidth. Low amplitude means that when $Y(t)$ interferes with the PWM input, the pulse-width modulator will not be driven into saturation. As in conventional analytical procedures from communication theory, the noise process is assumed to be stationary, meaning that the statistical properties are time invariant. Stationary Gaussian processes are known to be ergodic [6], meaning that when the random variable is mapped into a function of time, the statistics are preserved. Let $Z(t)$ represent the derivative of the noise signal. The signals $Y(t)$ and $Z(t)$ are uncorrelated in general. They have probability density functions $p_Y(y)$ and $p_Z(z)$, each with zero mean and with standard deviations σ_Y and σ_Z, respectively. Statistical independence of a noise signal and its derivative is true for Gaussian noise over extended time intervals, but it is well known that low-pass filtering creates short-term correlation. In practice, the bandwidth of the noise derivative is substantially wider than the bandwidth of the underlying noise, so the assumption of independence between $Y(t)$ and $Z(t)$ has little effect on accuracy of the analysis here.

The duty ratio, with a value between 0 and 1, will be represented as a random variable $X(t)$ with value x and probability density $p_X(x)$. The nominal duty ratio (with no noise present) will be denoted D_0. Let the PWM carrier be an ideal linear sawtooth with normalized amplitude of 1 and ramp slope $1/T$. The open-loop duty ratio is determined by comparing a signal to the sawtooth. Figure 7.23 [1] shows P_Y and P_Z, and also p_Y and p_Z for Gaussian white noise, band-limited through a two-pole rolloff at 10MHz, and then passed through a two-pole low-pass filter at 100kHz. The respective standard deviations at the output are $\sigma_Y = 0.1$V and $\sigma_Z = 61900$V/s. The standard deviation of the band-limited noise prior to the 100kHz filter was 1V.

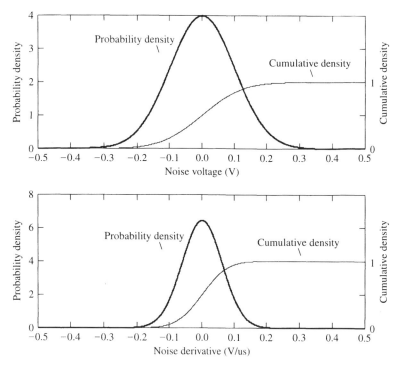

Figure 7.23 Cumulative probability density and probability density for band-limited Gaussian noise.

The latch in the PWM process biases the effects: a noise pulse that arrives prior to switching could be large enough to alter the timing, while a noise pulse that occurs after switching will have no effect. The switch will reset at a specific duty ratio X if and only if both of the following conditions are met:

1. The analog signal becomes equal to the sawtooth at duty ratio X (call this event A).
2. The switch did not turn off earlier in the cycle (call this event B).

To quantify the combination, we need conditional probability. Given two random events, A and B, the conditional probability $P(A \mid B)$ is the probability of random event A occurring given that event B has occurred. According to Bayes' Law [6,7], $P(A \mid B)$ is the ratio of the probability of both A and B occurring together to the probability of B occurring alone,

$$P(A \mid B) = \frac{P(A \wedge B)}{P(B)} \tag{7.25}$$

The probability of both A and B occurring in an incremental interval $(x, x + \Delta x)$ is the probability that the signal crossing occurs prior to $x + \Delta x$ less the probability that the crossing occurs prior to x, so $P(A \wedge B) = P_X(x + \Delta x) - P_X(x)$. The probability that B occurs alone $P(B)$ is the probability of no crossing occurring in the interval $(0, x)$, so $P(B) = 1 - P_X(x)$. The conditional probability $P(A|B)$ from (7.25) becomes

$$P(A \mid B) = \frac{P_X(x + \Delta x) - P_X(x)}{1 - P_X(x)} \tag{7.26}$$

Summary of Assumptions

Five assumptions have been identified so far:

1. External interference can be modeled as a noise process and noise derivative process that are Gaussian, stationary, and ergodic (conventional assumptions from communication theory).
2. The noise and its derivative each have bandwidth limits, which differ in general. (This is a property of a real power converter.)
3. The noise and its derivative are independent and uncorrelated (as discussed above).
4. The PWM process is ideal: the sawtooth is linear, and the set and reset processes take negligible time.
5. The noise level is low enough not to drive the PWM process to saturation limits.

In addition, since the statistics are Gaussian, it will be reasonable to terminate indefinite integrals after several standard deviations, such as 6σ, without loss of accuracy. These assumptions support closed-form analytical results.

Probability Density Function of Switch Timing

With Gaussian noise, a linear modulation method would result in a Gaussian distribution for the duty ratio density function $p_X(x)$. In normalized form with a sawtooth peak value of 1, we would expect $p_X(x) = p_Y(y)$. The cumulative probability density function of the duty ratio $P_X(x)$ is what we seek. A second equation is needed to eliminate $P(A|B)$ and support a solution. It is necessary to examine the noise effects in some depth to provide this second equation. The details can be found in [1]. To summarize, Figure 7.24 shows an expanded view of a narrow time interval. The switch has not yet reset, and the signal is higher than the sawtooth at the beginning of the interval. A reset will occur in this interval only if the control signal (plus noise) drops down to the sawtooth. Only certain values of noise and noise derivative Y and Z will result in a switch reset in a given interval, and the actual value of $P(A|B)$ is obtained by integrating

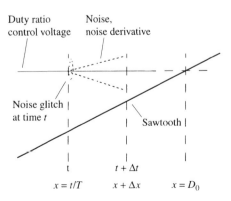

Figure 7.24 Expanded short-time interval near the turn-off time.

the appropriate probability density functions over these values. For example, in Figure 7.24, if the derivative causes the sum of signal and noise to rise more quickly than the PWM ramp, there will be no intersection. Thus values of the variable Z greater than the normalized ramp slope $1/T$ are not consistent with a switch reset in the interval of interest. A minimum value of Z can be assigned several standard deviations away. Limits on Y can be identified as well. The combination of conditions on y and z eliminates all but a triangular area in the yz plane. Any values that meet the constraints would be expected to cause a reset between x and Δx. Therefore, integration of $p_Y p_Z$ over a triangular area will give another expression for $P(A|B)$.

Integration of $p_Y p_Z$ involves the normal probability density function $p_Z(z)$, and has a closed-form solution. From [1], the ultimate result for the duty ratio distribution is

$$P_X(x) = 1 - [1 - P_Y(x - D_0)]^M \qquad (7.27)$$

Here M is a dimensionless *slope index*, a function of the ramp rate and the standard deviation of the noise derivative σ_Z given by

$$M = P_Z(1/T) + \frac{\sigma_Z T}{\sqrt{2\pi}} \exp\left(\frac{-1}{2\sigma_Z^2 T^2}\right) \qquad (7.28)$$

In effect, M is a measure of the high-frequency content of the noise normalized to the ramp rate. In the limiting case of zero noise derivative, $M = 1$, but any nonzero noise has nonzero derivative, and yields $M > 1$. Since the value of M will be greater than 1 for any noise level, the distribution function for the duty ratio, $P_X(x)$, always differs from the Gaussian represented by $P_Y(y)$. This distortion is a fundamentally nonlinear effect of the latch. It will be small for low levels of noise derivative, but for values of M significantly larger than 1, the duty ratio distribution is quite different from a Gaussian.

Implications of the Non-Gaussian Duty Ratio Distribution with Latch

Expression (7.27) represents a shift in the average duty ratio as well as a change in the distribution. Since the noise is ergodic, the time average value of duty ratio matches the expected value of the random variable X, given by

$$\langle D \rangle = E(X) = \int_{-\infty}^{\infty} x p_X(x) dx \qquad (7.29)$$

The probability density function is the derivative of (7.27) with respect to x. The duty ratio shift depends both on the value of σ_Y and on M. If the nonlinear effect of M were not present, there would be no shift in the average, and σ_X would match σ_Y. The actual average shift can be represented in normalized form relative to σ_Y,

$$\text{shift} = \frac{D_0 - \int_{-\infty}^{\infty} x p_X(x) dx}{\sigma_Y} \qquad (7.30)$$

This expression does not give rise to a closed-form expression, but evaluation at several values of σ_Y shows that the results of (7.30) depend only on M over a few orders of magnitude of σ_Y. The integral can be numerically evaluated, and results are given in Figure 7.25 for a case with $D_0 = 0.5$. The figure shows that the ac variation represented by σ_Y generates a dc shift as M increases above 1. When M reaches 2, the shift is more than half the standard deviation σ_Y. When M reaches 4, the shift is a full standard deviation. The behavior is approximately logarithmic. These results, and others from [1], can be summarized as follows:

- When the latch is controlled with a set-reset process, as M increases, average duty ratio decreases. This means that the distribution $p_X(x)$ is shifted to the left relative to $p_Y(y)$. (If the latch is controlled with a reset-set process, the average duty ratio will increase, in the same relative amount.)
- As M increases, the standard deviation decreases. The distribution becomes narrower and more peaked. The decrease is more than 10% for $M = 1.4$.
- There is skewness in the duty ratio distribution. Since $p_Y(y)$ has zero skew, this confirms that time asymmetry in the PWM process produces asymmetric noise effects.

All these effects are fundamentally nonlinear: any level of noise will shift the mean duty ratio and introduce skew. The effect is small when $M \approx 1$, but can be significant at realistic noise levels. The shift in average duty ratio can be explained qualitatively. As noise bandwidth increases, the number of uncorrelated noise samples increases. Thus, there is a higher chance of a reset event. Because of the latch, noise has no effect after a reset occurs, so it is more likely that noise will cause premature reset than delayed reset. Thus the average duty ratio should drop in the presence of noise.

In dc/dc converters, noise changes the average output dc value. In an inverter, the shift creates a dc imbalance, and can lead to saturation in the magnetics. Noise effects are biased in this sense, as a fundamental attribute of the PWM process. Interestingly, the average shift can provide a direct measure of the noise statistics in the PWM process. Since average shifts can be very precisely measured, Figure 7.25 can be used by a designer to infer noise standard deviation.

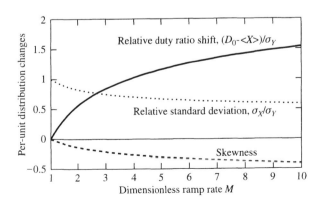

Figure 7.25 Duty ratio shift as a fraction of σ_Y for various values of M.

7.3.4 External Noise Effects in Closed-Loop DC/DC Converters

The Nature of Closed-Loop Noise Effects

Duty ratio shift, standard deviation compression, and skew are all open loop effects. When control loops are closed, noise effects propagate over time and introduce additional nonlinear behavior. Noise can alias with switch action to produce subharmonics. Voltage-mode and current-mode control techniques show significant differences in noise performance in dc/dc converters. Conventional small signal measures such as audio susceptibility [8,9,10] or control-to-output response [11] do not capture nonlinear noise effects. For example, designers have reported significant differences between noise susceptibility of current-mode and voltage-mode control [12]. Average current mode control [13] was developed in part to reduce the observed effects. In this section, we consider ways to evaluate closed-loop effects. Power spectra are used to show how nonlinear effects alter noise behavior.

In addition to the previous assumptions, for the purposes of closed-loop analysis, we will assume that $M \approx 1$. If $M > 1$, additional distortion occurs, as discussed above. Linear ripple of the state variables is also assumed. With these assumptions, a discrete-time approximation provides a way to follow noise propagation over time. None of these assumptions is unconventional.

Switching noise can complicate the issues further. In a realistic converter with feedback control, some switching noise will be fed back after a time delay. This tends to spread switching noise more uniformly through the interval, and broadens its effects to resemble those of random noise. Chaos is one example of a situation in which the closed-loop duty ratio becomes unpredictable.

The Closed-Loop Process and System Model

A discrete-time representation is proposed for closed-loop noise effects. In the nth switching cycle, in the absence of noise, the duty ratio can be written with a nominal value D_n. With noise, the duty ratio is a random variable X_n, and we can define a duty ratio error (equivalent to a timing error) in each cycle as $E_n = X_n - D_n$. In the analysis here, a conventional analog loop will be examined. Control loops can exhibit a *memory* effect that propagates noise effects from a given switching period to subsequent periods—perhaps with detrimental results. The feedback signal can be an inductor current or capacitor voltage, corresponding to current-mode or voltage-mode feedback, respectively.

For generality, consider the sawtooth as falling from a reference value V_{ref} with slope m_{RAMP}. In current-mode control, the current has a triangular waveform. These waveforms have slopes m_{ON} when the active switch is on and m_{OFF} when the switch is off (both slopes are used in their absolute value sense). The switch is reset when the current crosses a reference value. In peak current-mode control, the reference current is a dc quantity from an outer voltage loop, and $m_{\text{RAMP}} = 0$ as no control sawtooth is present. Figure 7.26 shows the waveforms in current-mode control with an external ramp. In the figure, a noise spike causes premature switch action in the second cycle shown. This alters the duty ratio in subsequent cycles; about three cycles are required for the return to steady state. For voltage-mode control, the output voltage is used for

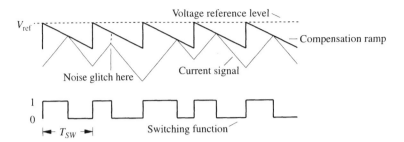

Figure 7.26 PWM waveforms when a noise glitch occurs.

feedback. The output voltage is approximately a constant dc quantity because of output filtering, and the PWM ramp has a slope opposite that in conventional current-mode control.

Time Domain Noise Analysis

During the nth switching cycle, the crossing point of the current feedback and sawtooth signals causes the switch transition to occur at time $t_{\text{off(n)}} = nT + D_n T$, where $D_n = D(nT)$ is the nominal duty ratio provided from the control loop during the nth cycle. A useful way to analyze the effects of noise is to consider the current at the moment of turn-off, then form recursion relations to determine how duty ratio evolves from cycle to cycle. When no noise is present, the current at the turn-off time is

$$i(t_{\text{off(n)}}) = V_{\text{ref}} - D_n T m_{\text{RAMP}} \tag{7.31}$$

Let Y_n represent the noise process on the current or current sensor signal at the switch turn-off point in the nth cycle. The duty ratio is the random variable X_n, with

$$i(t_{\text{off(n)}}) + Y_n = V_{\text{ref}} - X_n T m_{\text{RAMP}} \tag{7.32}$$

The difference between currents in cycle $n + 1$ and cycle n can be simplified to yield a recursive equation for the duty ratio random variable,

$$X_{n+1} = K X_n + \frac{Y_{n+1} - Y_n}{T(m_{\text{RAMP}} + m_{\text{ON}})} + \frac{m_{\text{OFF}}}{m_{\text{RAMP}} + m_{\text{ON}}} \tag{7.33}$$

where $K = (m_{\text{RAMP}} - m_{\text{OFF}})/(m_{\text{RAMP}} + m_{\text{ON}})$. The duty ratio error in the nth cycle, E_n, follows the recursion

$$E_{n+1} = K E_n + \frac{Y_{n+1} - Y_n}{T(m_{\text{RAMP}} + m_{\text{ON}})} \tag{7.34}$$

The duty ratio error E_n can be expressed in terms of the noise process Y injected in each previous cycle, and

$$E_{n+1} = \frac{Y_{n+1} + (K - 1)(Y_n + K Y_{n-1} + K^2 Y_{n-2} + \ldots)}{T(m_{\text{RAMP}} + m_{\text{ON}})} \tag{7.35}$$

The noise processes in the individual cycles $Y_n, Y_{n+1}, Y_{n+2}, \ldots$ arc statistically independent, each with standard deviation σ_Y. The standard deviation σ_X of the duty ratio X is the same as that of the duty ratio error E. Since the variance of a weighted sum of independent random variables equals the weighted sum of the variances (7.35) can be used to find σ_X through reduction to a geometric series in the slope ratio K. The variance sum becomes

$$\sigma_X^2 = \frac{\sigma_Y^2 + \sigma_Y^2(K-1)^2 + \sigma_Y^2(K-1)^2 K^2 + \sigma_Y^2(K-1)^2 K^4 + \dots}{T^2(m_{\text{RAMP}} + m_{\text{ON}})^2} \qquad (7.36)$$

The geometric series sum yields a simple form, and the standard deviation is

$$\sigma_X = \frac{\sigma_Y \sqrt{2}}{T(m_{\text{RAMP}} + m_{\text{ON}})\sqrt{1+K}} \qquad (7.37)$$

Now consider various control scenarios. Under voltage mode control in a dc/dc converter, the feedback signal is a nearly constant voltage. The slopes m_{ON} and m_{OFF} are small, which gives $K \approx 1$, and

$$\sigma_X = \frac{\sigma_Y}{V_{\text{RAMP}}} \qquad (7.38)$$

(or $\sigma_X = \sigma_Y$ for a 1V sawtooth). This is the ideal duty ratio variation that would be expected in the open-loop case when $M \approx 1$. In short, the duty ratio has the same statistical behavior as it would in the open-loop case when $K = 1$.

Under current-mode control, when the control ramp is matched to the off-state current rate of change (this is the ramp needed to support duty ratio values right up to 100%), $K = 0$, and the standard deviation σ_X is $\sqrt{2}$ times the value in the open-loop case. For current-mode control with no ramp, K reduces to a function of duty ratio, $-D_0/(1 - D_0)$. In this case, (7.37) becomes

$$\sigma_X = \frac{\sigma_Y}{T(m_{\text{RAMP}} + m_{\text{ON}})} \sqrt{\frac{2(1 - D_0)}{1 - 2D_0}} \qquad (7.39)$$

In the limit as $D_0 \to 1/2$, the standard deviation given by (7.39) increases without bound. This is not quite realistic. Consider the case of nominal 50% duty ratio, but complete loss of control (or perhaps chaotic operation) such that the duty ratio random variable X becomes uniformly distributed between 0 and 1. The uniform distribution over this interval has standard deviation of $\sqrt{1/12} = 0.289$. Therefore, we should truncate the result from (7.39) at this value, representing complete duty ratio randomness. Typically, this bound is not reached until $D_0 = 0.49$ or above.

Confirmation

Monte Carlo simulation is a useful way to confirm noise effects. In Figure 7.27, the duty ratio standard deviation from (7.37) is compared to statistical results from a Monte Carlo simulation of a buck converter switching at 10kHz, with a 250μH inductor, a stiff load with 0.5Ω in parallel with 100μF, and 10V input. In the figure, the input value of σ_Y has been set to 0.05V, corresponding to duty ratio standard deviation of 1% since a nominal 5V ramp is being used in the voltage-mode case. Each plotted point is the standard deviation for 3000 sample cycles, simulated for 10% nominal duty ratio steps and three types of controls. The duty ratio standard deviation is shown in dB relative to the scaled σ_Y value.

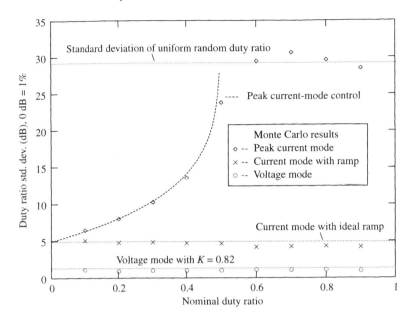

Figure 7.27 Normalized σ_X vs. D_0 for three dc/dc control methods.

For voltage-mode control, the value of K is a bit less than 1, since the slopes are not truly zero. In the voltage-mode converter used for Figure 7.27, the actual value of K is 0.82, and the predicted noise gain from (7.37) is 1.32dB. For current-mode control with a matched ramp, K is exactly zero. The predicted noise gain is 4.95dB for this buck converter. In both the voltage-mode and ramped-current-mode cases, the results are not a function of the nominal duty ratio. In the figure, the dotted and dashed lines show analytical results. Symbols show Monte Carlo results for various values of D_0. The Monte Carlo results match (7.37), and confirm that σ_X is substantially higher under current-mode control with a ramp than under voltage-mode control.

For peak current-mode control, the performance degrades as the nominal duty ratio increases. The mechanism is this: as D_0 approaches 50%, even slight noise will push the operation into the well-known subharmonic instability. At 50% and above, the standard deviation from Monte Carlo simulation is near the 0.289 level expected for a uniform random value of duty ratio. Subject to this ceiling on standard deviation, the theory matches the Monte Carlo results very closely.

Figure 7.27 confirms an amplification effect on noise under current-mode control. One implication is that the observed output ripple will be higher in the current-mode-control cases than in the voltage-mode case. Will this increased ripple be spread across the output spectrum, or will it appear at specific frequency values? A frequency domain analysis is needed to determine the actual spectral effects on converter output.

Frequency Domain Analysis

The energy spectral density of a random process can be obtained from the auto-correlation function of the process. The recursion relation developed in (7.33) suggests that there is some correlation between duty ratio variation in a cycle and duty ratio variation in previous cycles. This is true even though the noise values themselves are independent. A discrete spectrum can be identified through further analysis of (7.34) to obtain the autocorrelation in the discrete time domain; then a Fourier transform gives the energy spectral density in the frequency domain [6]. While these results use a discrete approximation to the actual continuous spectrum, they provide useful closed-form expressions for the converter duty ratio spectrum, and hence output ripple implications of noise.

The autocorrelation function $R_X(\tau)$ is the mean value of the product of a signal with a delayed version of itself. The function can be written as

$$R_X(\tau) = \langle x(t)x(t + \tau) \rangle \tag{7.40}$$

where the angle brackets indicate a time average. The autocorrelation of the error E_n with zero delay gives the signal RMS value, which is the same as the variance. From (7.37), this gives

$$R_E(0) = \frac{2}{1 + K} \frac{\sigma_Y^2}{T^2(m_{RAMP} + m_{ON})^2} = \sigma_X^2 \tag{7.41}$$

For a time difference of k cycles, the correlation of the duty ratio error E can be obtained from (7.35) as

$$R_E(kT) = \frac{\langle [Y_n + (K - 1)(Y_{n-1} + KY_{n-2} + \ldots)]}{T^2(m_{RAMP} + m_{ON})^2}$$
$$\times [Y_{n+k} + (K - 1)(Y_{n+k-1} + KY_{n+k-2} + \ldots)] \rangle \tag{7.42}$$

Since the noise terms Y are uncorrelated, the average values of all cross terms will be zero, and (7.42) can be reduced to a geometric series form. The series sum yields a closed-form expression for the correlation k cycles apart,

$$R_E(kT) = K^{k-1} \frac{K - 1}{K + 1} \frac{\sigma_Y^2}{T^2(m_{RAMP} + m_{ON})^2} \tag{7.43}$$

The power spectral density $S_E(f)$ is obtained by taking a Fourier transform of (7.43). This yields a close approximation to the continuous spectral density for values below the switching frequency,

$$S_E(f) = \frac{2\sigma_Y^2(1 - \cos\theta)}{(1 + K^2 - 2K\cos\theta)T(m_{RAMP} + m_{ON})^2} \tag{7.44}$$

where f is an arbitrary frequency and $\theta = 2\pi fT$. Details of the derivation can be found in [3]. The maximum of this spectrum occurs at half the switching frequency. The energy spectrum at this frequency is

$$S_E(f_{switch}/2) = \frac{4\sigma_Y^2}{(1 + K)^2 T(m_{RAMP} + m_{ON})^2} \tag{7.45}$$

Figure 7.28 shows the noise spectrum of the switching function (and hence the switch action prior to filtering) for three control methods. For Monte Carlo comparisons, converter simulations were performed over 5000 cycles for each control case and several values of nominal duty ratio. Each output switching function was analyzed to

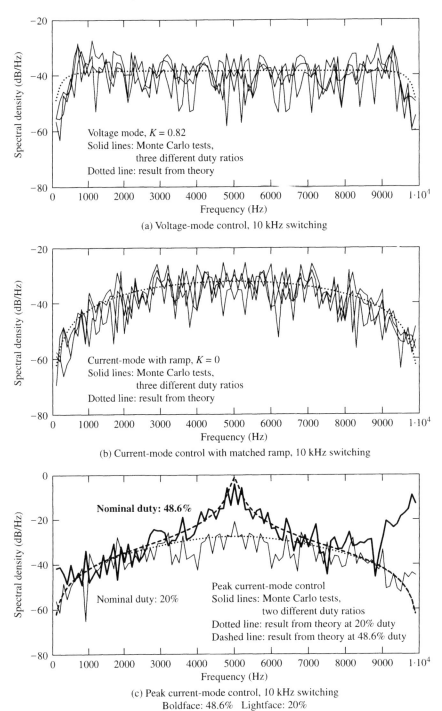

(a) Voltage-mode control, 10 kHz switching

(b) Current-mode control with matched ramp, 10 kHz switching

(c) Peak current-mode control, 10 kHz switching
Boldface: 48.6%   Lightface: 20%

Figure 7.28 Spectral density of noise, normalized to σ_y^2.

determine its Fourier coefficients at intervals of 100Hz. The random number generation yields input noise with a flat spectrum over the frequency range of interest.

The Monte Carlo simulation results are compared to (7.44) in the figure. Trace (a) corresponds to voltage-mode control with $K = 0.82$. (The theoretical spectrum would be flat for $K = 1$. The curving effects at 0 and at f_{switch} are caused by the lower value of K.) The input noise corresponds to 0.01 duty ratio standard deviation, and spectral density results are plotted in dB relative to the scaled σ_Y^2 value. The effects are independent of duty ratio. In trace (a), results are overlaid for nominal duty ratios of 20%, 50%, and 70%. Trace (b) has $K = 0$ and corresponds to current-mode control with optimal ramp. The spectrum is altered noticeably from the flat spectrum of the injected white noise. The peak at $f_{switch}/2$ is 6.2dB higher than in the voltage-mode trace. The Monte Carlo results are again plotted for 20%, 50%, and 70% duty ratios, and no duty ratio effect appears.

The bottom trace (c) corresponds to peak current-mode control. In this case, duty ratio alters the value of K. Results for nominal duty ratio of 0.2 (or $K = -0.25$) and $D_0 = 0.486$ (or $K = -0.946$) are shown. The peak at $f_{switch}/2$ reaches -1dB/Hz for the highest duty ratio. In the near-50% case, the actual spectrum shows an increase as the test frequency approaches the switching frequency as well. This represents a harmonic of the $f_{switch}/2$ component. It is not predicted by (7.44) because noise was assumed to be uncorrelated. When the strong subharmonic occurs, noise will alias against the subharmonic in addition to the switching frequency.

The results in Figure 7.28 confirm the analytical method used to obtain (7.44). They also suggest that very substantial subharmonic distortion will occur when noise is present in peak current-mode control. Even when the duty ratio is well away from the 50% level for subharmonic instability, peak current mode still shows subharmonic behavior in the presence of noise. These results have been experimentally confirmed [3].

7.3.5 Summary

Noise interacts with the PWM converter control process to alter noise distributions and spectra, as a fundamental property of PWM as it is applied in power electronic systems. In closed-loop systems, current-mode control methods in particular propagate noise effects into future switching periods. This aliases noise energy into subharmonic frequencies, and produces a noise amplification effect. The results provide specific quantitative confirmation of qualitative observations that current-mode controls are noisier than voltage-mode controls. Analytical formulas for the noise effects have been presented. The duty ratio variation was used as the basis for analysis. This provides a valid comparison of noise effects without regard to topology or filter implementation.

In peak current-mode control, the nonlinear PWM noise process produces significant noise amplification and unusual spectral effects. Even well away from subharmonic instability, injected noise produces subharmonic distortion in the switching function and thus in the output ripple. If the subharmonic instability point is exceeded, Monte Carlo results suggest that injected noise causes the duty ratio to become almost uniformly random.

Biases and subharmonic spectral effects of noise are fundamentally nonlinear, and occur even when noise levels are low. It is possible to develop tractable analytical methods for these nonlinear effects. The results here can be used to evaluate potential noise sensitivity at the design stage, and can help with analysis of unexpected behavior if it appears. The techniques introduced here provide methods for evaluating noise properties of open-loop and closed-loop converters.

REFERENCES

[1] P. Midya and P. T. Krein, Noise properties of pulse-width modulated power converters, Open-loop effects, *IEEE Trans. on Power Electronics*, vol. 15, 1134–1143, 2000.

[2] P. Midya and P. T. Krein, Closed-loop noise properties of pulse-width modulated power converters, *Rec. 1995 IEEE Power Electronics Specialists' Conf.*, pp. 15–21.

[3] P. Midya, Nonlinear control and operation of dc to dc switching power converters, PhD diss., University of Illinois at Urbana-Champaign, April 1995.

[4] S. Haykin, *Communication Systems*. New Delhi: Wiley Eastern Limited, 1985.

[5] R. E. Steele, *Delta Modulation Systems*. New York: Wiley, 1975.

[6] A. B. Carlson, *Communication Systems*. New York: McGraw-Hill, 1986.

[7] P. Z. Peebles, *Communication System Principles*. Reading, MA: Addison-Wesley, 1976.

[8] R. B. Ridley, B. H. Cho, and F. C. Y. Lee, Analysis and interpretation of loop gains of multiloop-controlled switching regulators, *IEEE Trans. on Power Electronics*, vol. 3, pp. 489–497, 1988.

[9] R. B. Ridley, A new, continuous-time model for current-mode control, *IEEE Trans. on Power Electronics*, vol. 6, pp. 271–280, 1991.

[10] D. M. Sable, R. B. Ridley, and B. H. Cho, Comparison of performance of single-loop and current-injection control for pwm converters that operate in both continuous and discontinuous modes of operation, *IEEE Trans. on Power Electronics*, vol. 7, pp. 136–142, 1992.

[11] R. Tymerski and D. Li, State-space models of current programmed pulsewidth-modulated converters, *IEEE Trans. on Power Electronics*, vol. 8, pp. 588–595, 1993.

[12] R. D. Middlebrook, Modelling current-programmed regulators, notes of short course presented at *IEEE Applied Power Electronics Conference*, San Diego, March 1987.

[13] L. Dixon, Average current-mode control of switching power supplies, in *Unitrode Power Supply Design Seminar Handbook*. Lexington, MA: Unitrode, 1990.

7.4 NONLINEAR PHENOMENA IN THE CURRENT CONTROL OF INDUCTION MOTORS

István Nagy
Zoltán Sütő

The three-phase full-bridge converter with six controlled switches is one of the most frequently applied static power processing units in power electronics and in electric drives. This converter with a closed-loop *hysteresis ac current controller* (HCC) can be applied as an ac/dc rectifier supplying power to the dc side, a dc/ac inverter delivering active and/or reactive power to the ac side, a reactive power compensator, an active filter suppressing the harmonics in line current or backing the voltage against short dips, or as a combination of these functions. The main differences among the above applications are the methods for calculating the reference ac current space vector for the HCC. After determining the reference ac current, the various applications have many common features. In most cases the reference current is obtained from a closed-loop controller, but our study concentrates on the steady-state behaviors of the system, when the reference current can be viewed as an open-loop parameter or as a constant. This section is concerned with the example of a dc/ac inverter supplying power to an induction motor and controlled by a somewhat sophisticated HCC with two concentric tolerance band circles. The main objective of this study is to show the most characteristic bifurcation phenomena in order to prepare the ground for finding ways of locking the trajectory to periodic orbit or inducing chaos under certain conditions. For tuning the process, a suitable parameter could be the radius r_{out} (or r_{in}) of the outer (or inner) tolerance band circle of the HCC.

7.4.1 System Model

The block diagram of the system studied is shown in Figure 7.29. It consists of a three-phase voltage source converter, a simple circuit modeling the ac side, a reference frame transformation, a comparator, and the hysteresis ac current controller.

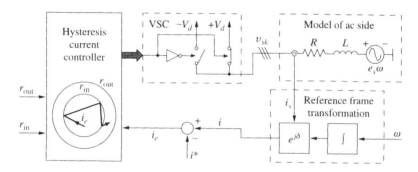

Figure 7.29 Voltage source converter with its hysteresis ac current controller.

Voltage Source Converter (VSC)

The VSC is modeled by a simple three-phase bridge connection of six controlled lossless switches. The dc side is assumed to be a constant and ideal voltage source with voltage $2V_d$. In each switching state, three switches are on and the other three are off, connecting each of the three ac legs of the VSC either to the positive dc voltage $+V_d$ or to the negative dc voltage $-V_d$. This results in $2^3 = 8$ possible switching states, with six active ones and two inactive[1] ones. The voltage space vectors v_{sk} of the VSC interfaced with the ac side are [1]

$$v_{sk} = \begin{cases} 2V_d \frac{2}{3} e^{j(k-1)\pi/3} & \text{for } k = 1, 2, \ldots 6 \quad \text{(active states)} \\ 0 & \text{for } k = 0, 7 \qquad \text{(inactive states)} \end{cases} \tag{7.47}$$

All other voltage and current vectors used in this chapter are three-phase space vectors. Space vectors with suffix s are written in natural or *stationary reference frame* (SRF). A space vector y_s is defined by the instantaneous values of the phase variables y_{sa}, y_{sb} and

---

1 The inactive states generate the same zero output voltages, but their switching patterns are different. Both zero vectors are used in optimizing the number of switchings.

y_{sc} as follows: $\boldsymbol{y}_s = 2(y_{sa} + \bar{a}y_{sb} + \bar{a}^2 y_{sc})/3$, where $\bar{a} = e^{j2\pi/3}$. For example, active voltage vector $\boldsymbol{v}_{s1}$ can be obtained by connecting phase a to the positive and phase b and c to the negative dc rail, that is, $\boldsymbol{v}_{s1} = 2(V_d - \bar{a}V_d - \bar{a}^2 V_d)/3 = 2(2V_d)/3$.

AC Side

The three-phase ac terminals of the VSC are connected through series $L - R$ components to voltage space vector $\boldsymbol{e}_s$ rotating with angular speed ω [1]. This simple circuit can model either an induction motor or the ac mains. Using this model the voltage balance equation for the ac side is

$$\boldsymbol{v}_{sk} = \boldsymbol{e}_s + L\frac{d\boldsymbol{i}_s}{dt} + R\boldsymbol{i}_s \tag{7.48}$$

There are some benefits of using a *rotating reference frame* (RRF) over SRF. The transformation of any space vector from SRF to RRF is as follows: $\boldsymbol{y} = \boldsymbol{y}_s e^{-j\delta}$ where $\boldsymbol{y}$ is the vector in RRF and $\delta = \delta_0 + \omega t$ is the angle of the real axis of RRF in SRF. In RRF all the vectors are stationary, except for the six active output voltage vectors $\{\boldsymbol{v}_k : k = 1, 2 \ldots 6\}$, which rotate clockwise with speed $\omega = d\delta/dt$. Space vectors without suffix s are in RRF. Rewriting the voltage balance equation in RRF, the following relation is obtained

$$\boldsymbol{v}_k = \boldsymbol{e} + L\frac{d\boldsymbol{i}}{dt} + \bar{Z}\boldsymbol{i} \tag{7.49}$$

where $\bar{Z} = R + j\omega L$. The reference ac current is denoted by $\boldsymbol{i}_s^*$, and may be thought of as being produced by a fictitious output voltage $\boldsymbol{v}_s^*$ that satisfies

$$\boldsymbol{v}_s^* = \boldsymbol{e}_s + L\frac{d\boldsymbol{i}_s^*}{dt} + R\boldsymbol{i}_s^* \tag{7.50}$$

This equation in RRF becomes

$$\boldsymbol{v}^* = \boldsymbol{e} + L\frac{d\boldsymbol{i}^*}{dt} + \bar{Z}\boldsymbol{i}^* = \boldsymbol{e} + \bar{Z}\boldsymbol{i}^* \tag{7.51}$$

where $\boldsymbol{i}^*$ is the reference signal of the HCC, and is usually the output signal of another outer control loop (e.g., torque and/or speed loop of an induction motor). Assuming constant load for the motor and balanced sinusoidal reference current with constant frequency, the vectors $\boldsymbol{i}_s^*$, $\boldsymbol{v}_s^*$ and $\boldsymbol{e}_s$ rotate with constant angular speed ω, so they are constant and stationary in RRF, that is, $d\boldsymbol{i}^*/dt = 0$, which explains the second equality in (7.51).

Subtracting (7.51) from (7.49) the state equation for the error current $\boldsymbol{i}_e = \boldsymbol{i} - \boldsymbol{i}^*$ is

$$\boldsymbol{d}_k = L\frac{d\boldsymbol{i}_e}{dt} = \boldsymbol{v}_k - \boldsymbol{v}^* - \bar{Z}\boldsymbol{i}_e \tag{7.52}$$

where the *direction vector* $\boldsymbol{d}_k$ is proportional to the derivative of the error current vector. It is well approximated by the difference vector $(\boldsymbol{v}_k - \boldsymbol{v}^*)$, as in most cases $|\boldsymbol{v}_k - \boldsymbol{v}^*| \gg |\bar{Z}\boldsymbol{i}_e|$. Equation (7.52) is a nonautonomous second-order system with known input $\boldsymbol{v}_k$. The time function of $\boldsymbol{v}_k$ is discontinuous whenever the trajectory of the state variable reaches the periphery of the tolerance band; however, it is continuous between two switchings. The solution of (7.52) can be explicitly written in switching state n between impact points $i_{e,n}$ and $i_{e,n+1}$

$$i_e(t) = e^{-\overline{\Omega}(t-t_n)}i_{e,n} + \frac{v_k}{R}\left(1 - e^{-\text{Re}\{\overline{\Omega}\}(t-t_n)}\right) - \frac{v^*}{\overline{Z}}\left(1 - e^{-\overline{\Omega}(t-t_n)}\right) \qquad (7.53)$$

where $t_n \leq t < t_{n+1}$, $\overline{\Omega} = (R/L) + j\omega$ and $i_e(t_{n+1}) = i_{e,n+1}$. Here t_n and t_{n+1} are the time at the beginning and at the end of switching state n, respectively. An iteration technique based on the Newton-Raphson method is applied to calculate the next impact point of $i_e(t)$ on the tolerance band, that is, the next switching time instant t_{n+1}, with a given accuracy [1,2].

Hysteresis Current Control (HCC)

First, the ac current i_s is transformed into RRF revolving with angular frequency ω. In RRF the reference current i^* is a stationary space vector. Whenever the error current i_e reaches the periphery of the *tolerance band* (TB) a new converter voltage v_k is switched. The ac voltage of VSC tailored from the eight v_k voltages by the HCC has to approximate the reference voltage v^*. The best approximation is obtained by applying the *adjacency principle*, that is, when voltages v_k are selected from the adjacent vector set consisting of the two neighboring or closest active vectors to v^* and the two zero vectors. The adjacent set is v_2, v_3 and v_0, v_7 in Figure 7.30(a).

Voltage v_k is selected using the *regular switching pattern* from the adjacent set. Regular switching, well known from the literature [3,4], can be understood from Figure 7.30(a). The TB is divided into three equal parts: arc(AB), arc(BC), and arc(CA), determined by the positions of the active voltage vectors of the adjacent set. In RRF both the active vectors v_k and the points A, B, and C are rotating in the clockwise direction while v^* is stationary. When v^* enters in the control region bounded by v_3 and v_4, the points A, B, and C jump suddenly back by $60°$ in the counterclockwise direction. Depending on the impact point of the error current trajectory along the periphery of TB, in part arc(AB) and arc(BC) and arc(CA) the direction vectors d_2 and $d_0 = d_7$ and

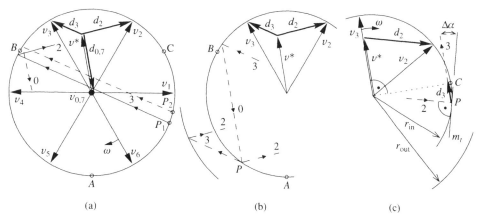

(a)   (b)   (c)

Figure 7.30 (a) Regular switching. (b) Elimination of double commutation using a second concentric tolerance band circle. (c) Elimination of fast switching.

d_3 (Figure 7.30(a)) must be selected, respectively. In part arc(BC) there is a choice between the two inactive states. This selection depends on the previous switching state and is aimed at avoiding simultaneous commutations in two legs of the VSC. Direction vector d_0 is selected if the suffix of the previous active state is odd (e.g., d_3), while in case of even suffix (e.g., after d_2) d_7 will be selected.

Although regular switching is a well-established technique, it has some shortcomings. The most important one, occurring rather frequently, is *double commutation.* Changing v_k from an inactive to an active state and keeping the suffix even or odd (e.g., from v_0 to v_2 at point P in Figure 7.30(b) or from v_7 to v_3) causes double commutation to occur. This increases the switching losses of the VSC. It can be eliminated by using a second concentric tolerance band circle with larger radius. Choosing the alternative voltage vector still from the adjacent set of v_k and permitting the error current to leave the inner circle, the double commutation is avoided. For example, in Figure 7.30(b) at impact point P, direction d_3 pointing out of the inner TB is selected in place of d_2. The other shortcoming of regular switching is that it generates occasionally nonuniform switching trains with periods of slow and *fast switchings.* Fast switchings occur when v^* is almost aligned with one of the active voltages and the radius of TB pointing to the impact point of i_e on the periphery of TB is almost $90°$ out of phase with the direction of v^*; that is, the impact point P of i_e is near the point C (Figure 7.30(c)). Close to point C on its clockwise side, d_3 is selected by regular switching after d_2 while on the counterclockwise side d_7 is selected. Both direction vectors are almost tangential to the TB, and a sequence of fast switchings develops. This situation can be avoided by applying the outer TB again. Let $\Delta\alpha$ denote the angle between the tangential line m_t of the circle at the impact point P of i_e and the direction vector d_k (e.g., d_3 in Figure 7.30(c)). Whenever $\Delta\alpha$ is less than a given minimum value, the outer TB is used (e.g., d_7 is selected). Note that fast switching can occur on the outer TB, but its probability is much smaller, as the number of impacts is much smaller. When neither fast switchings nor double commutation are predicted, the switching algorithm works in exactly the same way as regular switching does.

Poincaré Map

Unfortunately, an expression for the Poincaré map in closed form cannot be derived for this system. Numerical computation is the way to explore its behaviors. The Poincaré map is defined as the stroboscopic sampling of the error current vector i_e. To get rid of the transient phase, the first several points from $m = 1$ to $m = M_{tr}$ are dropped. M_{tr} is the estimated number of sampling points taken in transient state. The sampling period T_s is selected to be $T_p/6$, that is, a sixth of the period T_p of i_s^*, because of the symmetry of the six active and two inactive voltages of v_k (Figure 7.30(a)). The ac side of the VSC faces the same kind of v_k adjacent voltage set in every $T_p/6$ period. Only the switching patterns in the VSC generating the output voltages are different. The sampling points are taken at $\{i_e(mT_s + \varphi_0) : m > M_{tr}\}$.[2] Using an arbitrarily selected component of the Poincaré samples, a bifurcation diagram can be plotted. It can in fact be two diagrams in our case, showing the real or imaginary (absolute or phase) com-

---

2 Unless otherwise stated, $\varphi_0 = 0$ is used. Other values of φ_0 are usually chosen to reach a better appearance of the bifurcation diagrams by shifting the sample points along the system trajectory.

ponents of the sample points of i_e calculated as a function of one system parameter or variable. Qualitatively, there is no difference between bifurcation diagrams calculated for the real and imaginary (absolute and phase) components of the error current, therefore only one of them is used.

7.4.2 Nonlinear Phenomena

In this system the response can be either periodic, subharmonic, or chaotic, and it is rich in various kinds of bifurcations. In general, varying a parameter leads to a new steady state after an initial transient process, where the lengths of transients could be very different even if the change of parameter is very small. It is quite difficult to decide algorithmically whether the observed system state is chaotic or chaotic transient (see Sections 4.2, 5.2). In the latter case, the simulation time should be increased to reach the real final state, which could be periodic or even chaotic.

Sensitive Dependence on Initial Condition

One of the principal properties of chaotic systems is their sensitive dependence on initial conditions [5,6]. Consider the trajectory departing from point P_1 in Figure 7.30(a). If the initial condition is changed by the small value $\Delta i_e = \overline{P_2 P_1}$, the new trajectory follows a path that closely parallels the first one at the beginning. At the next switching instant, however, the two neighboring impact points in our example will belong to different segments of the TB circle (i.e., to the two sides of point B in Figure 7.30(a)). After these impact points, they continue their journey along two completely different orbits in the direction d_2 (continuous line) and in the direction d_0 (dotted line) respectively. This scenario highlights the strong nonlinear behavior of our system and its sensitive dependence on initial conditions. The information about the initial condition is quickly lost, implying an apparent randomness in the response and long-term unpredictability of the process.

Period Doubling Bifurcation

A bifurcation diagram is shown in Figure 7.31(a), presenting $\text{Re}\{i_e(mT_s)\}$ versus r_{out}/r_{in} at rated operation of the motor, while r_{in} is kept constant ($r_{in} = 0.05$). The system is chaotic in wide ranges of parameter values. On the other hand, there are infinitely many windows representing periodic states, where different kinds of period-doubling bifurcation can be identified. In a very wide periodic window, period doubling occurs around parameter value $r_{out}/r_{in} \cong 1.5322$. (Period-doubling bifurcation has been reported in [7,8] at $r_{in} = 0.2$). Increasing r_{out}/r_{in}, the length of the period of the motion is doubled only once, and after the second-order subharmonic state chaotic movement is abruptly generated.

Enlarging a not-so-wide window in Figure 7.31(a), a cascade of period-doubling bifurcations is presented in Figure 7.31(b). Proceeding from $r_{out}/r_{in} = 1.2207$ to the left, the length of the period of the motion is doubled at successively closer critical values of parameter r_{out}/r_{in}. In Figure 7.32 are plotted periodic, second-order subharmonic, and fourth-order subharmonic trajectories corresponding to Figure 7.31(b), where the periodic trajectory repeats itself every $T_p/6$ seconds. The corresponding switching instants are shifted relative to each other within one period in Figures 7.32(b) and (c), and finally the trajectory becomes chaotic and never repeats itself in Figure 7.33(a). The Lyapunov

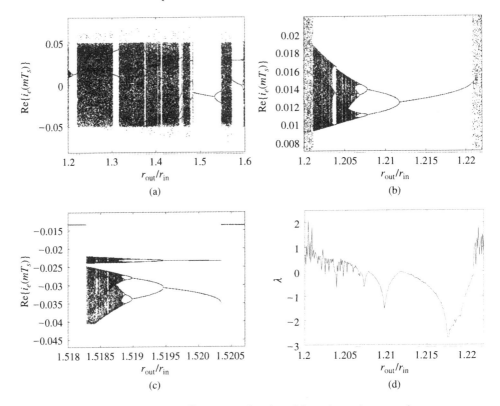

Figure 7.31 (a) Bifurcation diagram as a function of the ratio $r_{\text{out}}/r_{\text{in}}$ at rated operation of the motor. (b) Period-doubling cascade. (c) Coexisting period-doubling cascade and a periodic state. (d) Variation of the maximal Lyapunov exponent corresponding to (b).

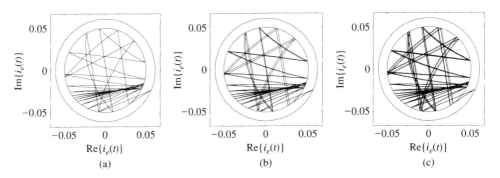

Figure 7.32 Error current trajectories in the period-doubling cascade. (a) Periodic state with period $T_p/6$ $r_{\text{out}}/r_{\text{in}} = 1.2207$. (b) Second-order subharmonic state with period $2T_p/6$, $r_{\text{out}}/r_{\text{in}} = 1.21$. (c) Fourth-order subharmonic state with period $4T_p/6$, $r_{\text{out}}/r_{\text{in}} = 1.207$.

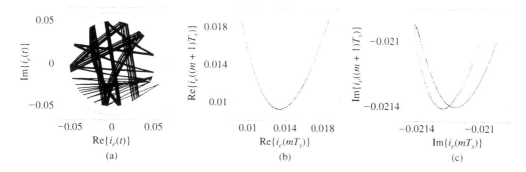

Figure 7.33 Chaotic state in the period-doubling cascade, $r_{out}/r_{in} = 1.2012$. (a) Error current trajectory near the periodic path in Figure 7.32 (a). (b) Return map for the real component of $i_e(mT_s)$. (c) Return map for the imaginary component of $i_e(mT_s)$.

exponent is calculated for the period-doubling cascade region in Figure 7.31(d); λ becomes zero at the period-doubling points. The chaotic trajectory (Figure 7.33(a)) developing after the cascade of period doubling also stays in a narrow segment of the state space, near the periodic and subharmonic paths in Figure 7.32. Return maps for the real and imaginary part of the error current samples $i_e(mT_s)$ are shown for this chaotic state in Figures 7.33(b) and (c). The period-doubling bifurcations occur in quite a smooth way. Decreasing r_{out}/r_{in}, the periods of the error-current trajectories are doubled, but the sequences of converter states remain periodic with period T_p in the periodic, subharmonic, and chaotic cases.

Intermittency

The period-doubling cascade in Figure 7.31(b) is embedded into a wide chaotic region, where the trajectories wander practically throughout the TB. The transition into these wide chaotic movements is abrupt, both on the left and on the right side of the diagram. Starting from the periodic state, and increasing r_{out}/r_{in} from 1.2207 to 1.2208, a critical parameter value is reached and in place of the periodic movement a chaotic one is suddenly generated (Figure 7.34(a)). This is a different kind of chaotic state from what we have seen in Figure 7.33. After several hundreds of periods the trajectory almost completely blacks out the inner TB. At first the trajectory stays in the neighborhood of the periodic orbit, but the distance between them increases. The switching pattern remains unchanged but the switching instants are shifting. Finally, at a critical state change of the VSC (see the neighborhood of point B in Figure 7.30(a)) the trajectory is continued in a completely different orbit from the one used in the periodic state (as with the case explained under the heading Sensitive Dependence on Initial Conditions). Unlike in the chaotic state in Figure 7.33, now the periodicity of converter states is also lost. However, the movement occasionally returns to the periodic path at apparently random time instants, and stays in its neighborhood for an unforeseeable interval. Approaching the critical value of r_{out}/r_{in}, the time spent near the periodic path is increased, and finally a periodic state develops from the critical value of r_{out}/r_{in}. Around the other end of the bifurcation diagram in Figure 7.31(b), the system state switches back and forth between two chaotic states (Figure 7.34(b)). The trajectory either is restricted to a narrow area in TB or it wanders in the whole TB. The Poincaré map for the latter situation is shown in Figure 7.34(c).

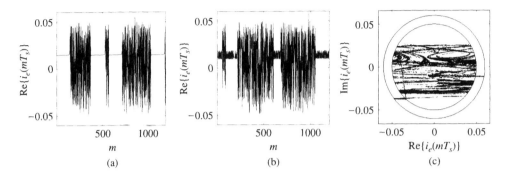

Figure 7.34 Chaotic states with intermittent behaviors; m is the number of iteration steps. (a) $r_{out}/r_{in} = 1.2208$. (b) $r_{out}/r_{in} = 1.2011$. (c) Poincaré section for (b).

Coexisting Attractors

The coexisting system states at the same parameter setting can further complicate the system behavior. For example, in a quite narrow segment between $r_{out}/r_{in} \cong 1.5183$ and 1.52035, system states different from the periodic one could develop, as seen in Figure 7.31(c). The attractor in Figure 7.31(c) cannot be seen in Figure 7.31(a). Increasing r_{out}/r_{in}, the periodic solution is reached at about 1.491 in Figure 7.31(a). If we keep increasing the control parameter r_{out}/r_{in} in small steps, the periodic state is maintained up to the value $r_{out}/r_{in} \cong 1.5322$, where a period-doubling bifurcation occurs (Figure 7.31(a)). However, the system can exhibit a state different from the one shown in Figure 7.31(a) by starting the operation from an appropriate basin of attraction (e.g., at $r_{out}/r_{in} = 1.52$). The bifurcation diagram for this other state is shown in Figure 7.31(c). A second-order subharmonic state (period-2 orbit) develops at and around $r_{out}/r_{in} = 1.52$. Decreasing r_{out}/r_{in}, a period-doubling cascade and finally a chaotic state develops. At both ends of the diagram in Figure 7.31(c), the original periodic state is suddenly restored; at the left side the chaotic state disappears, while at the right side the second-order subharmonic state disappears. This multifaceted behavior is quite usual for nonlinear systems, since the states developed can depend on the initial conditions.

7.4.3 Numerical Values

The parameters used in the calculations are for a medium-size *induction motor* (IM) rated about 10kW. Using a normalized synchronous speed of $\omega = 1$ and an impedance normalization derived from the rated stator voltage and current, the normalized (or per-unit) mutual inductance $L_m = 3$, stator and rotor leakage inductances $L_{sl} \approx L_{rl} = 0.1$, and stator and rotor winding resistances $R_r \approx R_s = 0.03$ are obtained. The magnitude of the reference stator current i^* and the magnitude of the voltage vector e are set to unity. The resistance and inductance of the model in Figure 7.29 are the stator winding resistance $R = 0.03$ and the transient inductance of the motor $L = L_s - L_m^2/L_r$

$\cong 0.2$ where $L_s = L_m + L_{sl}$, $L_r = L_m + L_{rl}$ are the stator and rotor inductances, respectively. Keeping in mind that the transient time constant $T = L/R$ is much longer than the average time between two consecutive switch operations, the switching transients do not affect the rotor flux Ψ_r. Assuming Ψ_r = constant and ω = constant, the voltage behind the transient inductance $e = j\omega L_m \Psi_r / L_r$ is constant as well. The simple model for the IM in Figure 7.29 is well justified. The values $\Psi_r = L_r / L_m$ and $e = j$ are used. Using the basic concept of field-oriented control of IM, the angle between e and i^* is $\varphi = \sin^{-1}(\Psi_r / L_m) \cong 20°$. The reference voltage v^* can then be obtained from (7.51). The magnitude of the active voltage vectors v_k of the VSC is selected to be $(2/\sqrt{3})v^*$. For the radius of the inner tolerance band circle, $r_{in} = 0.05$ is used, and the angle for predicting fast switching is set at $\Delta \alpha_{min} = 8°$.

7.4.4 Conclusions

The ac/dc voltage source converter with its hysteresis current controller can exhibit periodic, subharmonic, and chaotic response. Various bifurcation types can be discovered between the periodic and chaotic states, including the well-known period-doubling cascade. Either chaotic or periodic states can be achieved by changing parameters r_{out} and/or r_{in}. Usually, the periodic state is preferred because here the system becomes predictable, the stress of switching components can be calculated, and harmonic analysis and other conventional methods can be applied. Subharmonic states could cause problems if the frequency of the subharmonic variables comes close to either the mechanical resonance of the machine or the resonance of some electric circuit. The chaotic state might offer some benefits in *electromagnetic compatibility* (EMC) with its dispersed, broadband spectrum.

ACKNOWLEDGMENTS

The authors wish to thank the Hungarian Research Fund (OTKA F023753 and OTKA T029026) and the Control Research Group of the Hungarian Academy of Science for their financial support.

REFERENCES

[1] I. Nagy, Z. Sütő and L. Backhausz, Periodic states of hysteresis current control of I.M., *Proc. 29th Int. Power Conversion and Intelligent Motion Conference (PCIM'96)* (Nürnberg), pp. 605–619, May 21–23, 1996.

[2] Z. Sütő, I. Nagy, and E. Masada, Avoiding chaotic processes in current control of AC drive, *Proc. 29th Annual IEEE Power Electronics Specialists' Conference (PESC'98)*, vol. 1 (Fukuoka, Japan), pp. 255–261, May 17–22, 1998.

[3] D. M. Brod and D. W. Novotny, Current control of VSI-PWM inverters, *IEEE Trans. on Industry Applications*, pp. 562–570, 1985.

[4] A. Nabae, S. Ogasawara, and H. Akagi, A novel control scheme of current controlled PWM inverters, *IEEE Trans. on Industrial Applications*, pp. 697–701, 1986.

[5] F. C. Moon, *Chaotic and Fractal Dynamics*. New York: Wiley, 1992.

[6] R. C. Hilborn, *Chaos and Nonlinear Dynamics: An Introduction for Scientists and Engineers*. New York: Oxford University Press, 1994.

[7] Z. Sütő, I. Nagy, and Z. Jákó, Periodic responses of a nonlinear current controlled IM drive, *Proc. 7th European Conference on Power Electronics and Applications (EPE'97)*, vol. 3 (Trondheim, Norway), pp. 3.847–3.852, Sept. 8–10, 1997.

[8] Z. Sütő, I. Nagy, and K. Zabán, Nonlinear current control of three phase converter, *Proc. IEEE Int Symposium on Industrial Electronics (ISIE'98)*, vol. 2 (Pretoria, South Africa), pp. 353–358, July 7–10, 1998.

7.5 ANALYSIS OF STABILITY AND BIFURCATION IN POWER ELECTRONIC INDUCTION MOTOR DRIVE SYSTEMS

Yasuaki Kuroe

7.5.1 Introduction

Adjustable speed control of induction motors with the use of variable-frequency power sources such as PWM inverters are gaining momentum in industrial and domestic applications. It is well known that variable frequency induction motor drives become unstable at certain operating conditions, which causes unusual vibrations in the systems [1,2,3,4,5]. In this section we present the mechanism of such nonlinear phenomena [5,9,10]. We use the computer-aided methods which were explained in Section 4.5. In the analysis, the harmonics of the output voltage of variable-frequency power sources and the characteristics of mechanical loads are taken into consideration. First we derive the mathematical model of those systems and investigate the stability properties of their nominal steady-state solutions. Next the Poincaré map is defined in terms of the periodicity of the solutions and their stability is investigated by the method described in Subsection 4.5.3. Instability regions in the parameter space are determined. Furthermore, qualitative structures in the instability are investigated from the point of view of bifurcation phenomena. It is shown that some kinds of bifurcations occur in the system depending on the drive and load conditions and bifurcation values are determined by the method described in Subsection 4.5.6.

7.5.2 Model of Power Electronic Induction Motor Drive Systems

Figure 7.35 shows a schematic diagram of the power electronic induction motor drive system studied here. The system comprises a rectifier with filter, an inverter, a three-phase induction motor, and a mechanical load. The induction motor is supplied power through three symmetrical a, b, c stator windings. The rotor windings are short circuited and only two phases, α and β, are shown for convenience. The mathematical model of the system is given as follows.

Model of Induction Motor and its Mechanical Load

For induction motors, we employ the (dr, qr) and (ds, qs) axis in the synchronously rotating reference frame with the angular velocity ω_e for the rotor and the stator windings, respectively:

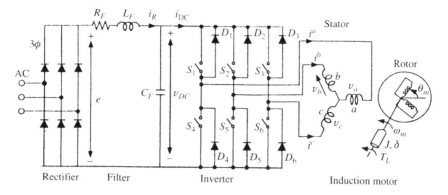

Figure 7.35 A power electronic induction motor drive system.

$$\begin{bmatrix} i^a(t) \\ i^b(t) \\ i^c(t) \end{bmatrix} = C_1 \begin{bmatrix} i^{ds}(t) \\ i^{qs}(t) \\ i^0(t) \end{bmatrix}, \quad \begin{bmatrix} v_a(t) \\ v_b(t) \\ v_c(t) \end{bmatrix} = C_1 \begin{bmatrix} v_{ds}(t) \\ v_{qs}(t) \\ v_0(t) \end{bmatrix} \tag{7.54}$$

$$\begin{bmatrix} i^\alpha(t) \\ i^\beta(t) \end{bmatrix} = C_2 \begin{bmatrix} i^{dr}(t) \\ i^{qr}(t) \end{bmatrix}, \quad \begin{bmatrix} v_\alpha(t) \\ v_\beta(t) \end{bmatrix} = C_2 \begin{bmatrix} v_{dr}(t) \\ v_{qr}(t) \end{bmatrix} \tag{7.55}$$

The transformation matrices C_1 and C_2 are defined by

$$C_1 = \frac{\sqrt{2}}{3} \begin{bmatrix} \cos\theta_e & -\sin\theta_e & \frac{1}{\sqrt{2}} \\ \cos\left(\theta_e - \frac{2}{3}\pi\right) & -\sin\left(\theta_e - \frac{2}{3}\pi\right) & \frac{1}{\sqrt{2}} \\ \cos\left(\theta_e - \frac{4}{3}\pi\right) & -\sin\left(\theta_e - \frac{4}{3}\pi\right) & \frac{1}{\sqrt{2}} \end{bmatrix}$$

$$C_2 = \begin{bmatrix} \cos(\theta - \theta_e) & \sin(\theta - \theta_e) \\ \sin(\theta - \theta_e) & -\cos(\theta - \theta_e) \end{bmatrix}$$

where θ is the electrical angle of the rotor windings and $d\theta_e/dt = \omega_e$. Note that $\theta = \lambda\theta_m$ where θ_m is the mechanical rotor angle and λ is the pole pair. Using (7.54) and (7.55), the model of induction motor is expressed as [6]:

$$\frac{d}{dt}\boldsymbol{i}(t) = -\boldsymbol{L}^{-1}\{\boldsymbol{R}\boldsymbol{i}(t) + \lambda\boldsymbol{G}\omega_m\boldsymbol{i}(t) - \boldsymbol{v}(t)\} \tag{7.56}$$

$$v_0(t) = R_s i^0(t) \tag{7.57}$$

where

$$
\boldsymbol{i}(t) := \begin{bmatrix} i^{dr}(t) \\ i^{qr}(t) \\ i^{ds}(t) \\ i^{qs}(t) \end{bmatrix}, \quad \boldsymbol{v}(t) := \begin{bmatrix} v_{dr}(t) \\ v_{qr}(t) \\ v_{ds}(t) \\ v_{qs}(t) \end{bmatrix}
$$

$$
\boldsymbol{R} := \begin{bmatrix} R_r & -L_r\omega_e & 0 & -M\omega_e \\ L_r\omega_e & R_r & M\omega_e & 0 \\ 0 & -M\omega_e & R_s & -L_s\omega_e \\ M\omega_e & 0 & L_s\omega_e & R_s \end{bmatrix}
$$

$$
\boldsymbol{L} := \begin{bmatrix} L_r & 0 & M & 0 \\ 0 & L_r & 0 & M \\ M & 0 & L_s & 0 \\ 0 & M & 0 & L_s \end{bmatrix}, \quad \boldsymbol{M} := \begin{bmatrix} 0 & L_r & 0 & M \\ -L_r & 0 & -M & 0 \\ 0 & 0 & 0 & 0 \\ 0 & 0 & 0 & 0 \end{bmatrix}
$$

and i^{dr} = rotor d-axis current; i^{qr} = rotor q-axis current; i^{ds} = stator d-axis current; i^{qs} = stator q-axis current; v_{dr} = rotor d-axis voltage; v_{qr} = rotor q-axis voltage; v_{ds} = stator d-axis voltage; v_{qs} = stator q-axis voltage; ω_m = rotor angular velocity; R_r = resistance of rotor windings; R_s = resistance of stator windings; L_r = self-inductance of rotor windings; L_s = self-inductance of stator windings; M = mutual inductance.

The model of the rotational motion is expressed by

$$
\frac{d}{dt}\omega_m(t) = -\frac{1}{J}\left\{\delta\omega_m(t) + T_L(\theta_m) - \lambda \boldsymbol{i}^t(t)\boldsymbol{G}\boldsymbol{i}(t)\right\} \tag{7.58}
$$

$$
\frac{d}{dt}\theta_m(t) = \omega_m(t) \tag{7.59}
$$

where J = moment of inertia; δ = damping coefficient; T_L = external load torque. The load torque T_L is given by the characteristics of the mechanical load which is connected to the rotor shaft. We consider here the following two cases: (1) T_L is constant and (2) T_L is periodic in θ_m:

$$
T_L(\theta_m) = T_L(\theta_m + 2\pi). \tag{7.60}
$$

There are a lot of examples of the mechanical loads which satisfy (7.60); typical examples are compressors.

Model of Inverter and Rectifier

We assume that the inverter in Figure 7.35 is the voltage-controlled type and lossless. Such inverters are usually implemented so that their output voltages v_a, v_b, and v_c are expressed as

$$
\begin{bmatrix} v_a(t) \\ v_b(t) \\ v_c(t) \end{bmatrix} = \begin{bmatrix} s_a(t) \\ s_b(t) \\ s_c(t) \end{bmatrix} v_{DC}(t) \tag{7.61}
$$

where v_{DC} is the voltage across the capacitor C_F of the filter. The functions s_a, s_b, and s_c represent the characteristics of inverters. They are usually given to be balanced and periodic with period $T_e = 2\pi/\omega_e$, and are given by

$$s_a(t) = s_a(t + T_e), \quad s_b(t) = s_b(t + T_e), \quad s_c(t) = s_c(t + T_e)$$

$$s_a(t) = s_b\left(t + \frac{1}{3}T_e\right) = s_c\left(t + \frac{2}{3}T_e\right)$$

$$s_a(t) + s_b(t) + s_c(t) = 0$$

where $f_e = 1/T_e$ corresponds to the drive frequency of the inverter. Figures 7.36 and 7.37 show examples of the waveform of s_a for a six-step voltage inverter and a pulse-width modulated inverter with sinusoidal modulation of the pulse width, respectively. Then $s_a(t)$, $s_b(t)$, and $s_c(t)$ can be expanded into the Fourier series in the form:

$$\begin{bmatrix} s_a(t) \\ s_b(t) \\ s_c(t) \end{bmatrix} = \sum_{n=1}^{\infty} K_n \begin{bmatrix} \cos(n\omega_e t + v_n) \\ \cos(n\omega_e t - \frac{1}{3}T_e + v_n) \\ \cos(n\omega_e t - \frac{2}{3}T_e + v_n) \end{bmatrix} \tag{7.62}$$

Let

$$\begin{bmatrix} s_{ds}(t) \\ s_{qs}(t) \\ 0 \end{bmatrix} := C_1^{-1} \begin{bmatrix} s_a(t) \\ s_b(t) \\ s_c(t) \end{bmatrix} \tag{7.63}$$

From (7.54), (7.61), and (7.62) we can obtain:

$$\begin{bmatrix} v_{ds}(t) \\ v_{qs}(t) \\ v_0(t) \end{bmatrix} = C_1^{-1} \begin{bmatrix} s_a(t) \\ s_b(t) \\ s_c(t) \end{bmatrix} v_{DC}(t) = \begin{bmatrix} s_{ds}(t) \\ s_{qs}(t) \\ 0 \end{bmatrix} v_{DC}(t)$$

$$= \frac{\sqrt{3}}{2} \left\{ K_1 \begin{bmatrix} \cos v_1 \\ \sin v_1 \\ 0 \end{bmatrix} + \sum_{n=2}^{\infty} K_n \begin{bmatrix} \cos(n\omega_e t + v_n) \\ \sin(n\omega_e t + v_n) \\ 0 \end{bmatrix} \right\} v_{DC}(t) \tag{7.64}$$

Then the input voltage of induction motor (7.56) is given by

$$v(t) = [0, 0, s_{ds}(t)v_{DC}(t), s_{qs}(t)v_{DC}(t)]^t \tag{7.65}$$

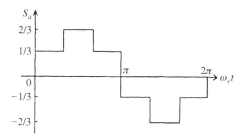

Figure 7.36 An example of waveform of s_a for six-pulse voltage inverter.

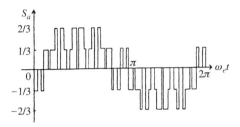

Figure 7.37 An example of waveform of s_a for PWM inverter with sinusoidal modulation of pulse-width.

Note that $s_{ds}(t)$ and $s_{qs}(t)$ defined in (7.63) are periodic functions with period T_e:

$$s_{ds}(t) = s_{ds}(t + T_e)$$
$$s_{qs}(t) = s_{qs}(t + T_e)$$

(7.66)

and if $K_n = 0$ for all $n \geq 2$ in (7.62), $s_{ds}(t)$ and $s_{qs}(t)$ are both constant. The model equation of the filter in Figure 7.35 is expressed as:

$$\frac{d}{dt} i_R(t) = -\frac{1}{L_F} \{ R_F i_R(t) + v_{DC}(t) - e(t) \}$$

(7.67)

$$\frac{d}{dt} v_{DC}(t) = \frac{1}{C_F} \{ i_R(t) - i_{DC}(t) \}$$

(7.68)

where $e(t)$ is the output voltage of the rectifier. Considering the conservation of power for the inverter, we have

$$i_{DC}(t) v_{DC}(t) = i^{ds}(t) v_{ds}(t) + i^{qs}(t) v_{qs}(t)$$

(7.69)

Substituting (7.65) into the right hand of (7.69),

$$i_{DC}(t) v_{DC}(t) = \{ i^{ds}(t) s_{ds}(t) + i^{qs}(t) s_{qs}(t) \} v_{DC}(t)$$

Hence the following equation is obtained:

$$i_{DC}(t) = i^{ds}(t) s_{ds}(t) + i^{qs}(t) s_{qs}(t)$$

(7.70)

Assuming that $e(t)$ is a constant voltage source $e(t) = E$ and substituting (7.70) into (7.68), we finally obtain:

$$\frac{d}{dt} i_R(t) = -\frac{1}{L_F} \{ R_F i_R(t) + v_{DC}(t) - E \}$$

(7.71)

$$\frac{d}{dt} v_{DC}(t) = \frac{1}{C_F} \{ i_R(t) - \left(i^{ds}(t) s_{ds}(t) + i^{qs}(t) s_{qs}(t) \right) \}$$

(7.72)

Equations (7.56), (7.58), (7.59), (7.71), and (7.72) represent the complete model by which we will investigate stability and bifurcation phenomena. They can be solved if E, the output voltage of the rectifier, and the functions $s_a(t)$, $s_b(t)$, and $s_c(t)$ are given. Note that the model derived here can deal with various kinds of inverters by choosing the functions $s_a(t)$, $s_b(t)$, and $s_c(t)$ appropriately. Let the state variable x be defined:

$$x = [i_R, v_{DC}, i, \omega_m, \theta_m]' \in R^8$$

(7.73)

Then (7.71), (7.72), (7.56), (7.58), and (7.59) can be written in the following nonlinear nonautonomous differential equation form:

$$\frac{d}{dt}x(t) = f(x(t), t) \tag{7.74}$$

where

$$f := \begin{bmatrix} -\frac{1}{L_F}\left\{R_F i_R(t) + v_{DC}(t) - E\right\} \\ \frac{1}{C_F}\left\{i_R(t) - \left(i^{ds}(t)s_{ds}(t) + i^{qs}(t)s_{qs}(t)\right)\right\} \\ -L^{-1}\left\{Ri(t) + \lambda G\omega_m(t)i(t) - v(t)\right\} \\ -\frac{1}{J}\left\{\delta\omega_m(t) + T_L(\theta_m) - \lambda i^t(t)Gi(t)\right\} \\ \omega_m(t) \end{bmatrix}$$

Due to the periodicity of the functions $s_{ds}(t)$ and $s_{qs}(t)$ and the load torque T_L, f has the following periodicity:

$$\begin{aligned} f(x_1, x_2, \ldots x_7, x_8, t) &= f(x_1, x_2, \ldots, x_7, x_8, t + T_e) \\ &= f(x_1, x_2, \ldots, x_7, x_8 + 2\pi, t) \end{aligned} \tag{7.75}$$

7.5.3 Poincaré Map and Periodicity of Steady States

We will now investigate instability and bifurcation phenomena of steady-state solutions of the system described by (7.74). According to the periodic properties (7.75), our discussions are divided into the following three cases.

Case I

We assume here that load torque T_L is constant, and $s_{ds}(t)$ and $s_{qs}(t)$ have the periodicity satisfying (7.66). In this case the last equation of (7.74) can be omitted for the analysis of stability and bifurcation. Hence the system equation is reduced to the following nonautonomous differential equation:

$$\frac{d}{dt}x(t) = f(x(t), t) \tag{7.76}$$

where $x = [i_R, v_{DC}, i, \omega_m]^T \in R^7$ and

$$f = \begin{bmatrix} -\frac{1}{L_F}\left\{R_F i_R(t) + v_{DC}(t) - E\right\} \\ \frac{1}{C_F}\left\{i_R(t) - \left(i^{ds}(t)s_{ds}(t) + i^{qs}(t)s_{qs}(t)\right)\right\} \\ -L^{-1}\left\{Ri(t) + \lambda G\omega_m(t)i(t) - v(t)\right\} \\ -\frac{1}{J}\left\{\delta\omega_m(t) + T_L - \lambda i^t(t)Gi(t)\right\} \end{bmatrix}$$

As per (7.66), f has the periodicity:

$$f(x, t) = f(x, t + T_e) \tag{7.77}$$

In this case the nominal steady-state solution $x^*(t)$ of (7.76) is periodic with the period $T_e : x^*(t) = x^*(t + T_e)$, which draws a closed orbit in the state space $\mathbb{R}^7$. The stability theorem of a periodic solution of the nonautonomous dynamical system is given in Subsection 4.5.2 (Theorem 1). According to the theorem, we introduce the

Poincaré map as follows. Let $\phi(t, t_0, x_0)$ be the solution of (7.76) at time t with the initial condition $x(t_0) = x_0$. The Poincaré map $P : \mathbb{R}^7 \to \mathbb{R}^7$ is defined by

$$P(x_0) = \phi(t_0 + T, t_0, x_0) \tag{7.78}$$

Note that the nominal steady-state solution of (7.76) corresponds to a fixed point of the Poincaré map P:

$$x_0^* - P(x_0^*) = 0 \tag{7.79}$$

where $x_0^* = x^*(t_0)$. By Theorem 1 in Section 4.5.2, the stability of the periodic solution $x^*(t)$ can be checked by evaluating the Jacobian matrix of the Poincaré map P with respect to x_0 at the point x_0^*. This is denoted as $DP(x_0^*)$. The theorem says that the periodic solution $x^*(t)$ is asymptotically stable if all the eigenvalues of the Jacobian matrix $DP(x_0^*)$ are inside the unit circle on the complex plane. The Jacobian matrix $DP(x_0^*)$ can be obtained by numerically solving the nonlinear equation (7.79) with respect to x_0^* by the use of the Newton-Raphson method. For details, see Subsection 4.5.3.

Case II

We assume here that T_L satisfies (7.60), and $s_{ds}(t)$ and $s_{qs}(t)$ are both constant. Note that the condition that $s_{ds}(t)$ and $s_{qs}(t)$ are both constant implies that the harmonics of the output voltage of the inverter are completely neglected. In this case the system equation (7.74) is reduced to the autonomous differential equation

$$\frac{d}{dt}x(t) = f(x(t)) \tag{7.80}$$

where $x = [i_R, v_{DC}, i, \omega_m, \theta_m]^t \in R^8$ and f has the periodicity

$$f(x_1, x_2, \ldots x_7, x_8) = f(x_1, x_2, \ldots x_7, x_8 + 2\pi) \tag{7.81}$$

In this case, the nominal steady-state solution $x^*(t)$ has the periodicity:

$$\begin{aligned}
x_i^*(t + T) &= x_i^*(t) \quad (i = 1, 2, \ldots 7) \\
x_8^*(t + T) &= x_8^*(t) + 2\pi
\end{aligned} \tag{7.82}$$

where T is the period. Note that the image of a solution with this periodicity does not proceed along a closed trajectory (limit cycle) in the usual state space R^8. Let us introduce a modified state space W defined by $W := R^7 \times S^1$, where S^1 is the unit circle on R^2:

$$W := \{(x_1, x_2, \ldots x_7, x_8) : x_i \in R^1 (i = 1, 2, \ldots 7), x_8 \in S^1\} \tag{7.83}$$

which is called the *hypercylindrical state space*. The circular component $S^1 = R(\text{mod} 2\pi)$ comes from the periodicity of f in $x_8 (= \theta_m)$. Note that a solution with the periodicity (7.82) is a closed orbit which encircles the hypercylindrical state space W. Such a periodic solution is usually referred to as a *periodic solution of the second kind* [7] as shown in Figure 7.38, whereas a limit cycle on the usual Euclidean state space is referred to as a *periodic solution of the first kind*.

The stability theorem of a periodic solution of the autonomous dynamical system is given in Subsection 4.5.2 (Theorem 2). Let $\phi(t, x_0)$ be the solution of (7.80) with the initial condition $x(0) = x_0$. The Poincaré map $P : W \to W$ is defined by

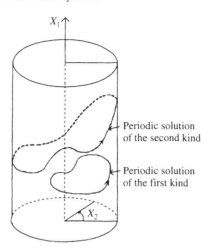

Periodic solution
of the second kind

Periodic solution
of the first kind

Figure 7.38 Cylindrical state space $W = R \times S^1$ and a periodic solution of the second kind.

$$P(x_0) = \phi(T, x_0) \tag{7.84}$$

The stability of the periodic solution $x^*(t)$ satisfying (7.82) can also be checked by evaluating $DP(x_0^*)$, the Jacobian matrix of the Poincaré map P with respect to x_0 at the point $x_0^* := x^*(0)$. In this case, the Jacobian matrix $DP(x_0^*)$ always has one eigenvalue of unity and the periodic solution of (7.80) is asymptotically stable if the magnitudes of the other seven eigenvalues of the Jacobian matrix $DP(x_0^*)$ are less than unity. Note that in this case the period T is unknown a priori. A method to compute the Jacobian matrix $DP(x_0^*)$ by using the Newton-Raphson method can also be found in Subsection 4.5.3.

Case III

Here we further simplify the system model and assume that the load torque T_L is constant and s_{ds} and s_{qs} are both constant. Similar to the Case I, the last equation of (7.74) can be omitted for the analysis. Then (7.76) is reduced to the autonomous differential equation:

$$\frac{d}{dt}x(t) = f(x(t)), \tag{7.85}$$

where $x = [i_R, v_{DC}, i, \omega_m]^t \in R^7$. In this case the nominal steady-state solution x^* is an equilibrium point of f: $f(x^*) = 0$. The stability of an equilibrium point x^* is investigated by evaluating the eigenvalues of the coefficient matrix of its linearized model at the point x^*: $\dot{\xi} = Df(x^*)\xi$, where $\xi = x - x^*$. Along this line, several studies on stability analysis of induction motor drive systems have been done [1,2,3,4].

Note that if both the load torque T_L and the inverter functions $s_{ds}(t)$ and $s_{qs}(t)$ are periodic, nominal steady-state solutions are not generally periodic any more.

7.5.4 Stability Analysis

As stated above, the stability of a periodic solution can be investigated by introducing the Poincaré map appropriately and evaluating the eigenvalues of its Jacobian matrix. The methods to compute the Jacobian matrix of the Poincaré map are discussed in Subsection 4.5.3, where it has been shown that solving the nonlinear equation corresponding to the fixed-point condition (e.g. (7.79)) of the periodic solution by the Newton-Raphson method gives us not only the nominal steady-state solution but also the Jacobian matrix of the Poincaré map.

We consider here a PWM inverter-fed induction motor drive with a compressor load [9] in which the load torque T_L is periodic in θ_m (satisfies (7.60)). It was observed by experiment that the system becomes unstable at some operating conditions. We investigate the effect of the control parameters V and f_e on its stability, where V is the effective value of the fundamental component of output line-voltage of the inverter and $f_e = \omega_e/2\pi$ is its drive frequency. Figure 7.39 shows the instability region in the f_e-V parameter plane obtained by the experiment. The points marked with o and × show stable and unstable operating points, respectively.

We investigate stability of the system by treating it as Case II in the previous subsection; that is, the functions $s_{ds}(t)$ and $s_{qs}(t)$ are assumed to be constant. The parameters are: $R_s = 0.5145\Omega$, $R_r = 0.2674\Omega$, $L_s = L_r = 0.09088\text{H}$, $M = 0.08953\text{H}$, $J = 0.55 \times 10^{-2}\text{Kgm}^2$, $\delta = 0.0\text{Kgm}^2/\text{s}$, $L_F = 0\text{mH}$, $C_F = 0\mu\text{F}$, $R_F = 0\Omega$. The first-order approximation model of the compressor is given by $T_L(\theta_m) = T_0 + T_1 \cos(\theta_m)$ where $T_0 = 10.83\text{Nm}$ and $T_1 = 13.98\text{Nm}$.

Figure 7.40 show the unstable region on the f_e-V parameter plane obtained by the method discussed in Subsection 4.5.3. In the figure, the inside of the contour line is the unstable region and the outside is the stable region. Figure 7.41 shows an example of the convergence behavior of the Newton-Raphson method to obtain the nominal steady-state solution at the unstable operating point ($V = 110\text{V}$, $f_e = 30\text{Hz}$). Note that the period T of the nominal steady-state solution is unknown a priori. In the figure the values of T and the motor speed ω_m versus the iteration number of the Newton-Raphson method are plotted. The figure clearly shows the Newton-Raphson method with quadratic convergence behavior, from which the Jacobian elements are also calculated. Table 7.1 shows examples of computational results of the eigenvalues of the Jacobian matrix. In the table the obtained eigenvalues and their absolute values at

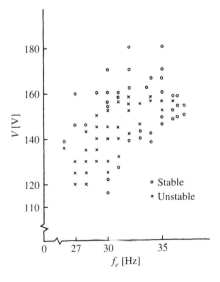

Figure 7.39 Instability region on the f_e-V parameter plane obtained by experiment.

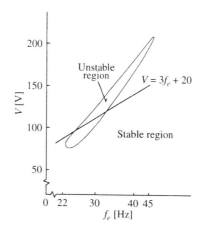

Figure 7.40 Obtained instability region on the f_e-V parameter space.

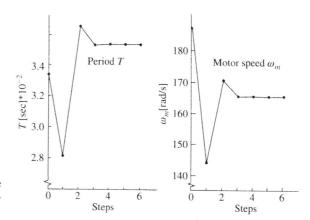

Figure 7.41 Convergence behavior of the Newton-Raphson method to locate the nominal steady-state solution.

TABLE 7.1 Computational results of eigenvalues of the Jacobian matrix of the Poincaré map.

	Stable Operating Point ($V = 160$V, $f_e = 30$Hz)		
	Eigenvalues		Absolute Values
$\lambda_1(=1)$	1.00005	$+j($ 0. $)$	
λ_2	-0.707593	$+j($ 0.533422 $)$	0.866130
λ_3	-0.707593	$+j(-0.533422)$	0.866130
λ_4	1.18223×10^{-2}	$+j($ 0. $)$	1.18223×10^{-2}
λ_5	3.36005×10^{-4}	$+j($ $1.48929 \times 10^{-4})$	3.36005×10^{-4}
λ_6	3.36005×10^{-4}	$+j(-1.48929 \times 10^{-4})$	3.36005×10^{-4}
	Unstable Operating Point ($V = 110$V, $f_e = 30$Hz)		
	Eigenvalues		Absolute Values
$\lambda_1(=1)$	1.00121	$+j($ 0. $)$	
λ_2	-1.11460	$+j($ 0. $)$	1.11460
λ_3	-0.80704	$+j($ 0. $)$	0.807035
λ_4	5.19397×10^{-2}	$+j($ 0. $)$	5.19397×10^{-2}
λ_5	8.77403×10^{-5}	$+j($ $5.84564 \times 10^{-5})$	1.05430×10^{-4}
λ_6	8.77403×10^{-5}	$+j(-5.84564 \times 10^{-5})$	1.05430×10^{-4}

the stable ($V = 160$V, $f_e = 30$Hz) and unstable ($V = 110$V, $f_e = 30$Hz) operating points
are shown. It is observed that the Jacobian matrix has one eigenvalue of unity at both
operating points, and at the unstable operating point the absolute value of the second
eigenvalue (λ_2) is greater than 1. Next we consider applying the well-known V/F control
to the system. Suppose that the voltage V is adjusted according to the control
$V = 3f_e + 20$. It can be found by drawing the line $V = 3f_e + 20$ in Figure 7.40 that
the instability occurs from $f_e \simeq 26$Hz to $f_e \simeq 33$Hz. Figure 7.42 shows the loci of the
dominant eigenvalues of the Jacobian matrix of the Poincaré map along the V/F control $V = 3f_e + 20$. It is observed that, as f_e is increased, a pair of the complex eigenvalues becomes real and then one of them goes out the unit circle.

Note that, as discussed in Subsection 4.5.5, the Jacobian matrix of the Poincaré
map can also be obtained by a transient simulator of power electronic induction motor
drive systems [8], which makes it possible to fully reflect the characteristics of power
electronic circuits (drive circuits).

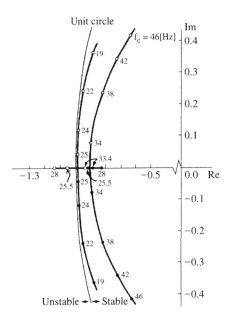

Figure 7.42 Loci of dominant eigenvalues of
the Jacobian matrix of the Poincaré map
along the control $V = 3f_e + 20$.

7.5.5 Analysis of Bifurcations

In this subsection, we analyze the instability in the power electronic induction
motor drive system from the point of view of bifurcation theory. Figure 7.43 shows
an example of waveforms of the motor speed ω_m obtained by the same experiment by
which the result in Figure 7.39 was obtained. In the figure the experimental waveforms
at two operating points are shown: (a) stable operating point ($f_e = 30$Hz, $V = 156$V)
and (b) unstable operating point ($f_e = 30$Hz, $V = 140$V). Note that both waveforms are
periodic, but the period in (b) is twice as long as that in (a). This implies that a period-
doubling bifurcation has occurred. Note also that in Figure 7.42 one of the eigenvalues
of the Jacobian matrix of the Poincaré map passes out the unit circle through the point
$(-1, 0)$, which also implies the occurrence of a period-doubling bifurcation.

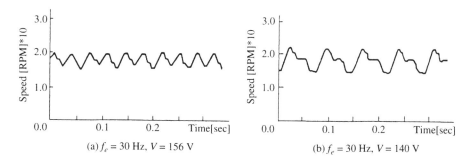

(a) $f_e = 30$ Hz, $V = 156$ V

(b) $f_e = 30$ Hz, $V = 140$ V

Figure 7.43 Experimental waveforms of motor speed ω_m of PWM inverter–fed induction motor drive with compressor loads.

In subsection 4.5.6 a computer method to determine bifurcation values, that is, values of system parameters corresponding to the occurrence of bifurcations, is described. By using the method we obtain the bifurcation values on the f_e-V parameter plane. In the method the Newton-Raphson algorithm is applied to solve a nonlinear equation, consisting of the fixed-point condition and the bifurcation condition, with respect to the parameter of interest μ and the state vector x. We find here two types of bifurcations: the Neimark bifurcation and the period-doubling bifurcation (for explanation, see Chapter 3). Figures 7.44 and 7.45 show the bifurcation set on the f_e-V parameter plane obtained by the algorithm, where we let $\mu = V$. In Figure 7.44, all the system parameters were the same values as those of the previous subsection, and in Figure 7.45 only J was changed to 1.4×10^{-2} Kgm^2. In these figures the solid and the dashed lines represent the Neimark bifurcation set and the period-doubling bifurcation set, respectively. It is observed that only the period-doubling bifurcations occur in Figure 7.44, whereas both the period-doubling and Neimark bifurcations occur in Figure 7.45. Figure 7.46 shows examples of the simulated waveforms of the motor speed ω_m and load torque T_L (a) before and (b) after the period-doubling bifurcation occurs in Figure 7.44. Note that both waveforms are periodic, but the period in (b) is

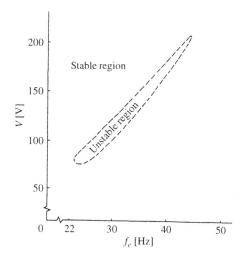

Figure 7.44 Bifurcation set in the f_e-V plane obtained by the Newton-Raphson algorithm (Case II).

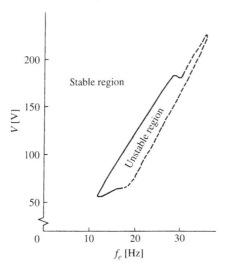

Figure 7.45 Bifurcation set on the f_e-V plane obtained by the Newton-Raphson algorithm (Case II, $J = 1.4 \times 10^{-2}$ Kgm^2).

twice as long as that in (a). Figure 7.47 shows examples of the simulated waveforms of the motor speed ω_m (a) before and (b) after the Neimark bifurcation occurs in Figure 7.45. It is observed that the waveform after the Neimark bifurcation is not periodic (but quasi-periodic).

Next we investigate bifurcation values of the system by treating it as Case I in Subsection 7.5.3. The model of the compressor load is simplified by using its averaged

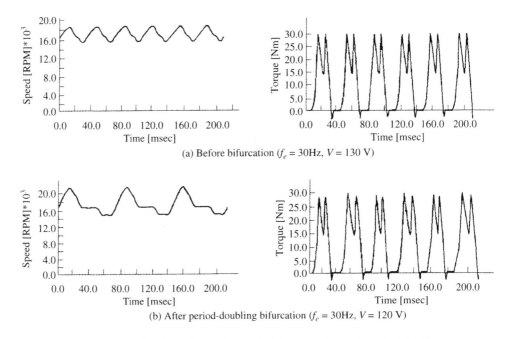

Figure 7.46 Simulated waveforms of ω_m and T_L before and after the period-doubling bifurcation.

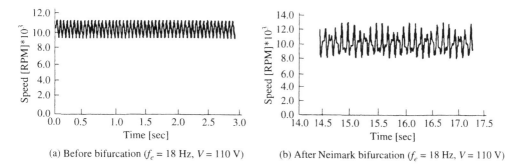

(a) Before bifurcation ($f_e = 18$ Hz, $V = 110$ V) (b) After Neimark bifurcation ($f_e = 18$ Hz, $V = 110$ V)

Figure 7.47 Simulated waveforms of ω_m and T_L before and after the Neimark bifurcation.

value; that is, T_L is assumed to be constant. The functions $s_{ds}(t)$ and $s_{qs}(t)$, which represent characteristics of the inverter, are in the form given by a sinusoidal PWM inverter with 27-mode carrier as shown in Figure 7.37. All the system parameters are the same as those of the previous subsection except that $J = 1.4 \times 10^{-2}$Kgm^2, $\delta = 3.043 \times 10^{-4}$ Kgm^2/s, $T_L = 10.83$Nm and the parameters of the filter are: $L_F = 0.8$mH, $C_F = 1500\mu$F, $R_F = 0\Omega$. Figure 7.48 shows the bifurcation set in the f_e-V parameter plane obtained by the algorithm. It is observed that only the Neimark bifurcations occur. We further simplify the model by letting $s_{ds}(t)$ and $s_{qs}(t)$ be constant, simplifying the system to Case III. In this case only the Hopf bifurcations were observed and the obtained bifurcation set in the f_e-V parameter plane was almost the same as that in Figure 7.48. This implies that the harmonics of the output voltage of the PWM inverter are small enough so that they can be neglected.

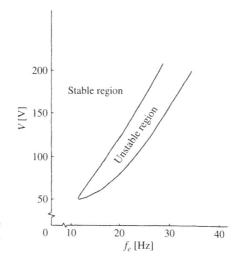

Figure 7.48 Bifurcation set on the f_e-V plane obtained by the Newton-Raphson algorithm (Case I).

REFERENCES

[1] T. A. Lipo and P. C. Krause, Stability analysis of a rectifier-inverter induction motor drive, *IEEE Trans. on PAS*, vol. PAS-88, no. 1, pp. 55–66, 1969.

[2] R. H. Nelson, T. A. Lipo, and P. C. Krause, Stability analysis of a symmetrical induction machine, *IEEE Trans. on PAS*, vol. PAS-88, no. 11, pp. 1710–1717, 1969.

[3] F. Fallside and A. T. Wortley, Steady state oscillation of variable-frequency inverter-fed induction-motor drives, *Proc. IEE*, vol. 116, No. 6, pp. 991–992, 1969.

[4] Y. Murai, I. Hosono, and Y. Tsunehiro, On system stability of PWM-inverter fed induction motor, *Trans. IEEJ*, vol. 105-B, no.5, pp. 467–474 (in Japanese).

[5] Y. Kuroe and T. Maruhashi, Stability analysis of power electronic induction motor drive system, *Proc. IEEE 1987 Int. Symposium on Circuits and Systems*, pp.1009–1013, 1987.

[6] P. C. Krause and C. H. Thomas, Simulation of symmetrical induction machinery, *IEEE Trans. on PAS*, vol. PAS-84, no. 11, pp. 1038–1053, 1965.

[7] N. Minorsky, *Nonlinear Oscillation*. New York: Van Nostrand, 1962.

[8] Y. Kuroe, H. Haneda, and T. Maruhashi, Computer-aided steady-state analysis of power electronic induction-motor drive systems, *Proc. IEEJ Int. Power Electronics Conference*, pp. 238–249, March 1983.

[9] Y. Kuroe, S. Hayashi, and T. Maruhashi, Stability analysis method for induction motor drive systems with periodic load, *Trans. IEEJ*, vol. D-107, no. 9, pp. 1175–1182, 1987 (in Japanese).

[10] Y. Kuroe and S. Hayashi, Analysis of bifurcation in power electronic induction motor drive systems, *Proc. IEEE Power Electronics Specialists' Conference*, pp. 923–930, June 1989.

Chapter 8

NONLINEAR CONTROL AND CONTROL OF CHAOS

8.1 CONVENTIONAL NONLINEAR CONTROLS IN POWER ELECTRONICS

Philip T. Krein

8.1.1 Introduction

In the most general sense, all power electronic controls are nonlinear. Since control is implemented with switches, and because switch action can occur at arbitrary times, the controls must take state, reference, and input information and translate it into timing. Switching makes the state derivatives discontinuous. The systems do not in general meet Lipschitz conditions [1], and indeed even the nature of a solution must be generalized to support any sort of analysis [2]. In fact, nonlinear systems of this type still pose tractability limits for mathematics and control theory. When chaos is added to the picture, it becomes even more difficult to consider the actual time-domain and frequency-domain behavior of a power electronic system. Poincaré maps and other tools provide a view of how the system evolves over time, but the action at any particular moment is essentially unpredictable.

Even though the underlying action is truly nonlinear, it is conventional in many contexts to use a linear framework for power electronics. A pulse-width modulation (PWM) process in a dc/dc converter, for instance, can be treated in an averaging sense as a simple gain block. This is true of many rectifier controllers as well, in which the average output is a linear function of the cosine of a phase angle. With this in mind, we can think of many converter controls as *linear*, with conventional PI loops or linear compensators applied around a linear model of a nonlinear block. Here we do not consider controls that can be classified in this way, as an extensive body of literature exists.

There are fundamental drawbacks to the use of linear techniques for power electronic system controls. First, the controls must be designed in light of approximations used in the system model. The performance becomes *model limited*, in the sense that a design should not be applied outside the range in which the linear model is valid. Second, not all phenomena or possible operating regimes can be addressed with linear controls, and there are many cases in which we would like to expand the range of techniques. A concern today is that many designers are beginning to confuse model limitations with performance limitations. In some papers, broad statements are made about what can or cannot be accomplished with power electronic controls, but the statements reflect limits of the averaging process rather than the true fundamental capabilities.

In this chapter, we consider a range of nonlinear controls that fall outside the conventional framework. These include both a set of nonlinear controls for power electronic circuits in conventional operating modes, and a suite of techniques that can be applied when chaotic operation occurs. The range of tools is such that each has its special applications. A few have broad application, and indeed some have been in use for decades.

There are a few examples with a long history of power electronic control approaches that are nonlinear in a fundamental sense. Three distinct classes of examples are:

1. Hysteresis controllers
2. Nonlinear modulation approaches, especially those used in cycloconversion
3. The application of multipliers within control loops, such as is conventional in active power-factor correction (PFC) rectifier circuits

In this section, these three will be reviewed.

8.1.2 Hysteresis Controllers

Hysteresis control is a long-established approach in which a power electronic circuit is controlled in a manner analogous to a thermostat. The power converter output is monitored. An active switch operates as the output crosses a threshold. The simplest technique is to compare the output to a reference waveform, switching on when the output is too low and off when it is too high. Section 8.2 examines many of the issues of hysteresis control, mainly in the context of dc/dc converters. The same technique is widely used for current control in ac drive systems: the motor line current is compared to the desired sinusoidal value, and switches drive the current high when the comparison shows it to be low and vice versa. It has proved to be convenient in *power-factor correction* (PFC) converters as well [3]. In the PFC case, the converter's input current is to be controlled to follow a full-wave rectified sinusoid. It is straightforward to use a direct comparator approach to turn the active switch on and off as the actual current crosses the reference.

Hysteresis control is inherently robust, since the switches operate to enforce a desired output, irrespective of time scale or line or load values. There are still fundamental limitations (a low-voltage input bus can force only a limited current slew rate on an inductor, for instance), but hysteresis can help keep a converter near any feasible operating condition. Unfortunately, the approach does not have general applicability (see Section 8.2) but is easy to use for many common topologies. In recent years, variations on this technique have become popular for low-power conversion applications in battery systems. Some approaches use *single-sided hysteresis*, in which either the switch turn-on or the turn-off is determined by a comparator, while the other operation is set by timing [4,5].

8.1.3 Nonlinear Modulation

Power converters are most often controlled with a modulation-based block such as pulse-width modulation (PWM). In PWM-based controls, a linear carrier waveform is almost universally used in the process. Although nonlinear modulation schemes have been discussed [6], linear PWM is the conventional approach in power electronics.

In ac/ac converters, phase modulation can be used as the basis of control. The most common ac/ac conversion topology, known as the *cycloconverter* or *line-commutated cycloconverter*, uses a pair of controlled rectifiers to supply the positive and negative load currents of an ac load, as in Figure 8.1. Control approaches for cycloconverters date from the 1930s [7]. The appropriate method can be represented as a nonlinear modulation method, in which the rectifier phase control angle α is given by

$$\alpha(t) = \cos^{-1}[k\cos(\omega_0 t)] \tag{8.1}$$

Here ω_0 is the desired output radian frequency and k is a gain value between 0 and 1. Since a rectifier has an average output proportional to $\cos\alpha$, a cycloconverter has its primary output frequency component proportional to $k\cos(\omega_0 t)$. To support both positive and negative current half-cycles, the inverse cosine is confined to different quadrants for each of the two controlled rectifiers. For the positive rectifier, the phase angle is held in the interval $[0, 180°]$, while the negative rectifier has an angle in the range $[-180°, 0]$.

This application of nonlinear modulation has been studied extensively [8,9]. The noise frequency components are a challenge to analyze, but the cycloconverter remains the most practical way to construct ac/ac converters at high power levels. There are other nonlinear modulating functions that will support cycloconversion [10], but the inverse cosine technique represents the approach used in practice.

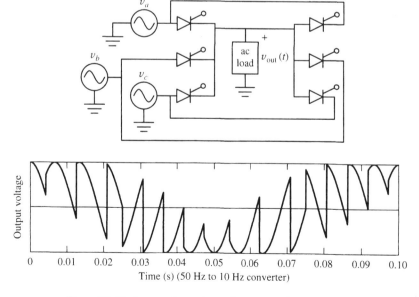

Figure 8.1 Cycloconverter and sample output voltage waveform.

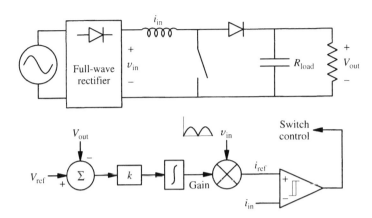

Figure 8.2 Power factor correction boost converter with multiplier gain control and current hysteresis control.

8.1.4 Multipliers in the Loop

It is perhaps natural that *power* might be used as the basis for control in power electronics. This implies the presence of multiplier blocks in a control loop, as voltage and current are processed to compute power. In fact, the use of power information is not typical (examples are given in later sections of this chapter). However, multiplier blocks do have application in conventional controls. The most typical use of a multiplier block is to implement a controllable gain for a converter stage.

Figure 8.2 shows a PFC circuit in which the *shape* of the input current is to match the shape of the input full-wave rectified sinusoid, but the current *amplitude* is uncertain since it depends on load. The circuit functions by measuring an output voltage error, then using the error magnitude to set a gain for the voltage waveform. This circuit converges to a steady-state operating condition in which the current shape matches the voltage, while the current amplitude is exactly that needed to supply the load power at the desired voltage. The use of the multiplier is common enough that it appears in most commercial integrated circuits that implement PFC control [11].

REFERENCES

[1] H. K. Khalil, *Nonlinear Systems.* New York: Macmillan, 1992.

[2] B. Lehman and R. M. Bass, Extensions of averaging theory for power electronic systems, *IEEE Trans. on Power Electronics*, vol. 11, no. 4, pp. 542–553, July 1996.

[3] C. Zhou, R. B. Ridley, and F. C. Lee, Design and analysis of a hysteretic boost power factor correction circuit, *IEEE Power Electronics Specialists' Conf. Rec.*, pp. 800–807, 1990.

[4] J. Scolio, Power conservation in 3V, 5V dual-supply systems, *Computer Design*, March 1996.

[5] M. Wilcox and R. Flatness, New LTC1148/LTC1149 switching regulators maximize efficiency from milliamps to amps, *Linear Technology*, vol. 3, no. 1, pp. 1, 10–12, February 1993.

[6] R. W. Erickson, *Fundamentals of Power Electronics.* New York: Chapman and Hall, 1997, p. 641.

[7] H. Rissik, *Mercury-Arc Current Converters*. London: Sir Isaac Pitman and Sons, 1935.

[8] P. Wood, *Switching Power Converters*. New York: Van Nostrand Reinhold, 1981.

[9] L. Gyugyi and B. R. Pelly, *Static Power Frequency Changers*. New York: Wiley, 1976.

[10] P. T. Krein, *Elements of Power Electronics*. New York: Oxford University Press, 1998, p. 265.

[11] M. Nalbant, Power factor calculations and measurements, *Proc. IEEE Applied Power Electronics Conf.*, pp. 543–552, 1990.

8.2 SLIDING MODE AND SWITCHING SURFACE CONTROL

Philip T. Krein

8.2.1 Introduction

Power electronic systems are an important class of systems that operate by *variable structure control*. Since they must act through switching, every control action changes the system structure. In control systems, *sliding mode control* is a type of variable structure control in which specific state dynamic behavior is imposed on a system through switching: system trajectories are drawn to, and remain on a surface in the state space on which the switching occurs (ideally at infinite frequency) [1,2,3]. Power electronic systems are more general variable structure control applications, since switch action is always used, whether or not the state dynamics are to be constrained in an explicit way. Sliding mode controls have been introduced in three-phase converters [4], in motor drives [5,6], and in a variety of other power electronics applications [7,8,9]. The term *switching surface control* will be used to refer to an extension of sliding mode control, suitable for broad use in power electronics. A general switching surface control framework was proposed by Burns [10], and later expanded by Bass [11]. The treatment given here follows [12]. A useful tutorial discussion of variable structure control in the context of switching can be found in [13]. In the general case, state-dependent switch action is represented using *switching surfaces* in state-space. The active switches toggle when surfaces are crossed.

Sliding mode control can be represented with a switching surface (which in the n-dimensional case is formally a hypersurface in state space) along which state action is to be constrained. The key issues are a *reaching condition*, to ensure an initial condition can be driven to the hypersurface, and a *sliding condition*, which must be met to keep the state on the hypersurface once it is reached. When the reaching condition is satisfied locally, the surface is said to be *attractive*. When reaching and sliding conditions are met, a *sliding surface* is said to be generated, and the system operates in a *sliding regime*. One drawback is that a sliding regime is associated with infinite switching frequency as switch action forces the state dynamics to follow the surface. This drawback can be avoided with the addition of timing constraints such as minimum on-time or off-time, or a hysteresis band can be added in the control. Sliding mode control is well-established as a useful tool in power converters [4,5,6,7,8,9,14,15,16]. Another widely used but rather limited related method is termed *hysteresis control*, in which a single specific state or output is used to make a decision about switch action.

Switching surface control is a generalization of sliding mode control, in which switching surfaces are not necessarily associated with sliding regimes. Switching surface

control is a direct large-signal method, meaning that it addresses the complete operation of a converter. Like other large-signal methods [17], switching surface methods do
not separate startup, steady-state, and protection modes. Hysteresis control is a simple
example of switching surface control, and can be illustrated with a buck converter. If
the transistor is turned on when the output voltage is too low, and then turned off when
it becomes too high, the converter will be constrained to operate close to a fixed output
voltage. The target output voltage defines a switching surface, and switch action is
taken when the actual output voltage crosses this surface. The general action does
not depend on initial conditions or even on circuit parameters.

   The systems to be considered in power electronics can be described with a nonlinear *network equation* of the form

$$\dot{\mathbf{x}} = f(\mathbf{x}, u, q) \tag{8.2}$$

where q is a vector switching function with $q_i \in \{0, 1\}$ and u is the external input. The
dimension of $\mathbf{x}$ matches the number of state variables in the converter, while the vector
switching function q has dimension equal to the total number of possible circuit configurations. When N switches are present, the dimension of q can be as high as 2^N,
although some configurations can be ruled out by circuit laws. By setting each switching
function q_i to a value of 1 (and setting values $q_j = 0$, $i \neq j$) in turn, state variable
expressions can be found for each configuration. Each will have an equilibrium
point. For switching surface control, we can define a surface through a constraint of
the form

$$\sigma(\mathbf{x}) = 0 \tag{8.3}$$

Flat surfaces are common, and will be used as the focus in this treatment. However,
curved surfaces and time-dependent surfaces $\sigma(\mathbf{x}, t) = 0$ are not ruled out and are a
topic of current research.

8.2.2 Hysteresis Control

   Hysteresis control, as the term is used in power electronics, operates the active
switch based on a comparison between the actual output voltage v_{out} and a pair of
reference values $V_{ref} + \Delta v$ and $V_{ref} - \Delta v$. Output voltage hysteresis control applies
only to buck-type converters. The time behavior of a buck dc/dc converter under
hysteresis control is shown in Figure 8.3. The figure illustrates the initial startup transient, followed immediately by steady-state behavior. It is usual to have separate high
and low switching surfaces (as in a thermostat), with $\sigma_{low} = v_{out} - V_{ref} - \Delta v = 0$ and
$\sigma_{high} = v_{out} - V_{ref} + \Delta v = 0$. The space between them defines a *hysteresis band*, in

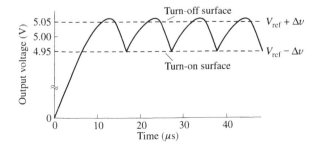

Figure 8.3 Output voltage of a buck converter under hysteresis control.

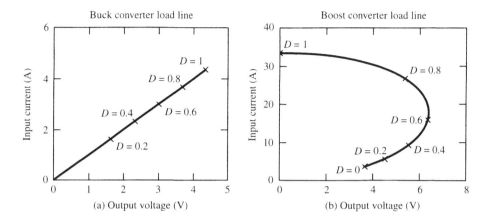

Figure 8.6 Operating point families for (a) buck and (b) boost converters. From [12], ©1998, Oxford University Press; used by permission.

figure shows that there are two duty ratio values that can provide a given output voltage, but duty ratios above 60% attempt to exceed the maximum power transfer capability of the converter and yield high losses. Thus the upper portion of the curve beyond 60% duty is not of interest. For any converter, the operating point family is the same as a *load line*, since the points are those that are physically consistent with the load resistance at various values of D.

Switching Surface-Based Control Laws

In switching surface control, a single switching surface $\sigma(\mathbf{x}) = 0$ governs switch action. The surface intersects the converter load line at the target operating point. For a straight line or flat surface,

$$\sigma(\mathbf{x}) = \mathbf{k}_\sigma \cdot (\mathbf{x} - \mathbf{x}_0) = 0 \qquad (8.4)$$

where $\mathbf{x}_0$ is the intended operating point and $\mathbf{k}_\sigma$ is a constant gain vector that defines the switching surface slope. The slope becomes a control parameter, and the switching surface divides the state space into two half-spaces. As the converter state follows a trajectory, switching takes place when the trajectory contacts the switching surface. In sliding mode control, the switching surface slope is selected to be attractive and to meet sliding conditions. In output hysteresis control, only one gain value is used and there is no explicit constraint on state action. In a two-state dc/dc converter, output hysteresis control can be represented by a vertical line at the intended V_{out}. For the buck converter, state behavior is summarized in Figure 8.7 during the transition from startup to steady-state operation. The transistor is on whenever the state is to the left of the surfaces, and off whenever the state is to the right of the surfaces. Since the vertical switching surface (in this case, a switching line) keeps the configuration equilibrium points separate, there is no chance that the state will somehow be pulled toward an incorrect operating point, and the control is globally stable.

For the boost converter, a vertical switching surface at a given V_{out} value greater than V_{in} will not work. If such a line is drawn on Figure 8.6, the switch action is ambiguous. If the state is to the left of the switching surface, turning the transistor on

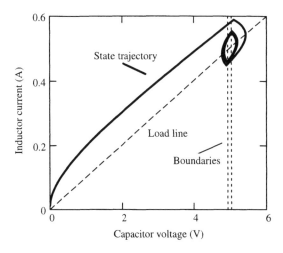

Figure 8.7 State evolution for buck converter under hysteresis control. From [12], ©1998, Oxford University Press; used by permission.

will move the system toward the point $(0, V_{in}/R_L)$, away from the switching surface. If the transistor is turned off instead, the trajectory will not intersect the switching surface from all possible starting points. Since neither configuration will ensure that the surface is reached from any possible starting point, a vertical switching surface does not lead to a practical control for a boost converter. More generally, notice that because both of the configuration equilibrium points are located to the left of the boundary, there is no global choice of switch action that avoids these points. A further dilemma in the boost converter is that a vertical switching surface intersects the load line at two points: the desired operating point and a second high-current high-loss point on the upper portion of the curve in Figure 8.6(b). Interestingly, if a starting point located above the target operating point is used for the boost converter, a sliding regime will be generated if we elect to switch the transistor off to the left side of the surface. However, this sliding regime moves away from the target point to the high-current operating point. Certain other switching surface choices, such as a horizontal switching surface, will satisfy a reaching condition and in fact will work as the basis of a switching surface control in a boost converter.

The undesired action of a boost converter with a vertical switching surface can also be understood in another way. It is well known [20] that most dc/dc converter topologies have nonminimum phase behavior and have small-signal models with right half-plane zeros. In the boost converter, when the active switch turns on, the output voltage falls; it recovers for a short time after the switch turns off. Thus the active switch action is only indirectly tied to output dynamics, and a vertical switching surface does not establish stable dynamics for the whole system. A switching surface that separates the configuration equilibrium points can avoid the problem caused by nonminimum phase behavior.

Necessary Conditions for Switching Surface Controls

A vertical switching surface yields a control that is not globally stable for a boost converter. This suggests certain conditions that must be met for any switching surface control:

- The switching surface must separate the equilibrium points of the two (or more) configurations.
- The switches must act in opposition to trajectory motion toward an equilibrium point. That is, the switch state at any given moment must be selected to force state dynamics toward the opposite configuration equilibrium point. This prevents the system from reaching any of the configuration equilibrium points.
- The switching surface must pass through the desired operating point to ensure that switch action can drive the system to that point.
- A hysteresis band or other switching frequency limit must be provided to avoid chattering.

The first two conditions are a verbal statement of the reaching condition. If each configuration of a converter has a stable equilibrium point, if the switching surfaces separate these points, and if the switch operates to move the converter toward the equilibrium point in the half-space opposite to that in operation, operation will always be forced toward the switching surface eventually. These are sufficient conditions for reaching the switching surface. In the sliding mode case, an additional condition is that the switching surface be attractive. This requires the trajectories to approach the switching surface from points in its neighborhood.

For the case of hysteresis controls, a single state variable is used, and the switching surfaces are either vertical or horizontal. Both such switching surfaces will meet the conditions for a buck converter. In a boost converter, input current (a horizontal switching surface) will work. In a half-bridge inverter, output current hysteresis will work. In a buck-boost converter, the inductor current provides a useful hysteresis control.

Sample Outputs and Hysteresis Design Approaches

Figure 8.8 shows results for a few choices of L and C for a 24V-to-5V buck converter with a 1.25Ω load, operating under hysteresis control with a 50mV hysteresis band. The results confirm the robustness with respect to converter parameters. For example, ripple changes only slightly over a factor-of-four change in capacitance and inductance. Indeed, the dynamics are such that a $1\mu F$ capacitor gives somewhat lower ripple than a $4\mu F$ capacitor for the same inductor. It is interesting that current-based hysteresis controls give much different operation. For this same converter, current hysteresis uses a 4A setting. Results with $L = 100\mu H$, $C = 2\mu F$, and a 1.25Ω load are shown in Figure 8.9. With these choices, a sliding mode control is achieved. Once the switching surface is reached, the system is constrained to remain at 4A, except only for the 1% hysteresis band. In Figure 8.9, the output voltage shows no overshoot, and the output ripple is determined entirely by the hysteresis band. In contrast, the ripple exceeded the hysteresis band in the voltage hysteresis case. This basic example demonstrates that sliding modes are special types of switching surface controls which are more effective than general switching surface controls at constraining dynamics.

For a voltage-source inverter, the output under current hysteresis control is a reference sine wave with a superimposed triangular ripple. An inverter must be able

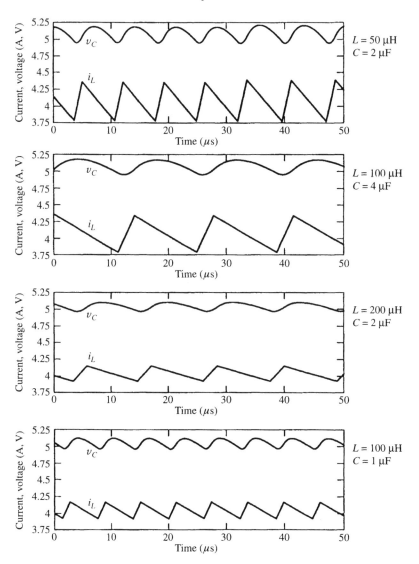

Figure 8.8 Behavior near operating point for 24V-to-5V buck converter, 1.25Ω load, with various L and C values. From [12], ©1998, Oxford University Press; used by permission.

to deliver a higher rate of current change than that in the reference waveform. If the inductor is too large or the input voltage is too small, the output will be unable to track the desired sinusoid, and distortion will appear.

Hysteresis control design is straightforward for any converter with just a few states. The design procedure follows along these lines:

1. Identify the converter specifications, including a nominal switching frequency and load.

2. Choose L and C values to enforce the desired ripple level under nominal conditions.

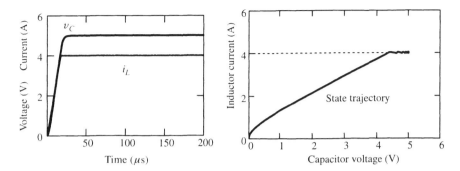

Figure 8.9 Behavior of buck converter with 4A hysteresis inductor current control. From [12], ©1998, Oxford University Press; used by permission.

3. Choose between vertical and horizontal switching surfaces. Check that the choice meets the key switching surface control requirements (equilibrium point separation and switch action that opposes motion toward equilibria).

4. For a dc/dc converter under current hysteresis, choose a hysteresis band consistent with the desired current ripple. The relationship is direct: a 1% hysteresis band yields 1% ripple. For voltage hysteresis, the ripple depends on the degree of over-shoot, and an iterative process is one way to select the right switching surface hysteresis band to enforce a specified ripple.

5. For an inverter, ensure that the maximum derivative determined by $v_L = L(di/dt)$ is high enough to follow the desired current change. If the current is $I_0 \sin(\omega t)$, then $v_L/L > \omega I_0$ is required. The inductor voltage does not match the input voltage, since $v_L = V_{in} - i(t)R_{load}$. An estimate motivated by the exponential behavior is the choice $V_{in}/L > 2.8\omega I_0$, which will provide a workable design.

An important consideration with many types of switching surfaces is the generation of reference values. A horizontal surface for a boost converter, for instance, presupposes that the correct value of inductor current is known. More commonly, only the target output voltage is known and the current value is a function of the unknown load. Multiloop control is well known in power electronics as a way to obtain a reference current value for an inner regulation loop [21]. For the outer loop, an integral loop or proportional-integral (PI) controller compares the output voltage to the reference value. The loop output is used in place of the current reference for the purposes of the "inner" switching surface controller. The outer loop must avoid any possibility of convergence to the undesired high-current value.

8.2.4 Global Stability Considerations

Successor Points

When a switching surface control satisfies the reaching condition, global stability can be assured if each successive intersection with the switching surface brings the operation closer to the target point. Based on [22,23], every subsequent intersection of the state trajectory and the switching surface $\sigma = 0$ after the first can be termed a *successor point*. For stability, it is necessary and sufficient to show that the distance from successor points to $\mathbf{x}_0$ diminishes with time. Successor points were used to evaluate stability of a boost converter as early as 1981 [24]. The successor point concept is closely related to a Poincaré return map.

Behavior Near a Switching Surface

Since switching occurs when a switching surface is crossed, it is helpful to consider the trajectory behavior at a switching surface. Points along $\sigma = 0$ can be classified based on the directions of *on* and *off* trajectories at the switching surface. There are three possibilities, shown in Figure 8.10:

1. *Refractive* points have state trajectories directed toward $\sigma = 0$ on one side and away from $\sigma = 0$ on the other.
2. *Attractive* points have state trajectories directed toward $\sigma = 0$ on both sides.
3. *Rejective* points have state trajectories directed away from $\sigma = 0$ on both sides.

When a switching surface exhibits refractive behavior, the system dynamics will act like those shown in Figure 8.7. The trajectories form a loop that tends to get smaller over time in the absence of a hysteresis band. Switching frequency limits or a hysteresis band will lead to a stable limit cycle in steady state. It is possible to show that voltage hysteresis control in dc/dc buck converters always yields refractive behavior when the load is resistive.

When attractive behavior is present at the switching surface, a sliding regime is generated. State action is constrained to follow the switching surface itself. In this case, a hysteresis band avoids chattering and directly enforces a specific ripple. Since state action remains very close to the switching surface, we can approximate the behavior as if the trajectory were in fact coincident with the switching surface—equivalent dynamics along a sliding surface that define a sliding regime. We can test for an attractive switching surface by examining the normal component of the state trajectories with respect to $\sigma(\mathbf{x})$; refer to Figure 8.11. Trajectory velocities are given by the state vector $\dot{\mathbf{x}}$, and the component of velocity normal to the switching surface should be such that the surface is approached from either side. This normal velocity turns out to be proportional to $\dot{\sigma}(\mathbf{x})$, so the trajectories will be directed toward $\sigma(\mathbf{x}) = 0$ if

$$\sigma(\mathbf{x})\dot{\sigma}(\mathbf{x}) < 0 \qquad (8.5)$$

is satisfied in the domain of interest. This condition does not apply to refractive boundaries but is necessary for a sliding regime to occur.

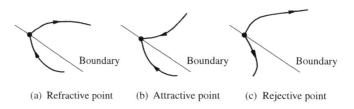

(a) Refractive point   (b) Attractive point   (c) Rejective point

Figure 8.10 Trajectory behavior at a switching surface crossing. From [12], ©1998, Oxford University Press; used by permission.

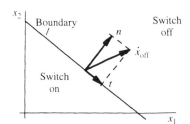

Figure 8.11 Switching surface with normal and tangential vectors.

Rejective behavior at the desired operating point implies that the converter is unstable, since the states will move away from the desired operating point.

A converter can make transitions among the three modes during a transient. However, only a few types of transitions are possible for a stable converter for which the various circuit configurations are linear and the load is resistive [25]. In particular, a transition from refractive mode to attractive mode can occur, but not the opposite. The implication is that once a sliding mode is established after a transition from refractive mode to attractive mode, it will be maintained right up to the target operating point [19]. A good strategy is to operate in refractive mode far from $\mathbf{x}_0$, then in attractive mode near $\mathbf{x}_0$. This tends to generate fast dynamics when large disturbances are encountered, and good small-signal behavior with no overshoot and with predictable sliding mode behavior near the operating point. Figure 8.12 shows the state space and time response for a buck dc/dc converter in which a high-performance switching surface controller is at work. From the initial condition at (0,0), the transistor turns on and ramps the inductor current. When the current reaches the switching surface (at about 8.5A in this example), the switch turns off, and refractive behavior drives the converter very close to its target operating point (5V, 5A) in a single switching cycle. This control provides excellent performance, although it is not robust since the switching surface that yields the desired switching point depends on system parameters.

Choosing a Switching Surface

The ideal switching surface provides global stability, good large-signal operation, and fast dynamics. Typically, a good choice for a switching surface has a sliding regime around the operating point and a refractive mode farther away. The nature of directional behavior near a switching surface can be tested with basic vector analysis. Each trajectory can be associated with a velocity field. As in Figure 8.11, the normal component tells whether a point of interest is refractive, attractive, or rejective. The tangential component tells the rate at which successor points approach or recede from the target operating point. For the best possible dynamic performance, the tangential velocity should be as high as possible, directed toward the operating point.

For applications such as inverters, the output reference is a function of time, and in general should be represented as a time-dependent surface $\sigma(\mathbf{x}, t) = 0$. Since the switching frequency is intended to be much higher than the output frequency in a practical inverter, the surface time variation is slow enough for it to be treated as quasi-steady, and no special problems are added.

Pulse-width modulation (PWM) is the conventional approach for fixed-frequency control in a power converter. Closed-loop PWM has an interesting switching surface

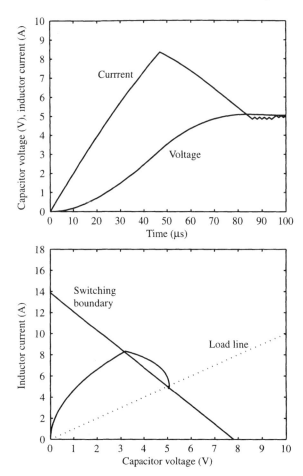

Figure 8.12 Time and state-space behavior of a high-performance switching surface controller for a buck converter, from [26].

interpretation: the triangle carrier waveform can be represented as a moving switching surface, with a slope dependent on feedback gains for voltage and current [11]. In this case, the surface dynamics are at about the same rate as the switching period. With a symmetric triangle waveform, the switching surface moves back and forth in state space. With a ramp, the switching surface moves in one direction, then jumps abruptly to its starting position. In most closed-loop converters, switch action takes place when the state values cross the moving switching surface. The concept of a moving switching surface can be used to model the conventional practice of adding a ramp in a hysteresis controller to enforce a fixed switching frequency [18]. It is also possible to relate converter action under pulse-width modulation directly to the behavior under sliding mode control [15].

Many practical power electronic systems require explicit current or voltage limits to protect sources or loads, or to keep components within ratings. In the context of switching surface control, limits can be enforced by adding additional surfaces. As an example, consider again a boost converter. The primary switching surface σ can be selected to separate configuration equilibrium points and produce the desired state dynamics. An additional horizontal surface can be added at the point $i_L = i_{max}$. The transistor switches off when the states move above this surface. The composite surface, now nonlinear, is a very practical way to control this converter.

Higher Dimensions

For converter systems with more than two states, switching surface control becomes only slightly more complicated. Most dc/dc converters with additional energy storage elements, for example, retain a single active switch and use only two configurations. Most converters in discontinuous mode add a configuration, but not another equilibrium point. A switching surface can be defined in n-space that separates the equilibrium points and meets the other requirements of switching surface control. In each half-space defined when the switching surface is in place, the switching configuration is chosen to drive toward the opposite equilibrium point so that the switching surface is reached from any initial state. The surface σ can be chosen for refractive behavior far from the operating point and attractive behavior nearby. While the geometry of the behavior is easy to visualize in two or three dimensions, the extension of sliding mode control to n-space is well established. The condition $\sigma\dot{\sigma} < 0$ still indicates attractive behavior.

In higher dimensions, the biggest challenge is to establish the appropriate switching sequence. In dc/dc converters, the switch action is unambiguous and is determined immediately by the choice of switching surface. In effect, the switch configuration is a static function of state. In contrast, a circuit such as a three-phase inverter has three pairs of switches. The switch configuration is a dynamic function of the states [5].

8.2.5 Summary

Switching surface control, which encompasses sliding mode control and hysteresis control, is a tool of variable structure systems well suited to power electronics. Switching surfaces can establish global stability, enforce state dynamics when in a sliding regime, or keep values within direct rating limits. Good converter performance can be obtained by choosing a surface with refractive behavior far from the desired operating point and attractive (sliding mode) behavior near the operating point. Switching surface controllers are robust to parameter variation and energy storage values within the converter. Outer-loop control can be used to make them robust to load variation. The applications of these controllers in power electronics continue to expand.

REFERENCES

[1] H. Sira-Ramirez, Nonlinear variable structure systems in sliding mode: The general case, *IEEE Trans. on Automatic Control*, vol. 34, no. 11, pp. 1186–1188, November 1989.

[2] V. I. Utkin, Application oriented trends in sliding mode control theory, *Proc. IEEE IECON*, pp. 1937–1942, 1993.

[3] K. D. Young, V. I. Utkin, and U. Ozguner, A control engineer's guide to sliding mode control, *IEEE Trans. on Control Sys. Tech.*, vol. 7, no. 3, pp. 328–342, May 1999.

[4] N. Sabanovic-Behlilovic, A. Sabanovic, and T. Ninomiya, PWM in three-phase switching converters – sliding mode solution, *IEEE Power Electronics Specialists' Conf. Rec.*, pp. 560–565, 1994.

[5] A. Sabanovic and D. B. Izosimov, Application of sliding modes to induction motor control, *IEEE Trans. on Industry Applications*, vol. IA-17, no. 1, pp. 41–49, January 1981.

[6] A. Damiano, G. Gatto, and I. Marongiu, A sliding mode control technique for direct speed control of induction motor drives, *IEEE Power Electronics Specialists' Conf. Rec.*, pp. 1106–1111, 2000.

[7] M. Carpita and M. Marchesoni, Experimental study of a power conditioning system using sliding mode control, *IEEE Trans. on Power Electronics*, vol. 11, no. 5, pp. 731–742, September 1996.

[8] G. Escobar and H. Sira-Ramirez, A passivity based sliding mode control approach for the regulation of power factor precompensators, *Proc. IEEE Conf. Decision and Control*, pp. 2423–2424, 1998.

[9] R. Venkataramanan, A. Sabanovic and S. Cuk, Sliding mode control of dc-to-dc converters, *Proc. IEEE IECON*, pp. 251–258, 1985.

[10] W. W. Burns III and T. G. Wilson, State trajectories used to observe and control dc-to-dc converters, *IEEE Trans. on Aerospace Electronic Systems*, vol. AES-12, pp. 706–717, 1976.

[11] R. M. Bass, Large-scale tools for power electronics: state space analysis and averaging theory, PhD diss., University of Illinois at Urbana-Champaign, March 1991.

[12] P. T. Krein, *Elements of Power Electronics*. New York: Oxford University Press, 1998. Portions of Chapter 17 are used by permission.

[13] R. DeCarlo, S. H. Zak, and G. P. Matthews, Variable structure control of nonlinear multivariable systems: A tutorial, *Proc. IEEE*, vol. 76, no. 3, pp. 212–232, March 1988.

[14] V. I. Utkin and A. Sabanovic, Sliding mode applications in power electronics and motion control systems, *Proc. IEEE Int. Symp. Industrial Electronics*, pp. TU22-TU31, 1999.

[15] H. Sira-Ramirez and M. Ilic, A geometric approach to feedback control of switch mode dc-to-dc power supplies, *IEEE Trans. on Circuits and Systems*, vol. 35, no. 10, pp. 1291–1298, October 1988.

[16] H. Sira-Ramirez, Switched motions in bilinear switched networks, *IEEE Trans. on Circuits and Systems*, vol. CAS-34, no. 8, pp. 919–932, August 1987.

[17] R. W. Erickson, S. Cuk and R. D. Middlebrook, Large-signal modelling and analysis of switching regulators, *IEEE Power Electronics Specialists' Conf. Rec.*, pp. 240–250, 1982.

[18] L. Malesani, P. Mattavelli, and P. Tomasin, Improved constant-frequency hysteresis current control of VSI inverters with simple feedforward bandwidth prediction, *IEEE Trans. on Industry Applications*, vol. 33, no. 5, pp. 1194–1202, September 1997.

[19] R. Munzert and P. T. Krein, Issues in boundary control, *IEEE Power Electronics Specialists Conf. Rec.*, pp. 810–816, 1996.

[20] D. Mitchell, *DC-DC Switching Regulator Analysis*. New York: McGraw-Hill, 1988.

[21] R. D. Middlebrook, Topics in multiple-loop regulators and current-mode programming, *IEEE Trans. on Power Electronics*, vol. 2, no. 2, pp. 109–124, April 1987.

[22] T. Vogel, Sur les systèmes déferlants, *Bulletin de la Société Mathématique de France*, 1953.

[23] N. Minorsky, *Nonlinear Oscillations*. Princeton: Van Nostrand, 1962.

[24] M. Kulawik, Stability analysis of boost converter with nonlinear feedback, *IEEE Power Electronics Specialists Conf. Rec.*, pp. 370–377, 1981.

[25] R. Munzert, Boundary control, applied to dc-to-dc converter circuits, Studienarbeit, Technische Hochschule Darmstadt, Germany. Also technical report UILU-ENG-95-2545, University of Illinois at Urbana-Champaign, July 1995.

[26] M. Greuel, R. Muyshondt, P. T. Krein, Design approaches to boundary controllers, *IEEE Power Electronics Specialists' Conf. Rec.*, pp. 672-678, 1997.

8.3 ENERGY-BASED CONTROL IN POWER ELECTRONICS

Alex M. Stanković
Gerardo Escobar
Romeo Ortega
Seth R. Sanders

8.3.1 Introduction

This section describes several power converter control techniques that are based on concepts of energy. We focus on control algorithms that are based on averaged models derived in Section 2.1. Later in this section we concentrate on dc/dc converters, and illustrate four fairly general control design approaches on a boost converter example. We also establish a link with control structures that aim to control switches directly, as in the case of sliding mode control. Many such policies may be based on measured or estimated energy in certain elements of a converter. For example, dynamical models expressed in terms of energy are very useful for the class of resonant converters that relies on bandpass property of a second-order circuit (the resonant tank) to enable power conversion. An example of this approach is [1], where the converter switching is directly based on the measured tank current and voltage, and the control goal is to minimize surges in tank energy. A related control idea is presented in [2], where the energy stored in the tank is linearized and used in discrete-time control design based on root locus.

In power electronic converters, as in most engineered systems, modeling tasks are often not separable from control tasks. Dynamical descriptions of energy processing systems in terms of power flows are typically simpler than alternatives. Such models may be regarded as coordinate transformations of original models expressed in terms of Kirchhoff's voltage and current laws, with the coordinate transformation being based on physical considerations. For example, model of a switching preregulator based on the square of the capacitor voltage is presented in [3, pp. 396–399], and results in a *linear time-invariant* discrete-time model (averaged over a line cycle). The same approach is extended to the case of fast preregulators in [4] where a *linear periodically varying* model is obtained (averaged over one switching cycle). Another instance when the resulting large signal model is linear in energy coordinates is provided by the up/down dc/dc converter in discontinuous conduction (Problem 12.5 in [3]). An energy-based averaging modeling procedure for dc/dc converters that includes parasitics is presented in [5].

Controllers designed from an energy flow standpoint often have a number of desirable properties in implementations [6]. These include: (1) ease in measuring or estimating key quantities in control laws; (2) downward compatibility with linear controllers (i.e., the possibility of deriving linear control laws while maintaining global stability); and (3) capability of globally stabilizing a more complex system in which the original converter is embedded.

The concepts of incremental energy [7] and scaled total energy are [8] pivotal in deriving control laws for switched-mode dc/dc power converters. Energy-based control, however, is not limited to switched-mode dc/dc converters. A useful application of the same general methodology to pulse-width modulated rectifiers (ac/dc converters) is presented in [9,10]. Models based on generalized averaging and dynamic phasors (see

Section 2.1) are well suited for use in energy-based control. Examples of this type include output feedback control of series resonant dc/dc converters [11], and control of unbalanced three-phase active filters (inverters) in [12].

8.3.2 Circuit-Theoretic Approaches

Consider the boost circuit shown in Figure 8.13; this is an example of the class of switched-mode dc/dc converters whose averaged dynamical model (introduced in Chapter 2) is of the form

$$\frac{d\mathbf{x}}{dt} = A\mathbf{x} + (B\mathbf{x} + \mathbf{b})d + \mathbf{f} \tag{8.6}$$

where $\mathbf{x}$ is the state vector (two-dimensional in this case), d is the duty ratio (i.e., the dc component of the switching function $q(t)$ averaged over a switching cycle), and A, B, $\mathbf{b}$ and $\mathbf{f}$ are matrices and vectors of appropriate sizes that depend on parameters such as the load, component values, and the input voltage. In the case of a boost converter, with the inductor current as the first component of the state vector $\mathbf{x}$ and the capacitor voltage as the second component,

$$A = \begin{bmatrix} 0 & \frac{-1}{L} \\ \frac{1}{C} & \frac{-1}{RC} \end{bmatrix}, \quad B = \begin{bmatrix} 0 & \frac{1}{L} \\ \frac{-1}{C} & 0 \end{bmatrix}, \quad \mathbf{f} = \begin{bmatrix} \frac{E}{L} \\ 0 \end{bmatrix}, \quad \mathbf{b} = 0$$

We denote the steady-state variables with a bar (i.e., $\bar{\mathbf{x}}$); these quantities satisfy

$$0 = A\bar{\mathbf{x}} + (B\bar{\mathbf{x}} + \mathbf{b})\bar{d} + \mathbf{f} \tag{8.7}$$

The typical control task is to maintain some variables (say output voltage) at a desired value $\mathbf{x_d}$ in the presence of parametric perturbations like load and input voltage changes. Following the standard practice in power electronics, we denote the duty ratio corresponding to the desired steady state $\mathbf{x_d} = [I_d \ V_d]^\top$ with $D = \bar{d}_d$. We will consider cases when either the full state or its subset (typically the output voltage) is available for measurements.

Basic Control

The approach introduced in [7] is to consider incremental quantities

$$\tilde{\mathbf{x}} = \mathbf{x} - \mathbf{x_d}$$

$$\tilde{d} = d - D$$

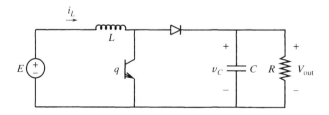

Figure 8.13 Circuit schematic of a boost converter.

In the basic formulation that assumes known parameters, the (scalar) energy (Lyapunov) function of interest is

$$V(\mathbf{x}) = \frac{1}{2}\tilde{\mathbf{x}}^\top Q \tilde{\mathbf{x}} \tag{8.8}$$

where Q is positive definite, for example

$$Q = \begin{bmatrix} L & 0 \\ 0 & C \end{bmatrix}$$

After evaluating the time derivative of $V(\mathbf{x})$,

$$\frac{d}{dt} V(\mathbf{x}) = \tilde{\mathbf{x}}^\top Q(A + DB)\tilde{\mathbf{x}} + \tilde{d}\big[\tilde{\mathbf{x}}^\top Q(B\mathbf{x}_d + B\tilde{\mathbf{x}} + b)\big] \tag{8.9}$$

Many control laws can be constructed to guarantee nonpositivity of the right-hand side of (8.9); a straightforward one defines the new output signal

$$y = \tilde{\mathbf{x}}^\top Q(B\mathbf{x}_d + B\tilde{\mathbf{x}} + b) \tag{8.10}$$

together with the control input selection

$$\tilde{d} = -\alpha y$$

where α is a positive scalar. Typically, the expression for $\tilde{d}$ simplifies considerably for an actual converter; e.g., for the boost converter

$$y = V_d x_1(t) - I_d x_2(t)$$

Note that this control law is linear with the idealized converter model that we use.

Before we can establish asymptotic stability of the power converter with the proposed control law, we have to look into two additional issues. The first deals with the fact that the duty ratio $d(t)$ is constrained to be between 0 and 1, which in turns places saturation limits on $\tilde{d}(t)$. It turns out, however, that at either saturation limit $d/dt V(\mathbf{x})$ is negative (and dependent on α), so that the system will quickly enter the unsaturated region. The second issue arises from the fact that the proposed control policy guarantees only the nonpositivity of $\frac{d}{dt}V(\mathbf{x})$, and not its negativity. Our intention is, of course, to use LaSalle's theorem [13, p. 115] to prove asymptotic stability, so we have to rule out the possibility that $\frac{d}{dt}V(\mathbf{x})$ is identically zero on a set (trajectory) outside the desired equilibrium. In the boost converter example this requires $y \equiv 0$ and $\frac{d}{dt}y \equiv 0$, which turns out to be equivalent [7] to $\mathbf{x} \equiv \mathbf{x_d}$, thus establishing the asymptotic stability.

We will compare this and all succeeding control schemes on the example taken from [14] with the following parameter values: $E = 15\text{V}$, $L = 20\text{mH}$, $C = 20\mu\text{F}$, $R = 30\Omega$; the desired operating point is $\bar{\mathbf{x}} = [3.125 \ 37.5]^\top$ corresponding to the steady-state duty ratio $D = 0.4$, and the initial state is $\mathbf{x_0} = [2 \ 25]^\top$. The disturbance scenarios are: P_1, where at $t_1 = 50\text{ms}$ the load is step changed to $R = 25\Omega$, and P_2, where at $t_2 = 80\text{ms}$ it is restored to $R = 30\Omega$. We want to stress that in all simulations the emphasis is on qualitative behavior, and not on the optimal tuning of controller parameters; in practical implementations, such tuning would critically depend on the quality of available measurements.

In Figure 8.14 we display the output voltage response of the system with the described controller for disturbances P_1 and P_2; note that the response is satisfactory when parameters (in this case load resistance) are known accurately, but deteriorates when the load varies (i.e., between t_1 and t_2).

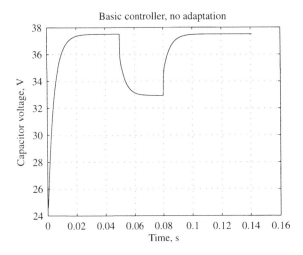

Figure 8.14 Output voltage with basic controller.

Adaptation

To make the proposed controller practical, we have to address the issue of parametric variations, in particular the inevitable changes in the load resistance R. The key issue in this extension is that the parameter variations (e.g., the load resistance, or equivalently, the load conductance $^1/_R$) enter the model *linearly*. In that case, we extend the model by gradient-type estimators for each varying parameter (simple integrators driven by weighted errors in measured quantities), and replace true parameters in (8.10) by their estimates. Note that this is a *certainty equivalence*–type controller, and we prove asymptotic stability of the overall system next.

We define a new energy function as

$$V(\mathbf{x}) = \frac{1}{2}\tilde{\mathbf{x}}^\top Q\tilde{\mathbf{x}} + \frac{1}{2}(\hat{\mathbf{x}}_d - \mathbf{x}_d)^\top K(\hat{\mathbf{x}}_d - \mathbf{x}_d) \qquad (8.11)$$

where $\hat{\mathbf{x}}_d$ is the estimate of the desired operating point $\mathbf{x}_d$. Note that in the case of a boost converter and load resistance uncertainty only the desired current I_d is affected, while the desired voltage V_d remains constant (and known). The estimates are of the gradient type

$$\frac{d}{dt}\hat{\mathbf{x}}_d = -K^{-1}Q(B\mathbf{x} + \mathbf{b}) \qquad (8.12)$$

The stability proof is now more technical, as boundedness of the estimate $\hat{\mathbf{x}}_d$ needs to be established; we refer the reader to [7] for details. All incremental quantities needed to evaluate the control output signal y in (8.10) are now calculated with respect to the corresponding estimates. With this control policy $^{dV}/_{dt}$ is again nonpositive, and we can conclude asymptotic stability as before.

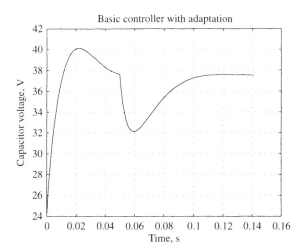

Figure 8.15 Output voltage with adaptive controller.

In Figure 8.15 we show the output voltage response to the load transient variation P_1, as described before. Note that this time load voltage recovers to its correct value.

Estimation and Output Feedback

The same energy function can be used in estimation, as shown in [15]. The estimator is a replica of the system with a correction term proportional to the difference between the measured system output y_m and the output predicted by the estimator:

$$\frac{d}{dt}\mathbf{z} = A\mathbf{z} + (B\mathbf{z} + \mathbf{b})d + \mathbf{f} + K[C(d)\mathbf{z} + F(d) - y_m]$$

$$y_m = C(d)\mathbf{x} + F(d)$$

(8.13)

Note that the duty ratio d is known in an estimation setup, rendering the system and estimation models linear. It is also required that the original system be observable from the measured output y_m, which can be checked by standard tools. In this case we are interested in the dynamics of error defined as $\mathbf{e} = \mathbf{x} - \mathbf{z}$, and we consider the candidate energy function

$$V(\mathbf{e}) = \frac{1}{2}\mathbf{e}^{\top}Q\mathbf{e}$$

(8.14)

and the feedback chosen as

$$K = -Q^{-1}C(d)^{\top}W(d)$$

(8.15)

for some positive definite $W(d)$, so that the time derivative of $V(\mathbf{e})$ is

$$\frac{d}{dt}V(\mathbf{e}) = \frac{1}{2}\mathbf{e}^{\top}\big[(QA + A^{\top}Q) + d(QB + B^{\top}Q) - 2C(d)^{\top}W(d)C(d)\big]\mathbf{e}$$

(8.16)

which is nonpositive because of the properties of matrices involved.

For our example of the boost converter, we display in Figure 8.16 the convergence of the output voltage open-loop estimate (8.13) during startup (assuming the controller (8.10) and accurate parameters).

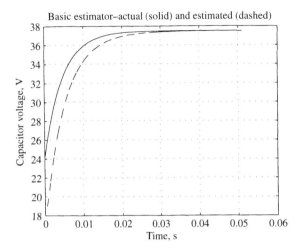

Figure 8.16 Actual (solid) and estimated (dashed) output voltage (with basic controller).

In addition, provided that it is sufficiently slow, parameter estimation can be added to estimation. Then an argument based on singular perturbation [15,16] can be used to establish convergence of state and parameter estimators.

At this point, we may be interested in combining state and parameter estimation with control. The main benefit of this type of control structure is in reducing the number of potentially costly measurements, while allowing for increased computations. While stability and convergence are established for the observer and the controller separately, and not jointly, the overall structure does lead to a satisfactory performance in our example, as illustrated in Figure 8.17 for the disturbance P_1.

8.3.3 Passivity-Based Control

Basic Controller

A closely related approach to the one described in the previous section is denoted as passivity-based control (PBC), as described in detail in [14]; this book also contains experimental comparisons of a number of different controllers, including those based

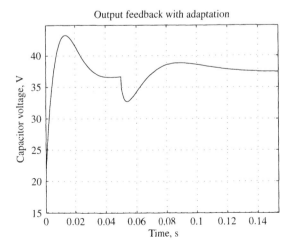

Figure 8.17 Adaptive output feedback controller.

on feedback linearization and sliding mode techniques. We first rewrite the boost converter model in the form (the same approach can be used for other converters as shown in [14]):

$$Q\frac{d}{dt}\mathbf{x} - (1 - d)J\mathbf{x} + G\mathbf{x} = \mathbf{u} \tag{8.17}$$

where the new symbols are

$$J = \begin{bmatrix} 0 & -1 \\ 1 & 0 \end{bmatrix}, \quad G = \begin{bmatrix} 0 & 0 \\ 0 & \frac{1}{R} \end{bmatrix}, \quad \mathbf{u} = \begin{bmatrix} E \\ 0 \end{bmatrix}$$

Consider first the *desired error dynamics*:

$$Q\frac{d}{dt}\tilde{\mathbf{x}} - (1 - d)J\tilde{\mathbf{x}} + G_t\tilde{\mathbf{x}} = \psi \tag{8.18}$$

where G_t contains added damping g_1 for the first state (i.e., the inductor current)

$$G_t = G + G_a = G + \begin{bmatrix} g_1 & 0 \\ 0 & 0 \end{bmatrix}$$

and

$$\psi = \mathbf{u} - \left(Q\frac{d}{dt}\mathbf{x_d} - (1 - d)J\mathbf{x_d} + G\mathbf{x_d} - G_a\tilde{\mathbf{x}}\right)$$

If we succeed (by choice of d and $\mathbf{x_d}$) to make $\psi = 0$, then the same energy function $V(\mathbf{x})$ as before (denoted as the *desired storage function* in [14]) will result in

$$\frac{d}{dt}V(\mathbf{x}) = -\tilde{\mathbf{x}}^\top G_t\tilde{\mathbf{x}} < 0 \tag{8.19}$$

The requirement that $\psi = 0$ translates into

$$L\frac{d}{dt}x_{1,d} + (1 - d)x_{2,d} - \frac{1}{g_1}(x_1 - x_{1,d}) = E$$
$$C\frac{d}{dt}x_{2,d} - (1 - d)x_{1,d} + \frac{1}{R}x_{2,d} = 0 \tag{8.20}$$

It is important to note that (8.20) defines the controller implicitly, as we need means to generate three functions ($x_{1,d}(t)$, $x_{2,d}(t)$, $d(t)$) that would satisfy (8.20). The stability of the controller is formally established in [14] by noting that the error dynamics (8.18) defines an output strictly passive map from ψ to $\tilde{\mathbf{x}}$.

To derive an explicit controller equation, we may try two routes: (1) direct regulation—fix the output voltage $x_{2,d}$ to the desired value, then calculate $x_{1,d}$ from the steady–state relationship (second line in (8.20)) as $x_{2,d}/R(1 - d(t))$ and then generate the duty ratio $d(t)$ to satisfy the first equation in (8.20); (2) indirect regulation—fix the input current to the desired value $E/R(1 - D)^2$, then calculate $x_{2,d}(t)$ from the first equation in (8.18), and finally generate the duty ratio $d(t)$ to satisfy the second equation in (8.18). It turns out that the direct regulation is not feasible in practice, as the associated zero dynamics (i.e., the dynamics of $d(t)$ with $\mathbf{x} = \mathbf{x_d}$) is not stable. The indirect regulation, however, results in stable zero dynamics, and this is the controller used in our simulations:

$$\frac{d}{dt}d = \frac{R(1-d)^2}{LV_d}\left[E - (1-d)V_d + \frac{1}{g_1}\left(i - \frac{V_d}{R(1-d)}\right)\right] \quad (8.21)$$

In Figure 8.18 we display the converter voltage following disturbances P_1 and P_2.

Adaptation

Parameters of any power electronic converter, and load in particular, are subject to variations. In that case, PBC has to be expanded to include adaptation. The key observation is that stable adaptation is possible for the case of parameters that enter *linearly* in the basic PBC. Thus the idea is to use the same control law as in PBC, and the parametric error will appear as an additive disturbance to the state error dynamics. The estimator will be of a gradient type, thus defining a passive operator from the state error to the parametric error. Given that the error dynamics defines an output strictly passive operator, and that a feedback interconnection of a strictly passive and a passive operator is stable (shown, e.g,. in [14]), we can establish stability of the overall adaptive control scheme.

In the case of a boost converter, the estimated parameter is the load conductance $\theta = 1/R$, and the desired output voltage $x_{2,d}$ is generated dynamically [14, p. 207]

$$\frac{d}{dt}\hat{\theta} = -\gamma x_{2,d}(x_2 - x_{2,d})$$

$$\frac{d}{dt}x_{2,d} = -\frac{\hat{\theta}}{C}\left[x_{2,d} - \frac{V_d^2}{Ex_{2,d}}\left[E + \frac{1}{g_1}\left(x_1 - \hat{\theta}\frac{V_d^2}{E}\right) + \gamma L \frac{V_d^2}{E}x_{2,d}(x_2 - x_{2,d})\right]\right]$$

$$d = 1 - \frac{1}{x_{2,d}}\left[E + \frac{1}{g_1}\left(x_1 - \hat{\theta}\frac{V_d^2}{E}\right) + \gamma L \frac{V_d^2}{E}x_{2,d}(x_2 - x_{2,d})\right] \quad (8.22)$$

This controller is derived using the energy function that is quadratic and contains state and parametric error terms; the actual proof is again technical in nature, as uniform continuity of the state error needs to be established; we refer the interested reader to [14,

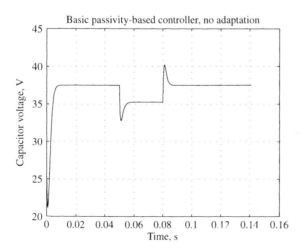

Figure 8.18 Basic passivity-based controller.

pp. 206–209]. In Figure 8.19 we display the response of the system controlled by the adaptive PBC following disturbances P_1 and P_2.

Hamiltonian Control

A novel way to control switched-mode dc/dc converters has recently been suggested in [17]. This methodology is based on the emerging theory of port-controlled Hamiltonian systems [18]. Its central feature is that it relies on energy-balancing principles instead of assigning quadratic incremental storage functions to the closed loop. This leads to designs which do not require an asymptotic inversion of the system dynamics—like the PBCs presented earlier—hence it applies directly to nonminimum phase systems.

We start by rewriting the system equations in the form

$$\frac{d}{dt}\mathbf{x} = [Jd - G]\frac{\partial H}{\partial \mathbf{x}}(\mathbf{x}) + \begin{bmatrix} 1 \\ 0 \end{bmatrix} E \tag{8.23}$$

where H is the total energy $H(\mathbf{x}) = 1/2\mathbf{x}^\top Q\mathbf{x}$.

The design objective, similarly to the previously described PBC, is to modify the energy function and modify the damping structure, but with two essential differences: (1) we want to preserve the Hamiltonian structure in closed loop; (2) the added damping need not be positive, but instead we will try to *distribute* the damping in the system so as to achieve better performance. That is, we want our closed-loop system to be of the form

$$\frac{d}{dt}\mathbf{x} = [Jd - G_d]\frac{\partial H_d}{\partial \mathbf{x}}(\mathbf{x}) \tag{8.24}$$

where

$$H_d(\mathbf{x}) = H(\mathbf{x}) + H_a(\mathbf{x})$$

is the desired energy function, $H_a(\mathbf{x})$ is the energy of the controller, and $G_d = G + G_a$ is the closed-loop damping, with G_a not necessarily positive semidefinite. It can be shown

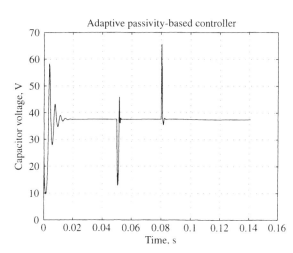

Figure 8.19 Adaptive passivity-based controller.

that stability will be assured if $H_d(\mathbf{x})$ has a minimum at the desired operating point $\mathbf{x}_d$ and $G_d \geq 0$. It can be shown that the control objectives will be achieved if and only if the following partial differential equation (which depends on the added damping G_a and control d)

$$[Jd - G_d]\frac{\partial H_a}{\partial \mathbf{x}}(\mathbf{x}) = G_a \frac{\partial H}{\partial \mathbf{x}}(\mathbf{x}) = gE \qquad (8.25)$$

is solvable for H_a.

The authors of [17] pose an additional output feedback constraint that the controller should use only measurements of the voltage x_2. Surprisingly, the solution turns out to be a nonquadratic energy function $H_d(\mathbf{x})$, and the controller does *not* depend on the load resistance:

$$d = D\left(\frac{x_2}{V_d}\right)^\alpha, \quad 0 \leq \alpha < 1$$

Notice that by selecting $\alpha = 0$ we recover the open-loop control $d = D$. Also, the standard notion of an incremental error $(x_2 - V_d)$ is conspicuous by its absence; instead the control action can be understood as a *modulation* of the open-loop control.

In Figure 8.20 we display the response of the system controlled by the adaptive PBC following disturbances P_1 and P_2.

8.3.4 Connections with Sliding-Mode Control

Many approaches to sliding-mode feedback control described in the literature disregard physical properties of the power converter and, in particular, ignore its energy dissipation characteristics. We will demonstrate in this section that an approach which combines the robustness of a sliding mode control with energy-based considerations is key to an improved control performance.

Sliding mode control strategies require a high switching frequency to guarantee good performance, and this condition may be difficult to satisfy in all cases. Furthermore, it may lead to *chattering* phenomena where high-frequency oscillations excite unmodeled system modes. It is well appreciated in the control community that

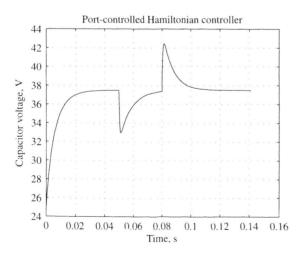

Figure 8.20 Port-controlled Hamiltonian controller.

sliding mode control belongs to the family of high gain controllers, which explains its robustness to parametric uncertainties in cases when parameters in question do not affect the definitions of sliding surfaces. On the other hand, this disregard for plant model will likely result in poor performance when the switching frequency is reduced.

We will study two modifications to sliding mode techniques. (1) We propose to incorporate into the sliding mode design the energy dissipation properties of the passivity-based control methodology. We show that with this new hybrid controller (PB + SMC) the regulated system may achieve a "smoother" behavior with enhanced performance. (2) With the goal of preventing the unnecessary switchings because of a control signal that may be the correct choice at the sampling instant, but not the right one during the whole sampling period, we suggest to use the average between the actual and the (one step ahead) errors. This strategy is referred as the *sliding mode + prediction*. This modification can be seen as an introduction of a *dynamical* sliding surface (i.e., the sliding surface is now represented by (first-order) stable dynamics).

Sliding-Mode Controller Revisited

While many sliding surfaces may be defined using linear combinations of both current and voltage, in a traditional SMC for the boost converter circuit a *current-mode* version is used in the inner control loop. Then the controller adopts as a sliding surface a desired *constant* inductor current corresponding to a desired equilibrium value of the capacitor voltage. Note that a sliding surface based on a desired constant equilibrium capacitor voltage leads to an *unstable* closed-loop dynamics [19]. This phenomenon is due to the underlying nonminimum phase zero dynamics associated with the capacitor voltage output. We first present the case of known parameters (treated in [20] in more detail), and later we analyze the more realistic unknown parameter case.

Consider the switching line $s = x_1 - V_d^2/RE$, where $V_d > 0$ is a desired constant capacitor voltage value, R is the load resistance, and E is the source. It has been shown in [20] that the switching policy given by

$$u = 0.5[\,1 - \text{sgn}\,(s)\,] = 0.5\left[\,1 - \text{sgn}\,(x_1 - V_d^2/RE)\,\right] \qquad (8.26)$$

locally creates a *stable* sliding regime on the line $s = 0$ with ideal sliding dynamics characterized by

$$\bar{x}_1 = \frac{V_d^2}{RE} \quad ; \quad \dot{\bar{x}}_2 = -\frac{1}{RC}\left[\bar{x}_2 - \frac{V_d^2}{\bar{x}_2}\right] \quad ; \quad u_{eq} = 1 - \frac{E}{\bar{x}_2} \qquad (8.27)$$

Moreover, the ideal sliding dynamic behavior of the capacitor voltage can be explicitly computed as

$$\bar{x}_2(t) = \sqrt{V_d^2 + \left[\bar{x}_2^2(t_h) - V_d^2\right]e^{-\frac{2}{RC}(t-t_h)}} \qquad (8.28)$$

where t_h stands for the reaching instant of the sliding line $s = 0$ and $\bar{x}_2(t_h)$ is the capacitor voltage at time t_h. The sliding regime is established on $s = 0$ provided $x_2 > E$, which is a well-known *amplifying* property of the *boost* converter.

In the case of *unknown* parameters consider the switching line $s = x_1 - x_1^*$ where x_1^* is the desired inductance current, necessary to guarantee the variation of the capacitor voltage toward its desired value $V_d > 0$ and to supply the power to the load (usually represented by an unknown resistor and/or a current source). It can be shown that the controller

$$u = 0.5\,[\,1 - \text{sgn}\,(s)\,] = 0.5\,[\,1 - \text{sgn}\,(x_1 - x_1^*)\,] \tag{8.29}$$

locally creates a *stable* sliding regime on the line $s = 0$. In this case and assuming that the dynamics of x_1 is much faster than the dynamics of x_2, that is, in a relatively short time x_1 converges toward its desired reference value x_1^* computed as follows

$$x_1^* = -K_p\big(x_2^2 - V_d^2\big) - K_i\phi \tag{8.30}$$

$$\dot{\phi} = \big(x_2^2 - V_d^2\big) \tag{8.31}$$

where parameters K_p and K_i are chosen relatively small to restrict the introduction of higher harmonics into the control loop.

Passivity-Based Sliding-Mode Controller

In the following development we utilize system dynamics written in terms of the auxiliary state vector, denoted by x_d. The basic idea is to create the sliding surface in terms of the auxiliary state vector x_d instead of using only the converter state vector x (note that x is used in generating x_d with the help of an auxiliary system). The feedback regulation, by means of a sliding mode, of the auxiliary state x_d toward the desired constant equilibrium value of the state x will in fact result in the specification of a *dynamical output feedback controller* for the original converter state. We will present the known parameter case for simplicity; a slight extension allows us to address the unknown parameter case as well.

Consider the switching line $s = x_{1d} - V_d^2/RE$, where $V_d > 0$ is a desired constant capacitor voltage value for the auxiliary variable x_{2d} and for the converter's capacitor voltage x_2. In [20] the following sliding surface which locally creates a sliding regime on the line $s = 0$ has been presented

$$u = 0.5\,[1 - \text{sgn}\,(s)\,] = 0.5\,[1 - \text{sgn}\,(x_{1d} - V_d^2/RE)\,] \tag{8.32}$$

This control drives the auxiliary dynamics toward the desired constant equilibrium state $(x_{1d}(\infty), x_{2d}(\infty)) = (V_d^2/RE, V_d)$ of the boost converter. The sliding mode locally exists on $s = 0$ provided $x_{2d} > E + R_1(x_1 - V_d^2/RE) > 0$.

Moreover, if the sliding-mode switching policy (8.32) is applied to *both* the converter and the auxiliary system, the converter state trajectory $x(t)$ converges toward the auxiliary state trajectory $x_d(t)$ and, in turn, $x_d(t)$ converges toward the desired equilibrium state, i.e.,

$$(x_1, x_2) \;\rightarrow\; (x_{1d}, x_{2d}) \;\rightarrow\; \left(\frac{V_d^2}{RE},\, V_d\right)$$

The ideal sliding dynamics is then characterized by

$$\bar{x}_{1d} = \frac{V_d^2}{RE} \; ; \; \dot{\bar{x}}_{2d} = -\frac{1}{RC}\left[\bar{x}_{2d} - \left(\frac{V_d^2}{E}\right)\frac{E + R_1(\bar{x}_1 - V_d^2/RE)}{\bar{x}_{2d}}\right]$$

$$u_{eq} = 1 - \frac{E + R_1(\bar{x}_1 - V_d^2/RE)}{\bar{x}_{2d}}$$

where $\bar{x}_1$ is the inductor current under sliding mode conditions. Furthermore, the ideal sliding dynamics for $\bar{x}_2(t)$ for $t \geq t_h$, can be explicitly computed in terms of the inductor current error signal $(x_1(t) - Vd^2/RE)$ as

$$\bar{x}_{2d}(t) = \sqrt{e^{-\frac{2(t-t_h)}{RC}}\bar{x}_{2d}^2(t_h) + \frac{2V_d^2}{RC}\int_{t_h}^{t} e^{-\frac{2(t-\sigma)}{RC}}\left[1 + \frac{R_1}{E}(\bar{x}_1(\sigma) - x_{1d}(\sigma))\right]d\sigma}$$

Simulations for this controller are presented in Figure 8.21 where we show the time response of the output capacitor voltage under the disturbances $P_1 - P_2$ described before, for the known-parameters case and also for the adaptive controller. Notice that while the former is nonrobust against disturbance in the load resistance since a steady-state error appears, the adaptive controller fully compensates for this type of disturbances.

Combining SMC with Prediction

As another extension to basic sliding mode control, we describe the case in which information about the system trajectory is included in selection of the switching pattern. In a straightforward implementation of SMC, a switching decision is based on sampled system states; this is actually quite reasonable with the assumed arbitrarily fast switching. In practice, however, switching frequency is limited, so the use of system model is likely to result in improved performance which avoids unnecessary switchings. We propose a controller that makes switching decisions based on the average between sampled and *one step ahead* predicted errors; the predicted error is estimated using the system model. This modification can be viewed as one involving a *dynamical* sliding surface represented by first-order stable dynamics.

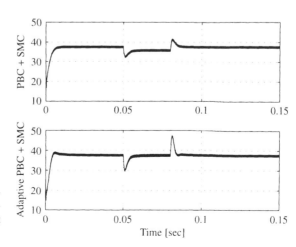

Figure 8.21 Output voltage $x_2(t)$: top trace—passivity-based sliding mode controller; bottom trace—adaptive passivity-based sliding mode controller.

We consider again the current control case; the control objective consists of selecting a control u from the set $\mathcal{U} = \{0,\ 1\}$ which minimizes the error signal

$$\sigma(u) = e_k + e_{k+1} \cong 2e_k + T\dot{e}_k(u)$$

where $e = x_{1,k} - x_{1,k}^*$, and $x_{1,k}^*$ is the desired inductance current reference to be defined later and we have used the following approximation

$$e_{k+1} \cong e_k + T\dot{e}_k(u)$$

with T the sampling period and $(\cdot)_k$ is the value of $(\cdot)$ at the instant $t = kT$. Notice that, $\dot{e}_k$, and thus σ, are functions of u, which is due the fact that the selected output has the relative degree 1. Since the set of possible controls is composed of only two elements, either 0 or 1, the switching policy can be rewritten in a form that is easier to implement

$$u(k) = 0.5[1 - \mathrm{sgn}\,(s)] \tag{8.33}$$

$$s = |\sigma(0)| - |\sigma(1)| \tag{8.34}$$

where $\sigma(0) = 2e + T\dot{e}(0)$ and $\sigma(1) = 2e + T\dot{e}(1)$. Notice that computations may be reduced if in the expressions above the reference signal x_1^* is assumed to be almost constant, i.e., $\dot{x}_1^* \cong 0$.

Let us now consider the outer control loop; we will assume that the load is represented by a power sink of unknown magnitude P_0. We again use the time scale separation between the states of the converter (i.e., after a relatively short time $x_1 = x_{1d}$). Under this assumption, the dynamics of x_2 reduces to

$$C\frac{d}{dt}\left(\frac{x_2^2}{2}\right) = -L\dot{x}_1^* x_1^* + Ex_1^* - P_0 \tag{8.35}$$

When x_1^* is slowly varying, and since the inductance L is typically quite small, we can neglect the effect of the above derivative term (i.e., $L\dot{x}_1^* x_1^* \cong 0$). Then the same proportional plus integral term proposed in (8.30) solves the regulation problem.

If the sampling frequency is high (100kHz in our simulations), there is no appreciable difference between the standard and predictive SMC, and traces are very similar to the ones presented in Figure 8.21. If, however, the sampling frequency is reduced to a value which is very small from a practical standpoint (10kHz in our simulations), we observe that excursions of the trajectories around the desired equilibria are smaller by approximately 40% when the prediction feature is included.

8.3.5 Conclusions

This section presented a cross-section of ideas in large signal control of power electronic converters centered around notions of energy. The control algorithms are based on physics of energy conversion, and aim to effectively utilize benign nonlinearities, instead of trying to cancel them. A growing number of references suggests that this not only leads to an economy in using the control inputs, as the plant is not swamped with the actuation, but also results in control structures that exhibit undeniable simplicity and elegance. We are thus convinced that energy-based control has an important role to play in emerging demanding applications across the spectrum of electronic power conversion.

REFERENCES

[1] R. Oruganti, J. J. Yang, and F. C. Lee, Implementation of optimal trajectory control of series resonant converter, *IEEE Trans. on Power Electronics*, vol. 3, no. 3, pp. 318–327, 1988.

[2] M. G. Kim and M. J. Youn, An energy feedback control of series resonant converter, *IEEE Trans. on Power Electronics*, vol 6, no. 3, pp. 338–344, 1991.

[3] J. G. Kassakian, M. F. Schlecht, and G. C. Verghese, *Principles of Power Electronics.* Reading, MA: Addison-Wesley, 1991.

[4] M. O. Eissa, S. B. Leeb, G. C. Verghese, and A. M. Stanković, A fast analog controller for a unity power factor ac/dc converter, *IEEE Trans. on Power Electronics*, vol. 11, no. 1, pp. 1–6, 1996.

[5] D. Czarkowski and M. K. Kazimierczuk, Energy-conservation approach to modeling PWM dc/dc converters, *IEEE Trans. on Aerospace and Electronic Systems*, vol. 29, no. 3, pp. 1059–1063, 1993.

[6] S. R. Sanders and G. C. Verghese, Lyapunov-based control for switched power converters, *IEEE Trans. on Power Electronics*, vol. 7, no. 1, pp. 17–24, 1992.

[7] S. R. Sanders, Nonlinear control of switching power converters. PhD diss., MIT, 1989.

[8] N. Kawasaki, H. Nomura and M. Masuhiro, A new control law of bilinear dc-dc converters developed by direct application of Lyapunov, *IEEE Trans. on Power Electronics*, vol. 10, no. 3, pp. 318–325, 1995.

[9] H. Komurcugil and O. Kukrer, Lyapunov-based control of three-phase PWM ac/dc voltage-source converters, *IEEE Trans. on Power Electronics*, vol 13, no. 5, pp. 801–813, 1998.

[10] G. Escobar, R. Ortega, and A. van der Schaft, A saturated output feedback controller for the three phase voltage sourced reversible boost-type rectifier, *Proc. IEEE IECON*, 1998.

[11] A. M. Stanković, D. J. Perreault, and K. Sato, Synthesis of dissipative nonlinear controllers for series resonant dc/dc converters, *IEEE Trans. on Power Electronics*, vol. 14, no. 4, pp. 673–682, 1999.

[12] P. Mattavelli and A.M. Stanković, Dynamical phasors in modeling and control of active filters, *IEEE Int. Symp. on Circuits and Systems*, vol. 5, pp. 278–282, 1999.

[13] H. K. Khalil, *Nonlinear Systems* (2nd Ed.). Upper Saddle River, NJ: Prentice Hall, 1996.

[14] R. Ortega, A. Loria, P.J. Nicklasson, and H. Sira-Ramirez, *Passivity-Based Control of Euler–Lagrange Systems.* Berlin, Germany: Springer, 1998.

[15] L. A. Kamas and S. R. Sanders, Parameter and state estimation in power electronic circuits, *IEEE Trans. on Circuits and Systems—I*, vol. 40, no. 12, pp. 920–928, 1993.

[16] B. Anderson, R. Bitmead, C. Johnson, P. Kokotovic, R. Kosut, I. Mareels, L. Praly, and B. Riedle, *Stability of Adaptive Systems: Passivity and Averaging Analysis.* Boston, MA: MIT Press, 1986.

[17] H. Rodriguez, R. Ortega, G. Escobar, and N. Barabanov, A robustly stable output feedback saturated controller for the boost dc-to-dc converter, *Systems and Control Letters*, vol. 40, pp. 1–8, 2000.

[18] G. Escobar, A. van der Schaft, and R. Ortega, A Hamiltonian viewpoint in the modeling of switching power converters, *Automatica*, vol. 35, no. 3, pp. 445–452, 1999.

[19] H. Sira-Ramírez and M. Ilic-Spong, A geometric approach to the feedback control of switchmode dc-to-dc power supplies, *IEEE Trans. on Circuits and Systems*, vol. 35, no. 10, pp. 1291–1298, 1988.

[20] H. Sira-Ramírez, G. Escobar, and R. Ortega, On passivity-based sliding mode control of switched dc-to-dc power converters, *Proc. 35th IEEE Conf. Decision Contr.* (Kobe, Japan), 1996.

8.4 RIPPLE CORRELATION CONTROL

Philip T. Krein

8.4.1 Background

Power electronic circuits and systems manipulate energy flows with switches. This makes switching power converters nonlinear large-signal systems. Switching action produces ripple, and it is well known that ripple cannot be avoided without a power loss penalty. The nonlinear character of the circuits complicates their analysis, particularly from a stability and control standpoint. The describing equations have discontinuous derivatives, and basic tools such as linearization do not directly apply.

In most commercial inverters and dc/dc converters, control is effected through a pulse-width modulation (PWM) block based on averaging. The conventional practice is to limit closed-loop dynamics to about one-tenth of the switching frequency to support averaged models. The advantage of averaging is that it smooths out switch action and avoids ripple and the discontinuities of switching. Average-based models lend themselves to conventional small-signal analysis and linearization. The theory of averaging is complicated, but its properties are known [1].

One of the key challenges in power electronics is the need for high performance. Applications ranging from 500MHz microprocessors to switch-based semiconductor automobile fuses demand the fastest possible response and small transient disturbances. Unfortunately, the model-based limitations of averaging do not lend themselves well to a wide range of high-performance objectives. The geometric methods of Section 8.2 offer an important alternative. The drawback of geometric methods is a tradeoff between dynamic performance and load regulation, but there is room for future progress.

Other methods that can act on time scales similar to the switching intervals include observer-based approaches [2], energy-based approaches (see Section 8.3), and approaches related to dead-beat controls [3]. Design tools are not readily available for these nonlinear methods. This discussion suggests that most controls for power electronics can be classified as either average-based (which often supports conventional linear control loop design) or geometric-based (as when sliding modes are used to give specific dynamics). In general, these methods are used for regulation, and controls that meet optimality objectives are rarely considered.

8.4.2 Ripple-Based Control

In power converters and their controls, ripple is at best a substitute for a switching clock (as in a hysteresis control) and at worst a nuisance and source of noise and interference. In sliding mode control, for instance, it is convenient to consider the infinite switching frequency case, in which ripple reduces to zero and the behavior follows an ideal sliding dynamic. In averaging, ripple is removed either by filtering or with synchronous sampling. In any event, ripple is not considered as a source of information.

It is important to remember, however, that ripple is inherent to the switching action. It represents a consistent perturbation signal. In oscillatory control [4,5], external perturbations are imposed on nonlinear systems with limited degrees of freedom. The perturbations themselves can be used as a way to stabilize a system in certain cases. A power electronic system is unusual in that an internal perturbation is always present.

It is natural to ask whether these internal signals can be a source of information or a basis for control. In this section, we explore a general approach to ripple-based control. It is shown that significant control objectives, such as cost-function minimization, can be addressed with a ripple correlation technique. Ripple correlation control opens a whole suite of new possibilities for converter action and for control loops. Power electronic systems are uniquely suited to this approach because of their *self-perturbed* internal activity.

One helpful viewpoint for understanding ripple-based control is in the frequency domain. Averaging methods focus intentionally on the dc behavior of a converter. Geometric methods focus on time-domain action, and are difficult to interpret in the frequency domain. In contrast, ripple reflects circuit behavior at a set of interesting frequencies: the switching frequency and its harmonics. In a dc/dc converter switching at 100kHz, for example, the ripple represents a continuous ac excitation that supports measurements on time scales of a few microseconds. The frequency-domain viewpoint also brings out one necessary condition for ripple correlation control: in order to obtain ripple information, the perturbation must be present. If switch action stops (such as when a control variable saturates), ripple would cease and useful converter function would not be obtained. This issue is examined further below.

8.4.3 Ripple Correlation

To prepare ripple signals for measurements or control applications, a plausible approach is to mix them with a switching-frequency signal to downconvert to baseband. This is analogous to conventional automatic gain control (AGC) systems, in which an ac signal is downconverted with a carrier signal to serve as the basis for amplitude control. In the case of ripple, the amplitude and phase convey useful information. It is proposed to use a switching function $q(t)$, with a value of 1 when the corresponding switch is on and 0 when it is off, as the basis of the mixing process. It is convenient to use the cross-correlation between the ripple and a switching function-based carrier waveform $q'(t)$,

$$\int r(t)q'(t)dt \tag{8.36}$$

in which $r(t)$ is a ripple voltage, current, or other waveform, and $q'(t) = 2q(t) - 1$ is a nonzero function based on $q(t)$. Equation (8.36) is a ripple correlation function, and its use in a control loop is *ripple correlation control* [6], as defined here.

A general approach to ripple correlation control summarized from [7] is given here. Consider an application in which a power electronic system has a cost function $J(x)$ to be maximized or minimized. In addition, a modified switching function $q'(t)$ affects this cost function. The switching function is determined through a PWM block or with a direct comparator, as is typical. A suitable PWM block can be represented as

$$q'(t) = sgn[d - tri(t)] \tag{8.37}$$

where $tri(t)$ is a triangular waveform or ramp at the switching frequency, d is a control parameter (equal to duty ratio in steady state), and sgn(x) is the signum function equal to +1 if the argument is positive and −1 if it is negative. The objective is to find the steady-state value of d that minimizes (or maximizes) J. Let the triangle fall between 0 and 1; logically d should also be constrained to this range. A crucial system constraint emerges:

Constraint: Ripple correlation controls require that the control parameter d lies in the open interval $(0, 1)$. The implication is that the switch will always change state once during each period. *It is a necessary condition for these controls that switching action does not cease.*

Now formulate a control law,

$$d(t) = -k \int_0^t q'(t)\dot{J}\,dt = -k \int_0^t q'(t)\frac{\partial J}{\partial d}\dot{d}\,dt \tag{8.38}$$

Here k is a gain value that can be selected to meet noise immunity and response time objectives. The time derivative of (8.38) can be taken and rearranged to give

$$\dot{d}\left(1 + kq'(t)\frac{\partial J}{\partial d}\right) = 0 \tag{8.39}$$

Consider a case in which we have reached the desired operating condition, such that $\partial J/\partial d = 0$. Then the solution to (8.39) requires that $\dot{d} = 0$ (i.e., the duty ratio will not change and converter operation will remain at a fixed condition).

Figure 8.22 shows waveforms from a dc/dc converter intended to convert maximum power from a solar panel (more information about this application is provided

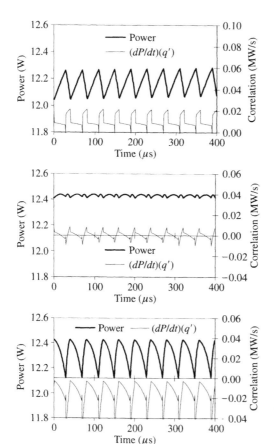

Figure 8.22 Sample waveforms for solar power converter controlled for maximum power delivery.

below). In the top plot, the duty ratio is lower than the value to provide maximum power. The power waveform exhibits ripple that shows power increase while the switch is on, and the correlation $\dot{p}q'(t)$ is positive on average. In the center plot, the duty ratio has been set to deliver maximum power. Now the power ripple excursion moves through the maximum value during the cycle, and the average correlation $\dot{p}q'(t)$ is zero. The bottom plot shows results when the duty ratio is too high, and the correlation is negative.

What if the cost function provides some other behavior, such that $\partial J/\partial d \neq 0$? Remember that the cost function should be chosen to have a dependence on $q(t)$, and therefore on d. Although ideally the result $\dot{d} = 0$ could still be a solution, the key is that switching is occurring. Switching will impose ripple, and the ripple will ensure that $\dot{J} \neq 0$ always. This will cause d to experience some change as well. Equation (8.38) indicates that as long as $\dot{J}$ has a nonzero correlation with $q'(t)$, the value of $d(t)$ will continue to change. The only solution when switching is active is $\partial J/\partial d = 0$, which in turn will maintain $d(t)$ fixed, at least over each switching cycle. Looking further, the control law in (8.38) will cause d to decrease if $\partial J/\partial d$ is positive and increase if $\partial J/\partial d$ is negative, which will drive the operation to a minimum of J.

Variables such as power, voltage, and current are of interest as cost functions within a power electronic circuit. As long as a variable experiences ripple, the ripple function $r(t)$ can be used in place of $\dot{J}$ in (8.38). The end result is that a dc quantity (such as a power) can be driven to a minimum or maximum based entirely on its ripple. A user could, for example, minimize a dc current flow based on information from an ac current transformer with this approach. It is possible in principle to adjust d nearly on a cycle-by-cycle basis to achieve the desired operating objective. It is also true that $\partial J/\partial d$ will not be exactly zero—but it will be as close as possible subject only to unavoidable ripple.

8.4.4 Some Application Examples

Adaptive Dead Time

One important high-performance application is low-voltage dc/dc conversion for advanced microprocessors. In this case, a dc bus level is stepped down through a buck converter to provide 2.5V or less for a fast CMOS VLSI processor. The load dynamics are fast, and high efficiency is an important attribute of the converter. To achieve high efficiency with outputs of 2.5V or less, it is necessary to use synchronous rectification, in which a power MOSFET substitutes for a diode in the output stage. Figure 8.23 shows a block view of such a circuit for 12V-to-2V conversion, with a ripple correlation controller in place.

A drawback of synchronous rectification is timing sensitivity. Ideally, the two switches in Figure 8.23 should operate exactly in complement, but this is not feasible. If the two separate switching functions $q_1(t)$ and $q_2(t)$ have any overlap, then there will be times when the input experiences a short circuit, and losses will be high. A dead time is essential to avoid this, but if it is too long, the efficiency benefits of synchronous rectification will not be fully realized. The ideal dead time is a function of load and temperature, among other external influences.

Ripple correlation control supports adaptive dead time. Assume that switching function $q_1(t)$ (V_{g1} in Figure 8.23) is controlled to provide fixed 2V output, through a PWM block and duty ratio command V_{d1}. The function $q_2(t)$ (or V_{g2}) can be used

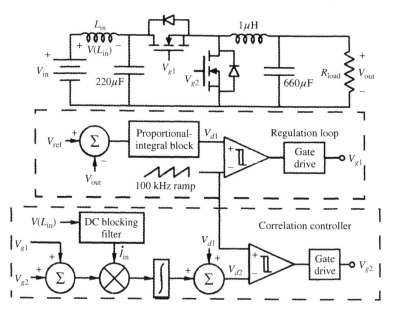

Figure 8.23 Synchronous buck converter with ripple correlation control for adaptive dead time.

with ripple correlation control to minimize the input current and therefore the input power. Details are given in [7]. It will perform this function dynamically without changing the average 2V output.

Solar Power Processing

In solar power applications, it is important to extract the maximum power from a solar cell array whenever this power can be used. In spacecraft systems, remote communication systems, and solar vehicle applications, a dc/dc converter interfaces the cells to a battery bus. The available power is a strong function of illumination and temperature, and a weaker function of aging and other parameters. Figure 8.24 shows the current-voltage and power-voltage characteristics of a single solar cell, and sets of current-voltage curves at various levels of illumination and temperature. The objective is to drive converter operation to the peak power point, with $\partial P/\partial v = 0$. Notice that temperature and illumination alter the curve in disparate ways, so passive strategies such as constant voltage, constant current, or others cannot be tuned to follow peak power as it changes.

The best conventional solar power processors use a *perturb-and-adjust* approach in which a duty ratio value is set for a time, then the average power draw is measured. The process is repeated at a slightly different duty ratio, and the sign of $\partial P/\partial d$ is determined from the two measurements. The controller then steps the duty ratio in the direction of higher power, and repeats the process. This is relatively slow (the best prior practice rates are about 1Hz), and requires memory and computation.

A ripple correlation controller for this battery interface application, based on a boost dc/dc converter, was presented in [8]. It was able to achieve a factor of 1000

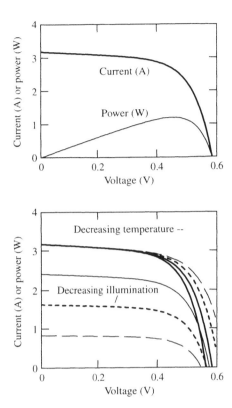

Figure 8.24 Solar cell current-voltage characteristic under various illumination and temperature conditions.

speedup in dynamic performance while achieving an improvement in power delivery. Stability was established by methods of Lyapunov.

Motor Power Minimization in Drives

Electronic motor drives for ac induction machines often are implemented with a dc link inverter to provide variable-frequency, variable-voltage operation. High-performance inverters use some form of *field-oriented control* (FOC) [9], and this can serve as a useful basis for the analysis of a ripple correlation application. In FOC, there are two independent command inputs: one for torque and one for flux. There is no loss of generality in using FOC for the analysis, since other motor controls have (perhaps indirectly) equivalent commands. Of interest here is the flux command. The torque command determines the force delivered to the motor load. Flux, on the other hand, influences power loss. High flux values support higher peak torque and faster dynamic response, but flux is limited by magnetic saturation and losses increase as saturation is reached. Low flux helps reduce power consumption, but limits torque production.

For optimum power consumption, the flux can be adjusted to minimize input power subject to meeting the torque needs. This issue is well known [10]. Existing implementations are awkward, requiring detailed motor measurements and tuning or slow perturb-and-adjust approaches. As a result, motor power minimization is rarely used. With ripple correlation control, the input power ripple can be used to make the flux adjustment. In a dc link drive, the dc bus provides a convenient place to sense the

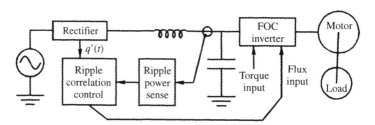

Figure 8.25 Motor power minimizer based on flux optimization.

power. Figure 8.25 shows a configuration that can be used. In this case, the flux command is set based on a ripple correlation between the bus power and a modified switching function,

$$\lambda = k \int \dot{p} q'(t) dt \qquad (8.40)$$

The power function $p(t)$ can be obtained by a chain rule, or directly with a Hall effect sensor.

8.4.5 Summary

Ripple correlation control is a unique approach that provides a direct basis for cost function minimization in systems with internal perturbation. In power electronic systems, ripple correlation control is especially useful because it uses unwanted ac ripple waveforms to drive the dc operating point to a cost function extremum. Applications with power minimization were explored with synchronous rectification and motors, and power maximization was discussed for solar power. Other cost functions related to minimum time or minimum energy transient behavior can be envisioned. Previous controls used in power electronics have not been suitable for cost function-based optimization. The ripple correlation method provides a direct way to address high-performance requirements in power electronic systems.

REFERENCES

[1] B. Lehman and R. M. Bass, Extensions of averaging theory for power electronic systems, *IEEE Trans. on Power Electronics*, vol. 11, no. 4, pp. 542–553.

[2] P. Midya, P. T. Krein, and M. Greuel, Sensorless current-mode control: An observer-based technique for converters, *Rec. IEEE Power Electronics Spec. Conf.*, pp. 197-202, 1997.

[3] K. M. Smedley, One-cycle controlled switching circuit, U.S. Patent 5,278,490, January 1994.

[4] S. M. Meerkov, Principle of vibrational control: Theory and applications, *IEEE Trans. on Automatic Control*, vol. AC-25, no. 4, pp. 755–762, 1980.

[5] B. Lehman, S. Weibel, and J. Baillieul, Open-loop oscillatory control, in *Encyclopedia of Electrical and Electronics Engineering* (J. G. Webster, ed.), New York, Wiley, 1999.

[6] P. Midya, P. T. Krein, and R. J. Turnbull, Self-excited power minimizer/maximizer for switching power converters and switching motor drive applications, U.S. Patent, 5,801,519, September 1998.

[7] J. W. Kimball, Application of nonlinear control techniques in low voltage dc-dc converters, MS Thesis, University of Illinois, 1996.

[8] P. Midya, P. T. Krein, R. J. Turnbull, R. Reppa, and J. Kimball, Dynamic maximum power point tracker for photovoltaic applications, *Rec. IEEE Power Electronics Spec. Conf.*, pp. 1710–1716, 1996.

[9] D. W. Novotny and T. A. Lipo, *Vector Control and Dynamics of AC Drives*. Oxford: Clarendon Press, 1996.

[10] P. Famouri and J. J. Cathey, Loss minimization control of an induction motor drive, *IEEE Trans. on Industry Applications*, vol. 27, no. 1, pp. 32–37, 1991.

8.5 CONTROL OF CHAOS

Mario di Bernardo
Gerard Olivar
Carles Batlle

8.5.1 Introduction

In recent years, the possibility of controlling nonlinear chaotic dynamical systems has been the subject of extensive investigation. Following the pioneering work by E. Ott, C. Grebogy, and J. A. Yorke, much research effort has focused on the possibility of solving the problem of *controlling chaos* (in the sense of suppressing chaotic regimes in a given system) by means of small, time-dependent parameter or input perturbations [1,2]. In particular, it has been pointed out that the many unstable periodic orbits (UPOs) embedded in a strange attractor can be used to produce regular behavior to the advantage of engineers trying to control nonlinear systems in which chaotic fluctuations are present but undesirable.

Over the years, many different strategies to control chaotic dynamics in nonlinear systems have been proposed. Recent surveys of some of the available methods for control of chaos can be found in [3] and [4]. In [5], a collection of some of the original papers can be found.

One class of strategies is based on the original method proposed by E. Ott, C. Grebogi, and J. A. Yorke [1,6], where control is achieved through small perturbations of an accessible parameter. These methods exploit the fact that, during its wandering within the strange attractor, the system will eventually come near the target UPO on a given Poincaré section. When this happens, and only then, a small perturbation is applied to the parameter so as to make the next Poincaré intersection land on the stable manifold of the target saddle fixed point. As a drawback, this method is not suitable for higher-dimensional systems, since a nontrivial computer analysis must be performed at each crossing of the Poincaré section. Also, small noise can drive the orbit away from the target orbit, and the control method must then wait for a while until the system comes near to the target orbit again.

An alternative method, called *time-delayed autosynchronization* (TDAS), was proposed by Pyragas in [7]. This strategy involves a control signal formed with the difference between the current state of the system and the state of the system delayed by one period of the UPO we want to stabilize. One variation to this method, called *extended TDAS* (ETDAS), proposed by Socolar et al. [8], uses a particular linear combination of signals from the system delayed by integer multiples of the UPO's period. Still another

modified approach was proposed by de Sousa Vieira et al. [9] and makes use of a
nonlinear function of the difference between the present state and the delayed state.
TDAS and its variants have the advantage that the only information needed about the
target orbit is its period and that no computer processing must be done to generate the
control signal. The method has recently been applied to systems described by partial
differential equations [10]. In general, the feedback gains which successfully stabilize the
orbit lie in a finite, and often narrow, orbit-dependent range. In the space of the feedback gains and the bifurcation parameters of the system, the region where the TDAS
can be applied with success is called the *domain of control*. In [11] a method was
proposed to compute the domain of control of a given system without having to
explicitly integrate the resulting time-delay equations, which is a nontrivial matter
due to the choice of initial conditions [12]. Essentially, the method reduces to the
computation of the index around the origin of a curve in the complex plane.

Finally, it is possible to identify a third class of control strategies, mostly developed
in the area of control engineering, which make use of state feedback controllers to solve
the problem of controlling chaos. These techniques, which include state feedback controllers (SFC), adaptive control schemes [13], and so on have been shown to be
extremely useful in applications to achieve the control goal while guaranteeing satisfactory robustness to noise, external disturbances, and parameter perturbations [3,14].

In what follows we will detail the possible application of some control techniques,
belonging to the three categories described above, to achieve the control of chaotic
dynamics in power electronic systems. Specifically, we will discuss in Section 8.5.2 the
application of OGY and Pyragas methods to power electronics. We will then show how
to achieve the suppression of chaos by controlling the occurrence of the *border-collision
bifurcations* in Section 8.5.3. Finally, in Section 8.5.4 the principles of time-delay-based
control will be presented and applied to the case of a current-controlled boost converter.

8.5.2 A Combination of OGY and Pyragas Methods

The OGY method was born in the discrete framework; the cornerstone of the
method is the perturbation, from its nominal value, of an available parameter p of
the system $\mathbf{x}_{n+1} = P(\mathbf{x}_n, p)$ in such a way that $\mathbf{x}_{n+1} = P(\mathbf{x}_n, p)$ lies in the stable manifold
of $(\mathbf{x}^*, p^*)$, where $(\mathbf{x}^*, p^*)$ is a saddle fixed point of the Poincaré map, representing an
unstable periodic orbit in continuous time. A detailed account of the OGY method can
be found in [5] and in Section 8.6 of this book, and will not be given here.

In [7], Pyragas proposed in fact two methods based on a special form of time-
continuous perturbation. A combination of feedback and a periodic external force was
used in the first method, while the second method is the already-mentioned TDAS.

The method presented in this section takes from OGY the philosophy and the
discrete framework, and from the first method of Pyragas the technique based on state
feedback. Its application to a simpler model of the current-controlled boost converter
can be found in [15].

Consider a system modeled by the discrete map

$$\mathbf{x}_{n+1} = P(\mathbf{x}_n, \xi) \qquad (8.41)$$

where $\mathbf{x} \in \mathbb{R}^n$ represents the state and $\xi \in \mathbb{R}^m$ are some accessible parameters of the
system. Assume that P has a fixed point $(\mathbf{x}^*, \xi^*)$ and P is smooth in a neighborhood of

$(\mathbf{x}^*, \xi^*)$. Then a linear stability analysis with $\mathbf{x} = \mathbf{x}^* + \tilde{\mathbf{x}}$ and $\xi = \xi^* + \tilde{\xi}$ leads to the following discrete system

$$\tilde{\mathbf{x}}_{n+1} = \left(\frac{\partial P}{\partial \mathbf{x}}\right)_{[\mathbf{x}=\mathbf{x}^*, \xi=\xi^*]} \tilde{\mathbf{x}}_n + \left(\frac{\partial P}{\partial \xi}\right)_{[\mathbf{x}=\mathbf{x}^*, \xi=\xi^*]} \tilde{\xi}_n \qquad (8.42)$$

that can be considered as a linear discrete control system where $\tilde{\mathbf{x}}_n$ are the states and $\tilde{\xi}$ the inputs.

We will apply a proportional state feedback

$$\tilde{\xi}_n(\tilde{\mathbf{x}}_n) = \mathbf{a}\tilde{\mathbf{x}}_n$$

where $\mathbf{a}$ is an $(m \times n)$ feedback gain matrix to force the eigenvalues of

$$\left[\left(\frac{\partial P}{\partial \mathbf{x}}\right)_{(\mathbf{x}^*, \xi^*)} + \left(\frac{\partial P}{\partial \xi}\right)_{(\mathbf{x}^*, \xi^*)} \mathbf{a}\right]$$

to be in the interior of the unit circle.

Application to the Current-Mode-Controlled Boost Converter

As we have said in the introduction, supression of chaos has an immediate appeal from an engineering point of view, since chaotic motion generally implies the wandering of the trajectory over a large zone in state space in an unpredictable way. For the dc/dc converters, chaotic motion imposes a bound on the values of the parameters such that the converter has a periodic output ripple. Any method that allows the converter to regulate for a larger range of the parameters (i.e., the input voltage for the buck converter or the reference current for the current-mode boost), has potential advantages.

We apply the method just described to the current-controlled boost converter whose equations in adimensional form we briefly summarize here.

The current-mode-controlled boost converter has been presented in Sections 2.1 and 5.1 of this book. We assume that the converter has ideal components and is operating in continuous-conduction mode (i.e., the inductor current never falls to zero). There are two circuit configurations, according to whether a switch S is closed or open (it is assumed that it is initially closed). The current i through the inductor L rises linearly until it reaches a reference value I_{ref}. During this phase, any pulse T-periodically provided by the clock is ignored (in case i takes more than T seconds to reach I_{ref}). When $i = I_{\text{ref}}$, S opens and remains open until the arrival of the next clock pulse, whereupon it closes again. Thus, at the beginning of a period T, S is always closed. In all the numerical results in this paper we assume the values used in [16] (i.e., $R = 20\Omega$, $L = 1\text{mH}$, $C = 12\mu\text{F}$, $T = 100\mu\text{s}$, $E = 10\text{V}$, and I_{ref} is varied from 0.5 to 5.5A, which acts as a bifurcation parameter). Although the details may vary, the general features of the analysis apply to any other set of values, provided that the converter is in continuous-conduction mode.

In a given cycle of S closed-open-closed (which can take any number of periods T) the system is described by

$$\frac{d}{dt}\begin{pmatrix} v \\ i \end{pmatrix} = \begin{pmatrix} -\frac{1}{RC} & 0 \\ 0 & 0 \end{pmatrix}\begin{pmatrix} v \\ i \end{pmatrix} + \theta(t - t_c)\begin{pmatrix} 0 & \frac{1}{C} \\ -\frac{1}{L} & 0 \end{pmatrix}\begin{pmatrix} v \\ i \end{pmatrix} + \frac{E}{L}\begin{pmatrix} 0 \\ 1 \end{pmatrix} \tag{8.43}$$

where t_c is the time when $i = I_{ref}$, i.e.,

$$t_c = \frac{L}{E}(I_{ref} - i(0)) \tag{8.44}$$

Thus (8.43) is a somewhat peculiar system of differential equations, where the right-hand side depends on an initial condition.

Introducing dimensionless variables

$$\hat{v} = v/E, \quad \hat{i} = iR/E, \quad \hat{I}_{ref} = I_{ref}R/E, \quad \hat{t} = t/T, \quad \hat{T} = T/T = 1$$

and defining

$$k = \frac{T}{2RC}, \quad \gamma = \frac{TR}{L}$$

equations (8.43) and (8.44) are converted into

$$\frac{d}{dt}\begin{pmatrix} v \\ i \end{pmatrix} = \begin{pmatrix} -2k & 0 \\ 0 & 0 \end{pmatrix}\begin{pmatrix} v \\ i \end{pmatrix} + \theta(t - t_c)\begin{pmatrix} 0 & 2k \\ -\gamma & 0 \end{pmatrix}\begin{pmatrix} v \\ i \end{pmatrix} + \begin{pmatrix} 0 \\ \gamma \end{pmatrix} \tag{8.45}$$

and

$$t_c = \frac{1}{\gamma}(I_{ref} - i(0)) \tag{8.46}$$

where we have dropped the hat for the sake of notational convenience. The reader has to assume that from now on all variables are dimensionless.

Notice that the new I_{ref} varies between 1 and 11.

The solution to (8.45) and (8.46) is explicitly given as follows:

- If $t < t_c$, $v(t) = e^{-2kt}v(0)$ and $i(t) = i(0) + \gamma t$.
- At $t = t_c = (I_{ref} - i(0))/\gamma$, $v(t_c) \equiv v_c = e^{-2kt_c}v(0)$, $i(t_c) = I_{ref}$.
- If $t > t_c$,

$$\begin{pmatrix} v(t) \\ i(t) \end{pmatrix} = \exp\left\{\begin{pmatrix} -2k & 2k \\ -\gamma & 0 \end{pmatrix}(t - t_c)\right\}\begin{pmatrix} v_c - 1 \\ I_{ref} - 1 \end{pmatrix} + \begin{pmatrix} 1 \\ 1 \end{pmatrix} \tag{8.47}$$

The exponential in (8.47) is given by

$$\exp\left\{\begin{pmatrix} -2k & 2k \\ -\gamma & 0 \end{pmatrix}t\right\} = e^{-kt}\begin{pmatrix} \cos\omega t - k/\omega \sin\omega t & 2k/\omega \sin\omega t \\ -\gamma/\omega \sin\omega t & \cos\omega t + k/\omega \sin\omega t \end{pmatrix} \tag{8.48}$$

where $\omega = +\sqrt{2k\gamma - k^2}$.

We take the reference current as the accessible parameter, $\xi = I_{ref}$. The Poincaré map with a dimensional return time 1 is given by

$$P((v, i), I_{\text{ref}}) = \left(f_1(1 - \frac{I_{\text{ref}} - i}{\gamma}, e^{-2k\frac{I_{\text{ref}} - i}{\gamma}}v), f_2(1 - \frac{I_{\text{ref}} - i}{\gamma}, e^{-2k\frac{I_{\text{ref}} - i}{\gamma}}v) \right) \qquad (8.49)$$

where $f_1(t, v) = e^{-kt}[(v - 1)(\cos \omega t - k/\omega \sin \omega t) + (I_{\text{ref}} - 1)2k/\omega \sin \omega t] + 1$, and $f_2(t, v) = e^{-kt}[-(v - 1)\gamma/\omega \sin \omega t + (I_{\text{ref}} - 1)(\cos \omega t + k/\omega \sin \omega t)] + 1$. Figure 8.26 shows the bifurcation diagram for the given values of the parameters.

The approximating linear map (8.42) around the fixed point $\mathbf{x}^* = (3.065799, 6.731273)$, $\xi^* = 8.0$, is given by

$$\begin{pmatrix} \tilde{x}_{1,n+1} \\ \tilde{x}_{2,n+1} \end{pmatrix} = \begin{pmatrix} 0.620941 & 1.160240 \\ -0.511134 & -2.392264 \end{pmatrix} \begin{pmatrix} \tilde{x}_{1,n} \\ \tilde{x}_{2,n} \end{pmatrix} + \tilde{\xi}_n \begin{pmatrix} -1.021537 \\ 3.339776 \end{pmatrix}$$

and the control law

$$\tilde{\xi}_n = a_1 \tilde{x}_{1,n} + a_2 \tilde{x}_{2,n}$$

where (a_1, a_2) are chosen in such a way that the coefficients of the characteristic polynomial $(1, \lambda_1, \lambda_0)$ of the controlled linear map satisfy the Jury criterion [17], namely

- $1 + \lambda_1 + \lambda_0 > 0$
- $1 - \lambda_1 + \lambda_0 > 0$
- $|\lambda_0| < 1$

The Jury criterion imposes conditions on the coefficients of a polynomial so that their roots lie in the interior of the unit circle. In our case this condition is that (a_1, a_2) is inside the triangle with vertices $\{(0.972817, 0.827929), (-0.146631, 1.084366),$ and $(-2.089465, -0.707576)\}$.

Figure 8.27 shows the result of the simulation for $v(t)$ corresponding to the former parameter values over a length of 120 periods. After 80 periods the feedback control is applied with a gain vector

$$(a_1, a_2) = (0., 0.75)$$

and, as it can be seen, it clearly stabilizes the UPO.

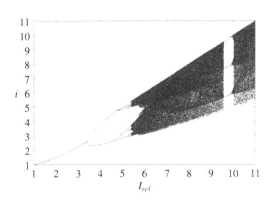

Figure 8.26 The bifurcation diagram of the system under consideration taking I_{ref} as the bifurcation parameter.

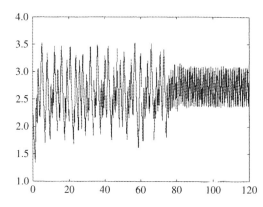

Figure 8.27 Simulation of the state feedback system for $I_{ref} = 8.0$, $a_1 = 0.0$, $a_2 = 0.75$. The voltage waveform is shown for 120 periods and the feedback gain starts to act after 80 periods.

8.5.3 Controlling Border-Collision Bifurcations

As reported in this book, many power electronic systems are naturally character-ized by switchings, impacts, and other nonsmooth functions. Therefore, the use of piecewise-smooth models for describing their dynamical behavior is of fundamental importance [18,19,20,21,22,23].

The occurrence of a novel class of nonsmooth bifurcations was discovered and outlined for this type of dynamical systems [24,25,26,27] and has been detailed in Section 3.4. Specifically, it has been shown that the system dynamical behavior under-goes dramatic changes when, as the system parameters are varied, a fixed point crosses one of the boundaries that divide the phase space into different regions associated with different (smooth) system equations. This type of bifurcation, called *border collisions* (or *C*-bifurcations in the Russian literature) has been shown to organize the dynamical behavior of several systems of relevance in power electronics. For instance, in [28] these bifurcations were shown to be the cause of much of the nonlinear phenomena exhibited by a current-controlled buck converter.

Thus, the control of these bifurcations can be very useful in applications and, as will be shown later, can be successfully applied in controlling the occurrence of chaotic behavior in power electronic circuits. Specifically, in this section we will describe a local strategy for the control and anticontrol of border-collision bifurcations in piecewise-smooth systems. The approach is to use an appropriately defined state feedback con-troller [29].

It is worth mentioning here that a possible classification method for border-col-lision bifurcations has been detailed in Section 3.4 and similar techniques are reported in [24,30,31,25,32], for both continuous-time and discrete-time dynamical systems. These classification strategies, in general, are based on a local analysis of the system dynamical evolution in a sufficiently small neighborhood of the border-collision bifur-cation point. In particular, it is possible to classify the system dynamical evolution after a border-collision bifurcation by studying the eigenvalues of an appropriately derived local map on each side of the border involved in the bifurcation in the phase space.

In what follows, we will assume that the system dynamics in the local neighbor-hood of a border-collision bifurcation can be described by a map of the form

$$\mathbf{x}_{n+1} = \begin{cases} \mathbf{A}_1 \mathbf{x}_n + \mathbf{c}\mu, & \text{if } \mathbf{L} \cdot \mathbf{x}_n \leq 0 \\ \mathbf{A}_2 \mathbf{x}_n + \mathbf{c}\mu, & \text{if } \mathbf{L} \cdot \mathbf{x}_n > 0 \end{cases} \tag{8.50}$$

where $\mathbf{x}_n \in \mathbb{R}^n$ and $\mathbf{L} \in \mathbb{R}^{1 \times n}$. This is typically the case in most of the systems of interest in applications [32,26].

Local Feedback Strategy

Suppose now that as μ is increased, an orbit of the system under investigation (i.e., a power electronic circuit), P_0, undergoes a border-collision bifurcation at $\mu = \mu^*$. Then locally the system dynamical evolution is described by (8.50) and, accordingly, the type of dynamical behavior exhibited by the system for $\mu > \mu^*$ depends on the eigenvalues of $\mathbf{A}_1$ and $\mathbf{A}_2$ in (8.50), as described in Section 3.3. Therefore, changing the eigenvalues of these dynamical matrices will cause a consistent change of the system dynamical behavior. In particular, if the matrices $\mathbf{A}_1$ and $\mathbf{A}_2$ are stabilizable, we can control the system at a border-collision by adding a state-feedback controller to system (8.50), through an appropriate matrix $\mathbf{B}$. Specifically, we have

$$\mathbf{x}_{n+1} = \mathbf{B}\mathbf{u}_n + \begin{cases} \mathbf{A}_1 \mathbf{x}_n + \mathbf{c}\mu, & \text{if } \mathbf{L} \cdot \mathbf{x}_n \leq 0 \\ \mathbf{A}_2 \mathbf{x}_n + \mathbf{c}\mu, & \text{if } \mathbf{L} \cdot \mathbf{x}_n > 0 \end{cases} \tag{8.51}$$

where we choose

$$\mathbf{u}_n = \begin{cases} -\mathbf{K}_1 \mathbf{x}_n, & \text{if } \mathbf{L} \cdot \mathbf{x}_n \leq 0 \\ -\mathbf{K}_2 \mathbf{x}_n, & \text{if } \mathbf{L} \cdot \mathbf{x}_n > 0 \end{cases} \tag{8.52}$$

so that locally

$$\mathbf{x}_{n+1} = \begin{cases} (\mathbf{A}_1 - \mathbf{B}\mathbf{K}_1)\mathbf{x}_n + \mathbf{c}\mu, & \text{if } \mathbf{L} \cdot \mathbf{x}_n \leq 0 \\ (\mathbf{A}_2 - \mathbf{B}\mathbf{K}_2)\mathbf{x}_n + \mathbf{c}\mu, & \text{if } \mathbf{L} \cdot \mathbf{x}_n > 0 \end{cases} \tag{8.53}$$

Using the *pole placement* technique, we can determine $\mathbf{K}_1$ and $\mathbf{K}_2$ such that the eigenvalues of the dynamical matrices in (8.51) satisfy one of the conditions described in Section 3.4. In so doing, we can select any of the possible dynamical evolutions following a border-collision bifurcation. Hence, the occurrence of a border-collision bifurcation can be used to control the system dynamical evolution.

Notice that the conditions on the eigenvalues of the map, as reported in [32], can also be given in a way that does not involve their exact numerical value; rather, only the sums of the eigenvalues that are greater than 1 or −1, respectively, are required to be either odd or even. Thus, the control objective can be achieved by a controller that is simpler than (8.53). This is because knowing the eigenvalues of the map on one side of the boundary, it is sufficient to control the position of the eigenvalues on the other side. Therefore, assuming the knowledge of the eigenvalues of $\mathbf{A}_1$, the control objective can be achieved, without loss of generality, by using the following controller:

$$\mathbf{u}_n = \begin{cases} 0, & \text{if } \mathbf{L} \cdot \mathbf{x}_n \leq 0 \\ -\mathbf{K}\mathbf{x}_n, & \text{if } \mathbf{L} \cdot \mathbf{x}_n > 0 \end{cases} \tag{8.54}$$

Using (8.54) we can select the eigenvalues of the system when $\mathbf{L} \cdot \mathbf{x}_n > 0$ and, thereafter, decide which of the conditions listed in Section 3.4 is satisfied.

An Example: A Two-Dimensional Map

We consider the map studied in [33], which is a suitable normal form for border-collisions of two-dimensional piecewise-smooth maps.

This map is of the form

$$
\mathbf{x}_{n+1} =
\begin{cases}
\begin{pmatrix} a_1 & 1 \\ b_1 & 0 \end{pmatrix} \mathbf{x}_n + \begin{pmatrix} 1 \\ 0 \end{pmatrix} \mu, & \text{if } (1 \quad 0)\mathbf{x}_n \leq 0 \\[2em]
\begin{pmatrix} a_2 & 1 \\ b_2 & 0 \end{pmatrix} \mathbf{x}_n + \begin{pmatrix} 1 \\ 0 \end{pmatrix} \mu, & \text{if } (1 \quad 0)\mathbf{x}_n > 0
\end{cases}
\tag{8.55}
$$

and has been used in [28] to study the dynamics of a current-controlled dc/dc buck converter. It has been shown [33] that this map undergoes a border-collision bifurcation when $\mu = 0$ and that the dynamics following this bifurcation are specified by the values of the parameters a_1, b_1, a_2, b_2 in the map (8.55). These correspond to the traces, a_1 and a_2, and the determinants $-b_1$ and $-b_2$, of the system matrices on both sides of the phase space boundary determined by $x_{n_1} = 0$.

For instance, by selecting $a_1 = 1.3$, $b_1 = 0.4$, $a_2 = 1.15$, and $b_2 = 0.3$ at the border-collision point (with $\mu = 0$), a stable equilibrium point existing on the left of the boundary will turn into a stable equilibrium on the right of the boundary. This is confirmed by Figure 8.28(a), where the bifurcation diagram of the map (8.55) is shown for these values of a_1, b_1, a_2, b_2. We can see that, as μ crosses zero, a branch of stable equilibria turns into another branch of a similar type.

We now apply a controller of the form (8.54), with $\mathbf{K} = (2 \quad 1)$, to the map. As shown in Figure 8.28(b), when the control is activated, the bifurcation diagram changes abruptly and the system orbit suddenly changes to a chaotic attractor at the border-collision point. In fact, with this choice of the feedback gain, the controlled map

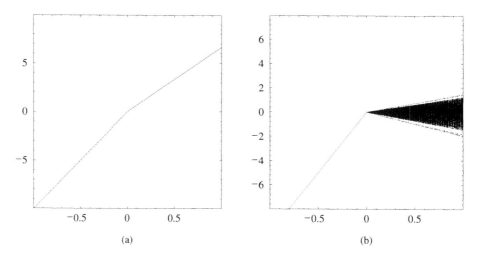

(a)   (b)

Figure 8.28 Bifurcation diagram of the map (8.55): (a) before and (b) after the feedback control is applied.

satisfies the conditions formulated in [34] that predict such a sudden transition to chaos to occur. Similarly, other types of dynamical behavior can also be obtained by simply tuning the feedback gain to different values.

This type of controller can be particularly effective in solving the problem of anticontrolling chaos [3,35]. In other words, given a system that is evolving along a periodic orbit, we are able to change it to a chaotic attractor by applying a simple state feedback control. In so doing, we have literally used the theory of border-collision bifurcation classification [32,33] as a controller design criterion.

8.5.4 Time-Delay Control of Chaos

Suppose we start with a dynamical system given by

$$\dot{\mathbf{x}} = f(\mathbf{x}, t), \quad \mathbf{x} \in \mathbb{R}^n, \quad f(\mathbf{x}, t + T) = f(\mathbf{x}, t) \tag{8.56}$$

We change the vector field, $f \to f_\eta$,

$$\dot{\mathbf{x}} = f_\eta(\mathbf{x}, t), \quad f_\eta(\mathbf{x}, t + T) = f_\eta(\mathbf{x}, t) \tag{8.57}$$

in such a way that $f_{\eta=0}(\mathbf{x}, t) = f(\mathbf{x}, t)$ and $f_\eta(\mathbf{x}^*(t), t) = f(\mathbf{x}^*(t), t)$ for $\mathbf{x}^*$ T-periodic. In particular, this can be implemented by means of TDAS and, from now on, we will assume that f and f_η differ by terms containing $\mathbf{x}(t) - \mathbf{x}(t - T)$.

A linear stability analysis of (8.57) yields a variational equation of the form

$$\dot{\xi}(t) = \mathbf{A}_0(t)\xi(t) + \mathbf{A}_1(t)\xi(t - T), \tag{8.58}$$

where both $\mathbf{A}_0(t)$ and $\mathbf{A}_1(t)$ are T-periodic. Since a time-delayed differential equation can be cast into an infinite-dimensional system of differential equations, we have here an infinite-dimensional Floquet problem [12] and we have to look for solutions of the form $\xi_\mu(t) = e^{\mu t}\mathbf{u}_\mu(t)$ with $\mathbf{u}_\mu(t)$ periodic. The equation for the periodic vector $\mathbf{u}_\mu(t)$ is

$$\dot{\mathbf{u}}_\mu(t) = \left(\mathbf{A}_0(t) - \frac{\mu}{T}\mathbf{I}\right)\mathbf{u}_\mu(t) + e^{-\mu}\mathbf{A}_1(t)\mathbf{u}_\mu(t) \tag{8.59}$$

which is an ordinary differential equation. A formal solution is given by

$$\mathbf{u}_\mu(t) = \mathbb{T} \exp\left\{\int_0^t \left(\mathbf{A}_0(t') + e^{-\mu}\mathbf{A}_1(t') - \frac{\mu}{T}\mathbf{I}\right) dt'\right\}\mathbf{u}_\mu(0) \equiv \mathbf{U}_\mu(t)\mathbf{u}_\mu(0) \tag{8.60}$$

where $\mathbb{T}$ means that the exponential has to be expanded in a Peano-Baker series [36]. Defining $z = e^{-\mu}$, the periodicity condition on $u_\mu(t)$ imposes

$$g(z) \equiv \det\left(\mathbf{U}_{-\log z}(T) - \mathbf{I}\right) = 0 \tag{8.61}$$

Notice that, due to the presence of $z\mathbf{A}_1(t')$ in (8.60), this is a transcendental equation with an infinite number of solutions in the complex plane, in concordance with the infinite dimensionality of the Floquet problem. In terms of z, the linear stability of $\mathbf{x}^*$ under the modified vector field is equivalent to the requirement that all the zeros of $g(z)$ be outside the unit circle. Since

$$\mathbf{U}_{\log z}(T) = z\, \mathbb{T} \exp\left\{\int_0^T (\mathbf{A}_0(t) + z\mathbf{A}_1(t))\, dt\right\} \tag{8.62}$$

has no poles inside the unit circle, the number of zeros of $g(z)$ inside the unit circle equals the index of the curve traced by $g(z)$ when z runs over the unit circle [11]. Thus, $\mathbf{x}^*$ is linearly asymptotically stable if and only if the index of $g(S^1)$ around the origin is zero. The function $g(z)$ cannot, in general, be obtained in closed form and one has to numerically integrate (8.58) for each value of the feedback gain η.

Some "no-go" theorems concerning the impossibility of stabilization of the orbit $\mathbf{x}^*$ in terms of the Floquet multipliers of $\mathbf{x}^*$ have been reported in the literature [37]. However, they do not apply to the current-mode boost converter, since its Floquet multipliers e^μ are both real but less than $+1$.

An Example: TDAS for the Current-Mode Boost Converter

Our strategy to stabilize the UPOs of the current-mode-controlled boost converter will consist of modifying the reference current with a term proportional to the difference between a linear combination of the present and past states of the system. Precisely, instead of comparing $i(t)$ to I_{ref}, we will compare it to

$$I_{ref} + \eta(\alpha(v(t) - v(t-1)) + \beta(i(t) - i(t-1))) \qquad (8.63)$$

where η is an overall feedback gain and α and β are relative weights. Notice again that, for a period-1 solution, the feedback signal vanishes. One must bear in mind that, although the mathematical computations to find out the range of parameters that stabilize the orbit can be quite imposing, once those values are known, the actual implementation requires only the knowledge of the period of the target orbit in order to form the feedback signal. For systems like the PWM-controlled converters the period of any orbit is a multiple of the period of the clock used to generate the pulses. More detailed information about the target unstable orbits can be obtained experimentally using, for instance, the techniques exposed in [38], and this can be used to numerically compute the parameter range mentioned above.

It is shown in [39] that for the current-mode boost converter $g(z)$ can be analytically computed as a function of t_c, v_c, i_c ($= I_{ref}$) and the feedback parameters. With this analytical expression we can numerically compute the index of $g(S^1)$ for several values of the weights α and β as a function of the bifurcation parameter I_{ref} and the feedback gain η. Figure 8.29 shows several of these diagrams, called *domains of control*. In every case, the black zone corresponds to index zero, while the gray and white ones correspond to index 1 and 2, respectively. Notice that for zero-feedback gain the stable zone always ends at $I_{ref} \sim 3.41$.

One can perform simulations of the system under the studied time-delay feedback gain in order to confirm the analytical results.

Close examination of Figure 8.29 predicts, when $I_{ref} = 8.0$ (which corresponds to a natural chaotic motion of the system) that, with $\alpha = -2.0$, $\beta = 2.0$, a feedback gain $\eta = -0.2$ should stabilize the UPO, while $\eta = -0.15$ should not. Figure 8.30 shows the result of the simulations for $v(t)$ corresponding to these two values of η over a length of 120 periods. In both cases the feedback is switched on after 80 periods have elapsed. For $\eta = -0.2$ (left) the feedback clearly stabilizes the UPO, while for $\eta = -0.15$ (right), it does not, although there is a certain decrease in the amplitude of the chaotic oscillations. In a practical implementation one should not expect this sharp behavior, and a "safe" value of η, well inside the domain of control, should be used.

Some simulations to test the robustness of the feedback scheme under changes in the parameters of the system can also be performed. Figure 8.31 shows the result of a simulation where the load R is changed, in steps of 5Ω, from 20Ω to 30 and then down

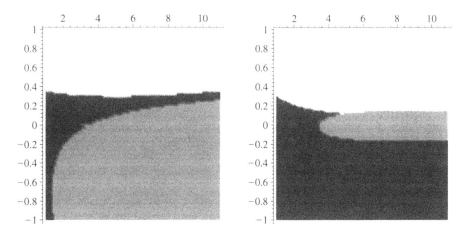

Figure 8.29 Domains of control for several values of α and β. Left: $(\alpha, \beta) = (2, 2)$. Right: $(\alpha, \beta) = (-2, 2)$. The black, gray, and white zones correspond to index 0, 1, and 2, respectively. Horizontal axis: I_{ref} from 1 to 11. Vertical axis: feedback gain η from -1 to 1.

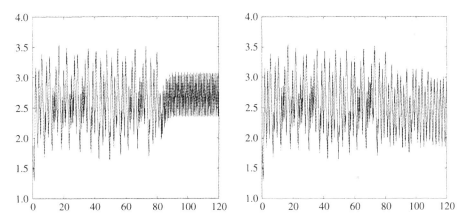

Figure 8.30 Simulations of the time-delay feedback system for $I_{\text{ref}} = 8.0$, $\alpha = -2.0$ and $\beta = 2.0$. The voltage waveform is shown for 120 periods and the feedback gain starts to act after 80 periods. Left: $\eta = -0.2$. Right: $\eta = -0.15$.

Figure 8.31 Simulation of the time-delay feedback system for $I_{\text{ref}} = 8.0$, $\alpha = -2.0$, $\beta = 2.0$, and $\eta = -0.5$. The voltage waveform is shown for 120 periods. The feedback starts after one period and the load is stepped first up and then down by 5Ω starting from 20Ω every 20 periods.

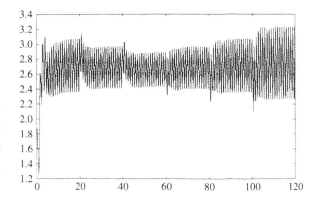

to 15, each phase lasting 20 periods. The simulation was performed with $I_{\text{ref}} = 8.0$, which for $R = 20\Omega$ corresponds to chaotic motion, and with $\alpha = -2$, $\beta = 2$, and $\eta = -0.5$ (this is deep inside the controllability zone for $R = 20\Omega$). Notice that the sudden changes in R introduce only a very short fluctuation in the output voltage. This shows that the domain of control does not vary too much under this kind of change and that a safe value of η can be used without regard to the exact value of the load, which is quite interesting from an implementation point of view.

8.5.5 Conclusions

We have seen that modern chaos control techniques can be indeed successful in controlling the occurrence of bifurcations and chaos in power electronic systems. Specifically, three different techniques to achieve this goal have been detailed and applied to circuits of relevance in applications. Much ongoing research is devoted to carrying out the experimental verification of some of these techniques and has shown very promising results [40]. This will soon help to identify the range of applications of the control strategies presented in this section, thus providing an additional approach for practitioners to tame chaos in power electronics.

REFERENCES

[1] E. Ott, C. Grebogi, and J. Yorke, Controlling chaos, *Phys. Rev. Lett.*, vol. 64, pp. 1196–1199, 1990.

[2] T. Shinbrot, C. Grebogi, E. Ott, and J. Yorke, Using small perturbations to control chaos, *Nature*, vol. 363, pp. 411–417, 1993.

[3] G. Chen and X. Dong, *From Chaos to Order: Methodologies, Perspectives and Applications*. Singapore: World Scientific, 1998.

[4] F. Chernousko and A. Fradkov (eds.), *Proc. 1st Int. Conf. on Control of Oscillations and Chaos*, IEEE, 1997. (St. Petersburg, Russia), August 1997.

[5] T. Kapitaniak, *Controlling Chaos*. San Diego, CA: Academic Press, 1996.

[6] F. Romeira, C. Grebogi, E. Ott, and W. Dayawansa, Controlling chaotic dynamical systems, *Physica D*, vol. 58, pp. 165–192, 1992.

[7] K. Pyragas, Continuous control of chaos by self-controlling feedback, *Phys. Lett.*, vol. A170, pp. 421–428, 1992.

[8] J. Socolar, D. Sukow, and D. Gauthier, Stabilizing unstable periodic orbits in fast dynamical systems, *Phys. Rev. E*, vol. 50, pp. 3245–3248, 1994.

[9] M. Sousa-Vieira and A. Lichtenberg, Controlling chaos using nonlinear feedback with delay, *Phys. Rev. E*, vol. 54, pp. 1200–1207, 1996.

[10] M. Bleich, D. Hochheiser, J. Moloney, and J. Socolar, Controlling extended systems with spatially filtered, time-delayed feedback, *Phys. Rev. E*, vol. 55, pp. 2119–2126, 1997.

[11] M. Bleich and J. Socolar, Stabilization of periodic orbits controlled by time-delay feedback, *Phys. Rev. E*, vol. 57, pp. 16–21, 1996.

[12] J. Hale and S. Verduyn-Lunel, *Introduction to Functional Differential Equations*. New York: Springer-Verlag, 1993.

[13] M. di Bernardo, An adaptive approach to the control and synchronization of continuous-time chaotic systems, *Int. J. Bifurcation and Chaos*, vol. 6, pp. 557–568, 1996.

[14] G. Chen and X. Dong (eds.), *Controlling Chaos and Bifurcations in Engineering Systems*. Boca Raton, FL: CRC Press, 2000.

[15] R. Santos and J. Marrero, Control of dc-dc converters in the chaotic regime, *Proceedings of the 1998 IEEE Int. Conf. Control Applications*, pp. 832–837, IEEE, 1999 (Trieste, Italy), September 1-4.

[16] J. H. B. Deane, Chaos in a current-mode controlled boost dc-dc converter, *IEEE Trans. on Circuits and Systems—I*, vol. 39, pp. 680–683, 1992.

[17] E. I. Jury, *Theory and Application of the Z-Transform Method*. Huntington, NY: Krieger, 1973.

[18] M. di Bernardo, A. R. Champneys, and C. J. Budd, Grazing, skipping and sliding: Analysis of the nonsmooth dynamics of the dc/dc buck converter, *Nonlinearity*, vol. 11, pp. 858–890, 1998.

[19] M. di Bernardo, F. Garofalo, L. Glielmo, and F. Vasca, Switchings, bifurcations and chaos in dc/dc converters, *IEEE Trans. on Circ. Syst.—I*, vol. 45, pp. 133–141, 1998.

[20] E. Fossas and G. Olivar, Study of chaos in the buck converter, *IEEE Trans. on Circ. Syst.—I*, vol. 43, pp. 13–25, 1996.

[21] K. Popp, N. Hinrichs, and M. Oestreich, Dynamical behaviour of friction oscillators with simultaneous self and external excitation, *Sadhana (Indian Academy of Sciences)*, vol. 20, pp. 627–654, 1995.

[22] M. Oestreich, N. Hinrichs, K. Popp, and C. J. Budd, Analytical and experimental investigation of an impact oscillator, *Proc. ASME 16th Biann. Conf. on Mech. Vibrations and Noise*, 1996.

[23] S. J. Hogan, On the dynamics of rigid-block motion under harmonic forcing, *Proc. Royal Society London A*, vol. 425, pp. 441–476, 1989.

[24] M. I. Feigin, Doubling of the oscillation period with c-bifurcations in piecewise continuous systems, *PMM*, vol. 34, pp. 861–869, 1970.

[25] M. I. Feigin, The increasingly complex structure of the bifurcation tree of a piecewise-smooth system, *J. Appl. Math. Mech.*, vol. 59, pp. 853–863, 1995.

[26] H. E. Nusse and J. A. Yorke, Border–collision bifurcations for piece-wise smooth one-dimensional maps, *Int. J. Bifur. Chaos*, vol. 5, pp. 189–207, 1995.

[27] H. E. Nusse and J. A. Yorke, Border-collision bifurcations including "period two to period three" for piecewise smooth systems, *Physica D*, vol. 57, pp. 39–57, 1992.

[28] G. Yuan, S. Banerjee, E. Ott, and J. A. Yorke, Border-collision bifurcations in the buck converter, *IEEE Trans. on Circ. and Syst.—I*, vol. 45, pp. 707–716, 1998.

[29] M. di Bernardo and G. Chen, Controlling bifurcations in nonsmooth dynamical systems, In *Controlling Chaos and Bifurcations in Engineering Systems* (G. Chen and X. Dong, eds.), Chapter 18, pp. 391–412. Boca Raton, FL, CRC Press 2000.

[30] M. I. Feigin, On the generation of sets of subharmonic modes in a piecewise continuous system, *PMM*, vol. 38, pp. 810–818, 1974.

[31] M. I. Feigin, On the structure of c-bifurcation boundaries of piecewise continuous systems, *PMM*, vol. 2, pp. 820–829, 1978.

[32] M. di Bernardo, M. Feigin, S. Hogan, and M. Homer, Local analysis of C-bifurcations in *n*-dimensional piecewise smooth dynamical systems, *Chaos, Solitons & Fractals*, vol. 10, no. 11, pp. 1881-1908, 1999.

[33] H. E. Nusse, E. Ott, and J. A. Yorke, Border collisions bifurcations: An explanation for observed bifurcation phenomena, *Phys. Rev. E*, vol. 49, pp. 1073–1076, 1994.

[34] C. Grebogi, E. Ott, and J. A. Yorke, Fractal basin boundaries, long-lived chaotic transients and unstable-unstable pair bifurcations, *Phys. Rev. Lett.*, vol. 50, pp. 935–938, 1983.

[35] G. Chen and D. Lai, Anticontrol of chaos via feedback, *Proc. 36th Conf. on Decis. Contr.*, pp. 367–372, (San Diego), December 1997.

[36] W. Rugh, *Linear System Theory*. Upper Saddle River, NJ: Prentice Hall, 1996.

[37] H. Nakajima and Y. Ueda, On the stability of delayed feedback control of chaos, in *Control of Oscillations and Chaos* (F. Chernousko and A. Fradkov, eds.), vol. 3, pp. 411–414, IEEE, 1997. (St. Petersburg, Russia), August 27–29.

[38] G. Poddar, K. Chakrabarty, and S. Banerjee, Control of chaos in dc-dc converters, *IEEE Trans. on Circ. and Syst.—I*, vol. 45, no. 6, pp. 672–676, 1998.

[39] C. Batlle, E. Fossas, and G. Olivar, Time-delay stabilization of periodic orbits of the current-mode controlled boost converter, in *Linear Time Delay Systems* (J. Dion, L. Dugard, and M. Fliess, eds.), pp. 111–116, IFAC, 1998 (Grenoble, France), July 6–7.

[40] G. Poddar, K. Chakrabarty, and S. Banerjee, Experimental control of chaotic behavior of buck converter, *IEEE Trans. on Circ. and Syst.—I*, vol. 42, no. 8, pp. 502–504, 1995.

8.6 CLOSED-LOOP REGULATION OF CHAOTIC OPERATION

José Luis Rodríguez Marrero
Roberto Santos Bueno
George C. Verghese

8.6.1 Introduction

In this section we examine how to design control schemes that *preserve* rather than eliminate operation of a power converter in the chaotic regime. The premise here is that operation in the chaotic mode may be desired for certain reasons, for example, because of the associated spectral broadening. The particular mode of operation we consider is current-mode control in dc/dc converters, specifically building on the development in Section 4.3 of Chapter 4.

We begin in Section 8.6.2 by presenting linearized averaged models that govern the dynamics of the converter—in both periodic and chaotic operation—and establishing that the simple first-order models which result can be used to design suitable feedback controllers. Section 8.6.3 presents experimental results with a real boost converter operating in the chaotic regime. In Section 8.6.4 we demonstrate that the approach of Ott, Grebogi, and Yorke [1] can be successfully used to synchronize the converter with a chaotic reference model that describes the (simplified) desired dynamics of the converter [2]. This synchronization allows feedback regulation of the inductor current and control of the spectral characteristics of the converter, as well as feedforward regulation of the output voltage.

8.6.2 Dynamics and Control

The dynamics of a boost converter under current-mode control can be conveniently and economically captured using averaged models [3]. We have demonstrated [4] that the same averaged models used in the periodic case are effective for the chaotic case. An accurate first-order continuous-time model for the dynamics of the boost converter under current-mode control was obtained in (2.11) of Chapter 2. Several transfer functions can be obtained by linearization of this averaged equation. In particular, let $\bar{v}_C(t) = V_{out} + v_{out}$, $i_P(t) = I_{ref} + i_{ref}$ and $\bar{v}_{in} = V_{in} + v_{in}$, where V_{out} is the steady-state average output voltage, I_{ref} is the nominal reference current, V_{in} is the nominal input voltage, and $V_{\text{out}}^2 = RI_{\text{ref}}V_{\text{in}}$. If we substitute these expressions in (2.11) and neglect second-order terms, we obtain

$$\frac{dv_{\text{out}}}{dt} + \frac{2}{RC}v_{\text{out}} = \frac{V_{\text{out}}}{V_{\text{in}}RC}\left(L\frac{di_{\text{ref}}}{dt} + v_{\text{in}} + \frac{V_{\text{in}}}{I_{\text{ref}}}i_{\text{ref}}\right) \tag{8.64}$$

The transfer function relating v_{out} to v_{in} can be obtained by setting $i_{\text{ref}} = 0$, while the transfer function relating v_{out} to i_{ref} is obtained by setting $v_{\text{in}} = 0$. For a streamlined representation that conveys useful insights, and invoking the notation established in Section 4.3 of Chapter 4, we choose dimensionless variables [5], expressing voltages with respect to a base value of V_{out}, currents with respect to a base value of I_{out}, and time with respect to the clock period T. With these choices, the transfer functions that describe the open-loop linearized averaged models of the boost converter become

$$\frac{\tilde{v}_{\text{out}}^*}{\tilde{v}_{\text{in}}^*}(s) = \frac{G}{2}\frac{1}{1 + \tau s/2} \tag{8.65}$$

$$\frac{\tilde{v}_{\text{out}}^*}{\tilde{i}_{\text{ref}}^*}(s) = \frac{1}{2G}\frac{1 - lG^2 s}{1 + \tau s/2} \tag{8.66}$$

where $\tilde{v}_{\text{out}}^*$, $\tilde{v}_{\text{in}}^*$, and $\tilde{i}_{\text{ref}}^*$ represent *small* perturbations of the averaged dimensionless variables from their values in the specified nominal operating condition, and

$$G = \frac{V_{\text{out}}}{V_{\text{in}}} \qquad l = \frac{L}{RT} \qquad \tau = \frac{RC}{T} \tag{8.67}$$

are new dimensionless parameters. The conditions

$$G > 2 \quad \text{and} \quad l > \frac{2G-1}{3}\frac{1}{G^2} \tag{8.68}$$

respectively guarantee that the converter is in the chaotic regime and in continuous conduction. The first condition in (8.68) follows from the fact that $G = 1 + \alpha$, and the chaotic regime exists for $\alpha > 1$; the second condition is obtained by combining (4.23) with the requirement that the inductor current falls from I_{ref} to 0 over the course of a single period at the transition from continuous to discontinuous conduction.

Classical feedback controllers in the configuration shown in Figure 8.32 can now be designed quite directly. For example, an integral controller with gain margin M is obtained by choosing the controller's small-signal transfer function to be

$$C(s) = \frac{2}{lG}\frac{1}{M}\frac{1}{s} \tag{8.69}$$

Figure 8.33 shows the (normalized) output voltage transient in response to an input voltage step of 20%, when an integral controller with $C(s) = 0.2/s$ is used.

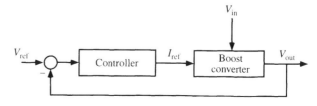

Figure 8.32 Closed-loop control configuration.

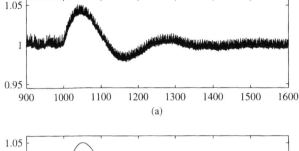

(a)

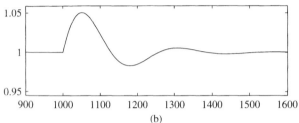

(b)

Figure 8.33 Dimensionless output voltage with integral control for a 20% change in v_{in}. (a) Simulation. (b) Response of averaged model.

Figure 8.33(a) is obtained from simulations (in SIMULINK) of the actual closed-loop converter, and Figure 8.33(b) is obtained using the transfer function of the closed-loop linearized averaged model. The system parameters used are $G = 3$, $l = 0.5$, and $\tau = 100$; note that the conditions in (8.68) are satisfied, so this indeed corresponds to chaotic operation with continuous conduction.

8.6.3 Experimental Results

We now explore experimentally the application of the foregoing theory to an actual boost converter introduced at the end of Section 4.3 and shown schematically in Figure 8.34, with nominal parameter values of $R = 195\Omega$, $L = 3.2\text{mH}$, $C = 100\mu\text{F}$ and $T = 50\mu\text{s}$. The converter operates under current-mode control. The reference current I_{ref} can be set at different levels, and a small-signal perturbation can also be imposed on I_{ref}.

Figure 8.35 shows the open-loop Bode magnitude and phase plots obtained experimentally (dots) and those given by (8.66) (solid line) for $V_{\text{in}} = 4\text{V}$ (we use $V_{\text{in,eff}} = 3.1\text{V}$ in the formulas to take into account the voltage drops across the inductor series resistance and the switches, as described in Section 4.3) and $V_{\text{out}} = 14.25\text{V}$. It can be seen that the agreement in the Bode phase plot is good, but the magnitude plot is off by a significant factor. Although this fact does not seriously affect the controller design, a better result can be obtained taking into account the equivalent series resistances

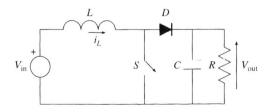

Figure 8.34 Boost converter circuit.

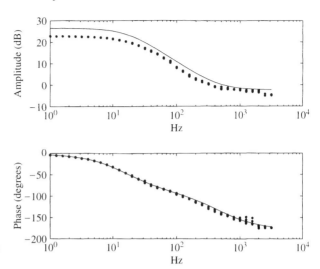

Figure 8.35 Bode plots of $v_{\text{out}}/i_{\text{ref}}$ for a boost converter in the chaotic regime.

(ESRs) of the inductor and capacitor. Inserting these ESRs, the following transfer functions are derived [4]:

$$\frac{\tilde{v}_{\text{out}}^*}{\tilde{v}_{\text{in}}^*} = \frac{G}{2} \frac{1 + \tau_2 s}{1 + (\tau_2 + \tau/2)s} \tag{8.70}$$

$$\frac{\tilde{v}_{\text{out}}^*}{\tilde{i}_{\text{ref}}^*} = \frac{1 - 2lG^2/l_2}{2G} \frac{\left(1 - \frac{lG^2}{1 - 2lG^2/l_2} s\right)(1 + \tau_2 s)}{1 + (\tau_2 + \tau/2)s} \tag{8.71}$$

In order to model the inductor ESR (r_L) and the capacitor ESR (r_C), two new dimensionless parameters appear:

$$l_2 = \frac{L}{r_L T} \quad \tau_2 = \frac{r_C C}{T} \tag{8.72}$$

Figure 8.36 shows the Bode magnitude and phase plots obtained experimentally (dots) and those given by (8.71) (solid line) using $r_L = 2.6\Omega$ and $r_C = 0.4\Omega$; these resistance values were obtained experimentally using an impedance analyzer. In this case we have used $V_{\text{in}} = 4V$ in the formulas, since the model takes care of the voltage drops in the resistances. The improved agreement between theory and experiment is evident.

Figure 8.37 shows the time response at the output of the chaotic boost converter when the input voltage changes from 2V to 2.5V. The first waveform is the measured output voltage and the second is the response predicted by the averaged model in equation (8.65). Again, measurements and theory are in good agreement.

8.6.4 The OGY Method

The approximate results of the first-order averaged model above are in excellent agreement with those obtained through computer simulations of the converter circuit model assuming ideal components [6,7]. However, computer simulations of the boost converter in closed loop in the chaotic regime show that the spectral characteristics (the *power spectrum*) of the converter are different from those in open loop. These changes

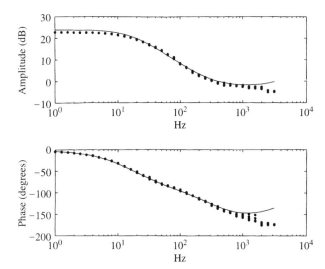

Figure 8.36 Bode plots of $v_{\text{out}}/i_{\text{ref}}$, including r_L and r_C, for a boost converter in the chaotic regime.

in the dynamics of the converter become even more apparent if the nonideal character-istics of the components of an actual converter are taken into account. The problem is that the map describing the dynamics of the converter in open loop no longer applies when the converter is operating in closed loop.

Here we propose a control technique [2] to produce a desired chaotic orbit by *synchronizing* the converter current samples with the samples generated by the one-dimensional map that approximately describes the dynamics of the converter in open-loop operation. Synchronization of the actual converter to a chaotic reference model will allow control of the spectral characteristics of the converter, since the inductor current and power spectrum of the converter will (almost) match those of the reference model. However, unlike the closed-loop case considered in the preceding section, there is now no direct feedback regulation of the output voltage. Nevertheless, feedforward control of the output voltage, for instance, to compensate for variations in the input voltage, can be conveniently implemented.

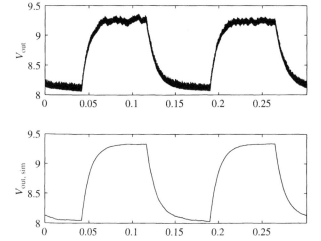

Figure 8.37 Time response of the boost con-verter for changes in V_{in} from 2V to 2.5V in the chaotic regime. (a) Measured output vol-tage waveform. (b) Predicted averaged output voltage.

Controlling chaos by parametric perturbation has been studied previously. The main objective in most previous studies has been the suppression of deterministic chaos, converting a chaotic attractor to any one of the possible attracting time-periodic motions by making small state-dependent perturbations of a system parameter, as suggested by Ott, Grebogi, and Yorke in [1] (hence the reference to the *OGY* method; see also [8,9]. A survey of recent references on control and synchronization of chaotic systems may be found in [10] and [11]. The method has also been used to generate desired aperiodic orbits or different chaotic trajectories [12]. We use the OGY approach in this section to control a dc/dc boost converter in the chaotic regime, synchronizing the dynamics of the converter to a reference model.

Review of the OGY Method

The OGY method requires the application of small state-dependent perturbations to one of the accessible parameters in a chaotic system so as to produce a desired periodic orbit in the system. The control is described in terms of its application to a k-dimensional discrete-time model or map [13], derived from the underlying (typically continuous-time) system via some form of sampling:

$$\mathbf{x}_{n+1} = F(\mathbf{x}_n, p) \tag{8.73}$$

where $\mathbf{x}_i \in R^k$ is a k-dimensional vector, p is some accessible parameter, and F is sufficiently smooth in both variables. It is assumed that the underlying nominal system (i.e., $p = \bar{p}$) contains a chaotic attractor. In the OGY method the parameter p is adjusted at each iteration (so instead of p we have p_n) in such a way that the dynamics of the map (8.73) converges to a fixed point, and correspondingly the underlying system converges to a desired periodic orbit.

Let $\mathbf{x}_*(\bar{p})$ denote an unstable fixed point of the map. For values of p close to $\bar{p}$ and in the neighborhood of the fixed point $\mathbf{x}_*(\bar{p})$, the map (8.73) can be *linearized*:

$$\mathbf{x}_{n+1} - \mathbf{x}_*(\bar{p}) = \mathbf{A}[\mathbf{x}_n - \mathbf{x}_*(\bar{p})] + \mathbf{B}(p_n - \bar{p}) \tag{8.74}$$

where $\mathbf{A}$ is the $k \times k$ Jacobian matrix

$$\mathbf{A} = \frac{\partial}{\partial \mathbf{x}} F(\mathbf{x}_*(\bar{p}), \bar{p})$$

and $\mathbf{B}$ is the k-dimensional vector

$$\mathbf{B} = \frac{\partial}{\partial p} F(\mathbf{x}_*(\bar{p}), \bar{p})$$

The state-dependence of the parameter p is assumed to be of the form

$$p_n - \bar{p} = -\mathbf{K}[\mathbf{x}_n - \mathbf{x}_*(\bar{p})] \tag{8.75}$$

The $1 \times k$ matrix $\mathbf{K}$ is to be determined so that the fixed point $\mathbf{x}_*(\bar{p})$ becomes stable. This choice of $\mathbf{K}$ determines the control law specifying p_n on each iteration. The point $\mathbf{x}_*(\bar{p})$ will be stable if $\mathbf{K}$ is chosen such that the eigenvalues of the matrix $\mathbf{A} - \mathbf{B}\mathbf{K}$ have modulus smaller than unity. This is well known from control theory and is called *pole placement by state feedback*.

Since the method is based on (8.74) and therefore only applies in the neighborhood of $\mathbf{x}_*(\bar{p})$, the control is activated only when $\mathbf{x}_n$ falls in a certain region, and is left at its nominal value $\bar{p}$ (or is not modified) when $\mathbf{x}_n$ is outside that region. The reason for leaving the control parameter p unchanged is that the ergodic nature of the chaotic dynamics ensures that the trajectory enters the region in which the control is activated. Once in that region, the control tries to keep the system near the fixed point.

The OGY method has also been used to generate any desired aperiodic orbit $\mathbf{r}_n$ in a chaotic system like (8.73) [12]. This is achieved by applying the method to a system that describes the dynamics of the error ($\mathbf{e}_n = \mathbf{r}_n - \mathbf{x}_n$) between the *desired* aperiodic motion (the reference orbit) $\mathbf{r}_n$ and the output of the original chaotic system $\mathbf{x}_n$. Under certain conditions, a map of the following form can be obtained:

$$\mathbf{e}_{n+1} = H(\mathbf{e}_n, p) \tag{8.76}$$

and the OGY technique can be used to stabilize the fixed point $\mathbf{e}_*(\bar{p})$ of (8.76). If this fixed point is close to zero, the error $\mathbf{e}_n$ can be made small too, leading to near-perfect generation of the reference orbit by the controlled chaotic system [12].

Controlling DC/DC Converters

In this section we use the OGY method to control the boost converter of Figure 8.34. The dynamics of the controlled current is approximately described by the following one-dimensional map:

$$i_{n+1} = \begin{cases} i_n + m_1 T & \text{if } i_n \leq I_{\text{ref}} - m_1 T \\ (\alpha + 1)I_{\text{ref}} - \alpha i_n - m_2 T & \text{if } i_n > I_{\text{ref}} - m_1 T \end{cases} \tag{8.77}$$

where $m_1 = V_{\text{in}}/L$, $m_2 = (V_{\text{out}} - V_{\text{in}})/L$, and

$$\alpha = \frac{V_{\text{out}} - V_{\text{in}}}{V_{\text{in}}} \tag{8.78}$$

Equation (8.77) can be directly obtained from Section 4.3 (equations (4.4) and (4.5)), and is in the form (8.73) if I_{ref} is taken as the governing parameter. The system (8.77) becomes chaotic for $\alpha > 1$ [6].

Since a modification of I_{ref} can only affect the value i_{n+1} if $i_n > I_{\text{ref}} - m_1 T$, the control will be applied only when this condition applies, leaving I_{ref} unchanged otherwise. Thus, we use the second equation in (8.77) as our model for the actual converter.

Setting

$$I_{\text{ref}} = I + p \tag{8.79}$$

in (8.77), where I is the nominal value obtained according to (4.23) of Chapter 4, namely

$$I = \frac{(1 + \alpha)^2 V_{\text{in}}}{R} + \frac{\alpha V_{\text{in}} T}{3L} \tag{8.80}$$

we obtain a map that constitutes our simplified one-dimensional model for the converter:

$$i_{n+1} = (\alpha + 1)(I + p) - \alpha i_n - m_2 T \tag{8.81}$$

Thus, the nominal value of the parameter is $\bar{p} = 0$.

The desired orbit r_n is generated by a model like (8.77), obtained by replacing i_n with r_n and I_{ref} with I:

$$r_{n+1} = (\alpha + 1)I - \alpha r_n - m_2 T \qquad (8.82)$$

Since $e_n = r_n - i_n$,

$$e_{n+1} = -\alpha e_n - (\alpha + 1)p \qquad (8.83)$$

The fixed point of (8.83) is

$$e_*(p) = -p \qquad (8.84)$$

and since the nominal value of the parameter is $\bar{p} = 0$, the nominal fixed point is zero. Because our simplified model in (8.81) contains unavoidable errors, we do not expect the difference between i_n and r_n to vanish even if the fixed point (8.84) is stabilized; however, these differences are expected to be small.

The matrix **A** and the vector **B** of (8.83) are both scalars:

$$A = -\alpha \; ; \; B = -(\alpha + 1)$$

Therefore the matrix $\mathbf{A} - \mathbf{B}K$ is a scalar (the system pole) given by:

$$A - BK = -\alpha + (\alpha + 1)K$$

The stability condition $|A - BK| < 1$ limits the values of K:

$$\frac{\alpha - 1}{\alpha + 1} < K < 1$$

A *dead beat* response (i.e., the system pole placed at the origin) is achieved by choosing

$$K = \frac{\alpha}{\alpha + 1}$$

The parameter p_n is thus varied according to (see (8.75)):

$$p_n = \frac{-\alpha}{\alpha + 1} e_n \qquad (8.85)$$

to stabilize the fixed point of (8.83)

Figure 8.38 shows the control scheme used for the simulations described here (experimental results are reported in [14]). In this figure, T represents a switch used to sample the inductor current to generate i_n. The box *Boost converter* is the circuit of Figure 8.34 with $R = 20\Omega$, $L = 1\text{mH}$, $C = 500\mu\text{F}$, $T = 100\mu\text{s}$, and $V_{\text{in}} = 10\text{V}$. The box *1st-order model* generates the sequence r_n according to (8.82). The control is activated when $V_{\text{out}} \approx V_{\text{ref}}$. It can be shown that in this case (8.83) gives an excellent approximation of the dynamics of the error signal. For values of V_{ref} far from V_{out} the control is

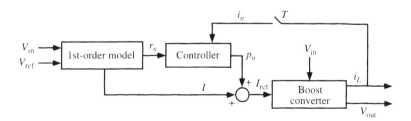

Figure 8.38 Control scheme.

not active, leaving the parameter I_{ref} unchanged, as explained in the description of the OGY method.

For a fixed V_{in}, Figure 8.39 shows that the samples of the inductor current are practically identical to the reference r_n once the control is activated ($t \approx 0.02$). Figure 8.40 shows the error signal versus time. Computer simulations show that this error is very small (less than 1%) as can be seen in Figure 8.41, which shows different maps e_{n+1} versus e_n obtained for the boost converter with the parameters given above. Figure 8.41(a) is obtained when no control is applied, while Figure 8.41(b) is obtained with the proportional control (8.85). Figure 8.41(c) is a detailed version of (b), showing some residual variation in e_n and a nonzero mean value. The offset in the mean value is due to the fact that our simplified one-dimensional model of the system contains unavoidable errors. (In the next subsection, an integral control is developed to make the mean value of this error zero.)

Feedforward regulation of the output voltage is achieved if the reference model that generates r_n is appropriately modified. Figure 8.42 shows the simulated response

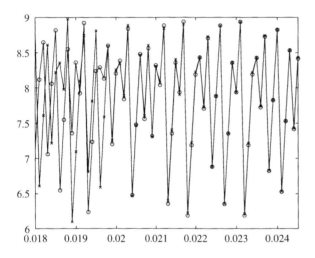

Figure 8.39 Samples of the inductor current (x) and reference (o) versus time.

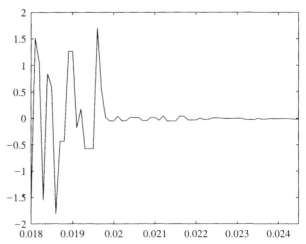

Figure 8.40 Error signal (reference−inductor current) versus time.

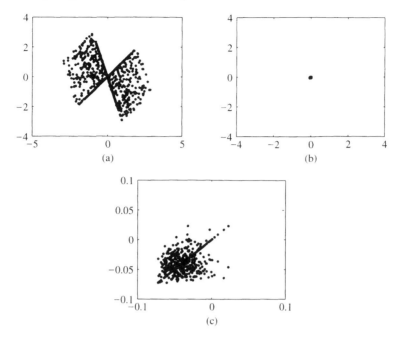

Figure 8.41 e_{n+1} vs. e_n. (a) No control. (b) With proportional control. (c) Magnified view of (b).

of the converter to a change in the output voltage reference V_{ref} from 40 to 35 volts at $t = 0.05$, which is implemented by adjusting α in accordance with (8.78) and then I in accordance with (8.80); these changes feed into (8.82). The feedback part of the control is disabled during the transient, and is activated again when $V_{out} \approx V_{ref}$. The response of the converter during the initial part of the above transient, as it approaches the new steady state, can be analyzed using the linearized averaged model of the open-loop boost converter in (8.66), and the predictions of this model are indicated by the circles in Figure 8.42. It is seen that the predictions are again very good.

Feedforward control can also be used to make the output voltage relatively insensitive to variations in the input voltage V_{in}. This is shown in Figure 8.43, where the converter input voltage changes from 10 to 12 volts at $t = 0.05$. If the change of V_{in} is not taken into account in the model, the output voltage follows the upper trace, settling to a value different from the desired one. However, if the changes in V_{in} are fedforward to the model, again modifying α and I as specified by the appropriate equations, the output voltage recovers to its original value after a brief transient (lower curve).

Integral Control

The proportional law in (8.85) produced a small error signal, but nevertheless one with a nonzero mean value. In this section we develop a procedure to obtain zero mean error. A discrete-time version of integral control is used for this purpose, with a new state variable f_n introduced to measure the accumulated error $\sum_0^{n-1} e_i$:

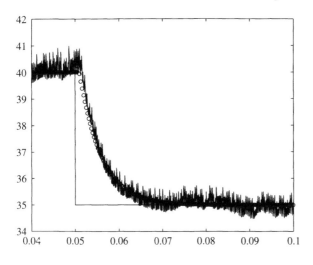

Figure 8.42 Output voltage versus time for a change in V_{ref} from 40 to 35V. Circles represent response of the average model of the converter.

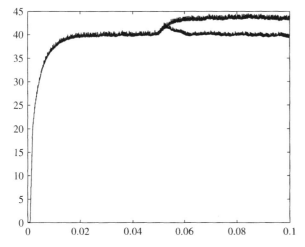

Figure 8.43 Output voltage versus time when V_{in} changes from 10 to 12V at $t = 0.05$. Start-up transient is also represented.

$$e_{n+1} = -\alpha e_n - (\alpha + 1)p$$
$$f_{n+1} = f_n + e_n \qquad\qquad (8.86)$$

The first equation is the method previously used, and the second describes the discrete time *integrator* or accumulator. Using state feedback to stabilize this equation causes e_n and f_n to settle to small bounded values if the external disturbance and errors are small; but since f_n is the integral of e_n, it can be bounded only if e_n has zero mean.

The matrix **A** and the vector **B** of (8.86) are now:

$$\mathbf{A} = \begin{pmatrix} -\alpha & 0 \\ 1 & 1 \end{pmatrix} \quad \mathbf{B} = \begin{pmatrix} -(\alpha + 1) \\ 0 \end{pmatrix}$$

With the control law

$$p_n = -K_1 e_n - K_2 f_n$$

a *dead-beat* response (i.e., both eigenvalues of $\mathbf{A} - \mathbf{BK}$ at zero) is obtained with the following values of K_1 and K_2.

$$K_1 = \frac{\alpha - 1}{\alpha + 1} \quad K_2 = \frac{-1}{\alpha + 1}$$

The resulting control law is

$$p_n = \frac{1 - \alpha}{\alpha + 1} e_n + \frac{1}{\alpha + 1} f_n \tag{8.87}$$

Figure 8.44 shows the map e_{n+1} versus e_n obtained in the boost converter with the parameters given before and with the integral control derived in this section. It can be seen that the error is still small, and that the mean value is now indeed essentially zero.

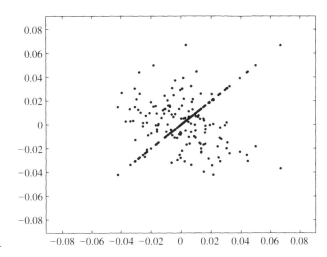

Figure 8.44 e_{n+1} vs. e_n with integral control.

8.6.5 Conclusions

In this section linearized averaged models that govern the dynamics of dc/dc converters under current-mode control have been used to design feedback controllers. Experimental results with a real boost converter operating in the chaotic regime compare favorably with our calculations and simulations.

Closing the feedback control loop around the controller changes the map that describes the dynamics of the converter, and correspondingly modifies the output spectral characteristics. The approach of Ott, Grebogi, and Yorke [1] can be successfully used to synchronize the converter with a chaotic reference model that describes the (simplified) desired dynamics of the converter.

ACKNOWLEDGMENT

José Luis Rodríguez Marrero is grateful for partial support by a grant from Comisión Interministerial de Ciencia y Tecnología of Spain (TIC97-0370).

REFERENCES

[1] E. Ott, C. Grebogi, and J. A. Yorke, Controlling chaos, *Physical Review Letters*, vol. 64, pp. 1196–1199, 1990.

[2] R. S. Bueno and J. L. R. Marrero, Control of a boost dc-dc converter in the chaotic regime, *IEEE Int. Conf. on Control Applications* (Trieste, Italy), pp. 832–837, 1998.

[3] J. G. Kassakian, M. F. Schlecht, and G. C. Verghese, *Principles of Power Electronics.* Reading, MA: Addison-Wesley, 1991.

[4] R. S. Bueno, J. P. Fuster, and J. L. R. Marrero, Dynamics and control of dc-dc converters: Dimensionless formulation, *SAAEI Seminario Anual de Automática y Electrónica Industrial* (Pamplona, Spain), pp. 537–540, 1998.

[5] E. Toribio, X. Odena, and L. Benadero, Región útil de funcionamiento del convertidor Boost PWM en un espacio de parámetros adimensionales 4D, *SAAEI Seminario Anual de Automática y Electrónica Industrial* (Valencia, Spain), pp. 133–138, 1997.

[6] J. L. R. Marrero, J. M. Font, and G. C. Verghese, Analysis of the chaotic regime for dc-dc converters under current-mode control, *IEEE Power Electronics Specialists' Conference* (Baveno, Italy), pp. 1477–1483, 1996.

[7] S. Banerjee, E. Ott, J. A. Yorke, and G. H. Yuan, Anomalous bifurcations in dc-dc converters: Borderline collisions in piecewise smooth maps, *IEEE Power Electronics Specialists' Conference* (St. Louis, MO), pp. 1337–1344, 1997.

[8] A. Y. Loskutov and A. I. Shishmarev, Control of dynamical systems behavior by parametric perturbations: An analytic approach, *Chaos*, vol. 4, no. 2, pp. 391–395, 1994.

[9] K. Chakrabarty, G. Poddar, and S. Banerjee, Control of chaos in the boost converter, *Electronics Letters*, vol. 31, pp. 841–842, 1995.

[10] G. Chen, Control and synchronization of chaotic systems: A bibliography, 1997. ECE Dept, Univ. of Houston, TX. Available from ftp.egr.uh.edu/pub/TeX/chaos.tex (login name: "anonymous" password: your email address).

[11] M. J. Ogorzalek, Overview of electronic chaos generation, control and applications, *Proc. SPIE* (Bellingham, WA), pp. 2–13, 1995.

[12] N. J. Mehta and R. M. Henderson, Controlling chaos to generate aperiodic orbits, *Physical Review A*, vol. 44, no. 8, pp. 4861–4865, 1991.

[13] F. J. Romeiras, C. Grebogi, E. Ott, and W. Dayawansa, Controlling chaotic dynamical systems, *Physica D*, vol. 58, p. 165, 1992.

[14] R. S. Bueno and J. L. R. Marrero, Application of the OGY method to the control of chaotic dc-dc converters: Theory and experiments, *IEEE Intl. Symp. Circuits and Systems*, (II) pp. 369–372 (Geneva, Switzerland), May 2000.

8.7 CONTROL OF BIFURCATION

Chi K. Tse
Yuk-Ming Lai

8.7.1 Background

Power electronics circuits are designed for stable operations. In most practical situations, the required stable operation is a period-1 operation. In dc/dc converters, for instance, the application of feedback control must guarantee stable operation at the switching frequency, and any subharmonic, quasi-periodic, or chaotic operation is regarded as being undesirable and should be avoided. It therefore transpires that the design objective must include the prevention of any bifurcation within the intended operating range. In fact, any effective design automatically has to avoid the occurrence of bifurcation for the range of variation of the parameters [1].

Bifurcations and chaos have been observed and analyzed throughout this book for various kinds of power electronics circuits. For systems that have been shown to bifurcate when a certain parameter is changed, the design problem is, in a sense, addressing the control of bifurcation. Note that the term *control* here should not be confused with the conventional usage, which refers to shaping dynamical response. In fact, the problem of *controlling bifurcation* is well tackled by power electronics engineers, whether they know it or not, since properly designed power electronics systems do not bifurcate within the chosen ranges of operation.

From the conventional viewpoint, the above-mentioned problem is a steady-state design problem if it does not take into account dynamical response of the system. Moreover, if dynamical response is considered in conjunction with the avoidance of bifurcation, then we are indeed dealing with a control problem from the conventional point of view. In the following we give two examples in which *control of bifurcation* involves both dynamical and steady-state considerations [2].

8.7.2 Controlling Bifurcation in Discontinuous-Mode Converters

Our first example illustrates how the design of a voltage feedback loop contributes to the control of bifurcation in a discontinuous-mode buck converter. In Chapters 3 and 5, we have seen the onset of a period-doubling bifurcation when the small-signal voltage gain has exceeded a certain bound. Also, in Chapter 5, we have derived an expression for the critical value of the feedback gain at which a discontinuous-mode buck converter begins to period-double. Denoting the small-signal output voltage (capacitor voltage) by Δv and the small-signal duty cycle by Δd, the feedback gain κ is defined as

$$\kappa = \frac{\Delta d}{\Delta v} \tag{8.88}$$

which is consistent with Chapters 3 and 5. The critical value of this gain, κ_c, is a function of the input voltage V_{in}, and for the buck converter it has been found as

$$\kappa_c = \frac{(1+\alpha)V_C^2 - \beta V_{in}^2 D^2}{2\beta V_{in} D V_C (V_{in} - V_C)} \tag{8.89}$$

where V_C is fixed, D is a function of V_{in} as stated in the relevant section of Chapter 5 on discontinuous-mode operation, and all symbols are as defined in Chapter 5. Hence, when we design the feedback system, we need to ensure that it remains stable for the entire range of input voltage used. In other words, we must keep the feedback gain below κ_c.

To see the control requirement, we plot the critical gain κ_c against the input voltage V_{in}, as shown in Figure 8.45. Clearly we observe that if the converter is to operate through a range of input voltage, say from 30V to 50V, then the critical gain ranges from 0.134 to 0.064 for $T/CR = 0.12$ and $RT/L = 20$, and from 0.22 to 0.087 when R is halved. To ensure stability for the ranges of load and input voltage, we must keep the gain below 0.064.

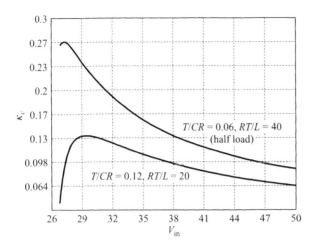

Figure 8.45 Critical feedback gain versus input voltage for $V_C = 25V$ and two loading conditions.

Obviously, a small feedback gain, while ensuring no bifurcation, slows down the response. Hence, if our design avoids bifurcation with excessive safe margin, then transient response will be sacrificed. It should now be apparent that we may develop a control scheme which continuously adjusts the gain according to the operating conditions, so as to maintain fast transient response while keeping the system clear of bifurcation.

8.7.3 Controlling Bifurcation in Current-Mode-Controlled DC/DC Converters

Use of Compensating Ramp for Controlling Bifurcation

Current-mode-controlled dc/dc converters are well known to be unstable when the duty cycle (designed steady-state value) exceeds 0.5. Thus, designers have to avoid operating current-mode-controlled converters beyond a duty cycle of 0.5, unless a compensating ramp is subtracted from the usual current reference. Such instability can in fact be examined in light of nonlinear dynamics. A handy starting point is the iterative function that describes the inductor current dynamics.

Consider the boost converter shown in Figure 8.46. Let I_{ref} be the reference current level which defines the peak inductor current. During a switching cycle, from $t = nT$ to $t = (n + 1)T$, the inductor current ramps up until it reaches I_{ref}, and then ramps down, as previously discussed in Chapter 5. We let i_n and i_{n+1} be the inductor current at $t = nT$ and $(n + 1)T$ respectively. Denote also the output voltage (voltage across the output capacitor) by v. By inspecting the slopes of the inductor current, we get, for $m_c = 0$ (uncompensated),

$$\frac{I_{ref} - i_{n+1}}{(1 - D)T} = \frac{v - V_{in}}{L} \quad \text{and} \quad \frac{I_{ref} - i_n}{DT} = \frac{V_{in}}{L} \tag{8.90}$$

Combining the above equations, we have the following iterative function:

$$i_{n+1} = \left(1 - \frac{v}{V_{in}}\right)i_n + \frac{vI_{ref}}{V_{in}} - \frac{(v - V_{in})T}{L} \tag{8.91}$$

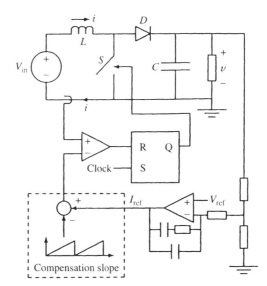

Figure 8.46 Boost converter under current-mode control.

If we are interested in the dynamics near the steady state, we may write

$$\delta i_{n+1} = \left(\frac{-D}{1-D}\right)\delta i_n + O(\delta i_n^2) \tag{8.92}$$

Clearly, the characteristic multiplier or eigenvalue, λ, is given by the slope

$$\lambda = \frac{-D}{1-D} \tag{8.93}$$

It is now apparent that the first period-doubling occurs when $D = 0.5$, which corresponds to $\lambda = -1$. Consistent with what is well known in power electronics, current-mode-controlled converters must operate with the duty cycle set below 0.5 in order to maintain a stable period-1 operation.

Remark: In the power electronics literature, the above criterion for stable operation (i.e., $D < 0.5$) is often obtained by comparing the upward and downward slopes of the inductor current. A simple graphical procedure demonstrates that stability is guaranteed if the magnitude of the upward slope is larger than that of the downward slope [3], and such requirement is equivalent to $|\lambda| < 1$.

In the application of current-mode control, the error signal derived from the output voltage is often used to modify I_{ref} directly (not the duty cycle as in the case of voltage-mode PWM control). It is thus helpful to look at the period-doubling bifurcation in terms of the current reference I_{ref}. Specifically we can express the *criterion of no bifurcation*, $D < 0.5$, in terms of I_{ref} by using the steady-state equation relating D and I_{ref}. For the boost converter, the equivalent criterion of no bifurcation is

$$I_{\text{ref}} < \frac{V_{\text{in}}}{R}\left[\frac{DRT}{2L} + \frac{1}{(1-D)^2}\right]_{D=0.5} \tag{8.94}$$

which can be derived from the power-balance equation

$$\left(I_{\text{ref}} - \frac{\Delta i}{2}\right)V_{\text{in}} = \frac{V_{\text{in}}^2}{(1-D)^2 R} \tag{8.95}$$

where $\Delta i = DTV_{\text{in}}/L$ and all symbols have their usual meanings. The critical value (upper bound) of I_{ref} for the uncompensated case is thus given by

$$I_{\text{ref},c} = \frac{V_{\text{in}}}{R}\left(\frac{RT}{4L} + 4\right) \tag{8.96}$$

Hence, period doubling occurs when I_{ref} exceeds the above-stated limit. To prevent period doubling, we must therefore control I_{ref} (i.e., the power level). Indeed, the use of a compensating ramp, as we will see, is to raise the upper bound of I_{ref}, thereby widening the operating range.

With compensation, the reference current is first subtracted from an artificial ramp before it is used to compare with the inductor current. Figure 8.47 illustrates the basic idea, which is well known in power electronics. By inspecting the inductor current waveform shown in Figure 8.47, we obtain the modified iterative function as

$$\delta i_{n+1} = \left(\frac{M_c}{1+M_c} - \frac{D}{(1-D)(1+M_c)}\right)\delta i_n + O(\delta i_n^2) \tag{8.97}$$

where $M_c = m_c L/V_{\text{in}}$ is the normalized compensating slope, and m_c is defined in Figure 8.47. Now, using (8.97), we get the eigenvalue or characteristic multiplier, λ, for the compensated system as

$$\lambda = \frac{M_c}{1+M_c} - \frac{D}{(1-D)(1+M_c)} \tag{8.98}$$

Hence, by putting $\lambda = -1$, the critical duty cycle, at which the first period doubling occurs, is obtained, i.e.,

$$D_c = \frac{M_c + 0.5}{M_c + 1} \tag{8.99}$$

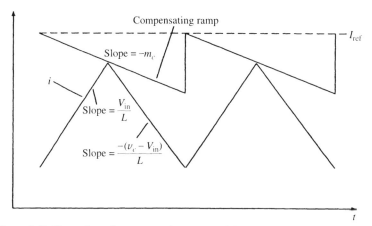

Figure 8.47 Illustration of current-mode control with compensating ramp showing inductor current in a boost converter. A compensating ramp of slope $-m_c$ is subtracted from the reference to give a new reference.

Note that $D_c = 0.5$ when $M_c = 0$ (i.e., uncompensated). Using (8.94) and the above expression for D_c, we get the critical value of I_{ref} for the compensated system as

$$I_{ref,c} = \frac{V_{in}}{R}\left[\frac{RT}{2L}\frac{M_c + 0.5}{M_c + 1} + 4(M_c + 1)^2\right] \qquad (8.100)$$

which increases monotonically as the compensating slope increases. Hence, it is obvious that compensation effectively provides more margin for the system to operate without running into the bifurcation region. Figure 8.48 shows some plots of the critical value of I_{ref} against R, for a few values of M_c.

Similar to the case of the discontinuous-mode buck converter discussed in the previous subsection, the choice of the magnitude of the compensating ramp constitutes a design problem which aims at avoiding bifurcation. In a likewise manner, we may

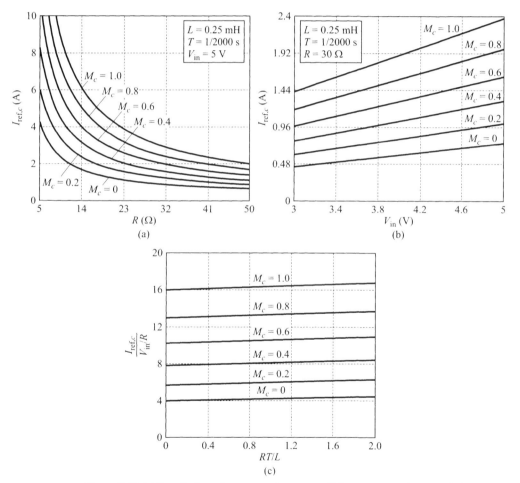

Figure 8.48 (a) Specific boundary curves $I_{ref,c}$ versus R for current-mode-controlled boost converter without compensation and with normalized compensating slope $M_c = 0.2, 0.4, 0.6, 0.8$ and 1; (b) specific boundary curves $I_{ref,c}$ versus V_{in} for current-mode-controlled boost converter without compensation and with compensation slope $M_c = 0.2, 0.4, 0.6, 0.8$ and 1; (c) boundary curves plotted with normalized parameters.

consider the input voltage variation and produce a similar set of design curves that provide information on the choice of the compensating slope for ensuring no bifurcation for a range of input voltage. This is shown in Figure 8.48(b). Also, for a general reference, the boundary curves in terms of normalized parameters are shown in Figure 8.48(c).

Effects on Dynamical Response

The transient response of a power converter can be compromised if bifurcation is kept too remote in order to give a large safe margin, especially when the operating range required is very wide, since guaranteeing *no bifurcation* for a wide range of parameter values would inevitably make the safe margin excessively large at one extreme end of the range.

It is therefore of interest to study the effect of the presence of a compensating ramp on the closed-loop dynamics of the *overall system*. We will take a simple averaging approach to derive the eigenvalues of the stable closed-loop system, mainly to reveal the transient speed for different values of the compensating slope. Specifically we can write down the normalized state equations for the boost converter as

$$\frac{dx}{d\tau} = -\gamma x + \gamma(1-d)y \tag{8.101}$$

$$\frac{dy}{d\tau} = \frac{-(1-d)x}{\zeta} + \frac{E}{\zeta} \tag{8.102}$$

where the normalized variables and parameters are defined by $x = v/V_{\text{ref}}$, $y = i/(V_{\text{ref}}/R)$, $E = V_{\text{in}}/V_{\text{ref}}$, $\tau = t/T$, $\gamma = T/CR$, and $\zeta = L/RT$. Here, we choose the steady-state output voltage as V_{ref}. The closed-loop control can be modeled approximately by (see Figure 8.47)

$$i + \frac{\Delta i}{2} \approx \overbrace{\left[\frac{P_o}{V_{\text{in}}} + I_{\text{offset}} - k(v - V_{\text{ref}})\right]}^{I_{\text{ref}}} - m_c dT \tag{8.103}$$

where P_o is the output power (i.e., $P_o = V_{\text{ref}}^2/R$), k is the voltage-feedback gain, and I_{offset} accounts for the steady-state shift due to the difference between the average and the peak value of the inductor current. This control equation can readily be translated, in terms of the normalized parameters, into

$$d = 1 - \frac{2\zeta\left(\frac{1}{E} - y - \kappa(x-1)\right) - 2M_c E}{x - E(1 - 2M_c)} \tag{8.104}$$

where $\kappa = kR$. Hence, putting (8.104) into the state equations, we get the closed-loop state equations which can then be used to study the closed-loop dynamics. Specifically, we can obtain the Jacobian matrix J_F as

$$J_F = \begin{bmatrix} \dfrac{2\zeta\gamma\left(2Y - \frac{1}{E}\right) + 2\gamma M_c E}{E(1 + M_c)} & \dfrac{\gamma\left(\frac{2\zeta}{E} + 2M_c E\right)}{E(1 + 2M_c)} \\[4mm] \dfrac{-2\kappa}{E(1 + 2M_c)} & \dfrac{-2}{E(1 + 2M_c)} \end{bmatrix}_{x=X, y=Y} \tag{8.105}$$

Note that in the steady state, $x = X = 1$ and $y = Y = 1/E$. Suppose the eigenvalues of the closed-loop system, λ_c, are complex. The real part of λ_c can be easily found as

$$\text{Re}(\lambda_c) = -\frac{E^2\gamma(1+2M_c) + 2E(1+\gamma M_c) - 2\kappa\gamma\zeta}{2E^2(1+2M_c)} \tag{8.106}$$

In practice, $E < 1$ and $\gamma \ll 1$. Also, for stable operation, κ has to be kept small enough so that $\text{Re}(\lambda_c) < 0$. Under such condition, we can readily show that $\frac{d}{dM_c}|\text{Re}(\lambda_c)| < 0$. In other words, the transient becomes slower as M_c increases. Some plots of $\text{Re}(\lambda_c)$ versus M_c are shown in Figure 8.49.

Consider the current-mode-controlled boost converter. We first observe that for a higher input voltage, the system is more remote to bifurcation. In fact, we can use (8.100) to determine if a current-mode-controlled boost converter, which is designed for a certain I_{ref} (corresponding to a given power level), may bifurcate and be chaotic for a given input voltage.

Remark: It should be reiterated that the overall dynamics is modeled by (8.101) and (8.102), while the inner current loop dynamics is described by (8.97). Inconsistent conclusions may be drawn from studying the two dynamical equations. Specifically, from (8.97), we observe that increasing M_c will make the inner loop dynamics "faster." However, the foregoing analysis of the overall system dynamics reveals that *for some range of parameter values,* the system actually becomes slower as M_c increases. Obviously, (8.97) is inadequate for the purpose of examining the overall system dynamics.

Experimental Measurements

An experimental prototype of a boost converter under current-mode control has been constructed, as schematically shown in Figure 8.46. The circuit parameters are: $L = 250\mu\text{H}$, $C = 440\mu\text{F}$, $R = 39\Omega$, $v = 8\text{V}$ (steady-state), and $T = 50\mu\text{s}$ (i.e., 20kHz). The feedback circuit has an appropriate integral control to adjust the steady-state level of I_{ref} in the event of a change in V_{in}. Such an arrangement is common in current-mode

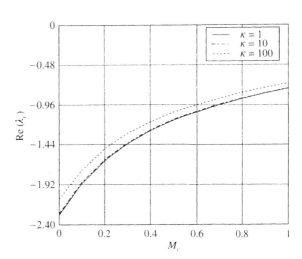

Figure 8.49 Plots of $\text{Re}(\lambda_c)$ versus M_c for $\zeta = 0.128$, $\gamma = 1/343$ (corresponding to $C = 440\mu\text{F}$, $L = 250\mu\text{H}$, $R = 39\Omega$ and $T = 1/20000\text{s}$) and $E = 3.5/8$.

control of dc/dc converters. Now, from $(1 - D)v = V_{in}$, we know that the uncompensated system will walk out of the stable region if V_{in} is reduced to below about 4V, since the output voltage is kept at 8V by the integral control. We may apply compensation to restore stability.

The results show that, without compensation (Figure 8.50(a)), the system becomes chaotic when the input voltage falls to 3.5V. Moreover, with compensation, the system remains stable. We further verify, from Figures 8.50(b) and (c), that excessive compensation lengthens the response time. Further elaboration will be given in the next subsection.

Variable Ramp Compensation

In order to keep bifurcation away while maintaining fast response, the control should incorporate a special function that dynamically adjusts the compensating ramp. The aim is to give just enough compensation under all input voltage conditions. Thus, the controller may contain, in addition to a conventional proportional-integral gain, a variable ramp generator providing necessary, but not excessive, compensation. For this simplified scenario (i.e., fixed load and output voltage), the compensating ramp need only be controlled according to

$$M_c(V_{in}) \geq \frac{v}{2V_{in}} - 1 \quad \text{or} \quad m_c(V_{in}) \geq \frac{v - 2V_{in}}{2L} \tag{8.107}$$

which is derivable from (8.99). Figure 8.51 shows the schematic of the variable ramp generator suitable for custom integration. In our experiment, this circuit has been constructed in discrete form. Figure 8.50(d) shows the measured waveforms for the boost converter under such control.

The series of waveforms shown in Figures 8.50(b) through (d) serve to illustrate the effect of applying ramp compensation to the system dynamics. Specifically, from (8.107), the value of M_c needed for a 3.5V input is about 0.14, and no compensation at all is needed for a 5V input. Thus, with $M_c = 0.8$ (Figure 8.50(c)), the system is overcompensated and hence suffers a slower transient compared with the case with $M_c = 0.2$ (Figure 8.50(b)). Furthermore, even for $M_c = 0.2$, the system is still overcompensated when the input is 5V. Thus, we can see a much faster response with the variable ramp compensation (Figure 8.50(d)) since it applies just enough compensation for the 3.5V input and none for the 5V input.

Remark: Due to its simplicity and monolithic design, the variable ramp generator shown in Figure 8.51 can be integrated with the standard current-mode-control IC to provide a simple yet effective solution to the compensation problem.

ACKNOWLEDGMENT

We wish to thank Mr. Velibor Pjevalica for his assistance in performing the experiments. This work has been inspired by a computer simulation work previously performed by Dr. William Chan in the study of bifurcations in dc/dc converters [4].

Acknowledgment

427

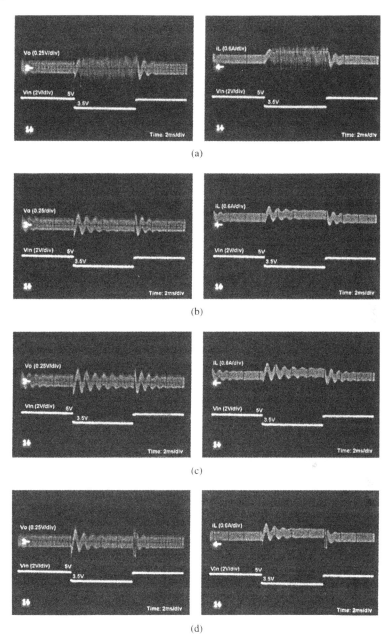

Figure 8.50 Measured output voltage v (or v_o) and inductor current i of boost converter under current-mode control: (a) without compensation showing chaos as input voltage drops to 3.5V; (b) with fixed ramp compensation $M_c = 0.2$ showing stabilized operation; (c) with fixed ramp compensation $M_c = 0.8$ showing stabilized operation but slow transient; (d) with variable ramp compensation showing improved transient for 5V input.

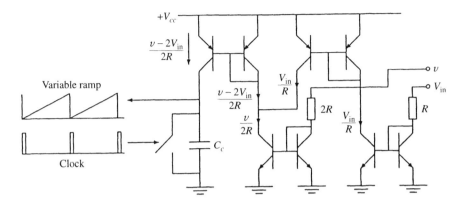

Figure 8.51 Schematic of variable ramp generator suitable for circuit integration.

REFERENCES

[1] T. Kapitaniak, *Controlling Chaos*. London: Academic Press, 1996.

[2] C. K. Tse and Y. M. Lai, Control of bifurcation in current-programmed dc/dc converters: A reexamination of slope compensation, *IEEE Int. Symp. on Circuits and Systems* (ISCAS'00), pp. I-671–674 (Geneva), June 2000.

[3] Y. S. Lee, *Computer-Aided Analysis and Design of Switch Mode Power Supplies*. New York: Marcel Dekker, 1993.

[4] W. C. Y. Chan, Study of chaos in dc/dc converters, PhD diss., Hong Kong Polytechnic University, 1999.

8.8 SYNCHRONIZATION OF CHAOS

Chi K. Tse

8.8.1 Background

One essential property of a chaotic system is that trajectories starting from nearby initial conditions diverge exponentially with time and quickly become uncorrelated. It is therefore nontrivial to show that two chaotic systems coupled through a chaotic signal can be in perfect synchronization. In Pecora and Carroll [1,2], such a surprising phenomenon was demonstrated. Since then, many investigations have been carried out to explore the properties and applications of chaos synchronization in a range of nonlinear circuits and systems [3]. Research in chaos synchronization has been further catalyzed by potential applications in communications [4,5,6].

The essential concept of synchronization lies in the deterministic nature of chaotic systems. As a deterministic system goes forward in time, the movements of the state variables are connected by a set of differential equations. Suppose we have two identical chaotic systems. Within each system, the state variables move according to the same differential equations. Moreover, their trajectories diverge if they start from two different sets of initial conditions, however close. Now we take a variable from one system, and the corresponding variable from the other system. If these two variables, one from each system, are *tied* together (i.e., their values are forced to become identical), there is a chance that the two systems will move in synchronization after a transient period. Of course it depends on the kind of the system (i.e., its dynamics) and the choice of the variables to be tied together (i.e., the coupling variable). In the following we will study the criterion for two chaotic systems to be synchronized. Our focus is on the classic *drive-response* configuration, and we will exemplify the phenomenon with power electronics circuits.

8.8.2 The Drive-Response Concept

Consider an n-dimensional ($n = m + k$) dynamical system which is described by a state equation of the form

$$\dot{\mathbf{x}} = f(\mathbf{x}(t)), \qquad (8.108)$$

where $\mathbf{x} = [x_1\ x_2\ \dots\ x_n]^T$. The system is divided into two subsystems in an arbitrary way. Accordingly the state vector $\mathbf{x}$ is partitioned as $\mathbf{x} = [\mathbf{x}_D\ \mathbf{x}_R]^T$, where $\mathbf{x}_D$ is an m-dimensional vector and $\mathbf{x}_R$ is a k-dimensional vector corresponding to a drive sub-system and a response subsystem, respectively (i.e., $\mathbf{x}_D = [x_1\ x_2\ \dots\ x_m]^T$ and $\mathbf{x}_R = [x_{m+1}\ x_{m+2}\ \dots\ x_n]^T$). Then, (8.108) can be rewritten as

$$\begin{cases} \dot{\mathbf{x}}_D = g(\mathbf{x}_D, \mathbf{x}_R) \\ \dot{\mathbf{x}}_R = h(\mathbf{x}_D, \mathbf{x}_R) \end{cases} \qquad (8.109)$$

where $g(\mathbf{x}) = [f_1(\mathbf{x}) \dots f_m(\mathbf{x})]^T$ and $h(\mathbf{x}) = [f_{m+1}(\mathbf{x}) \dots f_n(\mathbf{x})]^T$.

An identical copy of the system is constructed and driven by $\mathbf{x}_D$ taken from the above original system. We let the state variables of this new system be $\mathbf{x}'$, which is likewise partitioned (i.e., $\mathbf{x}' = [\mathbf{x}'_D\ \mathbf{x}'_R]^T$). The dynamics of this second system is thus described by

$$\begin{cases} \dot{\mathbf{x}}'_D = g(\mathbf{x}_D, \mathbf{x}'_R) \\ \dot{\mathbf{x}}'_R = h(\mathbf{x}_D, \mathbf{x}'_R) \end{cases} \qquad (8.110)$$

Since $\mathbf{x}'_D = \mathbf{x}_D$, we may ignore the dynamics of $\mathbf{x}'_D$, and hence describe the second system only by its response subsystem equation, i.e.,

$$\dot{\mathbf{x}}'_R = h(\mathbf{x}_D, \mathbf{x}'_R). \qquad (8.111)$$

Now, the original system (8.109) and the second response subsystem (8.111) constitute a complete *coupled* system.

It is readily shown that if all the Lyapunov exponents of the second response subsystem (8.111), also called *conditional Lyapunov exponents* (CLEs), are negative, then after the decay of the initial transient, $\mathbf{x}'_R$ will be exactly in step with $\mathbf{x}_R$. More precisely, under perfect synchronization, the difference between $\mathbf{x}'_R$ and $\mathbf{x}_R$ will tend to zero asymptotically, i.e.,

$$\lim_{t \to \infty} |\mathbf{x}'_R - \mathbf{x}_R| = 0 \qquad (8.112)$$

Alternatively, one may examine the error system that describes the dynamics of $(\mathbf{x}'_R - \mathbf{x}_R)$. Letting $\mathbf{e}_R$ be $(\mathbf{x}'_R - \mathbf{x}_R)$, we can write

$$\begin{aligned} \dot{\mathbf{e}}_R(t) &= h(\mathbf{x}_D, \mathbf{x}'_R) - h(\mathbf{x}_D, \mathbf{x}_R) \\ &= h_e(\mathbf{x}_D, \mathbf{x}_R, \mathbf{x}'_R) \end{aligned} \qquad (8.113)$$

The Lyapunov exponents of this system have been shown to be the same CLEs previously defined [3]. If h_e is linear, we can examine its eigenvalues and conclude that synchronization occurs if the real parts of all these eigenvalues are negative. Note that the real parts of these eigenvalues are also the CLEs. When h_e is nonlinear, we must resort to numerical procedure in order to calculate the CLEs.

For brevity, we will refer to the original system as *drive* system (with state vector **x**), and to the second system as *response* system (with state vector **x**′).

8.8.3 Synchronization in Chaotic Free-Running Ćuk Converters

We will demonstrate the phenomenon of *chaos synchronization* with an example. Our choice of the system is the *free-running Ćuk converter* studied earlier in Chapter 5. Readers may refer to Section 5.5 for details of the derivation of the autonomous describing equations, which we simply state as follows:

$$
\begin{cases}
\dfrac{di_{L2}}{dt} = -\dfrac{\mu i_{L2}}{2C} - \left(1 - \dfrac{\mu L}{CR}\right)\dfrac{v_{C1}}{2L} + \dfrac{v_{C2}}{2L} - \dfrac{E}{2L} \\[2mm]
\dfrac{dv_{C1}}{dt} = \dfrac{i_{L2}}{C} - \dfrac{v_{C1}}{CR} \\[2mm]
\dfrac{dv_{C2}}{dt} = -\dfrac{i_{L2}}{C} + \left(\dfrac{K - \mu v_{C1}}{2C}\right)\left(1 + \dfrac{\frac{\mu L}{C}i_{L2} - \left(1 + \frac{\mu L}{CR}\right)v_{C1} + E}{v_{C2}}\right)
\end{cases}
\tag{8.114}
$$

where μ and K are control parameters defined by

$$
i_{L1} + i_{L2} = K - \mu v_{C1} \tag{8.115}
$$

Note that the above representation is valid only when there is no saturation of the duty cycle d of the converter (i.e., $0 < d < 1$). However, in the real system, saturation does occur, especially when it is operating in chaotic regime. Thus, when we use (8.114) to calculate the corresponding CLEs, we must take into account the saturation of the duty cycle d.

Furthermore, to facilitate analysis, we prefer to express the system in dimensionless form by putting

$$
x_1 = \frac{Ri_{L2}}{V_{in}}, \quad x_2 = \frac{v_{C1}}{V_{in}}, \quad x_3 = \frac{v_{C2}}{V_{in}} \tag{8.116}
$$

$$
\tau = \frac{Rt}{2L}, \quad \xi = \frac{L/R}{CR}, \quad \kappa_1 = \mu R, \quad \kappa_0 = \frac{KR}{E} \tag{8.117}
$$

Hence, we obtain the dimensionless autonomous equations as

$$
\begin{cases}
\dfrac{dx_1}{d\tau} = -\xi\kappa_1 x_1 - (1 - \kappa_1\xi)x_2 + x_3 - 1 \\[2mm]
\dfrac{dx_2}{d\tau} = 2\xi(x_1 - x_2) \\[2mm]
\dfrac{dx_3}{d\tau} = -2\xi x_1 + \xi(\kappa_0 - \kappa_1 x_2)\left(1 + \dfrac{\kappa_1\xi x_1 - (1 + \kappa_1\xi)x_2 + 1}{x_3}\right)
\end{cases}
\tag{8.118}
$$

The above equations will be used for studying chaos synchronization in two chaotic free-running Ćuk converters.

We consider now two identical Ćuk converters arranged in the drive-response configuration described in Section 8.8.2. Let x_1, x_2, and x_3 be the state variables of the drive converter, and x_{1r}, x_{2r}, and x_{3r} be those of the response converter. In particular, we use x_3 (i.e., the dimensionless equivalence of capacitor voltage v_{C2}), as the driving signal.

The fifth-order coupled system is thus completely described by the set of drive converter equations [(i.e. (8.118)], and the set of response converter equations, which is

$$\begin{cases} \dfrac{dx_{r1}}{d\tau} = -\xi\kappa_1 x_{r1} - (1 - \kappa_1\xi)x_{r2} + x_3 - 1 \\[2mm] \dfrac{dx_{r2}}{d\tau} = 2\xi(x_{r1} - x_{r2}) \end{cases} \tag{8.119}$$

where subscript r denotes the response system variables. Figure 8.52(a) describes the interaction of the variables for the case of the isolated converter, and Figure 8.52(b) describes that for the case of the two converters being connected in the drive-response configuration.

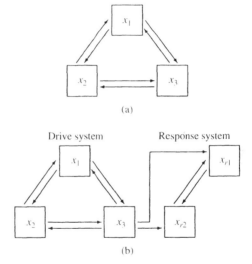

(a)

Figure 8.52 Interaction of state variables in (a) isolated converter; (b) coupled converters in drive-response configuration.

(b)

8.8.4 Derivation of the Conditional Lyapunov Exponents

As mentioned in Section 8.8.2, synchronization may be examined in terms of the error system [i.e. (8.8.6)]. For the system under study, the error system is linear. Letting $e_1 = x_1 - x_{r1}$ and $e_2 = x_2 - x_{r2}$, we can describe the error dynamics by

$$\begin{bmatrix} \dfrac{de_1}{d\tau} \\[2mm] \dfrac{de_2}{d\tau} \end{bmatrix} = \begin{bmatrix} -\xi\kappa_1 & -(1 - \kappa_1\xi) \\[1mm] 2\xi & -2\xi \end{bmatrix} \begin{bmatrix} e_1 \\[1mm] e_2 \end{bmatrix} \tag{8.120}$$

For brevity we let $\mathbf{e} = [e_1 \; e_2]^T$ and put (8.120) as

$$\dot{\mathbf{e}}(\tau) = A\mathbf{e}(\tau) \qquad (8.121)$$

where A can be extracted from (8.120). Our objective is to find the CLEs in order to determine if synchronization is possible. As mentioned in Section 8.8.2, the real parts of the eigenvalues of the error system are in fact the CLEs we need to find. Let λ_1 and λ_2 be the eigenvalues of A. For this linear system, we can readily show that

$$\lambda_{1,2} = -\frac{(2\xi + \kappa_1\xi) \pm \sqrt{(2\xi + \kappa_1\xi)^2 - 8\xi}}{2} \qquad (8.122)$$

Clearly, for $(2\xi + \kappa_1\xi)^2 \geq 8\xi$, both eigenvalues are negative. For $(2\xi + \kappa_1\xi)^2 < 8\xi$, moreover, the real part of both eigenvalues is negative for all positive values of κ_1 and ξ, i.e.,

$$\text{Re}(\lambda_i) < 0 \quad \text{for all } \xi, \kappa_1 > 0 \qquad (8.123)$$

Thus, by choosing suitable values of ξ, κ_1 and κ_0, the drive system can be set to operate in chaotic regime (characterized by one positive Lyapunov exponent) and the response system will also be driven to chaos when synchronization occurs.

8.8.5 Numerical Calculation of the Conditional Lyapunov Exponents

In the previous subsection, we have predicted that chaos synchronization is possible in the autonomous Ćuk converter. However, since the foregoing analysis is based on an averaged model of the converter which does not take into account the saturation of the duty cycle, the values of CLEs so obtained (i.e., those from (8.122)), remain to be verified.

In this subsection, we will numerically recalculate the CLEs, taking into account the saturation effect. In other words, we will consider the realistic case in which the switch can remain in the ON (or OFF) state continuously, as a result of the action of the control scheme. Essentially we need to include two extra subroutines to generate the flow corresponding to the cases $d = 1$ and $d = 0$, when the calculated duty cycle exceeds 1 and falls below 0, respectively.

Actual numerical calculation can be performed by using the INSITE software [7], which employs a Gram-Schmidt orthonormalization procedure for calculating Lyapunov exponents. In particular, we fix ξ at 0.1, which corresponds to a realistic practical choice, and for a range of κ_1 we have numerically calculated the corresponding CLEs using INSITE. The results are compared with those obtained from (8.122). As shown in Table 8.1, the numerical CLEs are all negative and reasonably close to those obtained from (8.122). We may thus conclude that chaos synchronization is possible in

TABLE 8.1 Comparison of CLEs from averaged model and by numerical calculation incorporating duty cycle saturation.

κ_1	$\text{Re}(\lambda_i)$ from (8.122)	$\text{Re}(\lambda_i)$ from numerical calculation
0.5	−0.125	−0.12
1.0	−0.15	−0.13
1.5	−0.175	−0.15
2.0	−0.120	−0.17

the free-running current-programmed Ćuk converter under study. Verification is yet to be sought using computer simulation of the actual converter circuits.

8.8.6 Computer Simulations

We will use "exact" time-domain simulations to verify the phenomenon. Our simulating model consists of two identical free-running current-mode-controlled Ćuk converters connected in the drive-response configuration, as previously described. Specifically, the drive system is a free-running current-mode Ćuk converter, exactly as described in Chapter 3, and the response system is constructed with the middle capacitor (the one connecting the two inductors) replaced by a voltage-controlled voltage source which copies the value of v_{C2} from the drive system.

Based on a piecewise-switched model, we simulate the cycle-by-cycle waveforms of the actual circuit, from which the phase portraits of the drive and response systems are obtained. The values of the parameters used are $L = 0.01\mathrm{H}$, $C = 1000\mu\mathrm{F}$, $R = 10\Omega$, $E = 12\mathrm{V}$, $\mu = 0.2$, and $K = 40$. These values correspond to $\xi = 0.1$, $\kappa_1 = 2.0$, and $\kappa_0 = 33.3$.

For ease of presentation, we use i_{L1}, i_{L2}, v_{C1}, and v_{C2} to denote variables of the drive circuit, and i_{L1r}, i_{L2r}, v_{C1r}, and v_{C2r} to denote those of the response circuit. We summarize the results as follows.

- Figure 8.53 shows the chaotic trajectories of the drive system and the response system after discarding the points in the transient period.
- Figure 8.54 shows the curves of i_{L2} versus i_{Lr2}, and of v_{C1} versus v_{Cr1}, which verify that chaos synchronization is achieved.
- For comparison, we have simulated the uncoupled system. The result is shown in Figure 8.55.

8.8.7 Remarks on Practical Synchronization

So far we have considered theoretical synchronization of two identical systems. However, in practice, slight deviation in parameter values is inevitable. Our question is:

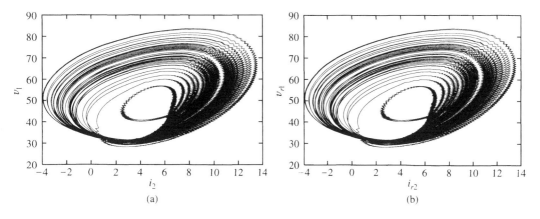

Figure 8.53 Chaotic trajectory from exact time-domain simulation of (a) drive system and (b) response system.

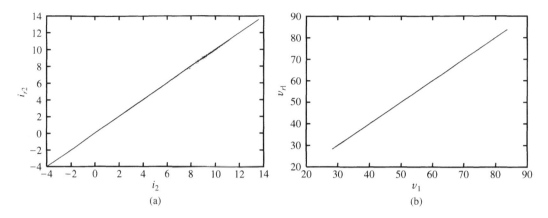

Figure 8.54 Graphical representation of synchronization from exact time-domain simulation: (a) i_{Lr2} versus i_{L2}; and (b) v_{Lr1} versus v_1.

Can two almost-identical chaotic systems be synchronized? This question can be translated to the study of two slightly different systems:

$$\dot{\mathbf{x}} = f(\mathbf{x}(t), \mathbf{p}) \tag{8.124}$$

$$\dot{\mathbf{x}} = f(\mathbf{x}(t), \mathbf{p} + \delta\mathbf{p}) \tag{8.125}$$

where $\mathbf{p}$ is a vector of parameters, and $\delta\mathbf{p}$ is a vector of the differences in the values of $\mathbf{p}$ between the two systems. Again we consider a drive-response configuration, and denote the variables for the response part of the two systems by x_R and x'_R, similar to what has been discussed in Section 8.8.2. However, the condition for synchronization (8.112) previously stated must be modified to

$$\lim_{t \to \infty} |\mathbf{x}'_R - \mathbf{x}_R| \le \epsilon \tag{8.126}$$

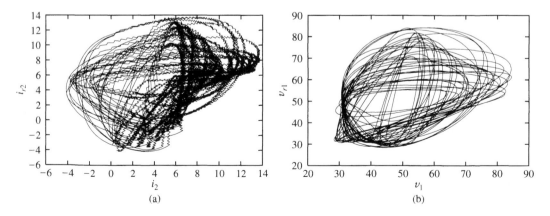

Figure 8.55 Plots of (a) i_{r2} versus i_2; and (b) v_{Cr1} versus v_{C1} for the uncoupled converters.

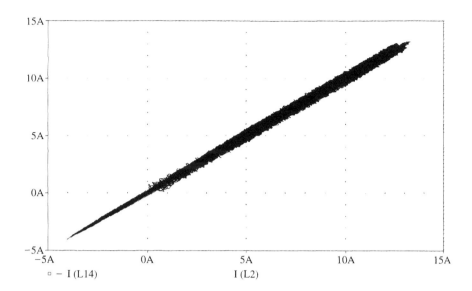

Figure 8.56 Graphical representation of synchronization from PSPICE simulation: i_{Lr2} versus i_{L2}.

where ϵ is a vector of small numbers ($\epsilon_i \ll 1$ for all $i = 1, 2, \ldots, n$). If (8.126) can be satisfied for a certain $\dot{p}$ and $\delta\dot{p}$, then synchronization is possible in practice, though not perfectly.

We have studied the practical synchronization in the case of the free-running Ćuk converter for which (8.126) can be shown to be true. Our verification tool this time is PSPICE simulation, which allows the use of practical MOSFET switches and drivers, and very handy change of parameters. The plot of i_{Lr2} versus i_{L2} shown in Figure 8.56 clearly demonstrates that synchronization is possible, though imperfectly, with a 1% deviation in the load resistor and inductor values.

REFERENCES

[1] L. M. Pecora and T. L. Carroll, Synchronization in chaotic systems, *Phys. Rev. Lett.*, vol. 64, pp. 821–824, Feb. 1990.

[2] L. M. Pecora and T. L. Carroll, Driving systems with chaotic signals, *Phys. Rev. A*, vol. 44, pp. 2374–2383, Aug. 1991.

[3] M. Hasler, Synchronization principles and applications, *IEEE Int. Symp. Circ. Syst. Tutorials*, pp. 314–329, May 1994.

[4] K. S. Halle, C. W. Wu, M. Itoh and L. O. Chua, Spread spectrum communication through modulation of chaos, *Int. J. Bifur. Chaos*, vol. 3, no. 2, pp. 469–477, 1993.

[5] G. Kolumbán, M. P. Kennedy, and L. O. Chua, The role of synchronization in digital communications using chaos—I: Fundamentals of digital communications, *IEEE Trans. on Circ. Syst.—I*, vol. 44, no. 10, pp. 927–936, Oct. 1997.

[6] U. Parlitz and L. Kocarev, Multichannel communication using autosynchronization, *Int. J. Bifur. Chaos,* vol. 6, no. 3, pp. 581–588, 1996.

[7] T. S. Parker and L. O. Chua, *Practical Numerical Algorithms for Chaotic Systems.* New York: Springer-Verlag, 1989.

INDEX

ABOUT THE EDITORS

Soumitro Banerjee was born in Calcutta in 1960, and earned his B.E. degree from Bengal Engineering College in 1981 and his M.Tech and Ph.D. degrees from the Indian Institute of Technology (IIT), New Delhi, in 1983 and 1987. He has been on the Electrical Engineering faculty of IIT Kharagpur since 1985, where he is now Associate Professor. His research and teaching interests are in the dynamics of physical systems, particularly energy systems, as well as in bifurcation theory and chaos.

Dr. Banerjee has focused in recent years on nonlinear phenomena in power electronics, demonstrating and studying many atypical bifurcation phenomena that have been observed only in the dynamics of switching circuits. He has been instrumental in developing a comprehensive theory of border collision bifurcation with which many nonlinear phenomena in power electronic systems can be analyzed. He has published over 40 papers—many of them in the area of this book—in refereed journals and conferences.

Dr. Banerjee is on the Editorial Board of the science magazine *Breakthrough*, published in India. He is also Editor of the popular Bengali science magazine *Prakriti*. He is a member of the IEEE, the Systems Society of India, and the Breakthrough Science Society (India).

George C. Verghese was born in Ethiopia in 1953, and received his B. Tech. from IIT Madras in 1974, his M.S. from the State University of New York at Stony Brook in 1975, and his Ph.D. from Stanford University in 1979, all in electrical engineering. Since 1979 he has been at the Massachusetts Institute of Technology, where he is Professor of Electrical Engineering in the Department of Electrical Engineering and Computer Science, and a member of the Laboratory for Electromagnetic and Electronic Systems.

Dr. Verghese's research interests and publications are in the areas of systems, control, estimation, and signal processing, with a focus on applications in power electronics, power systems, and electrical machines. He has served as an Associate Editor of *Automatica*, of the *IEEE Transactions on Automatic Control*, and of the *IEEE Transactions on Control Systems Technology*. He has also served on the AdCom and other committees of the IEEE Power Electronics Society, and was founding chair of its technical committee and workshop on Computers in Power Electronics. He is co-author (with J. G. Kassakian and M. F. Schlecht) of *Principles of Power Electronics*, Addison-Wesley, 1991. Dr. Verghese is a Fellow of the IEEE.

Printed in the USA/Agawam, MA
July 11, 2013

577629.069